D1195130

Nonsinusoidal Waves for Radar and Radio Communication

Advances in Electronics and Electron Physics

Edited by

L. MARTON CLAIRE MARTON

Smithsonian Institution
Washington, D.C.

SUPPLEMENTS

1. Electroluminescence and Related Effects, 1963 HENRY F. IVEY
2. Optical Masers, 1964 GEORGE BIRNBAUM
3. Narrow Angle Electron Guns and Cathode Ray Tubes, 1968 HILARY MOSS
4. Electron Beam and Laser Beam Technology, 1968 L. MARTON AND A. B. EL-KAREH
5. Linear Ferrite Devices for Microwave Applications, 1968 WILHELM H. VON AULOCK AND CLIFFORD E. FAY
6. Electron Probe Microanalysis, 1969 A. J. TOUSIMIS AND L. MARTON, EDS.
7. Quadrupoles in Electron Lens Design, 1969 P. W. HAWKES
8. Charge Transfer Devices, 1975 CARLO H. SÉQUIN AND MICHAEL F. TOMPSETT
9. Sequency Theory: Foundations and Applications, 1977 HENNING F. HARMUTH
10. Computer Techniques for Image Processing in Electron Microscopy, 1978 W. O. SAXTON
11. Acoustic Imaging with Electronic Circuits, 1979 HENNING F. HARMUTH
12. Image Transmission Techniques, 1979 WILLIAM K. PRATT, ED.
13. Applied Charged Particle Optics, 1980 (in two parts) A. SEPTIER, ED.

Nonsinusoidal Waves for Radar and Radio Communication

HENNING F. HARMUTH

DEPARTMENT OF ELECTRICAL ENGINEERING
THE CATHOLIC UNIVERSITY OF AMERICA
WASHINGTON, D.C.

1981

ACADEMIC PRESS
A Subsidiary of Harcourt Brace Jovanovich, Publishers
New York London Toronto Sydney San Francisco

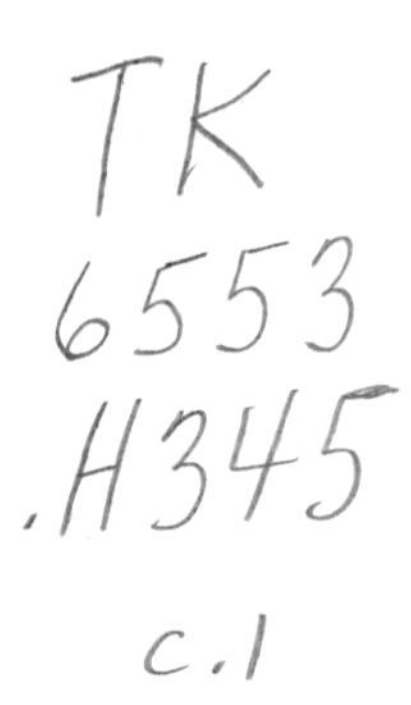

ACADEMIC PRESS, INC.
111 Fifth Avenue, New York, New York 10003

United Kingdom Edition published by
ACADEMIC PRESS, INC. (LONDON) LTD.
24/28 Oval Road, London NW1 7DX

LIBRARY OF CONGRESS CATALOG CARD NUMBER: 63–12814

ISBN 0–12–014575–8

PRINTED IN THE UNITED STATES OF AMERICA

81 82 83 84 9 8 7 6 5 4 3 2 1

To Richard B. Schulz
Editor of IEEE Transactions on Electromagnetic Compatibility
for publishing my papers

Contents*

* Equations are numbered consecutively within each of Sections 1.1–7.8. Reference to an equation in a different section is made by writing the number of the section in front of the number of the equation, e.g., Eq. (3.1-4) for Eq. (4) in Section 3.1.

Illustrations and tables are numbered consecutively within each of the Sections 1.1–7.8, with the number of the section given first, e.g., Fig. 1.2-1, Table 1.4-1.

References are characterized by the name of the author(s), the year of publication, and a lowercase Latin letter if more than one reference by the same author(s) is listed for that year.

4 Selective Receivers

5 Applications in Radar

Foreword

There is a dual pleasure that we feel with the publication of this volume by Dr. Harmuth. On the one hand, it is the third monograph that he has written to appear as part of *Advances in Electronics and Electron Physics*. On the other, the volume is unique in that this is the singular comprehensive presentation of this subject matter, so important in radar and other fields. We believe that the material covered here has a potential well beyond its immediate applications. We expect this volume to have a long and useful life, and Dr. Harmuth's efforts well deserve such success.

L. Marton
C. Marton

Preface

Twenty years have passed since the first papers on nonsinusoidal functions for information transmission and on carrier-free radar were published. On the academic level, this development has spread around the world, as shown by the volumes[1] from authors in England, Israel, Jordan, the Soviet Union, the United States, and West Germany. On the practical level, some 60 radars have been built and used from the tropics to the Arctic and the Antarctic.

Much has to be done before a new theory is accepted and routinely used, particularly if it calls for a radical change in one's thinking rather than for the more common step-by-step advance. One major requirement is that the theory has to be brought into a sufficiently simple and lucid form so that it can be mastered without undue investment of time. Another major requirement is that its useful applications have to be made clearly visible. This book attempts to comply with these two requirements for the field of radar and radio communication.

Opponents of sequency theory will have plenty of opportunity to point out all the problems that are not solved in this book. Very true. The perfect airplane never flies, and the perfect theory is never published. But a good theory grows as time marches on. Its final fate cannot be influenced, only its rate of development. Correct theories succeed, wrong theories die. Helping an unavoidable demise brings no honor, but speeding success does. This bias in favor of creation and advance makes science progress.

The author wants to thank the IEEE Electromagnetic Compatibility Society for many years of support. In particular, he wants to thank its officers: N. Ahmed, W. E. Cory, A. Farrar, J. R. Janoski, B. Keiser, H. Randall, G. R. Redinbo, G. F. Sandy, H. M. Schlicke, R. B. Schulz, R. M. Showers, L. W. Thomas, and J. C. Toler.

Much of the material in this book is based on papers that appeared in the *IEEE Transactions on Electromagnetic Compatibility*. Thanks are

[1] Ahmed and Rao (1974), Beauchamp (1975), Djadjunov and Senin (1977), Harmuth (1969, 1972, 1977a, 1979a), Karpovsky (1976), Maqusi (1981), and Trachtman and Trachtman (1975).

due to the Institute of Electrical and Electronics Engineers for the release of the copyright.

For many years the author hesitated to publish the military applications covered here, but was eventually convinced that he should do so through discussions, correspondence, and other means by the following scientists of the United States Department of Defense: L. C. Kravitz, Director of the Air Force Office of Scientific Research; J. O. Dimmock, Head of the Electronics and Solid State Science Program, Office of Naval Research; M. I. Skolnik, Superintendent of the Radar Division, Naval Research Laboratory; D. A. Miller, ELF Program Manager, Naval Underwater Systems Center; J. C. R. Licklider and C. H. Church of the Advanced Research Projects Agency; W. S. Ament, D. J. Baker, J. R. Davis, and J. P. Shelton of Naval Research Laboratory; and H. W. Mullany of the Office of Naval Research.

P. M. Gammell of the California Institute of Technology, Pasadena, helped with the proofreading and suggested numerous improvements; the author is greatly indebted to him.

Nonsinusoidal Waves for Radar and Radio Communication

1 Introduction

1.1 The Origins of Sinusoidal Waves in Radio Transmission

Sinusoidal waves have become so universal in radio communications that few people are aware it was not always so. Heinrich Hertz (1893) used a spark discharge to produce the electromagnetic waves for his experiments. These waves would be called colored noise today. Spark gaps and arc discharges between carbon electrodes were the dominant wave generators for about 20 years after Hertz's experiments, and radio signals consisted of short or long bursts of colored noise. The development of rotating high-frequency generators and the electronic tube eventually made the generation of sinusoidal currents and waves possible.

A strong incentive to use sinusoidal waves was provided by the need to operate several transmitters at the same time but to receive them selectively. Maxwell[1] (1891) had already studied what we now call the resonance of a circuit with coil and capacitor. Many people worked on the theoretical investigation and the practical implementation of this phenomenon, but the credit for the introduction of resonating filters using coils and capacitors for the selective reception of radio signals is usually given to Marconi on the strength of his patents[2] (Marconi, 1901, 1904). Apparently, no one ever raised the question seriously whether sinusoidal waves were the only ones for which the phenomenon of resonance existed. Hence, transmitters and receivers were developed on the basis of sinusoidal waves. Regulation followed common practice and brought the assignment of frequency bands for various radio services. However, a quotation from a textbook published in 1920 shows that nonsinusoidal waves were still used at that time, and that this was fully understood[3]

[1] Paragraph 779, "Combination of the electrostatic capacity of a condenser with the electromagnetic capacity of self-induction of a coil."

[2] Patent 763 772 introduces the term *tuning* and describes how a transmitter and a receiver can be tuned by proper choice of inductance and capacitance. Patent 676 332, applied for seven months later, uses the term *resonance* in the description of what we would now call an *LC* parallel circuit. The patenting of resonance for selective radio transmission created one of the worst controversies in the controversy-rich history of early radio.

[3] Figure 22 on page 22 shows a nonsinusoidal carrier for telegraphy, which was typical for the spark transmitters of early radio transmission.

(Edelman, 1920, p. 187): "The more advanced methods of wireless communication utilize continuous [sinusoidal] waves, produced either by an arc, quenched spark, or direct high-frequency generator. Inasmuch as these methods are quite likely to be developed into the ultimate perfect wireless system, some consideration of the theory together with experimental operation is worthy of attention."

Let us turn to the mathematical basis for the phenomenon of resonance. The homogeneous differential equation

$$(d^2v/dt^2) + \omega_0^2 v = 0 \tag{1}$$

has the general solution

$$v(t) = V_1 \sin \omega_0 t + V_2 \cos \omega_0 t \tag{2}$$

Let a force function[1] $\omega_0^2 v_f(t)$ replace the zero on the right-hand side of Eq. (1):

$$(d^2v/dt^2) + \omega_0^2 v = \omega_0^2 v_f(t) \tag{3}$$

The general solution of this inhomogeneous equation consists of the general solution of the homogeneous equation, given by Eq. (2), plus a particular solution of the inhomogeneous equation. The systematic way to find such a particular solution is by means of the method of *variation of the constant* or the *Laplace transform*. However, in simple cases one usually tries to shorten these methods by guessing a particular solution.

Let $v_f(t)$ be a sinusoidal function:

$$v_f(t) = V \sin \omega t \tag{4}$$

We guess that a particular solution of Eq. (3) has the form

$$v_p(t) = V_0 \sin \omega t \tag{5}$$

Insertion of $v_p(t)$ for v, and of the force function of Eq. (4) for $v_f(t)$ in Eq. (3), yields the value of V_0:

$$V_0 = \frac{\omega_0^2}{\omega_0^2 - \omega^2} V, \qquad v_p(t) = \frac{\omega_0^2}{\omega_0^2 - \omega^2} V \sin \omega t \tag{6}$$

Evidently, our guessed particular solution holds for all values of ω except for $\omega = \omega_0$, which is called the *resonance case*. We have to guess a new solution for this case:

$$v_p(t) = V_0 \omega_0 t \cos \omega_0 t + V_1 \omega_0 t \sin \omega_0 t \tag{7}$$

---

[1] We assume that t has the dimension of time, ω_0 that of inverse time, and v that of voltage. By using $\omega_0^2 v_f(t)$ rather than $v_f(t)$ we assure that $v_f(t)$ also has the dimension of voltage.

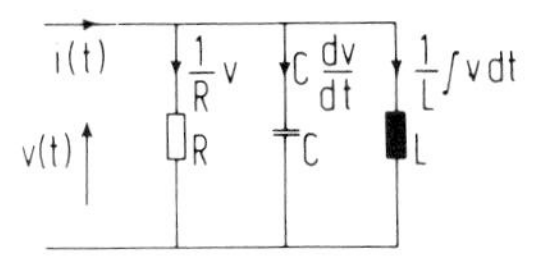

FIG. 1.1-1. Parallel resonant circuit for sinusoidal functions.

Insertion of $v_p(t)$ for v, and $V \sin \omega_0 t$ for $v_f(t)$, in Eq. (3) yields V_0 and V_1:

$$V_0 = -\tfrac{1}{2}V, \qquad V_1 = 0 \tag{8}$$

Hence, a particular solution for the resonance case equals

$$v_p(t) = -\tfrac{1}{2}V\omega_0 t \cos \omega_0 t \tag{9}$$

Let us now connect this purely mathematical concept of resonance with the resonance of a simple electrical circuit. Figure 1.1-1 shows a parallel resonant circuit with inductance L, capacitance C, and resistance R. The current $i(t)$ flowing into the circuit and the voltage $v(t)$ across the circuit are connected by the following differential equation:

$$\frac{1}{R} v + \frac{1}{L} \int v \, dt + C \frac{dv}{dt} = i(t) \tag{10}$$

Differentiation and reordering of the terms yields

$$\frac{d^2 v}{dt^2} + \frac{1}{RC}\frac{dv}{dt} + \frac{1}{LC} v = \frac{1}{C}\frac{di}{dt} \tag{11}$$

The two equations (3) and (11) become equal if R is sufficiently large, and if the relations

$$\omega_0^2 = \frac{1}{LC}, \qquad v_f(t) = \frac{1}{\omega_0^2 C}\frac{di}{dt} = L\frac{di}{dt}$$

are satisfied. The mathematical concept of resonance of a differential equation is thus connected with the concept of resonance of an electronic circuit.

The differential equations considered so far were linear and had constant coefficients. Such differential equations will only resonate with sinusoidal force functions.[1] Furthermore, a lumped circuit with linear, time-invariant components is always described by a linear differential equation with constant coefficients, and will thus resonate with sinusoidal functions only. However, the mathematical concept of resonance is more gen-

[1] More precisely, they will resonate only with the functions $\exp(-st)$, $\exp(-st) \sin \omega t$, and $\exp(-st) \cos \omega t$, where s is a nonnegative real number.

eral. In particular, it applies to linear differential equations with variable coefficients. They describe lumped electronic circuits with linear, time-*variable* components. Examples of such linear, time-variable components are switches, microphones, and modulators.[1] In the early days of radio communications there were no good time-variable components, and the phenomenon of resonance could thus be exploited for sinusoidal functions only. This situation was decisively changed by the advent of semiconductor technology. The switch is now one of the most desirable electronic components, and it is a linear, yet time-variable component. Let us then investigate some lumped circuits with linear, time-variable components that resonate with nonsinusoidal functions.

1.2 LUMPED, TIME-VARIABLE RESONANT CIRCUITS

In order to derive resonant circuits for nonsinusoidal functions we redesign first the parallel resonant circuit of Fig. 1.1-1 for implementation by operational amplifiers. To this end we integrate Eq. (1.1-11) twice,

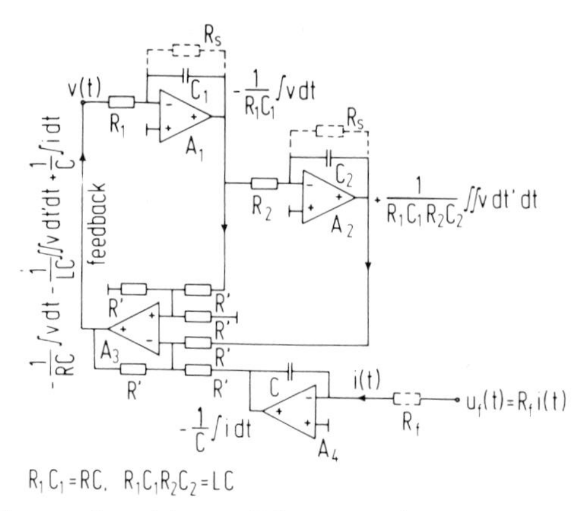

FIG. 1.2-1. Implementation of the parallel resonant circuit of Fig. 1.1-1 by means of operational amplifiers, resistors, and capacitors.

[1] Many textbooks call amplitude or frequency modulators nonlinear devices, and go on to discuss their use for the transmission of voice and music. There are two causes for this common error. First, no distinction is made between *linear, time-variable*, and *nonlinear*. Second, it is overlooked that a circuit with nonlinear components may have a linear relationship between input and output. For instance, a ring modulator with four square-law rectifiers transforms two input voltages u and v into the output voltage $(u + v)^2 - (u - v)^2 = 4uv$. The nonlinear terms u^2 and v^2 are canceled, and only the bilinear term uv remains.

$$v + \int \frac{1}{RC} v \, dt + \frac{1}{LC} \iint v \, dt' \, dt = \frac{1}{C} \int i \, dt \tag{1}$$

and rearrange the terms as follows:

$$v = -\frac{1}{RC} \int v \, dt - \frac{1}{LC} \iint v \, dt' \, dt + \frac{1}{C} \int i \, dt \tag{2}$$

Consider now the circuit of Fig. 1.2-1. The voltage $v(t) = v$ is fed in the upper left corner to the integrator[1] composed of the operational amplifier A_1, the resistor R_1, and the capacitor C_1. The integrated voltage $-(1/R_1C_1)\int v \, dt$ is produced. A second integration by A_2, R_2, and C_2 yields the voltage $+(1/R_1C_1R_2C_2)\iint v \, dt' \, dt$. Furthermore, the current $i(t)$ in the lower right-hand corner is integrated by the amplifier A_4 and the capacitor C to yield the voltage $-(1/C)\int i \, dt$. These three voltages are summed with proper signs to yield

$$-\frac{1}{R_1C_1} \int v \, dt - \frac{1}{R_1C_1R_2C_2} \iint v \, dt' \, dt + \frac{1}{C} \int i \, dt$$

If one chooses

$$R_1C_1 = RC, \qquad R_1C_1R_2C_2 = LC \tag{3}$$

one obtains just the right-hand side of Eq. (2). Since the right-hand side equals the voltage $v = v(t)$ on the left-hand side, we may close the feedback loop in Fig. 1.2-1. Hence, the circuit of Fig. 1.2-1 is a practical implementation[2] of Eq. (2).

The circuits of Figs. 1.1-1 and 1.2-1 will resonate with "periodic" sinusoidal functions, or functions that have sufficiently many periods. As the first generalization of such resonating circuits we will discuss one that resonates with sinusoidal or cosinusoidal *pulses* that have $i = 1, 2, \ldots$ periods and are zero outside the interval $-T/2 \leqq t \leqq +T/2$. The first few of these pulses are shown in Fig. 1.2-2, together with a rectangular pulse. These are the functions used for the Fourier series in a finite interval. They are defined by the equations

$$\begin{aligned} (d^2v/dt^2) + \omega^2 v &= 0, \qquad -T/2 \leqq t \leqq +T/2 \\ v &= 0, \qquad t < -T/2, \quad t > T/2 \end{aligned} \tag{4}$$

---

[1] It is assumed that the reader is familiar with the use of operational amplifiers for integration, summation, and differentiation. An excellent text for this field is the book by Graeme *et al.* (1971).

[2] The resistors R_s shown by dashed lines prevent the operational amplifiers from saturating due to drift. We do not discuss here the many technical refinements used in practical circuits with operational amplifiers, since we are interested in principles.

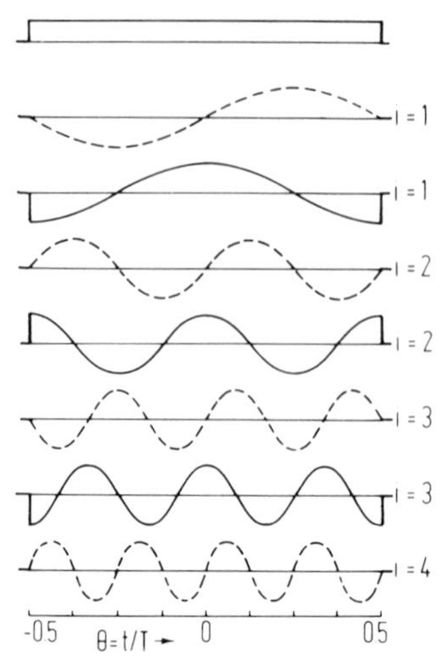

FIG. 1.2-2. The first eight functions of the Fourier series consisting of a constant, the sine functions $\sqrt{2}\sin 2\pi it/T$, and the cosine functions $\sqrt{2}\cos 2\pi it/T$.

We integrate the differential equation twice and add a force function $v_f(t)$ on the right-hand side:

$$v + \omega^2 \iint v \, dt' \, dt = v_f(t) \tag{5}$$

The terms are rearranged to make v the only term on the left-hand side:

$$v = -\omega^2 \iint v \, dt' \, dt + v_f(t) \tag{6}$$

Figure 1.2-3 shows the voltage $v(t)$ applied to the integrator consisting of amplifier A_1, resistor R_1, and capacitor C_1. The output voltage $-(1/R_1C_1)\int v(t)\,dt$ is integrated a second time to yield $+(1/R_1C_1R_2C_2)\iint v \, dt' \, dt$. This voltage is fed through a multiplier M that multiplies with $+1$ during the time interval $-T/2 \leqq t \leqq +T/2$ and with 0 otherwise. The summation of this voltage with a force function $v_f(t)$ according to Fig. 1.2-2 yields

$$-\omega^2 \iint v \, dt' \, dt + v_f(t)$$

in the interval $-T/2 \leqq t \leqq +T/2$ and zero otherwise. Closing of the feedback loop thus produces a circuit that implements Eq. (4).

In Fig. 1.2-1 we used the two resistors R_s to prevent saturation of the integrators due to drift. In Fig. 1.2-3 we use instead reset switches S_r that are always closed except during the interval $-T/2 \leqq t \leqq +T/2$. These

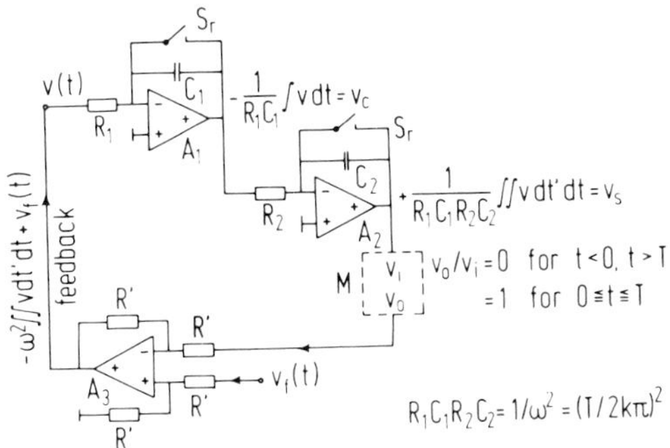

FIG. 1.2-3. Resonant circuit for the functions of Fig. 1.2-2. The choice of the time constant $R_1C_1R_2C_2$ determines the number $k = 1, 2, \ldots$ of the cycles of the sine or cosine pulse to which the circuit is tuned. A resonating *cosine* pulse with $i = k$ produces a large voltage v_c, and $v_s = 0$; a resonating *sine* pulse with $i = k$ produces a large voltage v_s, and $v_c = 0$. The reset switches S_r are closed at $t = \pm T/2$.

switches make the output voltages of both integrators zero outside this interval, which duplicates the effect of the multiplier M. Hence, the multiplier may be left out.

The voltages produced at the output terminals of the two integrators in response to a single cycle ($i = 1$) of a sinusoidal or cosinusoidal force function $v_f(t)$ is shown in Fig. 1.2-4. One may readily see that this circuit discriminates between sine and cosine pulses with the same period $2\pi/\omega$. The sinusoidal pulse (a) and the cosinusoidal pulse (d) were produced by digital circuits, which explains their steps.

Figure 1.2-5 shows oscillograms of the output voltage v_c of the first integrator if the circuit is tuned to $k = 128$ by the choice of the product $R_1C_1R_2C_2 = (T/2k\pi)^2$, whereas the force function $v_f = V\cos 2\pi it/T$ with $i = 128$, 129, and 130 is applied; this means that the circuit is tuned for the detection of a cosine (output v_c) or a sine pulse (output v_s) of 128 cycles, and that cosine pulses with 128, 129, or 130 cycles are fed to its input.

We obtained Eq. (5) by integration of Eq. (4) and addition of the force function $v_f(t)$. Another method is to make the following substitutions:

$$d^2v/dt^2 = u, \qquad dv/dt = \int u\,dt, \qquad v = \iint u\,dt'\,dt \tag{7}$$

One obtains from Eq. (4)

$$u + \iint u\,dt'\,dt = 0 \tag{8}$$

A force function

$$u_f(t) = d^2v_f/dt^2 \tag{9}$$

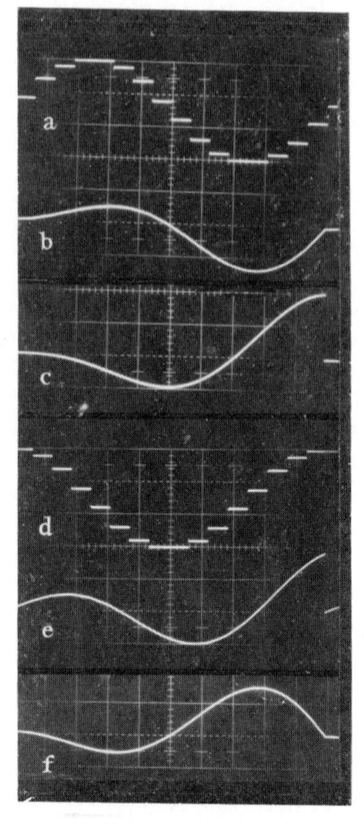

FIG. 1.2-4. Typical voltages of the circuit of Fig. 1.2-3: (a) force function $v_f(t) = V \sin 2\pi t/T$; (b) and (c) resulting voltages v_c and v_s; (d) force function $v_f(t) = V \cos 2\pi t/T$; (e) and (f) resulting voltages v_c and v_s. Horizontal scale: 15 msec/div. (Courtesy P. Schmid, R. Durisch, and D. Novak of Allen-Bradley Co., Milwaukee, Wisconsin.)

may be added on the right. The resulting equation is equal to Eq. (6), since the letters v are merely replaced by the letters u. The force function $u_f(t)$ in Eq. (9) differs only in sign from the force function $v_f(t)$ if $v_f(t)$ is a sinusoidal or cosinusoidal function.

The equality between integration and the substitutions according to Eq. (7) does not hold for differential equations with variable coefficients, to which we turn now. Figure 1.2-6 shows Legendre *pulses*. These pulses equal the Legendre polynomials $P_j(\theta)$ in the interval $-1 \leqq \theta \leqq +1$ and they are zero outside:

$$\begin{gathered} P_0(\theta) = 1, \qquad P_1(\theta) = \theta, \qquad P_2(\theta) = \tfrac{1}{2}(3\theta^2 - 1) \\ P_3(\theta) = \tfrac{1}{2}(5\theta^3 - 3\theta), \qquad P_4(\theta) = \tfrac{1}{8}(35\theta^4 - 30\theta^2 + 3) \\ -1 \leqq \theta \leqq +1, \qquad \theta = t/T \end{gathered} \tag{10}$$

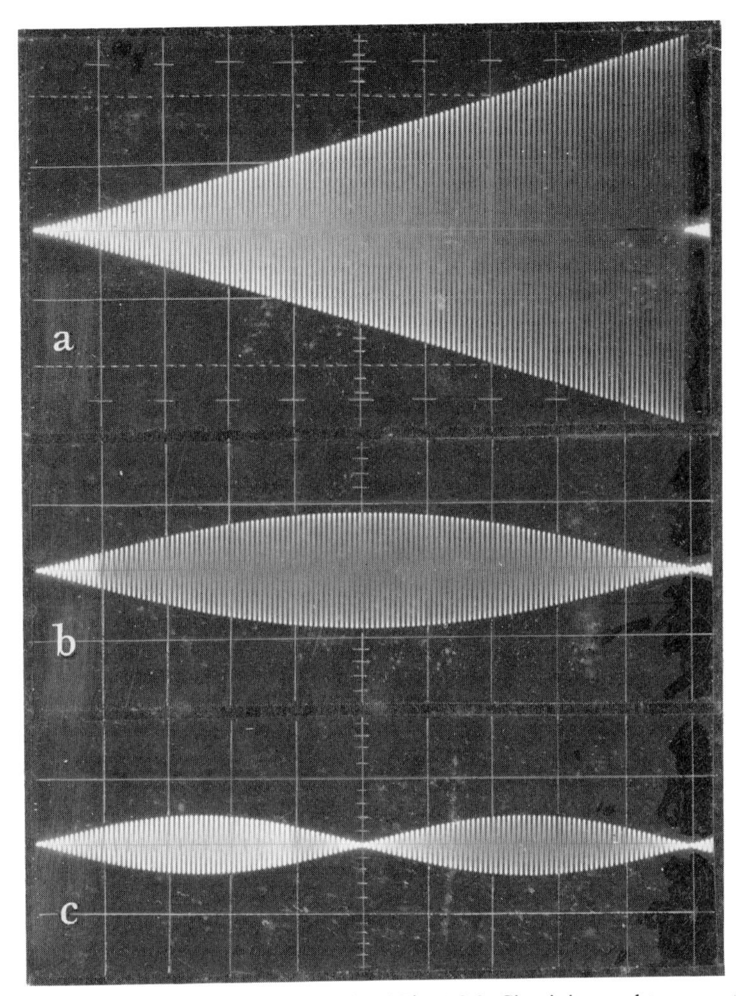

FIG. 1.2-5. Typical voltages of the circuit of Fig. 1.2-3. Circuit is tuned to resonate with sine and cosine pulses with $k = 128$ cycles. Output voltages v_c shown are caused by force functions $v_f(t) = V \cos 2\pi it/T$ with $i = 128$ (a), $i = 129$ (b), and $i = 130$ (c). Duration of the traces is $T = 78$ ms. (Courtesy P. Schmid, R. Durisch, and D. Novak of Allen-Bradley Co., Milwaukee, Wisconsin.)

The Legendre polynomials are defined[1] by the following differential equation:

$$(1 - \theta^2)\frac{d^2v}{d\theta^2} - 2\theta\frac{dv}{d\theta} + j(j + 1)v = 0, \qquad -1 \leqq \theta \leqq +1 \qquad (11)$$

[1] The mathematician says that the Legendre polynomials are the eigenfunctions of this differential equation, just as sinusoidal functions are the eigenfunctions of Eq. (4).

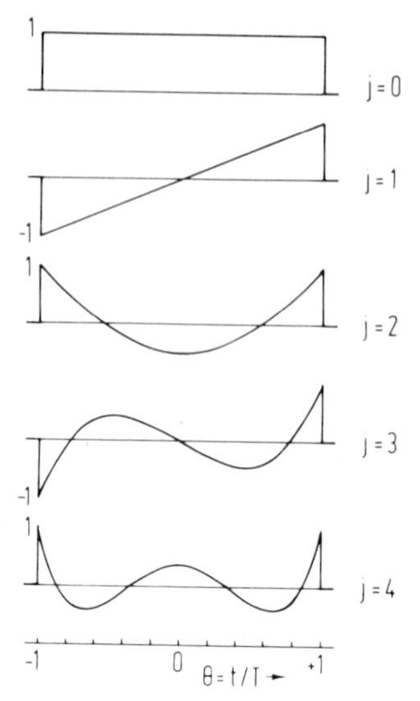

FIG. 1.2-6. Pulses of the form of the Legendre polynomials $P_j(\theta)$ in the interval $-1 \leqq \theta \leqq 1$ and zero outside.

We want equations that show the real time t rather than the normalized time $\theta = t/T$, and we thus substitute t/T for θ in Eq. (11):

$$(T^2 - t^2)\frac{d^2v}{dt^2} - 2t\frac{dv}{dt} + j(j+1)v = 0 \tag{12}$$

We use the substitutions of Eq. (7) and add a force function $u_f(t)$:

$$u - \frac{2t}{T^2 - t^2}\int u\,dt + \frac{j(j+1)}{T^2 - t^2}\iint u\,dt'\,dt = u_f(t) \tag{13}$$

Next we rearrange the terms so that only u stands on the left-hand side of the equation, and so that the time constant $1/T$ appears conspicuously with every integration:

$$u = 2\frac{tT}{T^2 - t^2}\frac{1}{T}\int u\,dt - j(j+1)\frac{T^2}{T^2 - t^2}\frac{1}{T^2}\iint u\,dt'\,dt + u_f(t) \tag{14}$$

A circuit implementing this equation is shown in Fig. 1.2-7. The voltage $u(t)$ is integrated twice to yield $\int u\,dt$ and $\iint u\,dt'\,dt$. These voltages are multiplied by the multipliers M_1 and M_2 with $tT/(T^2 - t^2)$ and $T^2/(T^2 - t^2)$. Further multiplications by 2 and by $j(j+1)$ are accomplished by the amplifiers A_3, A_4, and the resistors associated with them.[1] The resulting

[1] These two multiplication circuits can be combined with the amplifier A_5.

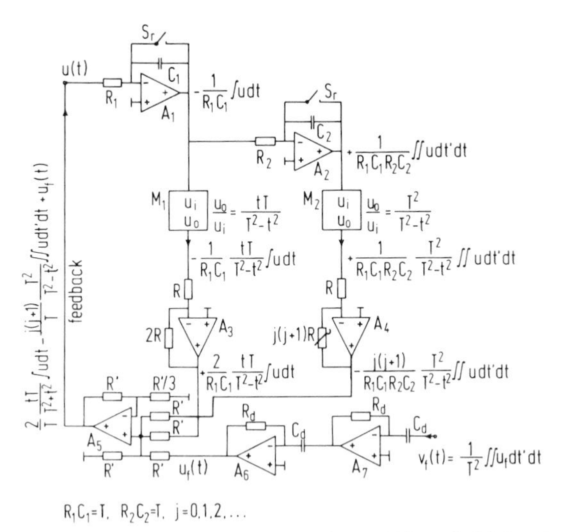

FIG. 1.2-7. Resonant circuit for the Legendre pulses of Fig. 1.2-6. The choice of the resistance $j(j+1)R$ in the feedback loop of the amplifier A_4 determines for which function $P_j(\theta)$ resonance occurs. The time constants R_1C_1 and R_2C_2 determine the time interval T of $\theta = t/T$.

voltages and the force function $u_f(t)$ are summed by the amplifier A_5, with attention to the proper positive or negative sign. The output voltage of A_5 equals the right-hand side of Eq. (14), and this voltage may thus be fed back to yield the input voltage $u(t)$ for the amplifier A_1.

The multipliers M_1 and M_2 have to produce the output voltages u_o from the input voltages u_i as functions of time. The ratios u_o/u_i as functions of t are shown in Fig. 1.2-8. In the days of the analog computer it was usual to implement such multipliers either by special, motor-driven potentiometers or by resistor networks and many switches. The modern way is to produce discrete values of the functions $tT/(T^2 - t^2)$ and $T^2/(T^2 - t^2)$ by means of a microprocessor, and feed the digital numbers to a digital–analog multiplier (Harmuth, 1979a, pp. 111–116).

Let us observe that circuits like the one in Fig. 1.2-7 were once routinely assembled for the simulation of differential equations by analog computers, but the advance of digital computers has made this almost a forgotten art (Korn and Korn, 1964; Johnson, 1963).

If the circuit of Fig. 1.2-7 is to resonate with Legendre pulses, we must feed the force function $v_f(t) = VP_j(t/T)$ into the circuit. However, our circuit implements the integral equation (14) for $u(t)$ rather than the differential equation (12) for $v(t)$. The force function $u_f(t)$ required to make it resonate is the second derivative of $v_f(t)$ according to Eq. (9). The circuit of

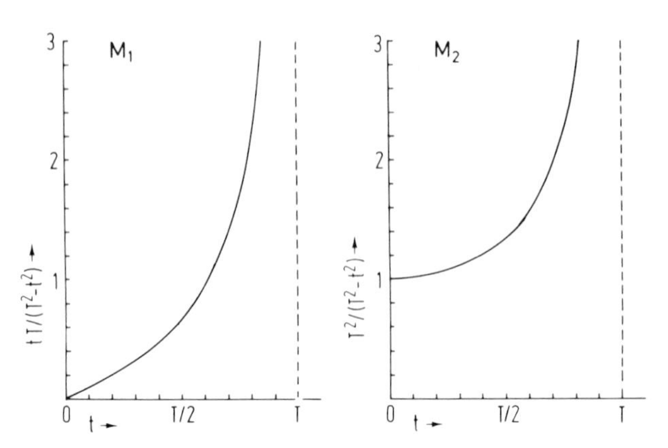

FIG. 1.2-8. Time variation of the ratio u_o/u_i of the two multipliers M_1 and M_2 in Fig. 1.2-7.

Fig. 1.2-7 shows at the bottom the voltage $v_f(t)$ fed through two differentiating stages (A_6, A_7) to yield $u_f(t)$. Such differentiations are undesirable. There are ways to avoid them, but we want to show here only how resonating circuits for nonsinusoidal functions can be devised, without advocating their construction and use.[1]

As the last example let us design a resonant circuit for Bessel functions. Figure 1.2-9 shows the first five functions $J_n(\theta)$. The functions $J_n(\theta)$ are defined by the following differential equation:

$$\theta^2 \frac{d^2v}{d\theta^2} + \theta \frac{dv}{d\theta} + (\theta^2 - n^2)v = 0, \qquad 0 \leqq \theta < \infty, \quad n = 0, 1, 2, \ldots \tag{15}$$

We replace θ by t/T, make the substitution of Eq. (7), add a force function $u_f(t)$, and rearrange the terms in analogy to Eq. (14):

$$u = -\frac{T}{t}\frac{1}{T}\int u\,dt - \left(1 - n^2\frac{T^2}{t^2}\right)\frac{1}{T^2}\int\!\!\int u\,dt'\,dt + u_f(t) \tag{16}$$

Figure 1.2-10 shows the circuit implementing the equation. The voltage $u(t)$ is integrated to yield $\int u\,dt$ and $\int\int u\,dt'\,dt$. The integrated voltages are multiplied with T/t and $1 - n^2T^2/t^2$ in the multipliers M_1 and M_2. The multiplied voltages and the force function $u_f(t)$ are summed by amplifier A_4, and the sum is fed back to amplifier A_1. The ratios u_o/u_i of the multipliers M_1 and M_2 are shown in Fig. 1.2-11. Note that the multiplier M_2 has

[1] One way is to find an integral equation rather than a differential equation that defines Legendre polynomials. Another is to use the circuit not for Legendre pulses, but for pulses obtained from the Legendre pulses by differentiating twice, e.g., $P_2''(\theta) = 3$, $P_3''(\theta) = 15\theta$, $P_4''(\theta) = (420\theta^2 - 60)/8$. More technical remedies are discussed by Graeme *et al.* (1971).

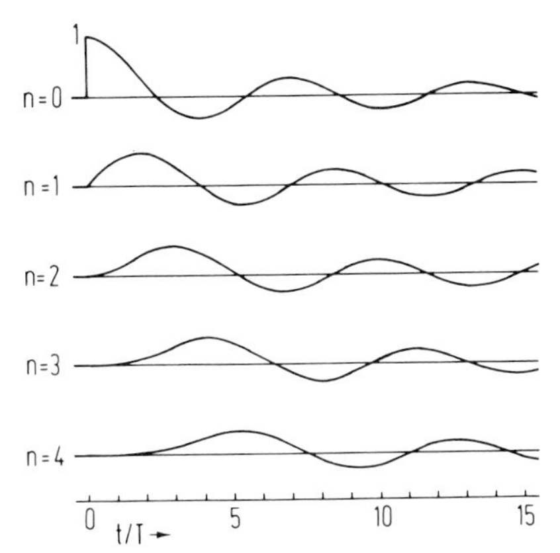

FIG. 1.2-9. Pulses of the form of the Bessel functions $J_n(\theta)$ in the interval $0 \leqq \theta < \infty$ and zero for $\theta < 0$.

to be changed for every value of $n = 0, 1, 2, \ldots$, while in Fig. 1.2-7 only a resistor needed changing for every value $j = 0, 1, 2, \ldots$. This causes no difficulty if the function $1 - n^2T^2/t^2$ is produced by a microprocessor, but it was a problem in bygone days when potentiometers and resistor networks had to be used.

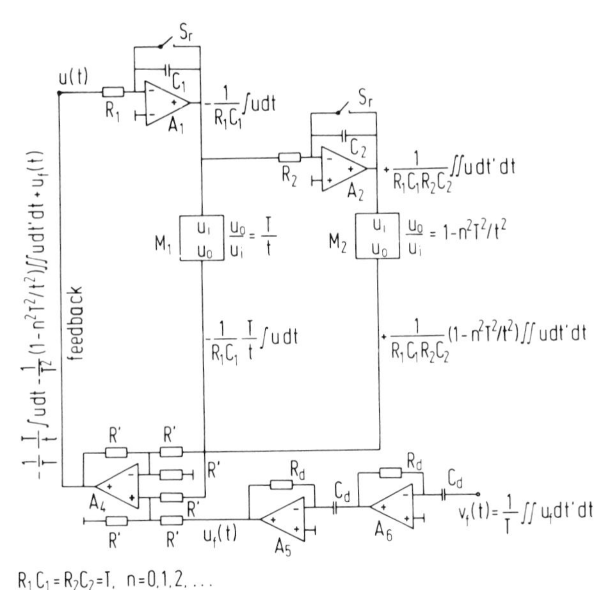

FIG. 1.2-10. Resonant circuit for the Bessel functions of Fig. 1.2-9. The choice of the ratio $u_o/u_i = 1 - n^2T^2/t^2$ of the multiplier $\mathbf{M_2}$ determines for which function $J_n(\theta)$ resonance occurs. The time constants R_1C_1 and R_2C_2 determine the time interval T of $\theta = t/T$.

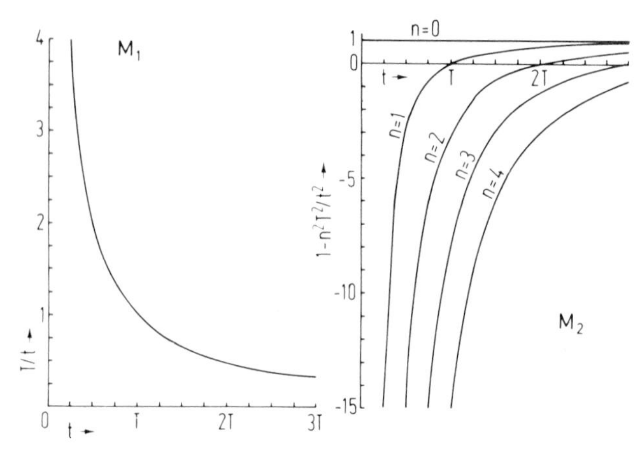

FIG. 1.2-11. Time variation of the ratio u_o/u_i of the two multipliers $\mathbf{M_1}$ and $\mathbf{M_2}$ in Fig. 1.2-10.

These examples should suffice to show that lumped resonant circuits are not restricted to sinusoidal functions. The examples also show that it is not easy, even today, to find resonant circuits that are as practical as the ones for sinusoidal functions. The circuits to be discussed later on will always be based on switches, sampled functions, and digital technology.

1.3 Distributed, Time-Invariant Resonant Circuits

A lumped circuit with linear, time-invariant components can only resonate with sinusoidal functions, while resonance with any other functions requires time-variable or nonlinear components. The reason is that lumped circuits are described by *ordinary* differential equations.

The same does *not* hold for distributed circuits and structures that are described by *partial* differential equations. To elaborate this statement we will study transmission lines, and derive the basic resonating circuit with time-invariant components for nonsinusoidal functions from them.

A transmission line is described by the following pair of partial differential equations:

$$\frac{\partial v}{\partial x} + L\frac{\partial i}{\partial t} + Ri = 0 \tag{1}$$

$$\frac{\partial i}{\partial x} + C\frac{\partial v}{\partial t} + Gv = 0 \tag{2}$$

The constants L, C, R, and G are the inductance, capacitance, resistance, and conductance per unit length of the line, while $v = v(x, t)$ and $i = i(x, t)$ are the voltage and current at the location x at the time t.

In order to separate the variables v and i one may introduce a new function:

$$v = \partial w/\partial x \tag{3}$$

Equation (2) then yields

$$\frac{\partial}{\partial x}\left(i + C\frac{\partial w}{\partial t} + Gw\right) = 0 \tag{4}$$

from which follows

$$i + C\,\partial w/\partial t + Gw = K \tag{5}$$

where K is independent of x. Without loss of generality one may choose $K = 0$, since $v = \partial w/\partial x$ remains unchanged if one adds to w a term that is independent of x. Hence, one obtains the following relation between i and w:

$$i = -C\,\partial w/\partial t - Gw \tag{6}$$

Substitution of Eqs. (3) and (6) into Eq. (1) yields the telegrapher's equation:

$$\frac{\partial^2 w}{\partial x^2} - LC\frac{\partial^2 w}{\partial t^2} - (LG + RC)\frac{\partial w}{\partial t} - RGw = 0 \tag{7}$$

To obtain its general solution one introduces a new function $u(x, t)$,

$$w(x, t) = e^{-at}u(x, t) \tag{8}$$

where a is a constant. Substitution of Eq. (8) into Eq. (7) yields

$$\frac{\partial^2 u}{\partial x^2} - LC\left(a^2u - 2a\frac{\partial u}{\partial t} + \frac{\partial^2 u}{\partial t^2}\right) - (LG + RC)\left(-au + \frac{\partial u}{\partial t}\right) - RGu = 0 \tag{9}$$

If one chooses

$$a = (LG + RC)/2LC \tag{10}$$

one eliminates the terms $\partial u/\partial t$ in Eq. (9) and obtains a simpler equation:

$$\frac{\partial^2 u}{\partial x^2} - LC\frac{\partial^2 u}{\partial t^2} + \frac{1}{2}(LG - RC)u = 0 \tag{11}$$

The term $(LG - RC)u$ vanishes if the condition

$$LG - RC = 0 \tag{12}$$

is satisfied. This is the case of the *distortion-free transmission line*. The product RC is generally much larger than LG for practical transmission lines. Thus in order to satisfy Eq. (12) one must either increase the inductance L, e.g., by means of materials with large magnetic permeability, or the conductance G, by using low-quality insulation. The increase of the inductance was common before the introduction of coaxial cables, whereas the increase of the conductance was never practical.[1] However, the length of the transmission lines we are interested in here is of the order of meters rather than kilometers, and we can well afford to increase the conductance G by means of poor insulation so that the condition for distortion-free transmission is satisfied. Equation (11) is in this case reduced to the wave equation:

$$\frac{\partial^2 u}{\partial t^2} = c^2 \frac{\partial^2 u}{\partial x^2}, \qquad c^2 = \frac{1}{LC} \tag{13}$$

Its general solution was found by d'Alembert in the eighteenth century:

$$u(x, t) = g_1(x - ct) + g_2(x + ct) \tag{14}$$

Substitution into Eq. (8) yields

$$w(x, t) = e^{-at}[g_1(x - ct) + g_2(x + ct)] \tag{15}$$

Using Eqs. (3) and (6), and writing f and g for the derivatives g_1' and g_2' with respect to their arguments $x - ct$ or $x + ct$, yields the voltage $v(x, t)$ and the current $i(x, t)$:

$$v(x, t) = e^{-at}[f(x - ct) + g(x + ct)] \tag{16}$$

$$i(x, t) = (1/Z)e^{-at}[f(x - ct) - g(x + ct)] \tag{17}$$

$$Z = (L/C)^{1/2}, \quad c = (LC)^{-1/2}, \quad a = (LG + RC)/2LC, \quad LG - RC = 0$$

[1] Communication links must always be "practically" distortion-free, or one could not transmit information. Radio links in vacuum are inherently distortion-free, whereas cable links are made so by means of compensating circuits, called *equalizers*, that are inserted at certain intervals. Power transmission lines, on the other hand, do not need to transmit information, and the distortion problem is solved by using sinusoidal currents and voltages rather than equalizers. One cannot use this expedient for communication links since information is always transmitted by nonsinusoidal currents, voltages, or field strengths. The distinction between power transmission and information transmission was not made for almost a century of development of electrical communications. Even today, the education of the electrical engineer is based on a double-think: He is presented a theory based on sinusoidal functions, and then told that sinusoidal functions transmit information at the rate zero; at this point he must switch to think in terms of nondenumerably many sinusoidal functions, and he is left to figure out which parts of the learned theory apply to this abstraction. A paper by Hartley (1928) marks the approximate time when the difference between power transmission and information transmission became more clearly recognized.

The function $f(x - ct)$ propagates with time in the direction of larger values of x, whereas the function $g(x + ct)$ propagates in the opposite direction. Let us consider the function $f(x - ct)$ alone. If it started propagating at $x = 0$ at the time $t = 0$ it will have reached the point $x = ct$ at the time $t = x/c$, and we may replace at by ax/c:

$$v(x, t) = e^{-ax/c}f(x - ct) \tag{18}$$

$$i(x, t) = (1/Z)e^{-ax/c}f(x - ct) \tag{19}$$

Consider the circuit of Fig. 1.3-1. It consists of a hybrid coupler HYC1, an amplifier AMP, a second hybrid coupler HYC2, and a transmission line DEL of length L and delay T. This transmission line is shown as a coaxial cable, but Eqs. (1) and (2) apply to many other types of transmission lines as well.

Let the amplification be set so that the feedback loop in Fig. 1.3-1 has essentially unit gain. Then let a periodic signal with period T,

$$f(t) = f(t - mT), \quad m = 0, 1, 2, \ldots \tag{20}$$

be fed to the input terminal of the hybrid coupler HYC1. It travels through the feedback loop and arrives after a delay time T at the feedback terminal of HYC1 as the signal $f(t)$. At this time the signal $f(t - T) = f(t)$ is fed again to the input terminal, and the sum of the new signal with the feedback signal produces the signal $f(t) + f(t - T) = 2f(t)$ with twice the amplitude. After n periods of the signal one has the signal $nf(t)$ with n times the amplitude. This is as good a resonance effect as the one shown in Fig. 1.2-5a, where the amplitude of the oscillations also increases propor-

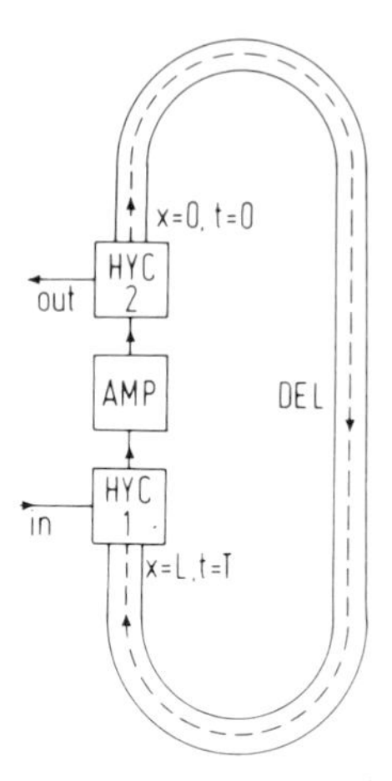

FIG. 1.3-1. Basic resonant circuit for general periodic functions with period T. HYC, hybrid coupler; AMP, amplifier; DEL, distortion-free delay line of length L and delay T.

tionate with the number of received periods of the sinusoidal signal. The difference is, of course, that any periodic function $f(t)$ will resonate with the distributed circuit of Fig. 1.3-1, whereas only sinusoidal functions will resonate with the lumped circuit of Figs. 1.2-1 and 1.2-3. Note that the circuit of Fig. 1.3-1 is linear and time-invariant. This circuit is as basic for nonsinusoidal waves as the LC circuit is for sinusoidal waves.

We have taken here great care to find the general solution for the distortion-free transmission line described by Eqs. (1), (2), and (12). Now we show that a completely different result is obtained by being not quite so careful. A typical method to solve a partial differential equation is to separate the variables by Bernoulli's product method. We apply this method to the telegrapher's equation (7):

$$w(x, t) = \varphi(x)\psi(t) \tag{21}$$

Substitution of $\varphi(x)\psi(t)$ into Eq. (7) yields two ordinary differential equations:

$$\begin{gathered} \frac{d^2\varphi}{dx^2} + (\mu - RG)\varphi = 0 \\ LC\frac{d^2\psi}{dt^2} + (LG + RC)\frac{d\psi}{dt} + \mu\psi = 0 \end{gathered} \tag{22}$$

Their general solutions are:

$$\varphi(x) = A_1 \cos(\mu - RG)^{1/2}x + A_2 \sin(\mu - RG)^{1/2}x \tag{23}$$

$$\psi(t) = B_1 e^{\gamma t} + B_2 e^{\delta t} \tag{24}$$

$$\begin{aligned} \gamma &= -a + (a^2 - \mu c^2)^{1/2}, & \delta &= -a - (a^2 - \mu c^2)^{1/2} \\ a &= (LG + RG)/2LC, & c &= (LC)^{-1/2} \end{aligned} \tag{25}$$

where μ is the eigenvalue to be determined by boundary conditions.

The function $w(x, t)$ now consists of sinusoidal and cosinusoidal terms multipled with an exponential function, e.g., $e^{\gamma t}\cos(\mu - RG)^{1/2}x$. The differentiation required to obtain the voltage v and the current i from $w(x, t)$ according to Eqs. (3) and (6) will produce additional terms of the same form. Hence, it appears that sinusoidal and exponential functions are distinguished by the telegrapher's equation. This is quite wrong. Bernoulli's product of Eq. (21) only yields particular solutions of the telegrapher's equation, not the general solution, and it is the method of solution rather than the equation that distinguishes the sinusoidal and exponential functions.

This is a very important point since the theory of electricity in general and the theory of electromagnetic radiation in particular—to the extent used in this book—are derived from Maxwell's equations. Most of the known solutions of Maxwell's equations use Bernoulli's product, and they must all be avoided if one wants to use nonsinusoidal signals or signals with a large relative bandwidth.

1.4 Relative Bandwidth

The relative bandwidth is usually defined as a quotient bandwidth/(carrier frequency). This definition is only applicable if there is a carrier. We will use the following more general definition for the relative bandwidth η,

$$\eta = (f_H - f_L)/(f_H + f_L) \tag{1}$$

where f_H is the highest and f_L the lowest frequency of interest. Typical radio signals used for communications or radar have relative bandwidths in the order of $\eta = 0.01$ or less, an amplifier specified from $f_L = 0.1$ GHz to $f_H = 2$ GHz yields $\eta = (2 - 0.1)/(2 + 0.1) = 0.9$, and an attenuator specified from dc to 2 GHz yields $\eta = 1$, which is the largest value permitted by the definition. The distortion-free lines discussed in the preceding section have $\eta = 1$. Many commercially available components, such as hybrid couplers or frequency-independent antennas, have relative bandwidths either close to 1 or at least much larger than the typical relative bandwidths of signals. Generally speaking, only circuits and structures designed to resonate with (almost) sinusoidal signals have a small relative bandwidth, since the phenomenon of resonance disappears with increasing relative bandwidth.

For an explanation of this statement refer to Fig. 1.4-1, which shows on top a parallel resonant circuit. The impedance Z presented by it to a sinusoidal current is given by the formula

$$Z = [1/R + j(\omega C - 1/\omega L)]^{-1} \tag{2}$$

which is usually written in a normalized form:

$$Z/Z_0 = [1 + jQ(\Omega - 1/\Omega)]^{-1} \tag{3}$$

$$Z_0 = R, \qquad \omega_0 = (LC)^{-1/2}, \qquad \Omega = \omega/\omega_0, \qquad Q = \omega_0 RC$$

The higher and lower half-power frequencies Ω_H and Ω_L follow from the condition $Q(\Omega - 1/\Omega) = 1$:

$$\Omega_H = 1/2Q + (1 + 1/4Q^2)^{1/2}, \qquad \Omega_L = -1/2Q + (1 + 1/4Q^2)^{1/2} \tag{4}$$

The difference and the sum of Ω_H and Ω_L define the half-power bandwidth and the sum frequency:

$$\Omega_H - \Omega_L = 1/Q, \qquad \Omega_H + \Omega_L = 2(1 + 1/4Q^2)^{1/2} \tag{5}$$

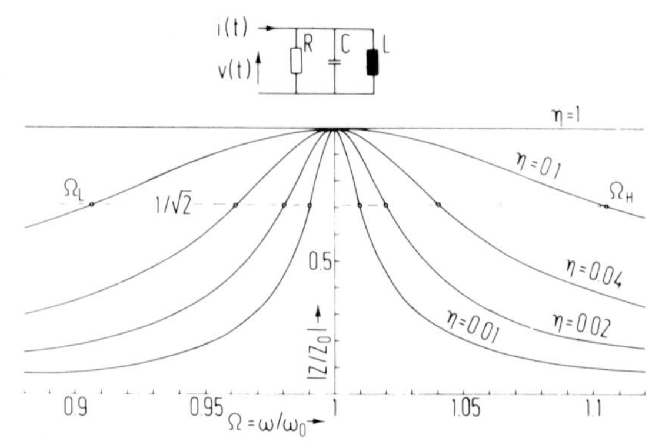

FIG. 1.4-1. The magnitude of the relative impedance Z/Z_0 of a parallel resonant circuit for various values of the relative bandwidth η as function of the normalized frequency $\Omega = \omega/\omega_0$.

The relative half-power bandwidth η is connected with the quality factor Q through Eq. (5):

$$\eta = \frac{\Omega_H - \Omega_L}{\Omega_H + \Omega_L} = \frac{\omega_H - \omega_L}{\omega_H + \omega_L} = \frac{f_H - f_L}{f_H + f_L} = \frac{1}{2Q(1 + 1/4Q^2)^{1/2}} \tag{6}$$

The magnitude of the normalized impedance and the half-power bandwidth can now be written in terms of η:

$$\left|\frac{Z}{Z_0}\right| = \left[1 + \left(\Omega - \frac{1}{\Omega}\right)^2 \frac{1 - \eta^2}{4\eta^2}\right]^{1/2} \tag{7}$$

$$\Omega_H - \Omega_L = 2\eta(1 - \eta^2)^{-1/2} \tag{8}$$

The normalized impedance $|Z/Z_0|$ and the bandwidths $\Omega_H - \Omega_L$ are shown in Fig. 1.4-1 for various values of the relative bandwidth η. One may readily see that η must be of the order of 0.01 or less to obtain an appreciable resonance effect. Similar results hold for other circuits or structures resonating with sinusoidal functions, in particular for the resonant dipole. This behavior is the main reason why our usual technology for radio transmission requires small relative bandwidths.[1]

The resonant circuits for sinusoidal pulses, Legendre pulses, and Bessel functions in Figs. 1.2-3, 1.2-7, and 1.2-10 do not have a small relative (fre-

[1] Radars have been built with a relative bandwidth η as high as 0.05, using the conventional technology based on sinusoidal functions. It is clear from Fig. 1.4-1 that the selectivity and noise rejection of such equipment cannot be good. We have here a classical example of how the lack of an adequate theory leads to enormous investments that eventually have to be written off.

quency) bandwidth since the Fourier transform of the signals for, e.g., $i = 1$ in Fig. 1.2-2, $j = 1$ in Fig. 1.2-6, and $n = 0$ in Fig. 1.2-9 contains important frequency components close to zero. However, this is so only because the usual Fourier transform and the usual concept of bandwidth are based on the periodic sinusoidal functions. One can generalize the Fourier transform and the concept of bandwidth by basing them on other systems of functions (Harmuth, 1969, 1972). The concept of relative bandwidth then becomes generalized too, and all resonant filters for a specific system of functions have a small relative bandwidth in terms of that system. This is not so for the resonant filter of Fig. 1.3-1, which resonates with the large class of periodic functions with period T.

The use of signals with a small relative bandwidth is obviously a restriction imposed on us by the current technology, but it was not felt as a restriction until very recently. Consider, e.g., a radio signal in the AM band from 535 to 1605 kHz. It has necessarily a small relative bandwidth, since our ears respond only to frequencies below about 16 kHz. A problem would have arisen if we had tried to transmit music with a radio carrier having a frequency of, e.g., 20 kHz, but there was not much demand for such a service.

The signals we want to transmit have usually a relative bandwidth close to 1. For instance, audible signals use the band from about 20 Hz to about 16 kHz. Their relative bandwidth is thus $\eta = (16{,}000 - 20)/(16{,}000 + 20) = 0.9975$. Even a low-quality telephone channel transmitting only the band $300 \text{ Hz} \leqq f \leqq 3000 \text{ Hz}$ has a relative bandwidth $\eta = 2700/3300 = 0.82$. Television signals start at dc and thus have a relative bandwidth equal to 1; the same holds true for rectangular pulses used for telegraphy, teletype or data transmission.

The transformation of the large relative bandwidth of these signals into a small relative bandwidth is typically done by means of amplitude modulation of a sinusoidal carrier. Consider the rectangular pulse on top of Fig. 1.2-2 with the duration T. Most of its energy is in the band from $f_\mathrm{L} = 0$ to $f_\mathrm{H} = 1/T = \Delta f$. Double sideband amplitude modulation of a sinusoidal carrier with frequency f_c shifts this energy to the band from $f_\mathrm{L} = f_\mathrm{c} - \Delta f$ to $f_\mathrm{H} = f_\mathrm{c} + \Delta f$. The relative bandwidth

$$\eta = \frac{(f_\mathrm{c} + \Delta f) - (f_\mathrm{c} - \Delta f)}{(f_\mathrm{c} + \Delta f) + (f_\mathrm{c} - \Delta f)} = \frac{\Delta f}{f_\mathrm{c}} = \frac{1}{f_\mathrm{c} T} \tag{9}$$

can in principle be made as small as one wants by choosing f_c sufficiently large compared with $1/T$. Let f_c equal $1/T$, $2/T$, $3/T$, ... The resulting sinusoidal carriers modulated by the rectangular pulse are the sinusoidal pulses with $i = 1, 2, 3, \ldots$ in Fig. 1.2-2. As i increases, the sinusoidal pulses look more and more like periodic sinusoidal functions. Hence, a small relative bandwidth means that a signal looks very similar to a periodic sinusoidal function.

The process of turning the generally large relative bandwidth of signals into a small relative bandwidth by means of a large carrier frequency f_c will fail, if conditions imposed by nature prevent us from choosing f_c as large as we want. We will investigate when this happens in the following section.[1]

The use of large relative bandwidths runs counter to current regulations based on CCIR recommendations. Tables 1.4-1 and 1.4-2 show the relative bandwidths of broadcasting and radio-location channels in the United States. Some of the broadcasting channels, and particularly the combination of adjacent channels, have fairly large relative bandwidths, but only two radio-location channels have a relative bandwidth slightly larger than 0.1. This seems to imply insurmountable regulatory obstacles. However, this is not so. CCIR Question 1A/29 asks specifically for a study of "what are technical criteria for sharing bandwidth expansion with conventional modulation systems and what are appropriate techniques and design factors conducive to improving the efficiency of spectrum utilization in

TABLE 1.4-1
RELATIVE BANDWIDTHS OF BROADCASTING BANDS ACCORDING TO THE UNITED STATES FREQUENCY ALLOCATIONS[a]

<table>
<tr><th>Service</th><th>Occupied band (MHz)</th><th colspan="4">Relative bandwidth</th></tr>
<tr><td>AM radio</td><td>0.535–1.605</td><td>0.5</td><td></td><td></td><td></td></tr>
<tr><td>TV channels 2–4</td><td>54–72</td><td>0.14</td><td rowspan="2">0.24</td><td></td><td rowspan="3">0.33</td></tr>
<tr><td>TV channels 5–6</td><td>76–88</td><td>0.07</td><td rowspan="2">0.17</td></tr>
<tr><td>FM radio</td><td>88–108</td><td>0.10</td><td></td></tr>
<tr><td>TV channels 7–13</td><td>174–216</td><td>0.11</td><td></td><td rowspan="3"></td><td rowspan="3"></td></tr>
<tr><td>TV channels 14–20</td><td>470–512</td><td>0.04</td><td rowspan="2">0.26</td></tr>
<tr><td>TV channels 21–69</td><td>512–806</td><td>0.22</td></tr>
</table>

[a] The relative bandwidth is shown for the individual bands as well as for certain combinations, e.g., 0.17 for TV channels 5–6 plus FM radio.

[1] One school of thought represented, e.g., by W. H. Kummer of Hughes Aircraft Co., holds that only the absolute bandwidth but not the relative bandwidth is significant, since the transmittable information is proportionate to the absolute bandwidth. This is correct within the realm of pure theory, where the properties of the transmission medium for the electromagnetic waves can be defined by the investigator to produce the desired result. The relative bandwidth is only significant for practical applications, where the features of the transmission medium are imposed on us by nature. The following Section 1.5 shows this for radars operating in the atmosphere. Similar restrictions hold for ice, fresh water, sand, clay, and other media encountered by the waves of the into-the-ground radar that will be discussed in Section 1.6. The restrictions imposed by saltwater will be discussed in Section 1.6 and in more detail in Chapter 7.

TABLE 1.4-2

RELATIVE BANDWIDTHS OF RADIO-LOCATION CHANNELS ACCORDING TO THE UNITED STATES FREQUENCY ALLOCATION

Band (MHz)	216–225	420–450	902–928			
Relative bandwidth	0.020	0.034	0.014			
Band (GHz)	1.215–1.4	2.3–2.5	2.7–3.7	5.25–5.925	8.5–10.55	13.4–14
Relative bandwidth	0.071	0.042	0.16	0.060	0.11	0.022
Band (GHz)	15.7–17.7	33.4–36	48–50	71–76	165–170	240–250
Relative bandwidth	0.060	0.037	0.020	0.034	0.015	0.020

shared frequency bands'' (CCIR, 1975). A number of reports have been written in response to this question (e.g., U.S.A., 1977a–c), and the CCIR recommendations will eventually be changed to permit frequency sharing services, which will allow the general use of large relative bandwidths. At the present time, the use of large relative bandwidths is quite legitimate if the radiation is sufficiently localized. This condition is met by the commercially available into-the-ground radars. Furthermore, the spreading of power over frequency bands with a width up to 10 GHz reduces the power per unit bandwidth—or the power density—to such low levels that the radiation cannot be detected by conventional monitoring equipment using a small relative bandwidth. There is, of course, no objection to radiation that is not detectable by conventional equipment.

1.5 Attenuation of Waves, Noise, and Distortions

Figure 1.5-1 shows the round-trip attenuation of sinusoidal waves due to rain or fog as function of the frequency for the frequency range from 1

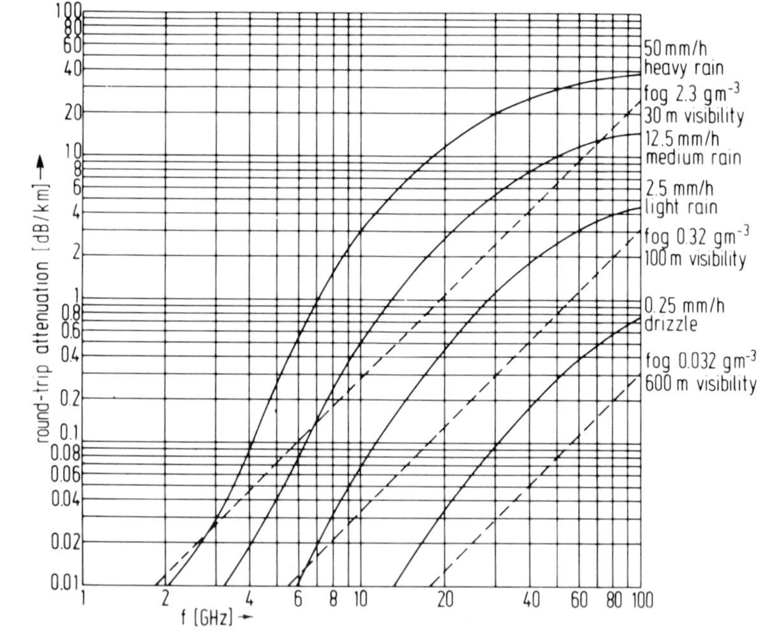

Fig. 1.5-1. Round-trip attenuation of a radar signal in decibels per kilometer distance to the target for the frequency range from 1 to 100 GHz. Curves of this type seem to have been obtained first by J. W. Ryde and D. Ryde (Ridenour, 1947, p. 61). They may be found in various presentations in many publications. The curves for fog were taken from p. 472 of the book by Barton (1964). The curves for rain were plotted from data computed by Setzer (1970); they differ slightly at low frequencies from corresponding curves published by Barton (1964, p. 472).

to 100 GHz. This illustration is specifically intended for radar. Round trip means that the waves travel the distance $2D$ if the target has the distance D from a radar.[1] The plot is not extended below 1 GHz since the attenuation becomes negligible there.

Figure 1.5-2 shows the noise temperature in the range from 100 MHz to 100 GHz for a radar with the elevation angles 0°, 1°, ..., 90°, where 0° means that the radar looks toward the horizon, and 90° means that it looks vertically up.

Figure 1.5-3 shows the round-trip attenuation of sinusoidal waves due to molecular absorption at sea level as a function of the frequency for the frequency range $0 \leqq f \leqq 130$ GHz. There is essentially no attenuation below about 10 GHz. Regions with low attenuation occur around 35 and 94 GHz. Peaks occur at 22.2, 60, and 118 GHz due to resonances of the H_2O and O_2 molecules.[2]

Let us use these three illustrations to determine when a signal should not be shifted to a higher frequency band in order to achieve a small relative bandwidth. Consider a radar using pulses with a duration of $T = 1\ \mu s$, which yield a range resolution of about 150 m. The pulses occupy the approximate band $0 \leqq f \leqq 1$ MHz. In order to turn the absolute bandwidth $\Delta f = 1$ MHz in the baseband into a relative bandwidth $\eta = 0.01$ we need a carrier frequency

$$f_c = 1/\eta T = \Delta f/\eta = 100 \text{ MHz}$$

---

[1] This illustration provides a good example of how the concept of small relative bandwidth enters our thinking. The attenuation of a pure sinusoidal wave between two points with distance D or $2D$ is a well-defined concept. There are no distortions, whereas the change of attenuation with frequency implies distortions for all other signals. The usual conclusion is that one should use signals with small relative bandwidth in order to keep the distortions low. It is rarely recognized that the price for no distortions is no range resolution, and the price for low distortions is poor range resolution.

[2] Data on molecular absorption may be found in many references (Altshuter *et al.* 1968; Falcone and Abreu, 1979; Moore 1979; U.S.A. 1977a,c). Some of these references also contain plots for rain and fog. The plots in Fig. 1.5-1 for fog correspond with data by Chen (1975) for *coastal fog;* the attenuation for *inland fog* is about half as large. Data for frequencies higher than used here are given by Zhevakin and Naumov (1967) and by Deirmendjian (1975). One should observe that the plots of Figs. 1.5-1–1.5-3 contain many simplifying assumptions that are needed to permit us to draw a few simple plots for "typical" conditions. For instance, the attenuation for rain in Fig. 1.5-1 depends not only on the rate of rainfall, but on the size of the drops, which may vary for the same rate of rainfall. The plots of Fig. 1.5-2 also depend on the reception angle of the antenna, the concentration of water vapor in the air, etc. The plot of Fig. 1.5-3 depends again on the concentration of water vapor in the air, on the air pressure, etc. The reader interested in such finer details is referred to the cited literature. The huge differences between attenuation and distortion in the bands $0.5 \text{ GHz} \leqq f \leqq 10.5$ GHz and $89 \text{ GHz} \leqq f \leqq 99$ GHz discussed here obviate the need to discuss these finer details.

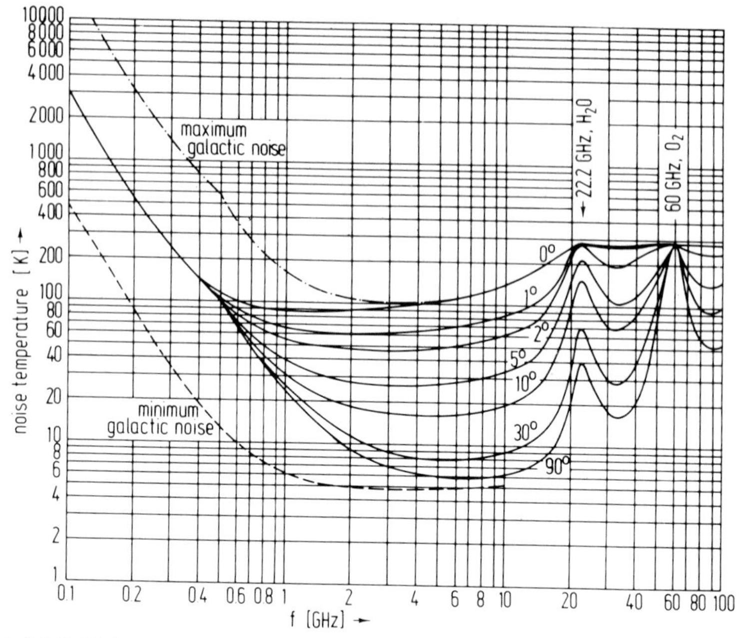

FIG. 1.5-2. Noise temperature as function of the frequency for the elevation angles 0°, 1°, ..., 90° of a radar. These plots are due to L. V. Blake of the Naval Research Laboratory. They may be found in a number of reports, papers, and books, e.g., in *Radar Handbook* (Blake, 1970).

according to Eq. (1.4-9). At 100 MHz, the noise temperature is still very high according to Fig. 1.5-2. One would much rather operate at a higher frequency, preferably in the range from 1 to 10 GHz. Hence, the requirement for a small relative bandwidth is overshadowed by the desire to operate at a higher frequency, where the noise temperature is lower. Furthermore, the radar dish assumes a more manageable size when one advances from 100 MHz to a higher frequency.

Let us next consider a radar that operates with pulses having a duration between 1 and 0.1 ns, which implies a range resolution between 15 and 1.5 cm. The baseband bandwidth Δf of the pulses is between 1 and 10 GHz. A relative bandwidth $\Delta f/f_c = 0.01$ calls for a carrier frequency between 100 and 1000 GHz. Figure 1.5-2 shows that the noise temperature at such high frequencies is well above the value in the band from 1 to 10 GHz. The real objection, however, does not come from the noise temperature but from the attenuation due to moisture and molecular absorption.

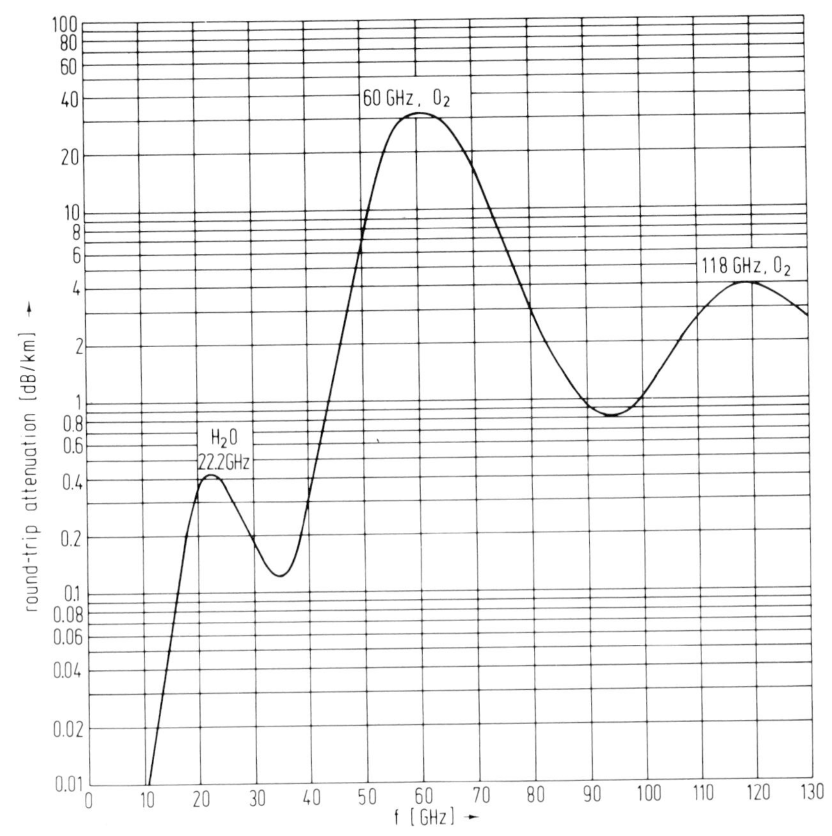

FIG. 1.5-3. Round-trip attenuation of sinusoidal waves due to molecular absorption at sea level. This plot is based primarily on data published by Zhevakin and Naumov (1967) and Blake (1970): The water vapor content of the air is assumed to be 7.5 g/m³.

Let us turn to Fig. 1.5-1 and consider a radar[1] using an absolute bandwidth of 10 GHz at the location 89 GHz $< f <$ 99 GHz. The relative bandwidth $\eta = 10/(89 + 99) \doteqdot 0.05$ is so large that there is only a very moderate resonance effect according to Fig. 1.4-1, and the noise rejection

[1] This reference system was originally advocated by D. K. Barton of Raytheon Co., a leading proponent of conventional techniques, as an example of what could be achieved by staying with small relative bandwidths. It now provides the best arguments for the use of large relative bandwidths. The pulse duration of 0.1 ns and the distance of 10 km are used because of the practical importance of these values for the low-elevation angle radar and the ICBM radar. More details on applications of radar operating at 35 or 94 GHz may be found in a publication by Wiltse (1979), and the many references given there.

by the filter is thus rather poor. The attenuation in fog of 30 m visibility for the signal is about 250 dB for a distance of 10 km to the target (Fig. 1.5-1 shows about 25 dB/km at $f = 100$ GHz). If, on the other hand, the signal with an absolute bandwidth of 10 GHz is at the location 0.5 GHz $< f <$ 10.5 GHz, one obtains a relative bandwidth $\eta = 10/(0.5 + 10.5) \doteqdot 0.9$. The attenuation at the upper band edge for a distance of 10 km is now 3 dB according to Fig. 1.5-1, and negligible at the lower band edge. The net gain due to the increase of the relative bandwidth is thus about[1] 247 dB.

Let us turn to Fig. 1.5-3. There is essentially no attenuation for the signal in the band 0.5 GHz $\leqq f \leqq$ 10.5 GHz, but the attenuation for the signal in the band 89 GHz $\leqq f \leqq$ 99 GHz equals more than 0.8 dB/km, or 8 dB for a distance of 10 km. Combined with the 247 dB due to fog we obtain an advantage for the large-relative-bandwidth radar over the conventional radar pushed to its limits of about 255 dB in the presence of heavy fog, and about 8 dB if the weather is fair.

It is worth contemplating the gain of 255 dB, since it means a factor of 3×10^{25}. The reader will probably not be able to come up with another example of a theoretical insight providing anywhere near such an improvement—be it in power or anything else—over an existing equipment or method.[2] To call this improvement astronomical would be an understatement, since we would have to give the distance to the farthest known celestial objects in meters or the known age of the universe in microseconds to obtain a number of the order of 10^{25}.

We return again to the question of noise. According to Fig. 1.5-2 the lowest noise temperature is in the region from about 1 to about 10 GHz, while the noise temperature in the band 89 GHz $< f <$ 99 GHz is significantly higher. However, the question is sometimes raised whether signals with large relative bandwidth are equally affected by noise as signals with small relative bandwidth. This question was answered by Kotel'nikov (1947) some 30 years ago. For additive thermal noise the time variation of the signal is quite unimportant, only the energy of the signal affects the error probability,[3] and the time variation is generally the only difference between any radio signals having the same polarization and power.

---

[1] Similar considerations apply, if the attenuation is due to chaff rather than water droplets. A pulse with a duration of 0.1 μs should penetrate chaff about like a pulse with a duration of 0.1 ns penetrates rain and fog, if no carrier is used.

[2] A frequency of 100 GHz implies a wavelength of 3 mm. The *millimeter-wave radar* is sometimes advocated for all-weather use (e.g., Whalen 1979). Millimeter waves are substantially less attenuated by fog than infrared or visible waves, but they are slightly more attenuated by rain and substantially more by molecular resonances.

[3] A derivation of this result using the same terms as this book was previously published (Harmuth 1972, pp. 292–299).

Another question often raised is that of signal distortions. In vacuum, the propagation of an electromagnetic wave is described by the wave equation, which implies distortion-free transmission. In the atmosphere, for transmission above the shortwave region, distortions may be caused by rain, fog, and molecular resonances. It was just discussed that a signal in the band 0.5 GHz $< f <$ 10.5 GHz suffers an attenuation of 3 dB at the upper band edge in heavy fog, whereas there is essentially no attenuation at the lower band edge. This implies a distortion that may be corrected, if one feels a need to do so, by an equalizer having 3 dB more attenuation at 0.5 GHz than at 10.5 GHz.

The signal in the band 89 GHz $\leqq f \leqq$ 99 GHz has an attenuation of about 20 dB/km at the lower band edge and of about 25 dB/km at the upper band edge according to Fig. 1.5-1. The difference equals 5 dB/km, or 50 dB for a distance of 10 km. Hence, the attenuation distortion increases from 3 to 50 dB if we replace the signal with large relative bandwidth with the one having a small relative bandwidth. This is just the opposite of what most communication engineers would expect.

The question of distortions due to molecular resonances can be answered with the help of Fig. 1.5-3. There is essentially no absorption and thus no distortion in the band 0.5 GHz $\leqq f \leqq$ 10.5 GHz. In the band 89 GHz $\leqq f \leqq$ 99 GHz the round-trip attenuation varies by about 0.2 dB/km, or 2 dB for a distance of 10 km. Hence, molecular resonances cause insignificant distortions in either case. The situation would be completely different for a radar operating in the band 40 GHz $\leqq f \leqq$ 50 GHz, where the attenuation varies by about 7 dB/km.

From all this it is clear that the region from about 0.5 GHz to somewhat above 10 GHz is ideal for an all-weather radar, and that signals with a bandwidth of about 10 GHz can be used even under the worst weather conditions. In fair weather, a large-relative-bandwidth radar can use a bandwidth close to 45 GHz, and have an attenuation below that of a narrow-band radar operating at 94 GHz according to Fig. 1.5-3.[1]

[1] The introduction of nonsinusoidal waves to radar aroused many objections from engineers and scientists involved in the manufacture and development of conventional radars. The scientific reasons for these objections have two sources: (a) It is not understood that Maxwell's equations do not distinguish sinusoidal waves, although solutions obtained with Bernoulli's product method do so, as discussed at the end of Section 1.3. (b) It is not recognized that most of the existing radar and radio equipment works best or even only with sinusoidal waves because it was designed to do so. In order to provide a public forum for a discussion of objections to nonsinusoidal waves, a panel discussion on "Radio Signals with Large Relative Bandwidth" was organized on 7 October 1980 at the IEEE International Symposium on Electromagnetic Compatibility in Baltimore, Maryland. Calls for opponents were published in *Microwave J.* **23** (No. 1, Jan. 1980, p. 10) and *IEEE Trans. Electromagn. Compat.* **EMC-22** (No. 1, Feb. 1980, p. 86). Many personal invitations were sent to known

1.6 When to Use Nonsinusoidal Waves

The frequency band from about 1 to about 10 GHz with a bandwidth of essentially 10 GHz has just been identified to be ideally suited for signals with large relative bandwidth, or *nonsinusoidal signals* for short. The typical application for this band is the line-of-sight, high-resolution, all-weather radar.

The conventional line-of-sight, all-weather radar cannot use signals with a bandwidth wider than about 100 MHz in the baseband if the highest frequency after modulation is restricted to about 10 GHz. The possible bandwidth of the radar with large relative bandwidth is about 100 times as large, which implies an increase in information acquisition by a factor of about 100. This gain consists inherently in better time resolution—the width of the used pulses can be decreased from 10 to 0.1 ns—but time resolution can be translated into range resolution, Doppler resolution, angular resolution, target signature, etc.

Radar is the major field for the use of very large absolute bandwidths. A lesser field is spread spectrum communications. Nonsinusoidal waves permit us to spread signals over a bandwidth of 10 GHz, stay below the resonance of the H_2O molecule, and avoid the attenuation by rain and fog.

To find other useful ranges in the frequency spectrum we must find instances where an increase of the frequency of a carrier is essentially impossible. The shortwave band is an example. Using a typical relative bandwidth of $\eta = 0.01$ and a carrier frequency $f_c = 15$ MHz in the middle of the shortwave band, we obtain from Eq. (1.4-9) the baseband bandwidth $\Delta f = \eta f_c = 150$ kHz. Since the shortwave band inherently permits the transmission of about 3 MHz wide signals before the frequency dependence of propagation becomes a problem, we use less than 10% of the bandwidth allowed by nature if we insist on a small relative bandwidth. Using techniques for large relative bandwidths, we may increase the absolute bandwidth of signals by about one order of magnitude compared with the typical case $\eta = 0.01$. The major use for this potential gain is for over-the-horizon radar, where the limited resolution has been a

---

opponents in industry and government, but none were accepted. The discussion panel, chaired by the author, consisted thus only of proponents: C. Bertram of Bertram Technology Inc., Merrimac, New Hampshire; J. C. Cook of Teledyne-Geotech, Dallas, Texas; A. Dean of U.S. Army Cold Regions Research and Engineering Laboratory, Hanover, New Hampshire; Fan Changxin of Northwest Telecommunication Engineering Institute, Xi'an, China; H. Lueg of Technische Hochschule Aachen, West Germany; J. V. Rosetta of Geophysical Survey Systems Inc., Hudson, New Hampshire. Two more proponents had planned to participate but could not come: J. R. Rossiter of Memorial University of Newfoundland, Canada; M. Zecha of Deutsche Akademie der Wissenschaften, Zentralinstitut für Kybernetik, Berlin, East Germany. A radar produced by Geophysical Survey Systems, Inc. was demonstrated at this symposium by J. V. Rosetta.

problem for many years. Note that the potential gain in absolute bandwidth for the over-the-horizon radar is only one order of magnitude, whereas it was two orders of magnitude for the line-of-sight radar.

The attenuation of an electromagnetic wave in seawater is given by the formula

$$\alpha(f) = 0.0345\sqrt{f} \quad [\mathrm{dB/m}] \tag{1}$$

where f is measured in hertz. For a penetration depth of 300 m, which is typically required for radio communication with deeply submerged submarines, one cannot use frequencies higher than some tens of hertz; e.g., for $f = 100$ Hz one obtains from Eq. (1) already an attenuation of 90 dB for a depth of 300 m. With the highest possible frequency so severely limited, the restriction of the relative bandwidth to values of about $\eta = 0.01$ would result in absolute bandwidths for a signal on the order of fractions of a hertz, and a corresponding small transmission rate of information. Hence, radio communications to and from deeply submerged submarines, and radar that penetrates seawater, are among the most intriguing applications of nonsinusoidal signals.

A very similar application is the into-the-ground radar. Inhomogeneities in the ground make the attenuation of electromagnetic waves increase with frequency very rapidly, just like raindrops and fog droplets make the attenuation in air increase very rapidly with frequency according to Fig. 1.5-1. In order to provide an acceptable resolution, such into-the-ground radars must use pulses with a duration of about 1 ns. The penetration depth currently achieved is a few meters in dry ground and at least 30 m in ice (Rossiter and Butt, 1979); the use of a carrier and a small relative bandwidth would result in no useful penetration. Some 60 such into-the-ground radars have been built so far, and they have been used in ten countries.[1] Their typical use is the probing of the ground in the course

[1] The following commercial organizations have developed into-the-ground radars: ENSCO Inc., Springfield, Virginia (J. Fowler, L. Davis); Geophysical Survey Systems Inc., Hudson, New Hampshire (J. Mann, J. V. Rosetta, D. F. Stanfill *et al.*); Morey Research Co., Nashua, New Hampshire (R. Morey); MPB Technologies Inc., Ste. Anne de Bellevue PQ, Canada (J. M. Keelty, S. Y. K. Tam); Teledyne-Geotech, Dallas, Texas (J. C. Cook); Terrestrial Systems Inc., Lexington, Massachusetts (J. Chapman). Academic developers and/or users are: Ohio State University Electronics Laboratory (C. W. Davis III, D. L. Moffatt, L. Peters *et al.*); Centre for Cold Ocean Resources Engineering, Memorial University of Newfoundland, Canada (J. R. Rossiter, K. A. Butt *et al.*); Department of Geophysics and Astronomy, University of British Columbia, Canada (G. K. C. Clarke, B. B. Narod); Institut für Technische Elektronik, Technische Hochschule Aachen, West Germany (H. Lueg). Government agencies that have made major contributions are: U.S. Army Cold Regions Research and Engineering Laboratory, Hanover, New Hampshire (J. Brown, A. Dean, A. Kovacs *et al.*); Geological Survey of Canada, Ottawa (A. P. Annan, J. L. Davis, W. J. Scott); Office of Naval Research, Washington, D.C. (H. Bolezalek).

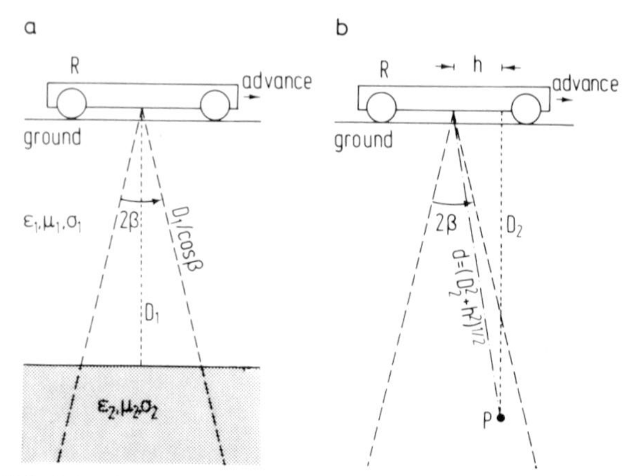

FIG. 1.6-1. Principle of the operation of an into-the-ground radar. (a) Probing of an extensive layer with discontinuity of ϵ, μ, σ; (b) detection of a pipe with small diameter: R, radar; P, pipe.

of excavations, since they will not only detect metallic pipes but also pipes made of concrete, clay or plastic. Other uses have been the detection of cavities in the ground, measuring the thickness of sea ice to assure a safe route for trucks or a landing site for aircrafts on the Arctic Ocean, etc.

The carrier-free into-the-ground radar was first proposed by J. C. Cook[1] (1960); a considerable number of papers on into-the-ground radar with and without carrier have been published since then.[2] For an explanation of the operation of this type of radar, refer to Fig. 1.6-1a. The radar is mounted on a cart that is pulled along the surface of the ground. A pulse of about 1 ns duration is radiated into the ground at intervals of the order of microseconds to milliseconds. A layer at the depth D_1 with a discontinuity of the dielectric constant ϵ, the magnetic permeability μ, or the conductivity σ will reflect the signal. The reflected signal will come from a

[1] By an incredible coincidence, this paper was published in the same year and in the same journal as the author's first papers on nonsinusoidal signals (Harmuth, 1960a,b).

[2] Annan and Davis (1976); Bertram *et al.* (1972); Campbell and Orange (1974); J. C. Cook (1970, 1973); Davis *et al.* (1976); Duckworth (1970); Harrison (1970); Kovacs and Abele (1974); Kovacs and Gow (1975, 1977); Kovacs and Morey (1978, 1979a,b,c); Moffat and Puskar (1976); Morey (1974); Morey and Kovacs (1977); Porcello *et al.* (1974); Rossiter and Butt (1979); Rossiter *et al.* (1979). More publications may be found referenced in the papers listed here. No papers on carrier-free radar or radio signals with large relative bandwidth were published by the more typical journals for radar and radio transmission, in particular, *IEEE Trans. Aerospace and Electronic Systems*, *IEEE Trans. Communications*, *IEEE Trans. Antennas and Propagation*, and *Microwave Journal*.

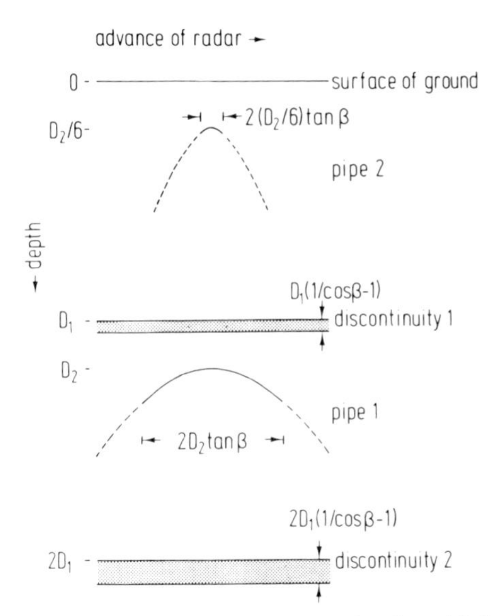

FIG. 1.6-2. Idealized recording of an into-the-ground radar with infinitely short pulses showing the ground profile along the line of advance of the radar. Layers with discontinuity of ϵ, μ, σ are shown at the depths D_1 and $2D_1$; pipes are shown at depths D_2 and $D_2/6$.

minimum distance D_1 and a maximum distance $D_1/\cos\beta$, if the beam angle of the radar equals 2β. Hence, the discontinuous surface is shown in the idealized recording of Fig. 1.6-2 as a band of width $D_1/\cos\beta - D_1$ at the depth D_1; a second discontinuous surface at the depth $2D_1$ produces a band that is twice as wide.

Consider next the detection of a pipe that runs vertically to the paper plane in Fig. 1.6-1 at a depth D_2. If the radar has the horizontal distance h from the pipe, the distance to the pipe will be $d = (D_2^2 + h^2)^{1/2}$. This signature is shown for a pipe 1 at a depth D_2, and for a pipe 2 at a depth $D_2/6$ in Fig. 1.6-2. Due to the beam angle 2β this signature will only be recorded for a maximum horizontal distance $h_{max} = D_2 \tan\beta$ or $h_{max} = (D_2/6)\tan\beta$ in both directions of the pipe. These sections of the pipe signatures are emphasized by solid lines in Fig. 1.6-2.

An actual recording of an into-the-ground radar is shown in Fig. 1.6-3. On top—i.e., at the surface of the ground—is the usual large signal caused by crosstalk from the transmitter to the receiver. It limits the minimum depth that can be probed, which explains why into-the-ground radars are often placed a considerable distance above ground. There are four bands indicating discontinuous surfaces, the deepest one at a depth of about 2.5 m being identified as the water table. The parabola-like signa-

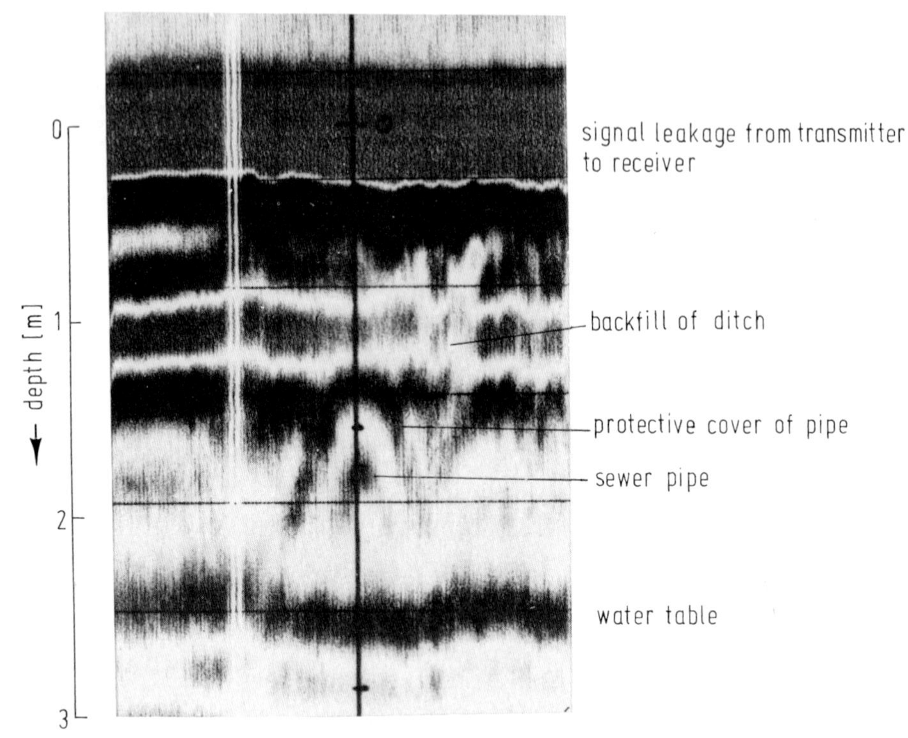

FIG. 1.6-3. Actual recording of an into-the-ground radar. The different signatures of extensive layers and the sewer pipe with its protective cover are conspicuous. (Courtesy J. Chapman of Terrestrial Instruments Inc., Lexington, Massachusetts.)

tures of the sewer pipe and a protective cover over it clearly have the shape shown in Fig. 1.6-2 for pipes. The disturbance above and to the right of the sewer pipe cover signature is caused by construction debris used to back-fill the ditch of the sewer pipe.

Figure 1.6-4 shows a radar profile of a peat deposit. The ground surface and the profile of the clay bottom are clearly visible. The echos just below the surface are caused by buried roots and tree trunks. Two buried trees can be recognized; one of them could be verified visually.

Figure 1.6-5 shows an into-the-ground radar mounted on a cart, as discussed in connection with Fig. 1.6-1, with some processing equipment. For instant displays of ground profiles, as in Figs. 1.6-3 and 1.6-4, one uses graphic recorders. Sometimes the data is stored on magnetic tape and computer processed at a later time.

A radar mounted on a sled and pulled by a tractor that contains the processing and display equipment is shown in Fig. 1.6-6. This radar was used

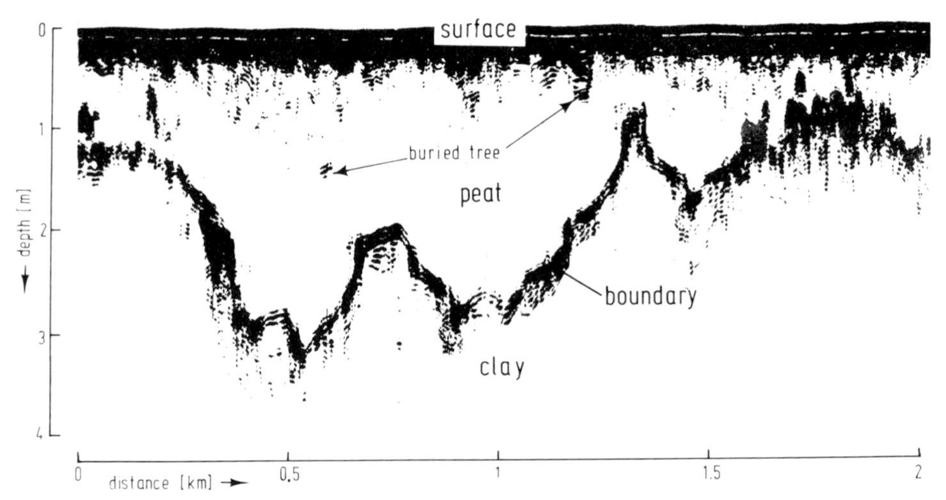

FIG. 1.6-4. Radar profile of a peat deposit. The boundary between the peat and the clay bottom is clearly visible. (Courtesy J. V. Rosetta and D. F. Stanfill of Geophysical Survey Systems Inc., Hudson, New Hampshire.)

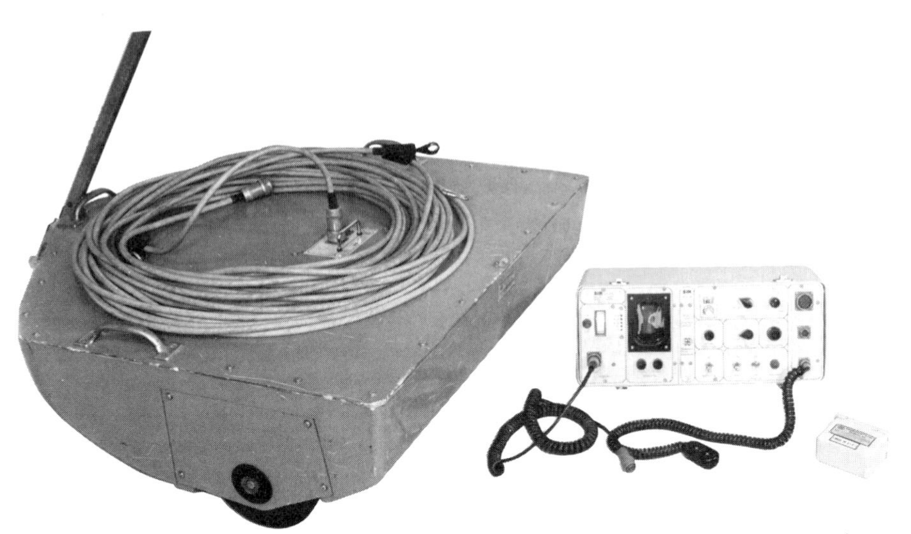

FIG. 1.6-5. Into-the-ground radar cart according to Fig. 1.6-1 and some processing equipment. (Courtesy J. V. Rosetta and D. F. Stanfill of Geophysical Survey Systems Inc., Hudson, New Hampshire.)

FIG. 1.6-6. Into-the-ground radar mounted on a sled used for surveying a safe truck route on the ice of the Arctic Ocean. (Courtesy J. V. Rosetta and D. F. Stanfill of Geophysical Survey Systems Inc., Hudson, New Hampshire.)

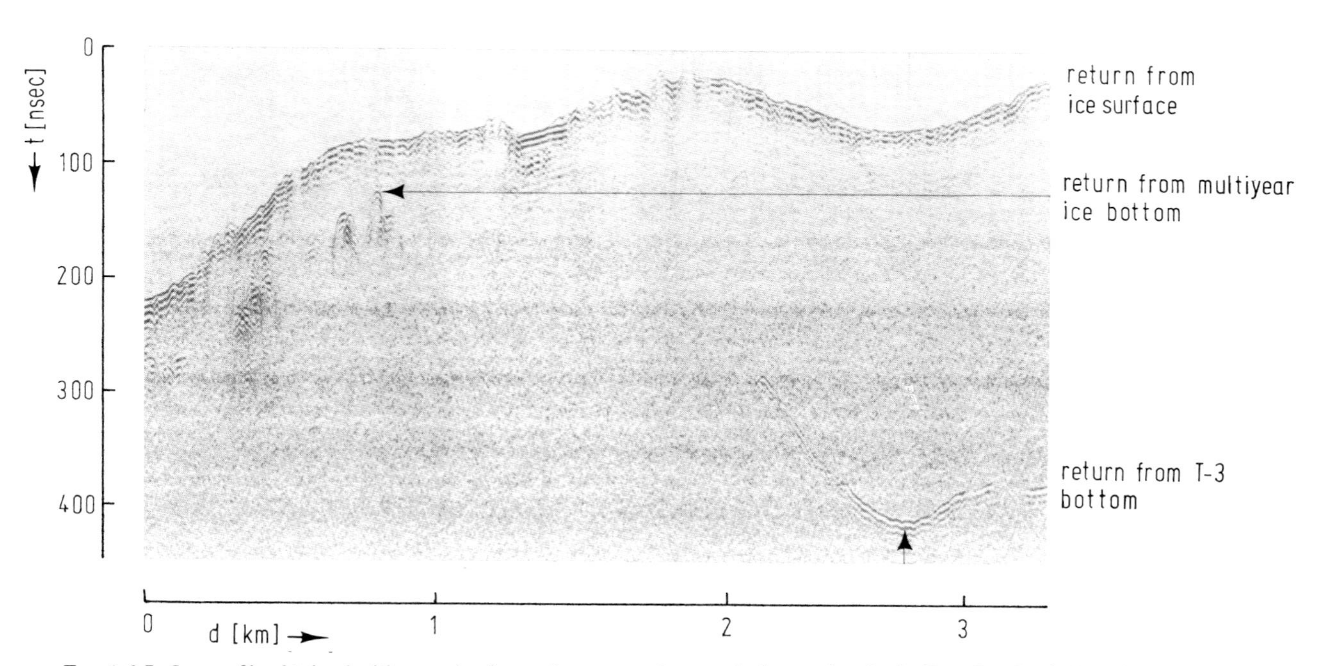

FIG. 1.6-7. Ice profile obtained with a carrier-free radar mounted on an airplane, taken in the Beaufort Sea in March, 1979. The airplane flew at a nominal altitude of 80 m with an airspeed of 70 m/s. The different thickness of single-year and multiyear ice is conspicuous. The return from the bottom of the ice island T-3 at the point indicated by the arrow comes from a depth of 26 m below the ice surface. (Courtesy J. R. Rossiter and K. A. Butt, Centre for Cold Ocean Resources Engineering, Memorial University of Newfoundland, St. John's.)

to probe the thickness of sea ice for the survey of a truck route on the Arctic Ocean.

Ice surveying is a major application of carrier-free radar. It is used either to find thick, safe ice or thin ice that is readily penetrated by ships. Figure 1.6-7 shows an ice profile obtained by an airborne radar in the Beaufort Sea (Rossiter and Butt, 1979). One may distinguish the different thicknesses of first-year ice, multiyear ice, and the ice island[1] T-3. The ice surface appears very uneven, but one must keep in mind that all distances were measured from the aircraft. The uneven surface of the ice indicates that the altitude of the aircraft was not constant along the surveyed line.

Figure 1.6-8 shows the radar with which the ice profile of Fig. 1.6-7 was obtained. This was the first successful use of a carrier-free radar mounted on a fixed-wing aircraft.[2] Helicopter-mounted carrier-free radars have

FIG. 1.6-8. Airplane with carrier-free radar used to obtain the ice profile of Fig. 1.6-7. (Courtesy J. R. Rossiter and K. A. Butt, Centre for Cold Ocean Resources Engineering, Memorial University of Newfoundland, St. John's.)

[1] For a description of the ice islands in the Arctic Ocean see Blyth (1953) or Thomas (1965).

[2] The previously mentioned J. C. Cook used an airborne carrier-free radar as early as 1966, but the beamwidth of the antenna was too large to obtain useful results (personal communication; see also Cook, 1970).

FIG. 1.6-9. Carrier-free radar mounted on a helicopter of the U.S. Coast Guard for ice survey. (Courtesy A. M. Dean, U.S. Army Cold Regions Research and Engineering Laboratory, Hanover, New Hampshire.)

been used since 1976; as far as the author is aware, the first users were A. Dean and A. Kovacs, both of U.S. Army Cold Regions Research and Engineering Laboratory, Hanover, New Hampshire, together with R. Morey of Morey Research Co., Nashua, New Hampshire.[1]

Figure 1.6-9 shows a typical installation of a carrier-free radar on a helicopter of the U.S. Coast Guard used for ice survey of shipping channels. Figure 1.6-10 gives a recording of the St. Marys River below Sault Ste. Marie—between Lake Superior and Lake Huron—obtained with this radar. The sheet ice on the left is clearly distinguishable from the brash

[1] Scientists of the U.S. Navy, Naval Research Laboratory, published a paper in 1979 that tried to prove that there were no useful applications for "carrier-free waveforms" (J. R. Davis *et al.*, 1979). This was three years after the U.S. Coast Guard in cooperation with the U.S. Army had started airborne ice surveys by carrier-free radar, nine years after Geophysical Survey Systems, Inc. had started the commercial production of such radars, and about 15 years after J. C. Cook had built the first experimental into-the-ground radar. Upward of 30 papers reporting experimental results had been published by 1979. We have here a well-recorded example of the time required for a basic new concept to become known and accepted in science.

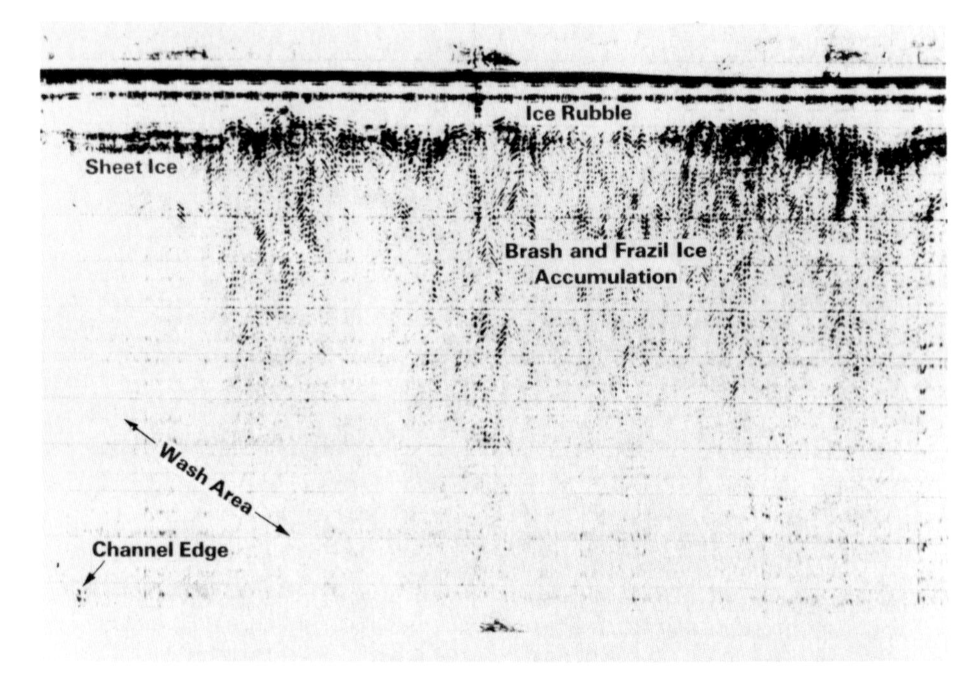

FIG. 1.6-10. Typical recording of the helicopter-borne radar of Fig. 1.6-9, showing a cross section of a shipping channel near Sault Ste. Marie. The ice rubble is about 300 m wide, the brash and frazil ice accumulation some 3–4 m deep. (Courtesy A. M. Dean, U.S. Army Cold Regions Research and Engineering Laboratory, Hanover, New Hampshire.)

and frazil ice accumulation caused by the ship traffic. In the lower left-hand corner one may recognize the channel edge at a depth of about 4 m (the ability of carrier-free radar to penetrate several meters of water, and give a useful resolution, has led to its use for profiling river bottoms from the air to find places that are sufficiently shallow for fording).[1]

The brash and frazil ice accumulation in Fig. 1.6-10 extends to a depth of about 4–5 m. This appears to contradict the depth of 4 m given previously for the channel edge. However, the channel edge in Fig. 1.6-10 is covered by almost 4 m of water—plus the sheet ice—and the electromagnetic wave has a significantly slower propagation velocity in the water than in the ice accumulation to its right. Hence, one must either know or

[1] In the Soviet Union, measurements of the thickness and salinity of sea ice as well as the salinity of seawater by means of airborne radar are carried out by scientists of the Institute of Radio Engineering and Electronics of the Academy of Sciences (Institut Radiotekniki i Elektroniki Ak. Nauk), Karl Marx Prospect 18, 103907 Moscow.

FIG. 1.6-11. Carrier-free dual radar used for probing massive ice in permafrost ground along the Alaska pipeline. (Courtesy A. Kovacs, U.S. Army Cold Regions Research and Engineering Laboratory, Hanover, New Hampshire, and R. Morey, Morey Research Co., Nashua, New Hampshire.)

measure the propagation velocity in addition to the propagation time of the radar pulse to make full use of ground-probing radars.

The problem of measuring simultaneously propagation time and velocity was neatly solved by means of the *dual radar* (Kovacs and Morey, 1979b). Figure 1.6-11 shows such a dual radar used for profiling regions of massive ice in the permafrost ground along the Alaska pipeline. The usual transmitting/receiving antenna is augmented by a second antenna that receives only. A recording of the ground profile is shown in Fig. 1.6-12. The numbers 218 ... 240 on top of the illustration identify the vertical support structures of the pipeline; two of these support structures are visible in Fig. 1.6-11. Two holes had to be drilled for each structure. The top and bottom of the massive ice penetrated by the holes on the west side of the pipeline are marked by circles in Fig. 1.6-12, whereas the stars mark top and bottom of the ice for the drill holes on the east side. The radar profile and the drill-hole data correlate quite well, major discrepancies being due to the change of thickness of the ice over short distances; e.g., at the vertical support structure 224 the drill hole on the west side (circles) shows the ice about 6 m thick, whereas the drill hole on the east side—which was only 3 m away—showed no ice at all.

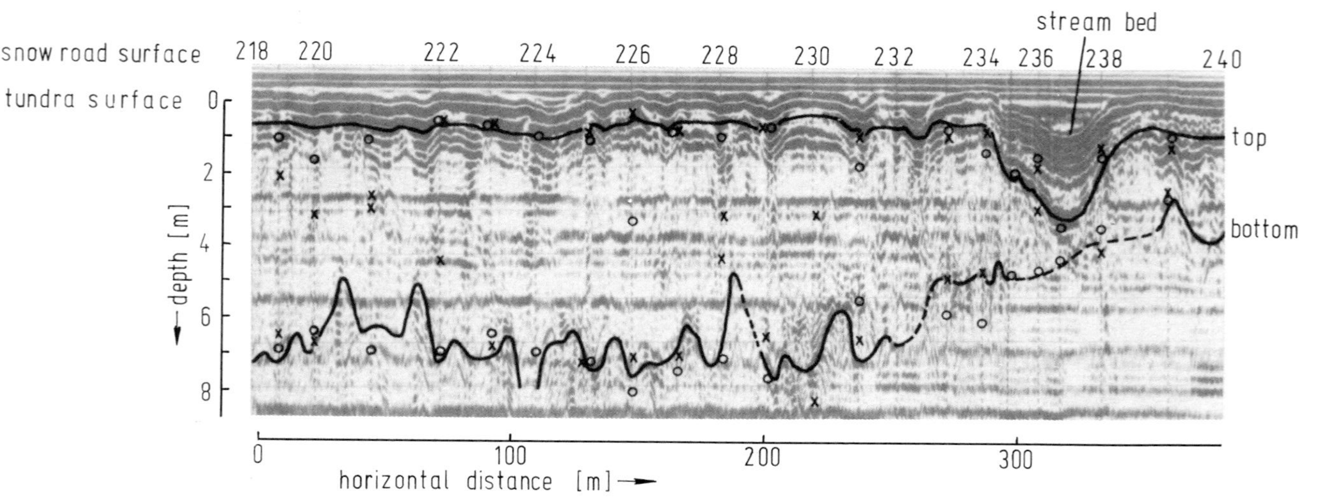

FIG. 1.6-12. Record of ground profile obtained by carrier-free dual radar and by drilling along the Alaska pipeline from vertical support structure 218 to 240. (Courtesy A. Kovacs, U.S. Army Cold Regions Research and Engineering Laboratory, Hanover, New Hampshire, and R. Morey, Morey Research Co., Nashua, New Hampshire.)

FIG. 1.6-13. Helicopter-mounted carrier-free dual radar used for surveys at Prudhoe Bay, Alaska. The man on the top of the helicopter is deicing the rotor. (Courtesy A. Kovacs, U.S. Army Cold Regions Research and Engineering Laboratory, Hanover, New Hampshire.)

Figure 1.6-13 shows a carrier-free dual radar with two receiving antennas mounted on a helicopter. This radar was used for surveys at Prudhoe Bay, Alaska.

A carrier-free dual radar probing the Ross Ice Shelf near McMurdo Sound in Antarctica is shown in Fig. 1.6-14. The great distance between the two antennas indicates that probing of substantially greater depths than in Figs. 1.6-11 and 1.6-13 is attempted.

A carrier-free radar designed for ground-probing from tunnels of coal mines is shown in Fig. 1.6-15. This radar produces rather long pulses of 25 ns duration with a peak power of 2 MW. The long pulses and the high power permit a probing distance of about 60 m in sandstone and about 9 m in shale, which is frequently encountered in coal mines. This radar was built by J. C. Cook, whose paper of 1960 we mentioned as the beginning of the carrier-free into-the-ground radar (J. C. Cook, 1975).

The use of carrier-free radar as airport ground control sensors and simi-

FIG. 1.6-14. Profiling of the Ross Ice Shelf at McMurdo Sound, Antarctica, with a carrier-free dual radar. (Courtesy A. Kovacs, U.S. Army Cold Regions Research and Engineering Laboratory, Hanover, New Hampshire.)

lar applications were described by Bennett, Nicholson, and Ross (Ross, 1974; Nicholson and Ross, 1975; Bennett and Ross, 1978).

In Chapters 2–4, we will briefly discuss basic concepts of implementation of radio transmission with large relative bandwidth. However, most of this book is devoted to principles and applications; they come first. Technology will be developed once sufficiently good principles and applications become widely known.

Experience has shown that many engineers have difficulties with the technology of radio signals with large relative bandwidth, since most of our technological advancement is done in small steps, and any radical change is thus not part of their experience. This is typical for a mature industry. There were few small but many radical changes when radio transmission got started around 1900. To obtain some idea of what one should expect, let us look at technology from a broader point of view.

The relative bandwidth η defined by Eq. (1.4-1) can have any value in the range $1 \geqq \eta \geqq 0$. Our current technology is based on a theory for the limit $\eta \to 0$. Both theory and technology that apply to the whole range

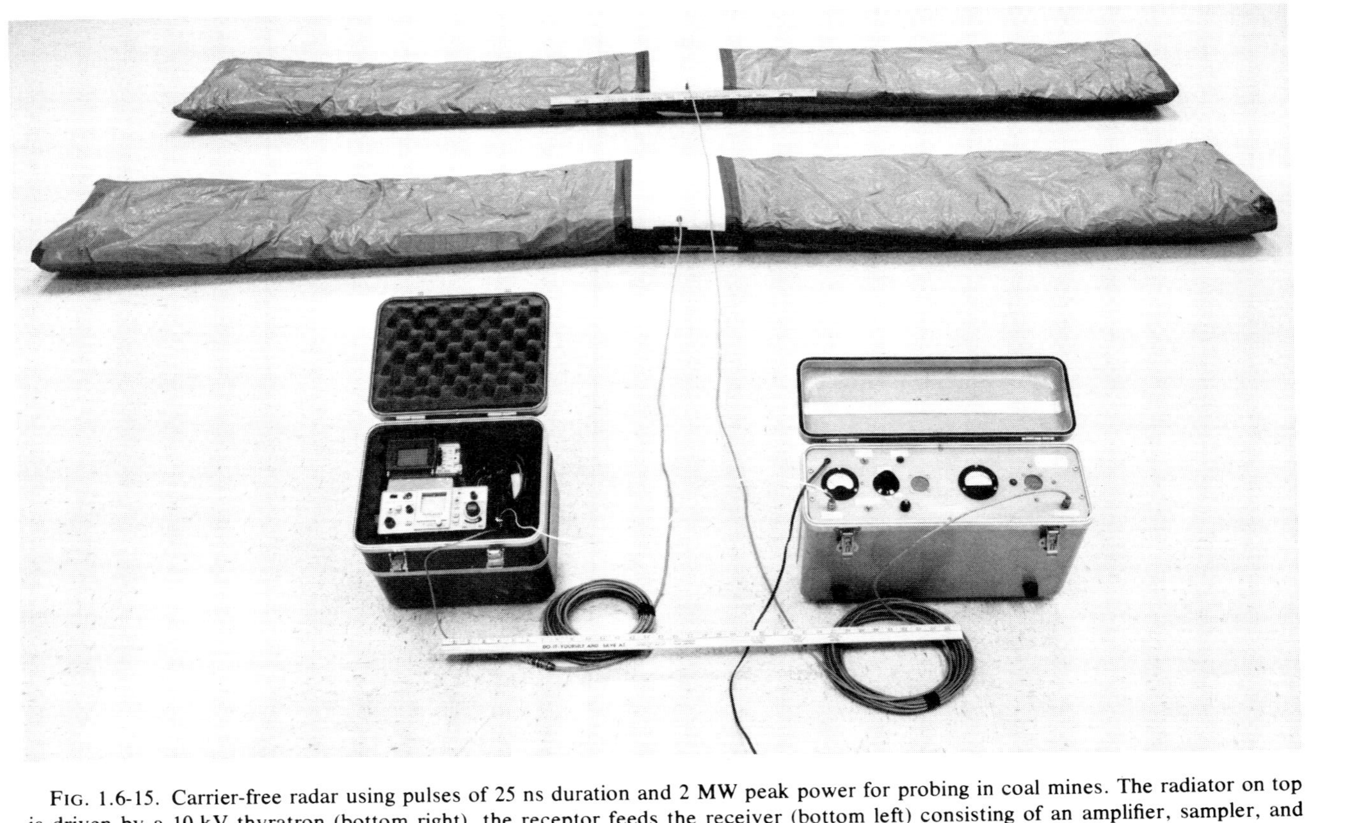

FIG. 1.6-15. Carrier-free radar using pulses of 25 ns duration and 2 MW peak power for probing in coal mines. The radiator on top is driven by a 10-kV thyratron (bottom right), the receptor feeds the receiver (bottom left) consisting of an amplifier, sampler, and oscilloscope. (Courtesy J. C. Cook, Teledyne-Geotech, Dallas, Texas.)

$1 \geqq \eta \geqq 0$ will have to be more general and more sophisticated than a theory and technology that apply to the limit $\eta \to 0$ only. Taking *Radar Handbook* (Skolnik, 1970) as a representative summary for the limit $\eta \to 0$ gives some idea of what one should reasonably expect for the general case $1 \geqq \eta \geqq 0$. It took 40 years of radar development, including several technology-advancing wars, before *Radar Handbook* could be compiled, and this indicates that a great deal of patience will be required before a comparable summary can be assembled for the general case.[1]

[1] As an example of an area of application that cannot be discussed in this book despite its potential, consider the problem of making an aircraft invisible to radar. At high frequencies, such as 35 or 94 GHz, one may use absorbing materials since the wavelength is in the range from 10 to 3 mm. At a frequency of 500 MHz, the wavelength is 60 cm, and the reduction of the backscattered wave by means of absorbing materials is hard to implement. This effect favors the use of low frequencies. On the other hand, the suppression of a backscattered wave by the radiation of an amplitude-reversed, but otherwise equal, wave is easy for a pure sinusoidal wave, but becomes increasingly difficult for signals with larger bandwidth. Furthermore, the larger the bandwidth of a signal the more information it can carry, which translates into a large radar signature. Both effects favor the use of large bandwidths. The radar most suited to overcome the problem of the invisible aircraft must thus use a large bandwidth and operate at low frequencies, which means its relative bandwidth should be close to 1.

2 Radio Signals with Large Relative Bandwidth

2.1 Basic Concepts

As the name implies, the carrier-free signals are radiated without the benefit of any kind of carrier. For the investigation of this type of transmission we summarize first some basic concepts:

(a) *Radiation of a Hertzian Electric Dipole.* The electric and magnetic field strengths produced by a current $i(t)$ flowing in a Hertzian dipole are given by the following formulas:

$$\mathbf{E} = Z_0 \frac{s}{4\pi c}\left[\frac{1}{r}\frac{di}{dt}\frac{\mathbf{r}\times(\mathbf{r}\times\mathbf{s})}{sr^2} + \left(\frac{c}{r^2}\,i + \frac{c^2}{r^3}\int i\,dt\right)\left(\frac{3(\mathbf{sr})\mathbf{r}}{sr^2} - \frac{\mathbf{s}}{s}\right)\right] \tag{1}$$

$$\mathbf{H} = \frac{s}{4\pi c}\left(\frac{1}{r}\frac{di}{dt} + \frac{c}{r^2}\,i\right)\frac{\mathbf{s}\times\mathbf{r}}{sr} \tag{2}$$

$$Z_0 = (\mu_0/\epsilon_0)^{1/2} \doteqdot 377\ \Omega \qquad c = (\mu_0\epsilon_0)^{1/2} \doteqdot 3\times 10^8\ \mathrm{m/s}$$

Here $i = i(t - r/c)$ is the dipole current $i(t)$ retarded by r/c; $\mathbf{r}$ is the location vector, s is the dipole vector, and $r = |\mathbf{r}|$, $s = |\mathbf{s}|$ are the magnitudes of these vectors (Harmuth, 1977a, p. 237).

At large distances the terms decreasing with $1/r$ will dominate; these are the wave zone or far zone components. The power transported by them through the unit area of a sphere around the dipole is given by Poynting's vector[1] $\mathbf{P}(1/r^2)$,

$$\mathbf{P}\left(\frac{1}{r^2}\right) = \mathbf{E}\left(\frac{1}{r}\right) \times \mathbf{H}\left(\frac{1}{r}\right) = Z_0\left(\frac{s}{4\pi rc}\right)^2\left(\frac{di}{dt}\right)^2 \sin^2\alpha\,\frac{\mathbf{r}}{r} \tag{3}$$

$$\sin\alpha = |\mathbf{s}\times\mathbf{r}|/sr$$

where α denotes the angle between the dipole vector s and the location vector $\mathbf{r}$.

(b) *Frequency—Band-Limited Signals.* Let a unit step function with amplitude A be applied to a *low-pass filter with infinite cutoff idealization,* which has no attenuation and linear frequency response of the phase in

[1] We use $\mathbf{E}(r, t)$, $\mathbf{H}(r, t)$ to denote the field strengths with all their components, while $\mathbf{E}(1/r, t)$, $\mathbf{H}(1/r, t)$ denote only the part of the field strengths that decreases like $1/r$, and $\mathbf{P}(1/r^2, t)$ denotes the part of Poynting's vector that decreases like $1/r^2$.

the band $0 \leqq f \leqq \Delta f$, but infinite attenuation for $f > \Delta f$. The output of the filter is shown by the function

$$A\left[\frac{1}{2} + \frac{1}{\pi}\,\mathrm{Si}(2\pi\,\Delta ft)\right]$$

in Fig. 2.1-1a, where $\mathrm{Si}(2\pi\,\Delta ft)$ denotes the sine-integral function. It is usual to approximate this complicated function by the one shown with a dashed line having the linear transient $1/2 + 2\,\Delta ft$ in the interval $-1/4\,\Delta f \leqq t \leqq 1/4\,\Delta f$, the value 0 for $t \leqq 1/4\,\Delta f$, and the value A for $t \geqq 1/4\,\Delta f$. A filter with this response to a unit step function is a *low-pass filter with linear transient idealization*. The output signals of this filter are not only easier to plot than those of an actual filter, but they are also much easier to produce practically.

Let the rectangular pulse of Fig. 2.1-1b be applied to a low-pass filter with linear transient idealization. The output signal is the triangular pulse of Fig. 2.1-1c. This pulse can readily be produced with high precision, even though the low-pass filter with linear idealization as well as the one with infinite cutoff idealization are difficult to approximate practically. One merely has to feed a rectangular pulse, like the one in Fig. 2.1-1b, to an integrator to produce the first half of the triangular pulse of Fig. 2.1-1c; the same rectangular pulse, amplitude-reversed and delayed by $t_0 = 1/2\,\Delta f$, fed to the integrator will produce the second half of the triangular pulse.

Both pulses of Fig. 2.1-1b and c occupy the infinite frequency band $0 \leqq f < \infty$, but the energy of the triangular pulse is better concentrated in the band $0 \leqq f \leqq \Delta f$. One might think that one could produce a pulse with all its energy in the band $0 \leqq f \leqq \Delta f$ by passing the rectangular pulse of Fig. 2.1-lb through a low-pass filter with infinite cutoff idealization, but this would lead to a pulse with infinite delay time, which is more of a defi-

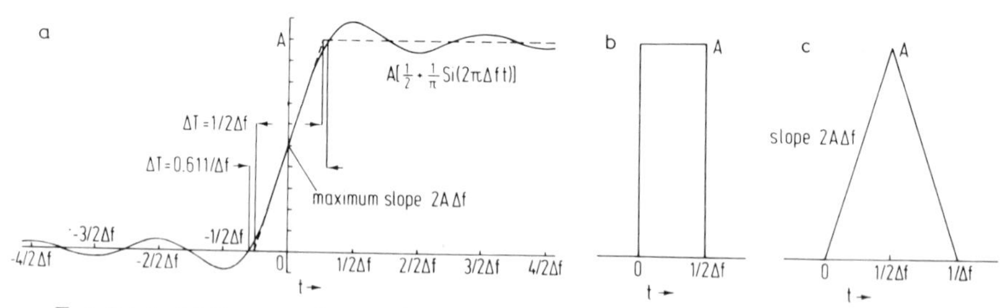

FIG. 2.1-1. (a) Output voltages of a filter with infinite cutoff idealization (solid line) and linear transient idealization (dashed line) due to a unit step input voltage. A rectangular input pulse (b) is transformed into a triangular output pulse (c) by a low-pass filter with linear transient idealization.

ciency[1] in radar than a pulse with only *most* of its energy in the band $0 \leqq f \leqq \Delta f$.

(c) *Antenna Current and the Field Strengths in the Far Zone.* Let us consider the relationship between the time variation of an antenna current and the electric and magnetic field strengths produced by it in the far zone. For a Hertzian electric dipole, the field strengths will vary like the first derivative of the current according to Eqs. (1) and (2). For a Hertzian magnetic dipole or a Hertzian electric quadrupole they will vary like the second derivative of the current[2] (Harmuth, 1977a, pp. 248–259). Other radiating antennas may have different relationships between the antenna current and the field strengths in the far zone, but in no case can these field strengths vary like the antenna current; the reason is that radiation is caused by the acceleration of charges, not by their constant motion, and the acceleration is proportionate to the first derivative of the current. A resonant electric dipole for sinusoidal waves can be viewed as many Hertzian electric dipoles that are all driven from one power source.[3] In this approximation, the electric and magnetic field strengths in the wave zone will vary like the first derivative of the current fed into the antenna.

Let a unit step current be fed through a low-pass filter with infinite cutoff idealization or linear transient idealization to a Hertzian dipole. The

[1] We are dealing here with an unavoidable deficiency of the Fourier description. The ordinary Fourier analysis is based on sine–cosine functions, which are distinguished by linear, time-invariant circuits and systems. The transmission rate of information is necessarily zero if everything is time-invariant, and Fourier analysis applies to this limiting case only. Whenever the transmission rate of information is greater than zero one runs into something beyond reality—such as infinite delays, infinite bandwidth, or an output signal preceding the application of an input signal—that needs correction by common sense. The mathematically correct way around this difficulty is to use an orthogonal system of functions defined in a finite interval, instead of the system of periodic sinusoidal functions in the infinite interval, as basis for a theory (Harmuth, 1969, 1972). However, this book was written to help advance results of the more rigorous theory from the academic to the industrial level. For this purpose it appears preferable to use concepts that are generally understood and accepted, even though this forces one to rely on the uncertain common sense.

[2] A Hertzian electric dipole is a very short dipole, a Hertzian magnetic dipole is a loop around a very small area, and a Hertzian electric quadrupole consists of two Hertzian electric dipoles with the currents flowing in opposite directions. A resonant electric dipole for sinusoidal waves has the length $\lambda/4$, $\lambda/2$, λ, 2λ, ... , where λ is the wavelength; it is much longer than a Hertzian electric dipole.

[3] This approximation holds if the energy in the standing wave of the antenna is large compared with the energy radiated or transformed into heat in the antenna during one period of the sinusoidal wave. One can build antennas consisting of many Hertzian dipoles driven by individual power sources to radiate more power than could a single Hertzian dipole. Hence, the following discussion based on the Hertzian electric dipole is quite close to what can be done practically.

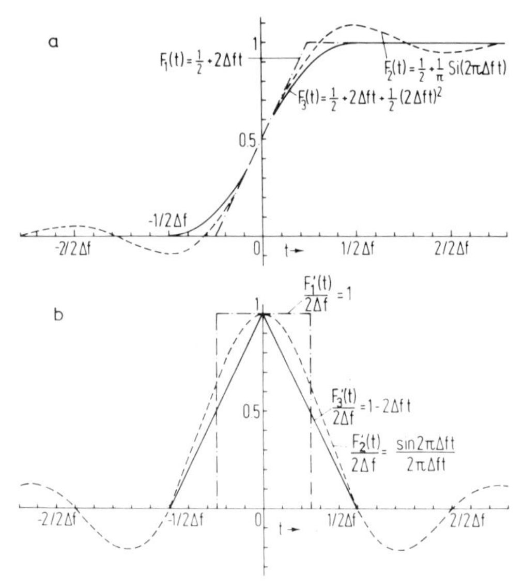

FIG. 2.1-2. (a) Output voltages of a low-pass filter with infinite cutoff idealization (dashed line), linear transient idealization (dash–dot line), and quadratic transient idealization (solid line). (b) Derivatives of the output voltages.

antenna currents are shown in Fig. 2.1-1a and repeated in Fig. 2.1-2a. The time variation of the electric and magnetic field strengths are shown by the differentiated functions in Fig. 2.1-2b.

If we want field strengths in the far zone that vary like the triangular pulse in Fig. 2.1-1c, we must pass the unit step current through a *low-pass filter with quadratic transient idealization*. The output voltage of such a filter is shown by the solid line in Fig. 2.1-2a; its derivative is the triangular pulse in Fig. 2.1-2b. The practical way to produce this field strength is, of course, *not* by means of an actual filter. Instead the triangular pulse in Fig. 2.1-2, which is best produced by integration of a positive rectangular pulse followed by a negative rectangular pulse, is integrated once more to produce an antenna current with the time variation of the solid line in Fig. 2.1-2a.

Summing up, we need an antenna current with linear transient idealization to produce field strengths with the time variation of a rectangular pulse, and an antenna current with quadratic transient idealization for field strengths with the time variation of a triangular pulse. The rectangular pulse is the easier one for the development of the theory, but it sometimes leads to paradoxical results since its energy is not sufficiently well concentrated in a finite bandwidth. In such a case one must use the trian-

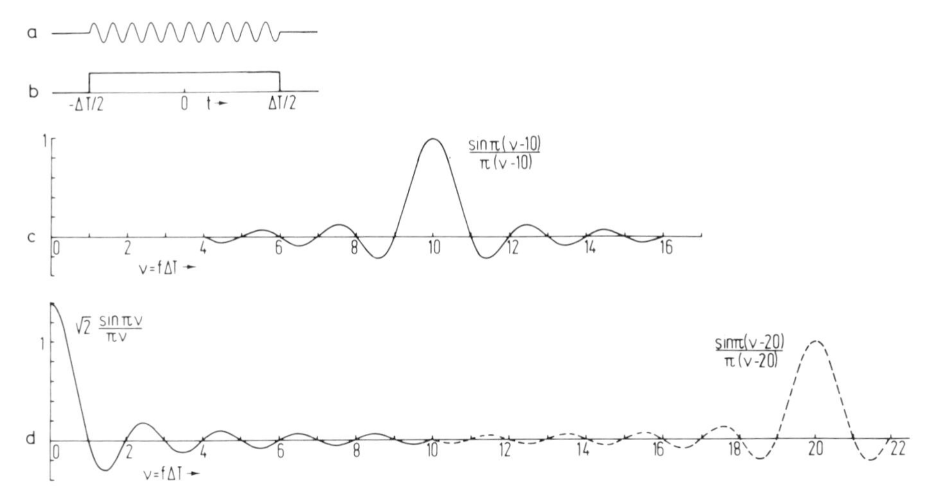

FIG. 2.1-3. Sinusoidal radar pulse (a) and rectangular pulse (b) of equal duration. (c) Fourier transform of the sinusoidal pulse. (d) Fourier transform of the rectangular pulse (solid line) and a sinusoidal pulse as shown by (a) but with twice as many cycles in the interval $-\Delta T/2 \leqq t \leqq \Delta T/2$ (dashed line).

gular pulse. The following example will show such a paradoxical result, what causes it, and how one can avoid it.

(d) *Radar Signal in a Practically Finite Band.* Consider a radio signal $f(t)$ that consists of i sinusoidal cycles in the interval $-\Delta T/2 \leqq t \leqq \Delta T/2$ and is zero outside this interval, as shown for $i = 10$ in Fig. 2.1-3a:

$$\begin{aligned} f(t) &= \sqrt{2}\sin(2\pi i t/\Delta T) \qquad &&\text{for} \quad -\Delta T/2 \leqq t \leqq \Delta T/2 \\ &= 0 &&\text{for} \quad t < -\Delta T/2, \quad t > \Delta T/2 \end{aligned} \tag{4}$$

The Fourier transform of the signal is shown in Fig. 2.1-3c:

$$\begin{aligned} a_s(i, \nu) &= 2\int_{-1/2}^{1/2} \sin(2\pi i\theta)\sin(2\pi\nu\theta)\,d\theta \\ &= \frac{\sin \pi(\nu - i)}{\pi(\nu + i)} - \frac{\sin \pi(\nu + i)}{\pi(\nu + i)}, \qquad \theta = t/\Delta T, \quad \nu = f\,\Delta T \end{aligned} \tag{5}$$

"Practically all" of the energy of this signal is concentrated in a band $(i-1)/\Delta T \leqq f \leqq (i+1)/\Delta T$, or $(i-2)/\Delta T \leqq f \leqq (i+2)/\Delta T$, etc., depending on the exact meaning of "practically all." The important point is that the bandwidth equals $2/\Delta T$, $4/\Delta T$, etc., which is independent of i.

The signal $f(t)$ shall be detected at the receiver by multiplication with two periodic local carriers $\sin(2\pi it/\Delta T + \alpha)$ and $\cos(2\pi it/\Delta T + \alpha)$ having the same period $\Delta T/i$ as $f(t)$ but an unknown phase shift α:

$$\begin{aligned} 2\sin(2\pi it/\Delta T)\sin(2\pi it/\Delta T + \alpha) &= \cos\alpha - \cos(4\pi it/\Delta T + \alpha) \\ 2\sin(2\pi it/\Delta T)\cos(2\pi it/\Delta T + \alpha) &= -\sin\alpha + \sin(4\pi it/\Delta T + \alpha) \end{aligned} \tag{6}$$

The constants $\cos\alpha$ and $-\sin\alpha$ are rectangular pulses as shown in Fig. 2.1-3b, but with amplitude $\cos\alpha$ or $-\sin\alpha$. The Fourier transform of such a pulse with amplitude 1 is shown in Fig. 2.1-3d:

$$a(0, \nu) = 2(\sin\pi\nu)/\pi\nu \tag{7}$$

The Fourier transform $a_s(2i, \nu)$ of a sinusoidal pulse $\sin(4\pi it/\Delta T)$ is given by Eq. (5) with $2i$ substituted for i. The transform $a_c(2i, \nu)$ of a cosinusoidal pulse has the same form, except that the two terms in Eq. (5) are summed rather than subtracted. The term $[\sin\pi(\nu + i)]/\pi(\nu + i)$ is very small for positive values of the frequency ν. Hence, one may ignore it in a first approximation. The Fourier (cosine) transform of the pulse $\cos(4\pi it/\Delta T + \alpha)$ and the Fourier (sine) transform of the pulse $\sin(4\pi it/\Delta T + \alpha)$ in Eq. (6) then become equal:

$$a_c(2i, \nu) \doteqdot a_s(2i, \nu) \doteqdot (\sin\alpha + \cos\alpha)\frac{\sin\pi(\nu - 2i)}{\pi(\nu - 2i)}, \qquad \nu \geqq 0 \tag{8}$$

This function is plotted for $i = 10$ in Fig. 2.1-3d.

The Fourier transforms of Fig. 2.1-3d suggest that one may suppress the high-frequency terms $\cos(4\pi it/\Delta T + \alpha)$ and $\sin(4\pi it/\Delta T + \alpha)$ in Eq. (6), particularly if $i = 10$ is replaced by $i = 100$ or $i = 1000$, which is more typical for a radar pulse. The dc terms $\cos\alpha$ and $-\sin\alpha$ may then be squared and added,

$$\begin{aligned} \cos^2\alpha + \sin^2\alpha &= 1 \qquad \text{for} \quad -\Delta T/2 \leqq t \leqq \Delta T/2 \\ &= 0 \qquad \text{for} \quad t < -\Delta T/2, \quad t > \Delta T/2 \end{aligned} \tag{9}$$

to yield the rectangular pulse of Fig. 2.1-3b. The infinitely fast rise time of this pulse implies an infinitely good resolution of the radar. This result is paradoxical but not absurd if one keeps in mind that the sinusoidal pulse of Fig. 2.1-3a occupies an infinite frequency band. Even though the Fourier transform of Fig. 2.1-3c shows that "almost all" energy of the pulse is concentrated in a finite, sufficiently wide band, there is always enough energy outside this finite band to make it nonnegligible. The use of triangular pulses according to Fig. 2.1-2b avoids this problem, since its Fourier transform decreases like $1/\nu^2$ for large values of ν, whereas that of a rectangular pulse according to Fig. 2.1-2b or a sinusoidal pulse according to Fig. 2.1-3a decreases only like $1/\nu$. One should note that a decrease like $1/\nu^2$ is not always fast enough. For instance, a filter with a frequency response of the output amplitude proportionate to ν or, more subtly, equivalent processing in the time domain, would make it impossible to treat the triangular pulse as "practically band limited."

In the following sections the results derived here will be used to investigate three types of radio signals: carrier-free signals, which are signals with a relative bandwidth η close to 1; (b) signals using a sinusoidal carrier but with a relative bandwidth much larger than the value $\eta = 0.01$ of conventional signals; and (c) signals using two-valued rather than sinusoidal carriers.

2.2 Carrier-Free Signals

The pulses in Fig. 2.1-2b all have a dc component, and such a component cannot be transmitted. A straightforward way to produce a radio signal in the band $f_L \leqq f \leqq f_H$, where f_L is not zero, is to filter the antenna current by a band-pass rather than a low-pass filter as in Figs. 2.1-1a and 2.1-2a. Such a band-pass filter may be considered to consist of two low-pass filters, one with the passband $0 \leqq f \leqq f_H$ and the other with the passband $0 \leqq f \leqq f_L$. A unit step pulse passed through the low-pass filter $0 \leqq f \leqq f_H$ minus a unit step pulse passed through the low-pass filter $0 \leqq f \leqq f_L$ will yield the same output as a unit step pulse passed through the band-pass filter $f_L \leqq f \leqq f_H$. Figure 2.2-1 shows an example.

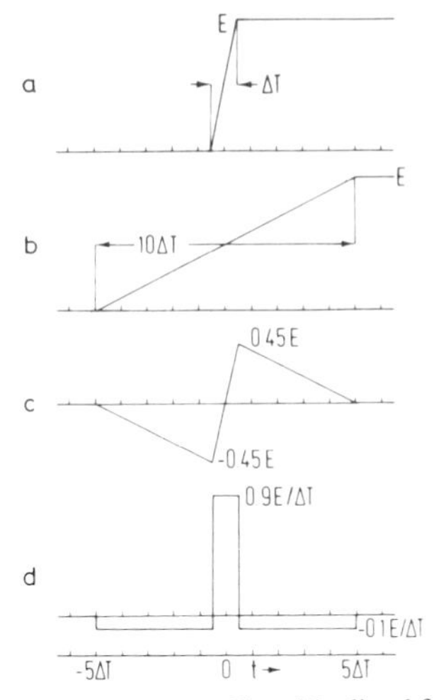

FIG. 2.2-1. The output signal of a band-pass filter idealized for a linear transient response to a step function. The nominal ratio of the upper to lower cutoff frequency equals $f_H/f_L = 10$; the relative bandwidth equals $\eta = (10 - 1)/(10 + 1) = 0.82$. (a) Output of a low-pass filter with transient time ΔT due to a step function with amplitude E. (b) Output of a low-pass filter with transient time $10\,\Delta T$ due to a step function with amplitude E. (c) Output of the band-pass filter equals (a) minus (b). (d) First derivative of the signal (c).

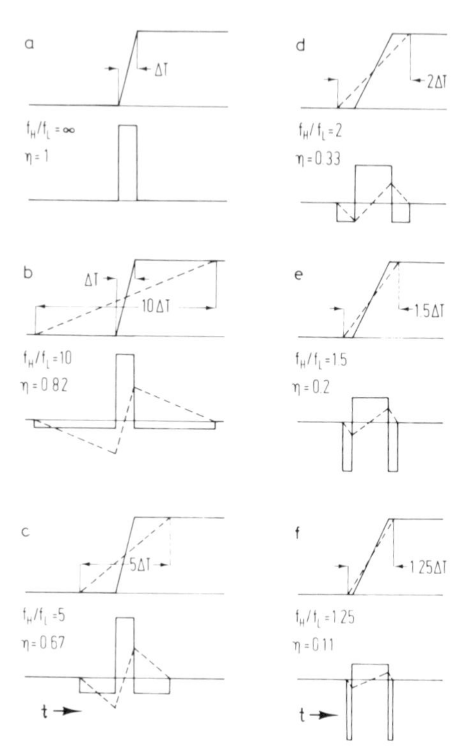

FIG. 2.2-2. Currents flowing in a Hertzian electric dipole and field strengths in the far zone produced by them as functions of time. (a) Low-pass filter with linear transient idealization according to $F_1(t)$ and $F_1'(t)$ in Fig. 2.1-2. (b) Repetition of Fig. 2.2-1 in compressed form. (c)–(f) Currents and field strengths in analogy to (b), but with decreasing ratios f_H/f_L and relative bandwidths η.

A unit step pulse with amplitude E applied to a low-pass filter with linear transient idealization, having the nominal cutoff frequency $f_H = 1/2\,\Delta T$, produces an output signal as shown in Fig. 2.2-1a. An equal step pulse passing through an equal low-pass filter, but having a nominal cutoff frequency $f_L = f_H/10 = 1/20\,\Delta T$, produces the output signal of Fig. 2.2-1b. The signal of Fig. 2.2-1a minus the signal of Fig. 2.2-1b yields the output signal of a band-pass filter with linear transient idealization, having the nominal passband $f_L \leqq f \leqq f_H$ shown in Fig. 2.2-1c. The first derivative is shown in Fig. 2.2-1d. This is the time variation of the electric and magnetic field strength in the far zone, if a current according to Fig.

2.2-1c is fed into a Hertzian electric dipole.

The practical realization of such a filter would be difficult, but there is no difficulty in generating the pulse of Fig. 2.2-1d, and integrating it to obtain the antenna current of Fig. 2.2-1c.

Figure 2.2-2 shows six examples of antenna currents and the field strengths in the far zone produced by them. Figure 2.2-2a is a repetition of the functions $F_1(t)$ and $F_1'(t)$ of Fig. 2.1-2. The band-pass case of Fig. 2.2-1 is repeated in Fig. 2.2-2b; one may readily recognize how the four plots of Fig. 2.2-1 are combined to save space. Figures 2.2-2c–f show the corresponding plots for decreasing relative bandwidths η.

The shape of the rectangular pulse in Fig. 2.2-2a is reasonably well preserved in Fig. 2.2-2b, and somewhat less so in Fig. 2.2-2c. The change of the shape becomes more conspicuous as the relative bandwidth η decreases in Figs. 2.2-2d–f.

Accordingly, a theory based on the current and the field strength shown in Fig. 2.2-2a will be acceptable for relative bandwidths η close to 1, but not for values of η close to 0.1. The conventional theory applies to signals with relative bandwidths on the order of $\eta = 0.01$ or less. Hence, we distinguish between carrier-free signals with η close to 1, conventional signals with η close to 0, and intermediate signals that are not satisfactorily represented by either one of these limiting cases. A line-of-sight, high-resolution, all-weather radar operating in the frequency range $1 \text{ GHz} \leqq f \leqq 10 \text{ GHz}$ or an into-the-ground radar operating in the range $100 \text{ MHz} \leqq f \leqq 1 \text{ GHz}$ have relative bandwidths $\eta = (10 - 1)/(10 + 1) = 0.82$. This value is close to 1, and one may thus use carrier-free signals. An over-the-horizon radar operating in the band $8 \text{ MHz} \leqq f \leqq 10 \text{ MHz}$ yields $\eta = (10 - 8)/(10 + 8) = 0.11$. This case is represented by Fig. 2.2-2f. The distortion of the field strengths compared with Fig. 2.2-2a is too large. Hence, over-the-horizon radar is a typical case where neither the conventional nor the carrier-free signal is satisfactory, and a generalization for intermediate signals is called for.

Figure 2.2-3 shows the extension of Fig. 2.2-2 to low-pass and band-pass filters with quadratic transient idealization. Figure 2.2-3a is a repetition of the pulse $F_3'(t)$ of Fig. 2.1-2; the antenna current $F_3(t)$ is not shown. Figure 2.2-3b shows the same pulse (dash–dot line) and one due to a current passed through the same kind of low-pass filter but with the nominal cutoff frequency reduced from f_H to $f_L = f_H/10$ (dashed line). The solid line shows the difference between the two pulses.

Figure 2.2-3 confirms for triangular pulses what Fig. 2.2-2 showed for rectangular pulses. A theory developed for $\eta = 1$ will be usable for $\eta = 0.82$ and perhaps still for $\eta = 0.67$, but not for $\eta = 0.11$.

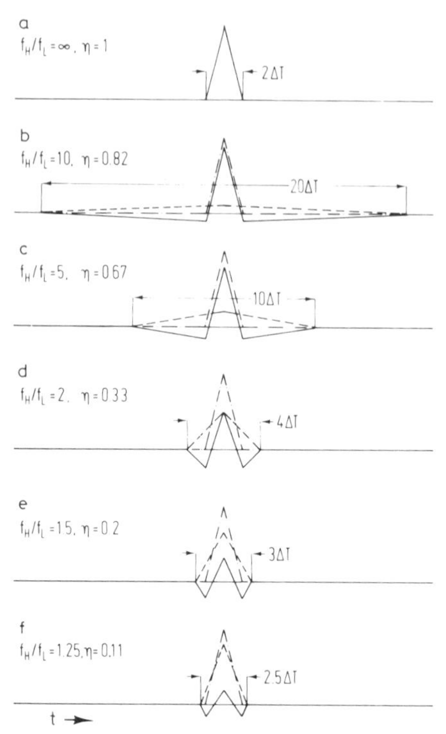

FIG. 2.2-3. Field strengths in the far zone as functions of time due to currents in a Hertzian dipole produced from step functions passed through low-pass and band-pass filters with quadratic transient idealization. The illustration is analogous to Fig. 2.2-2, except that the derivative $F_3'(t)$ instead of $F_1'(t)$ in Fig. 2.1-2 is used. The currents are not shown in order to simplify the illustration.

2.3 SINUSOIDAL CARRIERS

2.3.1 *Amplitude Modulation of a Sinusoidal Carrier*

The amplitude modulation of a sinusoidal carrier $\sin \omega_0 t$ by a baseband signal $F(t) = 1 + MG(t)$, where M is the modulation index, yields the function

$$f(t) = F(t) \sin \omega_0 t \tag{1}$$

An antenna current $i(t)$ with this time variation

$$i(t) = IF(t) \sin \omega_0 t \tag{2}$$

fed into a Hertzian electric dipole, yields the time variation di/dt for the electric and magnetic field strengths in the far zone:

$$di/dt = I[\omega_0 F(t) \cos \omega_0 t + F'(t) \sin \omega_0 t] \tag{3}$$

The field strength produced by a resonant dipole for the sinusoidal wave with the frequency $f_0 = \omega_0/2\pi$ will have the same time variation with very good approximation.

If the term $F'(t) \sin \omega_0 t$ in Eq. (3) is negligibly small compared with the term $\omega_0 F(t) \cos \omega_0 t$, one obtains the expression generally used in communications engineering[1] for the electric field strength at the receiving antenna or the input voltage to a receiver in the far zone:

$$e(t) = E\omega_0 F(t) \cos \omega_0 t \tag{4}$$

The condition "$F'(t) \sin \omega_0 t$ negligibly small compared with $\omega_0 F(t) \cos \omega_0 t$" must be made more specific before we can write it as a mathematical formula. There are several ways of doing so. We chose to specify that the largest absolute value of $F'(t) \sin \omega_0 t$ shall be small compared with the largest absolute value of $\omega_0 F(t) \cos \omega_0 t$:

$$|F'(t) \sin \omega_0 t|_{\max} \ll |\omega_0 F(t) \cos \omega_0 t|_{\max} \tag{5}$$

An important reason for choosing the condition in this particular form is, of course, that a simple physical explanation can be derived from it without much mathematical effort. To do so, we first assume that $F(t)$ assumes values between $+1$ and -1 only, $|F(t)| \leqq 1$; this normalization implies no reduction of generality, since Eq. (2) still contains the factor I. As a result, we can simplify the right-hand side of Eq. (5):

$$|F'(t) \sin \omega_0 t|_{\max} \ll \omega_0 \tag{6}$$

In general, a carrier and the signal modulating the carrier are not synchronized. This means that the modulating signal voltage does not begin to differ from zero, nor has its largest positive value, nor some other particular value at the moment when the sinusoidal carrier voltage is zero, but at an arbitrary point of the carrier period. If $F'(t)$ and $\sin \omega_0 t$ are not synchronized, it can happen that the largest absolute value of $F'(t)$ coincides with the values $+1$ or -1 of $\sin \omega_0 t$. Equation (6) must in this case be replaced by the following, more restrictive, condition:

---

[1] Scientific books and papers show that the field strengths vary like the first derivative of the current (Stratton, 1941, pp. 440 and 455; Becker and Sauter, 1964, p. 198; Harmuth, 1977a, pp. 237 and 395; Spetner, 1974), but textbooks on electrical communication usually treat the amplitude modulation of a radio carrier like the amplitude modulation used for frequency multiplexing. No differentiation is required for multiplexing, since the differentiation is due to the radiation process rather than the modulation process. The reason why the differentiation is generally overlooked may readily be seen from Stratton (1941). The antenna current $i(t)$ is assumed to have the form $u(\xi) \exp(-i\omega t)$ or $I_0 \sin(k\xi - \alpha) \exp(-i\omega t)$. A differentiation di/dt means a multiplication by $i\omega$. Such a factor is hard to trace through a lengthy calculation. Becker and Sauter (1964), Harmuth (1977a), and Spetner (1974) make no assumption about the time variation of the antenna current, and the differentiation thus shows up conspicuously in the form di/dt or $\ddot{\mathbf{p}}$.

$$|F'(t)|_{\max} \ll \omega_0 \tag{7}$$

Let us assume that $F(t)$ is a signal that has passed through one of the three types of low-pass filters defined by the step impulse response of Figs. 2.1-1a or 2.1-2a, which have a nominal passband $0 \leqq f \leqq \Delta f$. The largest possible value[1] of $F'(t)$ equals $2\,\Delta f$. Hence, Eq. (7) assumes the form

$$2\,\Delta f \ll \omega_0 = 2\pi f_0 \tag{8}$$

The antenna current of Eq. (2) will occupy the nominal band $f_0 - \Delta f \leqq f \leqq f_0 + \Delta f$, with the relative bandwidth

$$\eta = \frac{(f_0 + \Delta f) - (f_0 - \Delta f)}{(f_0 + \Delta f) + (f_0 - \Delta f)} = \frac{\Delta f}{f_0} \tag{9}$$

Equation (8) thus requires that the relative bandwidth be small if we want to ignore $F'(t) \sin \omega_0 t$ in Eq. (3) and use the approximation of Eq. (4).

We have seen in Section 1.4 that our usual methods for the implementation of resonant circuits and resonant structures require a small relative bandwidth, and we see now that the same requirement is also imposed by our usual method of radio transmission of signals by means of amplitude modulation of a carrier. We must find methods to modify amplitude modulation to make the use of general radio signals with a large relative bandwidth possible.

Three methods are readily apparent from Eq. (3). First, one may use synchronous demodulation. Insertion of $F(t) = 1 + MG(t)$ into Eq. (3) yields

$$di/dt = I\{\omega_0[1 + MG(t)] \cos \omega_0 t + MG'(t) \sin \omega_0 t\} \tag{10}$$

The term $\omega_0 \cos \omega_0 t$ can be extracted by a tracking filter and used for synchronous demodulation of $MG(t) \cos \omega_0 t$ and—after a phase shift—$MG'(t) \sin \omega_0 t$. The demodulated term $MG'(t)$ may be integrated to yield $MG(t)$. Hence, the signal power in both terms $MG(t) \cos \omega_0 t$ and $MG'(t) \sin \omega_0 t$ is used.

This method is not applicable to radar signals, since the carrier is only received when a returned radar pulse is received, and there is not enough time for a tracking filter to lock onto the phase of $\cos \omega_0 t$ if the signal does not have a small relative bandwidth.

The second method is to integrate the signal $f(t)$ of Eq. (1) and feed a current proportionate to this integral into the antenna:

$$i(t) = I \int f(t)\, d(t/\Delta T) = I \int F(t) \sin \omega_0 t\, d(t/\Delta T) \tag{11}$$

[1] The condition $|F(t)| \leqq 1$ implies $A \leqq 1$ in Fig. 2.1-1a.

The field strengths in the far zone then have the desired time variation:

$$di/dt = IF(t) \sin \omega_0 t \tag{12}$$

The third method is to use an antenna current according to Eq. (2), but integrate the input voltage at the receiver. According to Eqs. (3) and (4), this voltage has the following form:

$$e(t) = E[\omega_0 F(t) \cos \omega_0 t + F'(t) \sin \omega_0 t] \tag{13}$$

The integral yields

$$v(t) = \int e(t)\, d(t/\Delta T) = VF(t) \sin \omega_0 t \tag{14}$$

The difference between the second and third method is not only a mathematical one. The efficiency of power conversion from dc to the currents $i(t)$ of Eqs. (2) or (11) is not necessarily the same; the efficiency of radiation may also be different, and the frequency bands occupied during transmission may be different.

Consider next single-sideband modulation. We will use the phase shift method, since it is much easier to investigate for the signals of Figs. 2.2-2 and 2.2-3 than any method that requires filters.[1]

Let $F(t)$ be the triangular pulse of Fig. 2.3-1a. Its Fourier transform,

$$\begin{aligned} a_c(\nu) &= \sqrt{2} \int_{-\infty}^{\infty} F(\theta) \cos 2\pi\nu\theta \, d\theta = 2\sqrt{2} \int_0^1 (1 - \theta) \cos 2\pi\nu\theta \, d\theta \\ &= \sqrt{2}[(\sin \pi\nu)/\pi\nu]^2, \qquad \theta = t/\Delta T, \quad \nu = f\,\Delta T \end{aligned} \tag{15}$$

permits one to write $F(t)$ in the form

$$F(t) = 2 \int_0^{\infty} [(\sin \pi\nu)/\pi\nu]^2 \cos 2\pi\nu\theta \, d\nu \tag{16}$$

and the phase-shifted function $F^*(t)$ in the form

$$F^*(t) = 2 \int_0^{\infty} [(\sin \pi\nu)/\pi\nu]^2 \sin 2\pi\nu\theta \, d\nu \tag{17}$$

The function $F^*(t)$ is plotted in Fig. 2.3-1d.

To produce a single-sideband signal, we need carriers like $\cos(2\pi t/\Delta T)$ and $\sin(2\pi t/\Delta T)$ of Fig. 2.3-1b and e. Their amplitude modulation with the baseband signals $F(t)$ and $F^*(t)$ yields the double-sideband radio signals of

---

[1] The phase shift method is described in many books (Taub and Shilling, 1971, p. 102; Schwartz *et al.*, 1966, p. 188). The discussion here follows the one the author is most familiar with (Harmuth, 1969, p. 134, particularly Figs. 60 and 61; 1972, p. 198, Figs. 133 and 134).

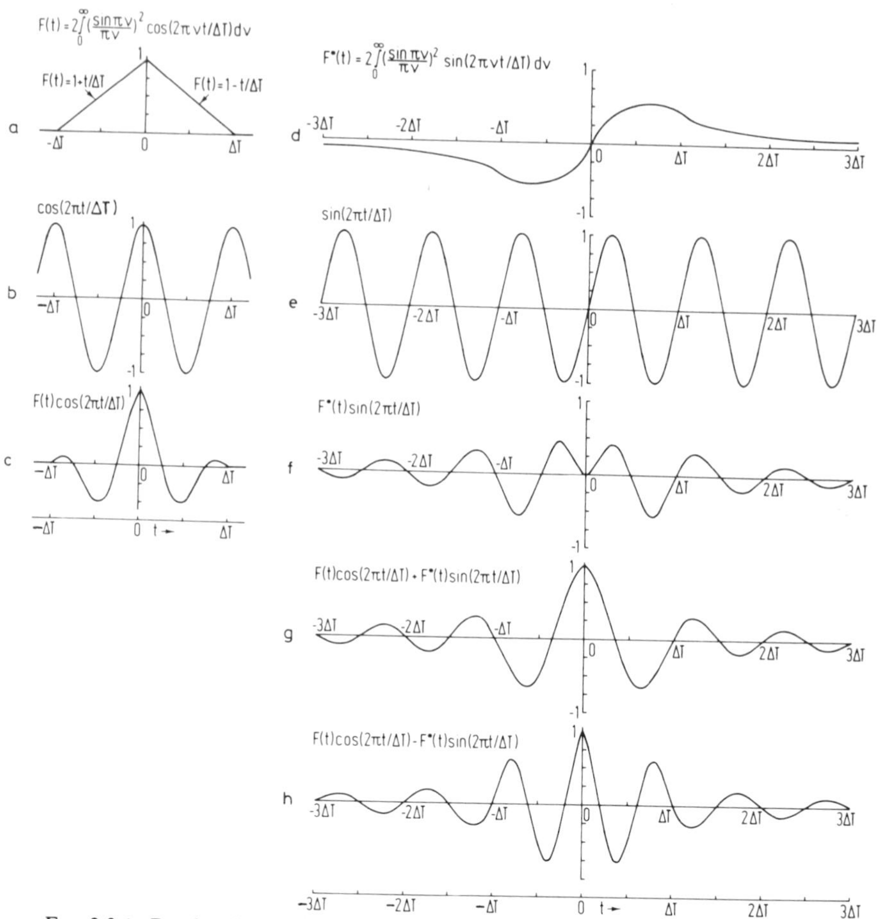

FIG. 2.3-1. Baseband signal (a), cosinusoidal carrier (b), and the carrier amplitude modulated by the baseband signal (c). The baseband signal phase shifted (d), a sinusoidal carrier (e), and the carrier amplitude modulated by the phase-shifted baseband signal (f). Lower (g) and upper (h) sideband signals derived from the signals (c) and (f).

Fig. 2.3-1c and f. A lower-sideband signal $F_{1L}(t)$ is produced by their sum,

$$F_{1L}(t) = F(t)\cos(2\pi t/\Delta T) + F^*(t)\sin(2\pi t/\Delta T) \tag{18}$$

and an upper-sideband signal $F_{1U}(t)$ by their difference:

$$F_{1U}(t) = F(t)\cos(2\pi t/\Delta T) - F^*(t)\sin(2\pi t/\Delta T) \tag{19}$$

These two signals are shown in Fig. 2.3-1g and h.

The radio signals of Fig. 2.3-1c, f, g, and h are completely different from

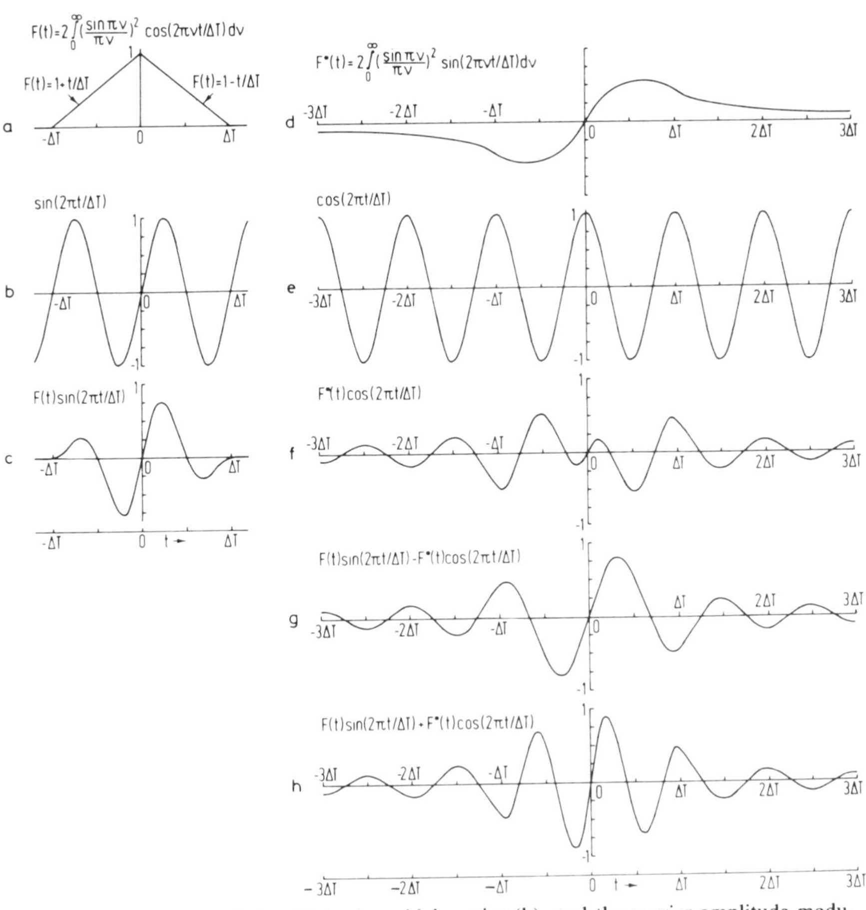

FIG. 2.3-2. Baseband signal (a), sinusoidal carrier (b), and the carrier amplitude modulated by the baseband signal (c). The baseband signal phase shifted (d), a cosinusoidal carrier (e), and the carrier amplitude modulated by the phase-shifted baseband signal (f). Lower (g) and upper (h) sideband signals derived from the signals (c) and (f).

sinusoidal signals. This is generally so for signals with a large relative bandwidth, whether a sinusoidal carrier is used or not. As pointed out before, a small relative frequency bandwidth implies that a signal looks very similar to a sinusoidal function, but this similarity disappears with increasing relative bandwidth.

The signal of Fig. 2.3-1c has no dc component, since the period of the carrier in Fig. 2.3-1b was chosen to avoid a dc component. It is not evident that the signals of Fig. 2.3-1f, g, and h have no dc component either.

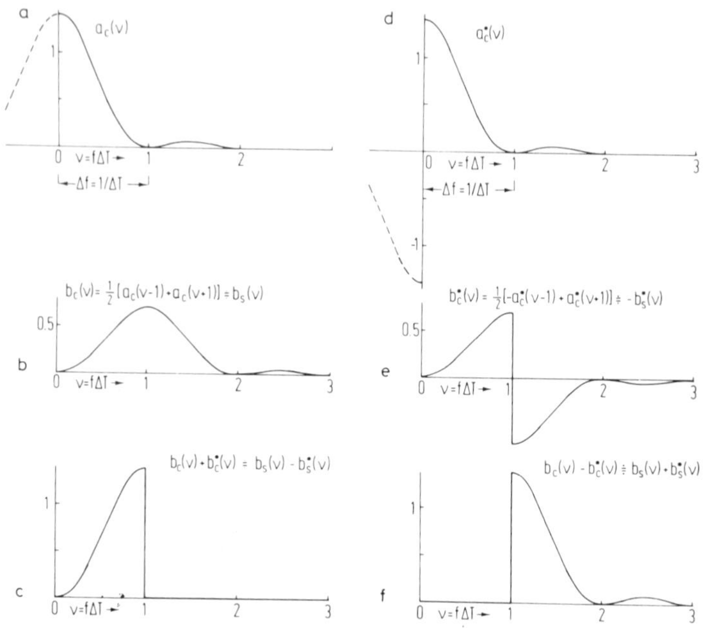

FIG. 2.3-3. Fourier transforms of the signals of Fig. 2.3-1. (a) Transform $a_c(\nu)$ of $F(t)$. (b) Transform $b_c(\nu)$ of $F(t)\cos(2\pi t/\Delta T)$. (c) Transform of $F(t)\cos(2\pi t/\Delta T) + F^*(t)\sin(2\pi t/\Delta T)$. (d) Transform $a_c^*(\nu)$ of $F^*(t)$. (e) Transform $b_c^*(\nu)$ of $F^*(t)\sin(2\pi t/\Delta T)$. (f) Transform of $F(t)\cos(2\pi t/\Delta T) - F^*(t)\sin(2\pi t/\Delta T)$.

This is a general feature of amplitude modulation of a cosinusoidal carrier with an even function, $F(t) = F(-t)$, or a sinusoidal carrier with an odd function, $F(t) = -F(-t)$, but the dc component can be neglected for signals with small relative bandwidth. One cannot do so for large relative bandwidths, particularly if one uses a current for the radiating antenna proportionate to the integral of the signal.

In order to avoid a dc component, one may either choose the period ΔT of the carriers in Fig. 2.3-1 carefully, or one may use a sinusoidal carrier for the even function $F(t)$ and a cosinusoidal carrier for the odd function $F^*(t)$. This approach is shown in Fig. 2.3-2. A lower-sideband signal $F_{2L}(t)$ is now produced by the difference of the double-sideband signals in Fig. 2.3-2c and f.

$$F_{2L}(t) = F(t)\sin(2\pi t/\Delta T) - F^*(t)\cos(2\pi t/\Delta T) \tag{20}$$

while an upper-sideband signal $F_{2U}(t)$ is produced by their sum:

$$F_{2U}(t) = F(t)\sin(2\pi t/\Delta T) + F^*(t)\cos(2\pi t/\Delta T) \tag{21}$$

These two signals are shown in Fig. 2.3-2g and h.

Let us look at the signals of Figs. 2.3-1 and 2.3-2 in the frequency domain. The Fourier transform $a_c(\nu)$ of the triangular pulse of Fig. 2.3-1a is defined by Eq. (15) and plotted in Fig. 2.3-3a. This function is defined for $\nu \geqq 0$ only; however, by plotting it symmetrically for $\nu \leqq 0$, we can express the fact that all the Fourier components have the phase of cosine functions. The function $F^*(t)$ of Fig. 2.3-1d has the same Fourier transform for $\nu \geqq 0$, but all its Fourier components have the phase of sine functions. This is expressed in Fig. 2.3-3d by plotting the Fourier transform skew symmetrically for $\nu \leqq 0$.

The Fourier transforms of the modulated carriers in Fig. 2.3-1c and f follow readily from the transforms $a_c(\nu)$ and $a_c^*(\nu)$ of the baseband signals:

$$\begin{aligned} b_c(\nu) &= \sqrt{2} \int_0^\infty [F(\theta) \cos 2\pi\theta] \cos 2\pi\nu\theta \, d\theta \\ &= \frac{1}{2} \left[\sqrt{2} \int_0^\infty F(\theta) \cos 2\pi(\nu - 1)\theta \, d\theta \right. \\ &\quad \left. + \sqrt{2} \int_0^\infty F(\theta) \cos 2\pi(\nu + 1)\theta \, d\theta \right] \\ &= \tfrac{1}{2}[a_c(\nu - 1) + a_c(\nu + 1)] \end{aligned} \tag{22}$$

$$\begin{aligned} b_c^*(\nu) &= \sqrt{2} \int_0^\infty [F^*(\theta) \sin 2\pi\theta] \cos 2\pi\nu\theta \, d\theta \\ &= \tfrac{1}{2}[-a_c^*(\nu - 1) + a_c^*(\nu + 1)] \end{aligned} \tag{23}$$

$$a_c^*(\nu) = \sqrt{2} \int_{-\infty}^\infty F^*(\theta) \sin 2\pi\nu\theta \, d\theta, \qquad \theta = t/\Delta T, \quad \nu = f\,\Delta T$$

The functions $b_c(\nu)$ and $b_c^*(\nu)$ are plotted in Fig. 2.3-3b and e. Their sum and difference, which are the Fourier transforms of the lower- and upper-sideband signals in Fig. 2.3-1g and h, are shown in Fig. 2.3-3c and f.

The Fourier transforms of the modulated carriers in Fig. 2.3-2c and f follow in analogy to Eqs. (22) and (23):

$$\begin{aligned} b_s(\nu) &= \sqrt{2} \int_0^\infty [F(\theta) \sin 2\pi\theta] \sin 2\pi\nu\theta \, d\theta \\ &= \frac{1}{2} \left[\sqrt{2} \int_0^\infty F(\theta) \cos 2\pi(\nu - 1)\theta \, d\theta \right. \\ &\quad \left. - \sqrt{2} \int_0^\infty F(\theta) \cos 2\pi(\nu + 1)\theta \, d\theta \right] \\ &= \tfrac{1}{2}[a_c(\nu - 1) - a_c(\nu + 1)] \end{aligned} \tag{24}$$

$$b_s^*(\nu) = \sqrt{2} \int_0^\infty [F^*(\theta) \cos 2\pi\theta] \sin 2\pi\nu\theta \, d\theta$$
$$= \tfrac{1}{2}[a_c^*(\nu - 1) + a_c^*(\nu + 1)] \quad (25)$$

In a first approximation, $b_s(\nu)$ equals $b_c(\nu)$ and $b_s^*(\nu)$ equals $-b_c^*(\nu)$ for $\nu \geqq 0$; this is indicated in Fig. 2.3-3b and e. The Fourier transforms of the signals of Fig. 2.3-2g and h are thus, in a first approximation, represented by the plots of Fig. 2.3-3c and f for $\nu \geqq 0$. For $\nu \leqq 0$, one must continue $b_c(\nu)$ and $b_c^*(\nu)$ as symmetric functions, while $b_s(\nu)$ and $b_s^*(\nu)$ are to be continued as skew-symmetric functions.

2.3.2 Demodulation

Consider a lower-sideband signal $F_{2L}(t)$ according to Eq. (20), but with the general frequency f_c for the carrier instead of the particular value $f_c = 1/\Delta T$. Let the antenna current of the transmitter be proportionate to the integral of $F_{2L}(t)$:

$$e(t) = E[F(t) \sin 2\pi f_c t - F^*(t) \cos 2\pi f_c t] \quad (26)$$

The field strength in the wave zone will be proportionate to di/dt, and the voltage $e(t)$ at the receiver input terminal will also be proportionate to di/dt. In order to produce the voltage of Eq. (26), one must thus feed a current into the radiating antenna that is proportionate to the integral of Eq. (26):

$$i(t) = I \int [F(t) \sin 2\pi f_c t - F^*(t) \cos 2\pi f_c t] \, d(f_c t) \quad (27)$$

The antenna current $i(t)$ is not necessarily a single-sideband signal, even though we have a single-sideband signal on the transmission path and at the receiver input terminal.

For demodulation of the received signal let us multiply the input voltage $e(t)$ with local carriers $\sin(2\pi f_c t + \alpha)$ and $\cos(2\pi f_c t + \alpha)$, produced at the receiver; they have the same frequency as the carrier used at the transmitter, but an arbitrary phase shift α:

$$e(t) \sin(2\pi f_c t + \alpha) = \tfrac{1}{2}E\{F(t)[\cos \alpha - \cos(4\pi f_c t + \alpha)] - F^*(t)[\sin \alpha + \sin(4\pi f_c t + \alpha)]\} \quad (28)$$

$$e(t) \cos(2\pi f_c t + \alpha) = \tfrac{1}{2}E\{F(t)[-\sin \alpha + \sin(4\pi f_c t + \alpha)] - F^*(t)[\cos \alpha - \cos(4\pi f_c t + \alpha)]\} \quad (29)$$

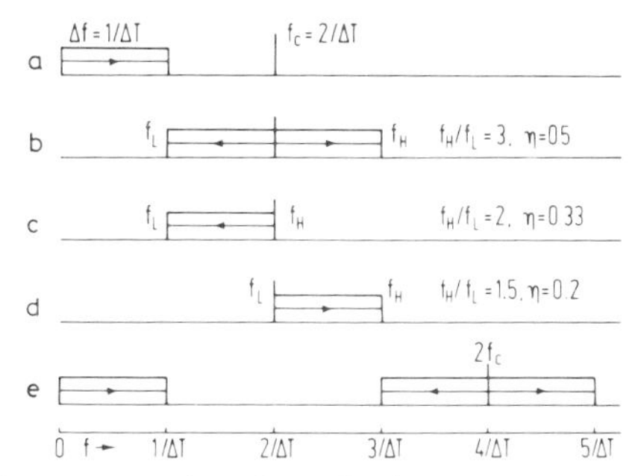

FIG. 2.3-4. Symbolic representation of certain signals in the frequency domain. (a) Baseband signal with bandwidth $\Delta f = 1/\Delta T$ and carrier with frequency $f_c = 2/\Delta T$. (b) Double sideband modulation signal. Lower (c) and upper (d) sideband modulation signal. (e) Demodulated signal with baseband and high-frequency components.

For signals with small relative bandwidths, one can readily ignore the "high-frequency components"

$$F(t)\cos(4\pi f_c t + \alpha), \qquad F(t)\sin(4\pi f_c t + \alpha)$$

$$F^*(t)\cos(4\pi f_c t + \alpha), \qquad F^*(t)\sin(4\pi f_c t + \alpha)$$

but this is not necessarily so for signals with large relative bandwidth.

Consider a signal in the frequency domain with the nominal bandwidth $\Delta f = 1/\Delta T$ as shown, e.g., in Fig. 2.3-3a and d. These signals are shown again in an easier to plot form in Fig. 2.3-4a by a Fourier transform of rectangular shape and the bandwidth $\Delta f = 1/\Delta T$. Also shown is a carrier with frequency $f_c = 2/\Delta T$. The frequency band occupied by the amplitude-modulated signal is shown in Fig. 2.3-4b, while the bands occupied by the lower and upper sideband signals are shown in Fig. 2.3-4c and d. The demodulation according to Eqs. (28) and (29) yields the high-frequency components in the band $3/\Delta T \leqq f \leqq 5/\Delta T$ in Fig. 2.3-4e, while the desired baseband components $F(t)\cos\alpha$, $F(t)\sin\alpha$, $F^*(t)\cos\alpha$, and $F^*(t)\sin\alpha$ are in the baseband $0 \leqq f \leqq 1/\Delta T$. This result holds for double-sideband modulation as well as for lower- or upper-sideband modulation, since Eqs. (28) and (29) include double-sideband modulation if $F^*(t)$ is replaced by zero, and upper-sideband modulation if the sign of $F^*(t)$ is changed. The high-frequency components in Fig. 2.3-4e can readily be suppressed by a filter, since a large frequency gap $1/\Delta T \leqq f \leqq 3/\Delta T$ exists between the baseband $0 \leqq f \leqq 1/\Delta T$ and the high-frequency component band $3/\Delta T \leqq f \leqq 5/\Delta T$.

Let us turn to Fig. 2.3-5, which is a repetition of Fig. 2.3-4, but the carrier frequency has been reduced to $f_c = 1/\Delta T$. We can see that there is no longer a gap in Fig. 2.3-5e between the baseband and the high-frequency

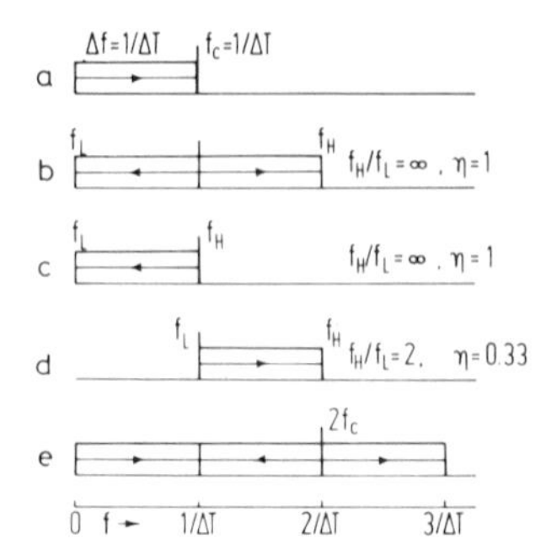

FIG. 2.3-5. Symbolic representation of certain signals in the frequency domain. (a) Baseband signal with bandwidth $\Delta f = 1/\Delta T$ and carrier with frequency $f_c = 1/\Delta T$. (b) Double sideband modulation signal. Lower (c) and upper (d) sideband modulation signal. (e) Demodulated signal with baseband and high-frequency component.

component band $1/\Delta T \leqq f \leqq 3/\Delta T$. Hence, the theoretical limit for the relative bandwidth η equals 1 for double-sideband and lower-sideband modulation, but only $\eta = 0.33$ for upper-sideband modulation.

If synchronous demodulation can be used, one may reduce α to zero in Eqs. (28) and (29). After suppression of the high-frequency terms, one then obtains $F(t)$ from Eq. (28) and $-F^*(t)$ from Eq. (29). Both contain, of course, the same information. However, in theory one can phase shift $-F^*(t)$ to yield $F(t)$, and add this $F(t)$ to the one obtained from Eq. (28). The two signals would sum to $2F(t)$, while the noise would sum statistically, so that an improvement of the signal-to-noise ratio is obtained. Hence, the energy used for the transmission of $F^*(t)$ not only reduces the occupied frequency band, but also increases the signal-to-noise ratio.

In radar, it is generally impossible to demodulate synchronously. The next best thing one can do is to suppress the high-frequency components in Eqs. (28) and (29) with a filter, square the baseband components, sum the squares, and take the square root:

$$\tfrac{1}{2}E\{[F(t)\cos\alpha - F^*(t)\sin\alpha]^2 + [F(t)\sin\alpha + F^*(t)\cos\alpha]^2\}^{1/2}$$
$$= \tfrac{1}{2}E[F^2(t) + F^{*2}(t)]^{1/2} \qquad (30)$$

If $F^*(t)$ is not transmitted, this formula yields the triangular pulse $F(t)$ of Fig. 2.3-1a. Twice the bandwidth of $F(t)$ is required in the radio channel, since the suppression of $F^*(t)$ means that double-sideband modulation is used.

The need for twice the width of the baseband in the absence of a knowledge of the carrier phase is generally accepted in communications. The basic argument is that the transmittable information is proportionate to the number of orthogonal—or linearly independent—functions that can be transmitted in a certain time interval (Harmuth, 1969, p. 245, or 1972, p. 316). The functions sin ωt and cos ωt are orthogonal to each other, if the phase is known, but undistinguishable without a knowledge of the phase. Hence, twice as many functions sin ωt are needed if the functions cos ωt cannot be distinguished from the functions sin ωt.

Such a doubling of the bandwidth did not occur in Fig. 2.2-3. However one might define the bandwidth of the pulse of Fig. 2.2-3a, one could hardly claim that the pulses of Fig. 2.2-3b or c need twice that bandwidth. Since we have used amplitude modulation solely to shift a signal from the band $0 \leqq f \leqq 1/\Delta T$ to a higher frequency band, we must be able to do so without increasing the bandwidth and without needing a phase tracking loop any more than we would need it for the pulses of Fig. 2.2-3.

One of the possible ways is to avoid demodulation and to use directly a sliding correlator for detection of the single-sideband signal. For an explanation refer to Fig. 2.3-1. The triangular pulse a would conventionally be modulated onto a carrier b with 100–1000 cycles during the interval $-\Delta T \leqq t \leqq \Delta T$, occupied by the triangular pulse, rather than the two cycles shown. For reception, the signal is first demodulated and the resulting triangular pulse is then fed into a sliding correlator that produces its cross-correlation function with a stored sample pulse. Instead of designing the sliding correlator for the triangular pulse a, one may design it for the single-sideband signals g or h. In this case, detection by cross-correlation with a sample function is performed directly, and demodulation is avoided. This process requires that the triangular pulse a and the carrier b be synchronized with each other. A second, practical, requirement is that the carrier have only a few cycles during the duration $-\Delta T \leqq t \leqq \Delta T$ of the signal, which means the modulated signal must have a large relative bandwidth.

This detection of signals with large relative bandwidth by means of a sliding correlator is currently done in the Doppler radar. Let the triangular pulse of Fig. 2.3-1 be amplitude-modulated onto a carrier with a frequency of $f_c = 2$ GHz, which is a typical radar frequency according to Table 1.4-2. A target approaching or *closing* with a velocity of $v = 300$ m/s produces a frequency shift Δf of the returned signal

$$\Delta f = 2\frac{v}{c}f = 2\frac{300}{3 \times 10^{10}} 2 \times 10^9 = 4 \text{ kHz}$$

We denote the original modulated signal with $EF(t) \sin 2\pi f_c t$, and the frequency-shifted signal with $EF(t) \sin 2\pi(f_c + \Delta f)t$. Synchronous demodulation with the carriers $\sin(2\pi f_c t + \alpha)$ and $\cos(2\pi f_c t + \alpha)$ according to Eqs. (28) and (29) yields

$$EF(t)\sin 2\pi f_c t\,\sin(2\pi f_c t + \alpha) = \tfrac{1}{2}EF(t)[\cos\alpha - \cos(4\pi f_c t + \alpha)] \quad (31)$$

$$EF(t)\sin 2\pi f_c t\,\cos(2\pi f_c t + \alpha) = \tfrac{1}{2}EF(t)[-\sin\alpha + \sin(4\pi f_c t + \alpha)] \quad (32)$$

Suppressing the high-frequency terms, squaring and summing the remaining terms, and taking their square root yields, in analogy to Eq. (30), the original signal $F(t)$:

$$[\tfrac{1}{4}E^2F^2(t)(\cos^2\alpha + \sin^2\alpha)]^{1/2} = \tfrac{1}{2}EF(t) \quad (33)$$

Let us now go through the same process for the Doppler-shifted signal $EF(t)\sin 2\pi(f_c + \Delta f)t$:

$$EF(t)\sin 2\pi(f_c + \Delta f)t\,\sin(2\pi f_c t + \alpha) = \tfrac{1}{2}EF(t)\{\cos(2\pi\,\Delta f t - \alpha) - \cos[2\pi(2f_c + \Delta f)t + \alpha]\} \quad (34)$$

$$EF(t)\sin 2\pi(f_c + \Delta f)t\,\cos(2\pi f_c t + \alpha) = \tfrac{1}{2}EF(t)\{\sin(2\pi\,\Delta f t - \alpha) + \sin[2\pi(2f_c + \Delta f)t + \alpha]\} \quad (35)$$

The low-frequency term of Eq. (34) is rewritten as follows:

$$\tfrac{1}{2}EF(t)\cos(2\pi\,\Delta f t - \alpha) = \tfrac{1}{2}E[F(t)\cos(2\pi\,\Delta f t)\cos\alpha + F(t)\sin(2\pi\,\Delta f t)\sin\alpha] \quad (36)$$

If $F(t)$ is the triangular pulse in Figs. 2.3-1a and 2.3-2a, and we chose $\Delta f = 1/\Delta T$, the term $F(t)\cos 2\pi\,\Delta f t$ is represented by Fig. 2.3-1c and the term $F(t)\sin 2\pi\,\Delta f t$ by Fig. 2.3-2c. A frequency shift $\Delta f = 4$ kHz calls for $\Delta T = 1/4000$ sec $= 250\ \mu$s. To detect the signal of Eq. (36) one needs two sliding correlators, one for $F(t)\cos 2\pi\,\Delta f t$ and the other for $F(t)\sin 2\pi\,\Delta f t$. At the moment when the signal of Eq. (36) sliding through the correlators coincides with these two sample signals one obtains the following outputs:

$$\frac{1}{2}E\int_{-\Delta T}^{\Delta T}[F(t)\cos(2\pi\,\Delta f t)\cos\alpha + F(t)\sin(2\pi\,\Delta f t)\sin\alpha]F(t) \times \cos(2\pi\,\Delta f t)\,d(t/2\,\Delta T)$$
$$= \frac{1}{2}E\cos\alpha\int_{-\Delta T}^{\Delta T}F^2(t)\cos^2(2\pi\,\Delta f t)\,d(t/2\,\Delta T) = K_1\cos\alpha \quad (37)$$

$$\frac{1}{2}E\int_{-\Delta T}^{\Delta T}[F(t)\cos(2\pi\,\Delta f t)\cos\alpha + F(t)\sin(2\pi\,\Delta f t)\sin\alpha]F(t) \times \sin(2\pi\,\Delta f t)\,d(t/2\,\Delta T)$$
$$= \frac{1}{2}E\sin\alpha\int_{-\Delta T}^{\Delta T}F^2(t)\sin^2(2\pi\,\Delta f t)\,d(t/2\,\Delta T) = K_2\sin\alpha \quad (38)$$

Forming the expression

$$\left[K_1^2 \cos^2 \alpha + \left(\frac{K_1}{K_2} K_2\right)^2 \sin^2\alpha\right]^{1/2} = K_1 \tag{39}$$

yields an output K_1 that does not depend on the phase angle.[1] One may readily recognize that the core of the method of processing of Doppler-shifted signals is the transformation of a signal $F(t)\sin 2\pi f_c t$ with small relative bandwidth into a signal $F(t) \cos 2\pi \Delta f t$ with large relative bandwidth, which can practically be detected by a sliding correlator.

Since this example indicated a difference between Doppler-shifted signals with small or large relative bandwidth, we will discuss the topic in more detail. Consider a radar signal of duration T on a sinusoidal carrier with period $\tau = T/n$, where $n \gg 1$. This is a signal with small relative bandwidth. A Doppler shift will change T and $n\tau$ into $T(1 + 2v/c)$ and $n\tau(1 + 2v/c)$. The term $2Tv/c$ is insignificant compared with T, while the term $2n\tau v/c$ is insignificant compared with $n\tau$ but not compared with τ. We obtain a significant Doppler effect for the carrier, since the resonant circuits respond to the period τ, and not to $n\tau$, but the baseband signal remains practically unchanged. For signals with large relative bandwidth, we do not have many carrier periods τ during the duration T of the signal. For instance, the duration of the triangular pulse in Fig. 2.3-1a equals $T = 2\,\Delta T$, while the period of the carrier in Fig. 2.3-1b equals $\tau = \Delta T$. Hence, we have $T/\tau = n = 2$, and a Doppler shift will be insignificant for the carrier if it is insignificant for the baseband signal. If one wants to use the Doppler effect for signals with large relative bandwidth, one must use it in the form of the moving target indicator or the pulse Doppler effect rather than the frequency shift of the Doppler effect. We will come back to the Doppler effect repeatedly in later chapters.

2.3.3 Frequency Modulation of a Sinusoidal Carrier

Frequency modulation in radar is used in the form of the chirp radar. It is also a possible method to produce spread-spectrum signals. Hence, we will briefly discuss frequency modulation for radio signals with large relative bandwidth.

---

[1] If the received signals do not coincide with the sample signals, one has to replace t by $t - \tau$ in the terms in brackets in Eqs. (37) and (38), and one obtains functions $K_1(\tau)$ and $K_2(\tau)$. Equation (35) can be used to distinguish between a closing target $(+\Delta f)$ and an opening target $(-\Delta f)$. It is usual to shift a received signal without Doppler shift not to dc, as in Eqs. (31) and (32), but to use carriers for demodulation with frequencies $f_c + f_0$, where f_0 is on the order of 10 kHz. A closing target then produces a frequency larger than 10 kHz, while an opening target produces a frequency smaller than 10 kHz. In this way one avoids the need to use Eq. (35), but the sliding correlators have to work at higher frequencies.

Consider an antenna current $i(t)$ frequency modulated by a signal $F(t)$:

$$i(t) = I \sin \left\{ \omega_0 \left[t + \frac{\Delta\omega}{\omega_0} \int F(t')\, dt' \right] \right\} \tag{40}$$

The electric and magnetic field strengths in the far zone are proportionate to the first derivative of the antenna current:

$$\frac{di}{dt} = I\omega_0 \left[1 + \frac{\Delta\omega}{\omega_0} F(t) \right] \cos \left\{ \omega_0 \left[t + \frac{\Delta\omega}{\omega_0} \int F(t')\, dt' \right] \right\} \tag{41}$$

Let $F(t)$ be normalized to the range $-1 \leqq F(t) \leqq +1$. This implies no reduction of generality, since one may still choose the value of the *modulation factor*[1] $\Delta\omega/\omega_0$. A small value $\Delta\omega/\omega_0$ makes di/dt vary like $i(t)$, except for a phase shift of $\pi/2$. A value of $\Delta\omega/\omega_0$ that is not small compared with 1, on the other hand, produces an amplitude modulation in addition to the frequency modulation in Eq. (41). As an example, consider the frequency modulation of the carrier $\sin(2\pi t/\Delta T)$ by the triangular pulse $F(t)$ in Fig. 2.3-6. One obtains:

$$\int F(t')\, dt' = \int_{-\Delta T}^{t} (1 + t'/\Delta T)\, dt'$$

$$= t + t^2/2\,\Delta T + \Delta T/2, \qquad -\Delta T \leqq t \leqq 0 \tag{42}$$

$$\int F(t')\, dt' = \int_{-\Delta T}^{0} (1 + t'/\Delta T)\, dt' + \int_{0}^{t} (1 - t'/\Delta T)\, dt'$$

$$= t - t^2/2\,\Delta T + \Delta T/2, \qquad 0 \leqq t \leqq \Delta T \tag{43}$$

$$\frac{di}{dt} = I\omega_0 \left[1 + \frac{\Delta\omega}{\omega_0} (1 + t/\Delta T) \right]$$

$$\times \cos \left\{ \omega_0 \left[t + \frac{\Delta\omega}{\omega_0} (t + t^2/2\,\Delta T + \Delta T/2) \right] \right\}$$

$$-\Delta T \leqq t \leqq 0 \tag{44}$$

---

[1] Note that $\Delta\omega/\omega_0$ is *not* the modulation index. If $F(t)$ were a sinusoidal function $\sin \omega_1 t$, the modulation index would be defined as $\beta = \Delta\omega/\omega_1$. No specific terminology seems to exist for nonsinusoidal signals $F(t)$ that transmit information at a rate larger than zero, even though the transmission of such signals is discussed in many books and papers (Schwartz *et al.*, 1966, p. 226; Middleton, 1960, p. 604). We have here one of the many relics of power engineering in communication engineering. Sinusoidal voltages and currents are ideal for the transmission of power and energy, since deviations from the sinusoidal form are due only to undesirable or auxiliary effects, such as lightning strokes or switching operations. In communication engineering, on the other hand, the nonsinusoidal form of voltages, currents, and field strengths is a basic necessity for the transmission of information at a rate larger than zero.

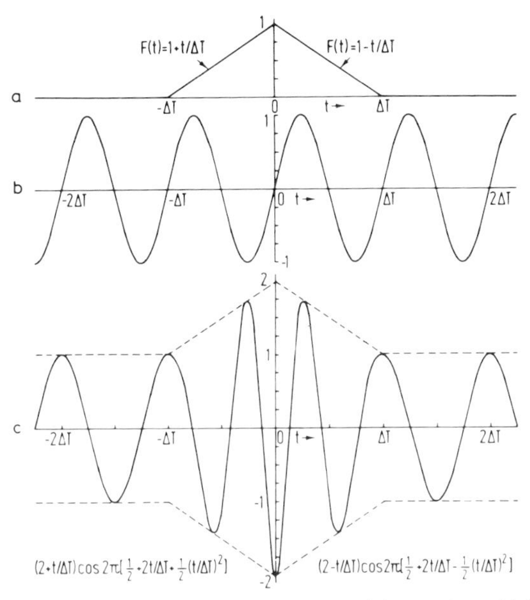

FIG. 2.3-6. Triangular pulse (a) used to frequency-modulate a sinusoidal antenna current (b) produces electric and magnetic field strengths in the far zone (c) that are amplitude-modulated if the modulation factor $\Delta\omega/\omega_0$ is not negligible compared with 1.

$$\frac{di}{dt} = I\omega_0\left[1 + \frac{\Delta\omega}{\omega_0}(1 - t/\Delta T)\right]$$
$$\times \cos\left\{\omega_0\left[t + \frac{\Delta\omega}{\omega_0}(t - t^2/2\,\Delta T + \Delta T/2)\right]\right\}$$
$$0 \leqq t \leqq \Delta T \quad (45)$$

The function $(I\omega_0)^{-1}\, di/dt$ is plotted in Fig. 2.3-6c for a carrier frequency $\omega_0 = 2\pi/\Delta T$ and a modulation factor $\Delta\omega/\omega_0 = 1$.

The important feature of frequency modulation, to yield a radio signal with constant amplitude, is lost if the condition

$$(\Delta\omega/\omega_0)|F(t)| \ll 1 \quad (46)$$

or

$$\Delta\omega/\omega_0 \ll 1 \qquad \text{for} \quad -1 \leqq F(t) \leqq +1 \quad (47)$$

is not satisfied. Note that this condition is independent of the bandwidth of the baseband signal $F(t)$. The peak amplitude of $F(t)$ in Eq. (46) is only apparently important, since we can choose $F(t)$ as in Eq. (47) and adjust

$\Delta\omega$ accordingly. Hence, the fact that the electric and magnetic field strengths in the far zone vary proportionately to the first derivative of the antenna current becomes important if $\Delta\omega/\omega_0$ is not small compared with 1. Of course, the modulation factor $\Delta\omega/\omega_0$ determines the bandwidth of the radio signal—which depends in a complicated way on the modulation factor—and the condition $\Delta\omega/\omega_0 \ll 1$ thus calls for a small relative bandwidth of the radio signal even though the bandwidth of the baseband signal $F(t)$ is not mentioned (Hund, 1942; Cuccia, 1952, p. 266). In order to radiate signals with constant amplitude, one has to make the antenna current proportionate to the integral of Eq. (40):

$$i(t) = I\omega_0 \int^t \sin\left\{\omega_0\left[t'' + \frac{\Delta\omega}{\omega_0}\int^{t''} F(t')\,dt'\right]\right\} dt'' \tag{48}$$

2.4 Nonsinusoidal Carriers

2.4.1 Amplitude Modulation of a Nonsinusoidal Carrier

The use of carrier-free signals or of signals with large relative bandwidth using a sinusoidal carrier will usually be sufficient for practical cases. The investigation of nonsinusoidal carriers is primarily required for theoretical completeness. However, such carriers are of practical interest for spread-spectrum signals and for efficient conversion of dc power into radio signal power.

Let us consider first the use of nonsinusoidal carriers for the generation of spread-spectrum signals.[1] A current with the time variation shown in Fig. 2.4-1a or b fed to a Hertzian dipole will produce electric and magnetic field strengths that vary like the first derivative of the current. Hence, we use the two-valued function with finite transition time ΔT of line c. Its first derivative is shown in line d. The short rectangular pulses approach Dirac pulses, if the transition time ΔT approaches zero.

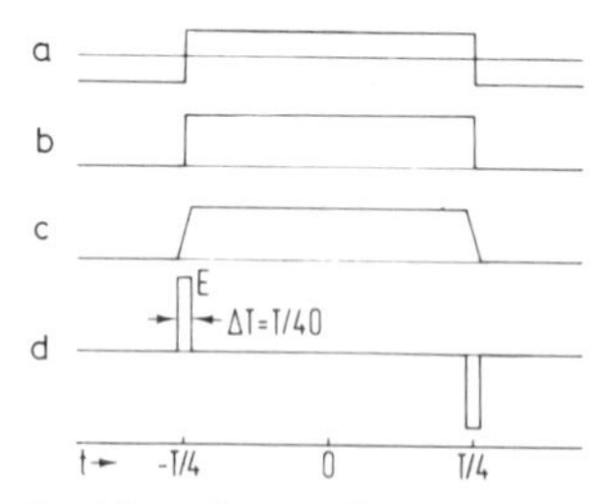

Fig. 2.4-1. Idealized two-valued functions (a, b). Two-valued function with a finite transition time (c), and its derivative (d).

[1] See also Schreiber (1976).

The Fourier transform a(ν) of the field strengths according to Fig. 2.4-1d,

$$a(\nu) = -2E \int_{1/4-\Delta T/2T}^{1/4+\Delta T/2T} \sqrt{2} \sin 2\pi\nu\theta \, d\theta$$

$$= -2\sqrt{2}\, E \frac{\Delta T}{T} \sin \frac{\pi\nu}{2} \frac{\sin \pi\nu\, \Delta T/T}{\pi\nu\, \Delta T/T} \tag{1}$$

$$\theta = t/T, \qquad \nu = fT$$

is shown for $\Delta T/T = 1/40$ in Fig. 2.4-2. The relative bandwidth equals 1 since $a(\nu)$ starts at $\nu = 0$. A periodic repetition of the function of Fig. 2.4-1d produces the coefficients $a(i)$, $i = fT = 1, 3, 5, \ldots$ of a Fourier series that are represented by the dash–dot lines in Fig. 2.4-2. The energy is now concentrated in discrete spectral lines that are distributed over the same frequency band as before. The distance between two adjacent lines equals $fT = 2$ in normalized notation and $f = 2/T$ in nonnormalized notation. In essence, the periodically repeated signal of Fig. 2.4-1c provides us with many sinusoidal carriers with the frequencies $f = i/T$, $i = 1, 3, 5, \ldots$.

Let us see how the periodically repeated signal of Fig. 2.4-1c can be used for spread-spectrum transmission. Figure 2.4-3 shows a typical block diagram for frequency spreading. The baseband signal—which may be a voice signal with a frequency band of 3 kHz, a radar signal with a frequency band of 3 MHz, etc.—is multiplied with a binary spreading carrier. This spreading carrier is typically produced by a digital shift register with a number of feedback loops. Such carriers and their generation are discussed in the literature on coding under headings like *maximum length binary sequences* and *Goldcodes*. An important parameter of the spreading carrier is the minimum pulse duration T_B, shown in Fig. 2.4-3. If T_B equals 1 μs, the baseband signal will be spread over a band of about

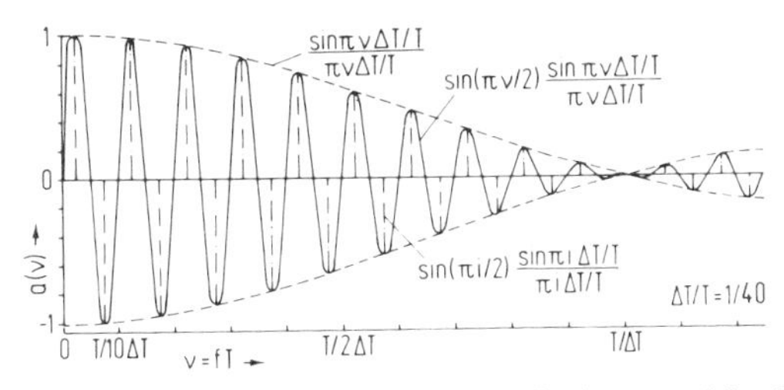

FIG. 2.4-2. The Fourier transform of the function of Fig. 2.4-1d in the infinite interval (solid line), and the coefficients of the Fourier series in the finite interval $-T/2 \leqq t \leqq T/2$ or the periodically repeated function (dash–dot lines).

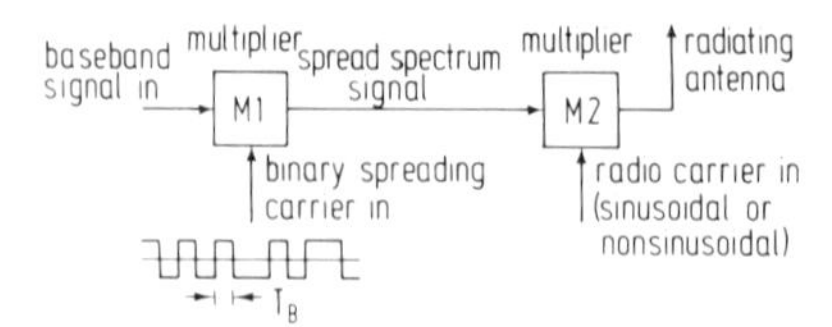

FIG. 2.4-3. Frequency spreading of a baseband signal by means of a binary spreading carrier, and radiation of the spread signal by means of a sinusoidal or a nonsinusoidal carrier.

1 MHz. The technical limit at the present time are values of T_B between 10 and 100 ns, and a spreading over a band of about 100 MHz.

The spread spectrum signal is then fed to another multiplier or amplitude modulator, and multiplied with a sinusoidal radio carrier. The second multiplication shifts the spread spectrum signal to the desired location in the frequency spectrum; it also doubles the occupied frequency band due to the double-sideband modulation feature of sinusoidal carriers.

Let now the sinusoidal carrier be replaced by a periodically repeated current with the time variation of Fig. 2.4-1c, producing electric and magnetic field strengths according to Fig. 2.4-1d. Choosing $T = 10$ ns, we obtain spectral lines in Fig. 2.4-2 at $f = i/T = 100, 300$ MHz, ... ; choosing further $\Delta T/T = 1/40$ we obtain $\Delta T = 0.25$ ns. The frequency $\nu = T/\Delta T$, where the envelope in Fig. 2.4-2 has its first zero, becomes $f = \nu/T = 1/\Delta T = 4$ GHz. Double-sideband modulation of a sinusoidal carrier with a 100-MHz-wide signal would have produced a signal spread over 200 MHz, which is 1/20 of 4 GHz.

One may readily verify that there are 20 spectral lines (dash–dot lines) in Fig. 2.4-2 in the band $0 \leqq \nu \leqq T/\Delta T$, and one is tempted to say that double-sideband modulation of 20 sinusoidal carriers with the frequencies 100, 300, ..., 3900 MHz would have produced essentially the same spread-spectrum signal in the band $0 \leqq f \leqq 4$ GHz. This is not so if one uses the conventional methods of modulation and filtering, since the relative bandwidth would not be small if a signal with a bandwidth of 100 MHz were modulated onto carriers with frequencies of 100, 300 MHz, ... One could, in principle, do it with the methods discussed for large relative bandwidths in Section 2.3, but one would hesitate to build 20 frequency-, phase-, and amplitude-stable transmitters, and also 20 such receivers.[1]

[1] The reader will get the impression that the difference between carrier-free signals, signals with large relative bandwidth produced by amplitude modulation of a sinusoidal carrier, or by amplitude modulation of a nonsinusoidal carrier, is one of practical implementation only. This is quite correct for signals with large relative bandwidth. For instance, the signals of Fig. 2.3-1c, f, g, and h could be produced in various ways without the help of the cosinusoidal and sinusoidal carriers in lines b and e, and they would then be called carrier-free signals.

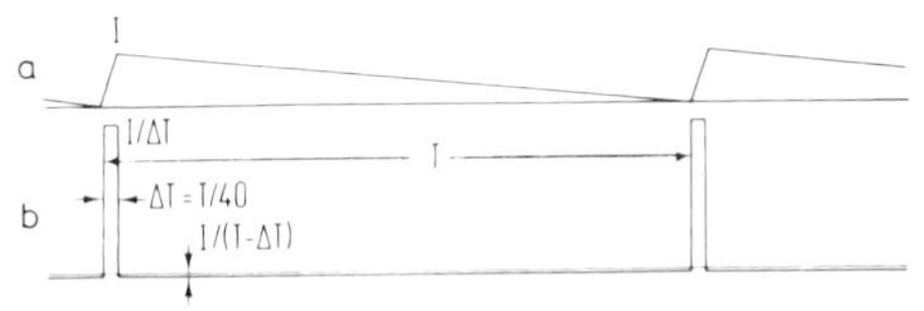

FIG. 2.4-4. Sawtooth current as function of time (a), and its first derivative (b).

Let us derive a figure that characterizes the possible improvement. Useful applications of spread-spectrum transmission range from below 10 Hz to above 10 GHz, and no single criterion can be applied to this whole range. However, at high frequencies the overriding criterion is the bandwidth Δf that a signal can have without exceeding a maximum frequency $f_{\max}$, since the attenuation by rain and fog becomes excessive above 10 GHz according to Fig. 1.5-1, and there is little point in having a radar or communication system that works in fair weather only. From the definition of the relative bandwidth,

$$\eta = (f_H - f_L)/(f_H + f_L)$$

follows the relation:

$$f_H - f_L = \Delta f = \eta(f_H + f_L) = \eta(2f_{\max} - \Delta f), \qquad \Delta f = 2\frac{\eta}{1+\eta} f_{\max} \quad (2)$$

For a transmission with large relative bandwidth, η is essentially 1 and $\eta/(1 + \eta)$ equals 0.5. A typical value of η for transmission with a sinusoidal carrier equals about 0.01, which yields $\eta/(1 + \eta) = 0.01$. Hence, the possible bandwidth for the nonsinusoidal signal is about 50 times as large as for the conventional signal, which implies a reduction of the signal power density by $-10 \log 50 = -17$ dB. This large reduction opens the possibility of using spread-sprectrum techniques for radar signals.

The waveform of Fig. 2.4-1d is one of the basic forms of nonsinusoidal waves. A second one, which we will encounter repeatedly, is produced by a sawtooth-shaped antenna current as shown in Fig. 2.4-4a. Its first derivative in line b consists of short positive pulses with large amplitude, and long negative pulses with small amplitude. The Fourier transform $a(\nu)$ of a single short pulse of Fig. 2.4-4b in the infinite interval,

$$a(\nu) = E\int_{-\Delta T/2}^{\Delta T/2} \sqrt{2}\cos 2\pi\nu\theta \, d\theta = \sqrt{2}\,\frac{\sin \pi\nu\,\Delta T/T}{\pi\nu\,\Delta T/T}$$
$$E = T/\Delta T, \qquad \theta = t/T, \qquad \nu = fT \quad (3)$$

is shown by the solid line in Fig. 2.4-5, while the dash–dot lines show the coefficients $a(i)$, $i = 1, 2, 3, \ldots$ of the Fourier series of the periodically re-

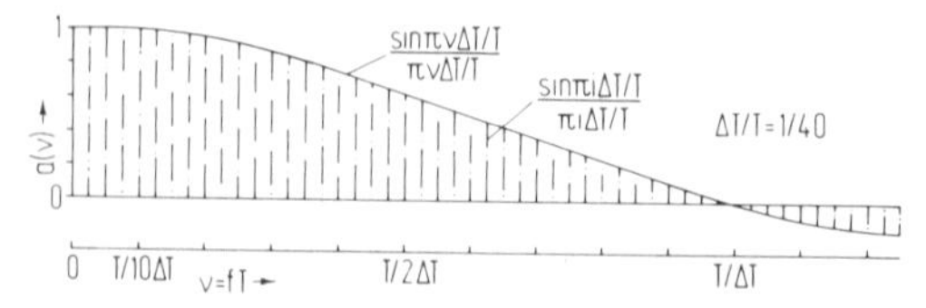

FIG. 2.4-5. The Fourier transform of a single pulse in the infinite interval according to Fig. 2.4-4b (solid line), and the coefficients of the Fourier series in the finite interval $-T/2 \leqq t \leqq T/2$ of the periodically repeated pulse (dash–dot lines). The negative pulses with amplitude $I/(T - \Delta T)$ in Fig. 2.4-4b are neglected.

peated pulse with period T. Most of the energy is in the band $0 \leqq f \leqq 1/\Delta T$.

The Fourier transforms of Figs. 2.4-2 and 2.4-5 show a rather poor distribution of the energy. One would like a fairly equal distribution over a certain frequency band, and little energy outside this band. We will investigate in the next section, how one can control the energy distribution in a practical way.

2.4.2 *Control of Energy Distribution*

The distribution of the energy versus frequency of the pulses in Figs. 2.4-1d and 2.4-4b can be controlled by means of low-pass and band-pass filters. However, we cannot send these pulses themselves through filters, since they represent electric and magnetic field strengths in the far zone. We must shape the radiator current by a filter, to produce the desired time variation of the field strengths in the far zone. The processes of time-invariant filtering and differentiation with respect to time can be interchanged, since both are linear operations. Hence, the same filter that would give the pulses of Figs. 2.4-1d and 2.4-4b the proper shape, if they could be passed through it, can be used to shape the radiator currents of Figs. 2.4-1c and 2.4-4a. Let us demonstrate this with the help of an example.

Figure 2.4-6a shows a unit step function $U(t)$. It becomes the function[1]

$$\frac{1}{2} + \frac{1}{\pi} \operatorname{Si}(2\pi\, \Delta f t)$$

of Fig. 2.4-6b after passing through a low-pass filter with infinite cutoff idealization, as discussed in connection with Figs. 2.1-1 and 2.1-2, hav-

[1] The use of the sine-integral function $\operatorname{Si}(2\pi\, \Delta f t)$ is justified here despite the nonrealizability of such filters, since we make a strictly mathematical point. The interchangeability of time differentiation and time-invariant filtering must hold for practical filters too.

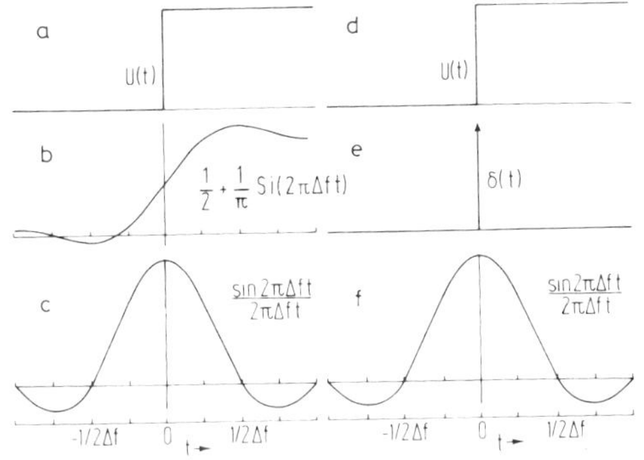

FIG. 2.4-6. Unit step function (a) transformed into the function $1/2 + (1/\pi)\ \mathrm{Si}(2\pi\ \Delta ft)$ by a low-pass filter with infinite cutoff idealization (b), and its derivative (c). The same unit step function (d), transformed into a Dirac delta function by differentiation (e), and into the function $\sin(2\pi\ \Delta ft)/2\pi\ \Delta ft$ by a low-pass filter with infinite cutoff idealization (f).

ing a cutoff frequency Δf. Differentiation yields the function $(\sin 2\pi\ \Delta ft)/2\pi\ \Delta ft$ of Fig. 2.4-6c.

Let now the processes of filtering and differentiation be performed in reversed order. Differentiation of the unit step function $U(t)$ in Fig. 2.4-6d yields the Dirac delta function $\delta(t)$ of line e. The same idealized low-pass filter as before transforms the delta function into the function $(\sin 2\pi\ \Delta ft)/2\pi\ \Delta ft$ of line f.

The periodically continued pulses of Figs. 2.4-1d and 2.4-4b are possible nonsinusoidal carriers. If one wants a system of carriers rather than a single carrier, one must construct an orthogonal system of functions. In particular, if one wants nonsinusoidal carriers with equal positive and negative pulses as in Fig. 2.4-1d, one must construct an orthogonal system of two-valued functions with linear transients. There are many ways of doing so, but the best known such system is derived from the Walsh functions (Walsh, 1923; Wallis *et al.*, 1972; Harmuth, 1969, 1972; Fritzsche, 1977). Figure 2.4-7, column a, shows the first eight Walsh functions with linear transients instead of the usual infinitely short jumps between positive and negative values. Differentiation of these functions yields the functions of column b. The use of a bandpass filter instead of a low-pass filter would replace the transients in column a by functions according to Fig. 2.2-1c. Their derivatives—for a ratio $f_H/f_L = 3$ between upper and lower cutoff frequency—are shown in column c of Fig. 2.4-7.

Let us note a few readily recognizable facts in Fig. 2.4-7. The periodically repeated functions $\mathrm{cal}(i, \theta)$ and $\mathrm{sal}(i, \theta)$, $i = 1, 2, \ldots, \theta = t/T$, differ by a time shift only for any particular value of i, and the same applies also to the functions in columns b and c of Fig. 2.4-7. In mobile com-

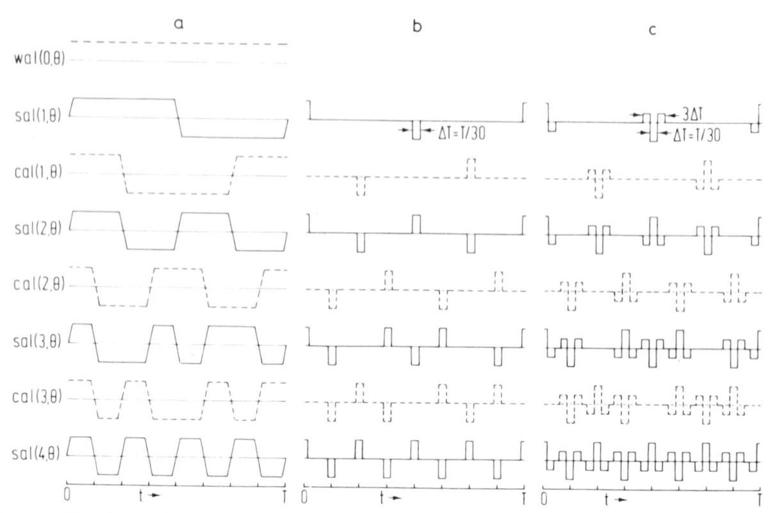

FIG. 2.4-7. Derivation of a set of radio carriers from an orthogonal system of two-valued functions.

munications, we have usually no control or knowledge of the absolute time of arrival of a signal, due to varying distances between transmitter and receivers. Hence, we can use only carriers derived either from the functions cal(i, θ) or sal(i, θ), but not from both.[1] This corresponds to being able to use only the sinusoidal carrier sin ωt or the cosinusoidal carrier cos ωt for a certain frequency, but not both. The first function wal(0, θ) of the Walsh system can be used no more than the first function of the Fourier system in Fig. 1.2-2, since its derivative is zero everywhere. The number of possible carriers in Fig. 2.4-7 depends on the ratio $T/\Delta T$ = period/(transient time). For a fixed transient time ΔT, a longer period T will permit more carriers, but the transmission rate of information varies like $1/T$, since each period of each carrier can carry at most one digit.[2]

2.4.3 *Demodulation of General Carriers*

One can use rectification and synchronous demodulation for the demodulation of a sinusoidal carrier. Both methods also work for nonsinusoidal carriers.

---

[1] We can use both sets of functions cal(i, θ) and sal(i, θ)—or cos ωt and sin ωt—if they come from the same transmitter, since they will be synchronized at any distance from the transmitter.

[2] One cannot say how much information each digit carries, since the digits do not have to be binary digits. The possible number of amplitudes of the digits—i.e., of the carrier during one period—depends on the signal-to-noise ratio.

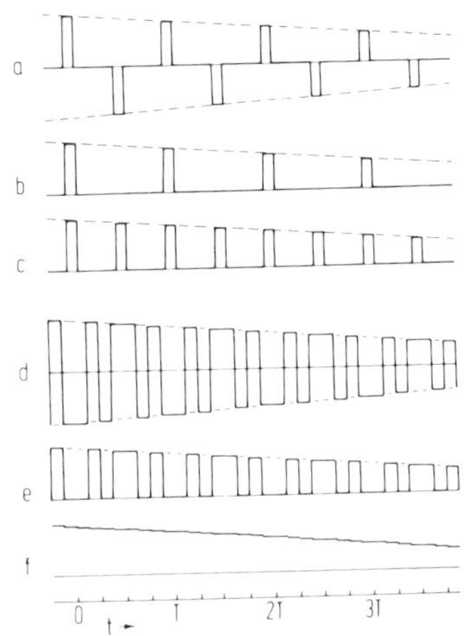

FIG. 2.4-8. Amplitude modulation of the periodically repeated pulses of Fig. 2.4-1d by a ramp function (a); half-wave (b) and full-wave (c) rectification of the modulated signal. Lines (d)–(f) show the amplitude modulation and rectification for the Walsh carrier cal(3, θ) of Fig. 2.4-7 shifted by $T/2$.

Refer to Fig. 2.4-8a. It shows a carrier consisting of the periodically repeated pulses of Fig. 2.4-1 amplitude-modulated by a ramp function. Half-wave rectification yields the signal of line b, and full-wave rectification the signal of line c. The circuits required for the rectification are exactly the same as for a sinusoidal carrier. The original signal is obtained by averaging the rectified carrier, just like in the case of sinusoidal carriers.

Figure 2.4-8d shows the carrier[1] cal(3, θ) of Fig. 2.4-7 amplitude-modulated by a ramp function. Half-wave rectification yields the signal of line e, full-wave rectification the signal of line f. There is no difference in principle between lines a–c and d–f but the rectified signal of line f will generally be more desirable than the other ones.

Consider synchronous demodulation. Let a baseband signal $F(t)$ be amplitude-modulated onto a carrier $g(t)$. Demodulation is accomplished by dividing the product $F(t)g(t)$ by $g(t)$:

$$F(t)g(t)/g(t) = F(t) \tag{4}$$

In the absence of noise, this division works for all values of $g(t)$ except for $g(t) = 0$. One could eliminate these points by suppressing the output at the times when $g(t) = 0$. In the presence of noise, the demodulation ac-

[1] Actually the carrier $-\text{cal}(3, \theta)$ is shown. However, $-\text{cal}(3, \theta)$ and sal(3, θ) differ from $+\text{cal}(3, \theta)$ only by time shifts $-T/2$ and $-T/4$.

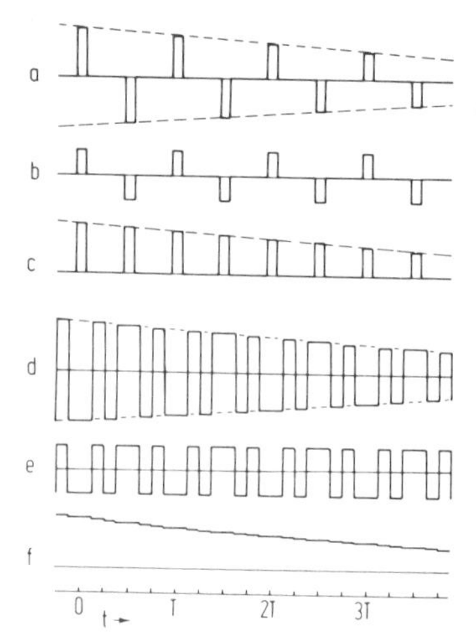

FIG. 2.4-9. The periodically repeated pulses of Fig. 2.4-1d amplitude-modulated by a ramp function (a), the same carrier without amplitude modulation (b), and the product of amplitude-modulated and not modulated carrier (c). Lines (d)–(f) show the same for the Walsh carrier cal(3, θ) of Fig. 2.4-7 shifted by $T/2$.

cording to Eq. (4) is poor whenever $g(t)$ as well as $F(t)$ are small. One avoids this problem if one replaces the division in Eq. (4) by a multiplication:

$$F(t)g(t) \times g(t) = F(t)[C + h(t)] \tag{5}$$

In order for this demodulation process to work one must have a carrier $g(t)$ whose square $g^2(t)$ has the form $C + h(t)$, and the term $F(t)h(t)$ must be easily separable from the term $CF(t)$. This condition is satisfied for $g(t) = \sqrt{2}\cos\omega t$,

$$(\sqrt{2}\cos\omega t)^2 = 1 + \sin 2\omega t \tag{6}$$

if the frequency $f = 2\omega/2\pi$ is larger than the Fourier component of $F(t)$ with the highest frequency.

The method is also applicable to the carriers of Fig. 2.4-8a and d. Their synchronous demodulation is shown in Fig. 2.4-9. The carriers $g(t)$ amplitude-modulated by a ramp function $F(t)$ are shown in lines a and d. The carriers $g(t)$ required for the multiplication of Eq. (5) are shown in lines b and e. The demodulated signals, representing the right-hand side of Eq. (5), are shown in lines c and f. These demodulated signals are equal to the full-wave rectified signals in Fig. 2.4-8c and f. This equality still holds if noise is added to the signal with a Walsh function[1] carrier in line d, but not if noise is aded to the signal in line a. Synchronous demodulation will suppress the noise whenever the carrier $g(t)$ in Fig. 2.4-9b equals zero, but this is not so for rectification in Fig. 2.4-8b and c.

[1] In the Soviet Union the Walsh functions are often called Trachtman functions, in recognition of the older A. M. Trachtman (Trachtman, 1973; Trachtman and Trachtman, 1973, 1975). A. M. Trachtman is associated with the Institute for Problems of Information Transmission of the Academy of Sciences (Institut Problemi Peredazi Informatsii Akad. Nauk), Ul. Ermolovoi 19, 103051 Moscow. See also Bljumin and Shtirin (1974, 1977), Bljumin and Trachtman (1977), Djadjunov and Senin (1977), Jaroslavskij (1979), Sitnikov (1976), Sobol (1969), and Soroko (1976) for recent work on Walsh functions and sequency theory in the Soviet Union. In the People's Republic of China work on Walsh functions and sequency theory is centered at the Northwest Telecommunication Engineering Institute in Xi'an, Shensi (Fan Changxin, Hu Zheng, and Yang Youwei). (See also list of references for China on p. 388.)

3 Radiators and Receptors

3.1 Frequency-Independent Antennas

Let us substitute a sinusoidal current,

$$i = I \sin[\omega(t - r/c)] = I \sin[(2\pi c/\lambda)(t - r/c)] \tag{1}$$

into Eq. (2.1-3) for Poynting's vector of the Hertzian dipole:

$$\mathbf{P}\left(\frac{1}{r^2}, t\right) = \frac{1}{4} Z_0 I^2 \left(\frac{s}{\lambda}\right)^2 \frac{1}{r^2} \cos\left[\omega\left(t - \frac{r}{c}\right)\right] \sin^2 \alpha \frac{\mathbf{r}}{r} \tag{2}$$

The factor $(s/\lambda)^2$ implies that the radiated power is a function of the wavelength, and the Hertzian dipole is thus called a frequency-dependent antenna. A frequency-independent antenna is one whose radiated power does not depend on the wavelength, at least for a large range of wavelengths (Rumsey, 1966).

Practical frequency-independent antennas have achieved radiation over a band with a ratio $f_H/f_L = 40$ between the highest and the lowest frequency. The practical relative bandwidth of such antennas equals

$$\eta = \frac{f_H - f_L}{f_H + f_L} = \frac{40 - 1}{40 + 1} = 0.975 \tag{3}$$

which is more than sufficient to achieve practically distortion-free radiation according to Figs. 2.2-2 and 2.2-3.

Consider a pulse of about 0.1 ns duration. An antenna with frequency-independent radiation in the band 1 GHz $\leqq f \leqq$ 10 GHz will permit essentially distortion-free radiation of this pulse. The wavelength at the upper band edge equals 3 cm, the one at the lower band edge equals 30 cm. The size of the antenna is determined by the longest wavelength, the acceptable tolerances of its construction by the shortest wavelength. The values in our case, 30 cm and 3 cm, are very reasonable. For longer pulses and lower frequencies one obtains rather large dimensions, but frequency-independent antennas covering the shortwave band from 6 to 30 MHz are still well within the state of the art (Smith, 1966).

We will briefly discuss the biconical antenna, the planar spiral antenna, and the log-periodic dipole antenna as examples of frequency-independent antennas. Many more types, and instructions for their design, may be found in the literature (Jasik, 1961; Rhodes, 1974; Rumsey, 1966; Smith, 1966; Smith *et al.*, 1979).

Consider first the biconical antenna shown in Fig. 3.1-1 (Schelkunoff, 1943, 1952). It consists of two circular cones with a common axis, and the angle α between this axis and the surface of the cone. The apexes of the two cones are assumed to be infinitesimally close together, the cones shall be infinitely long and perfect conductors. A particular solution of Maxwell's equations for the field strength has the following form:

$$E_r = 0, \qquad E_\theta = E_0 e^{-j2\pi r/\lambda}/r \sin\theta, \qquad E_\varphi = 0 \tag{4}$$

$$H_r = 0, \qquad H_\theta = 0, \qquad H_\varphi = H_0 e^{-j2\pi r/\lambda}/r \sin\theta \tag{5}$$

$$E_0 = Z_0 H_0, \qquad Z_0 \doteqdot 377\ \Omega$$

In order to obtain a connection between the field strengths and the current $i(t)$ fed into the antenna, as well as the voltage $v(t)$ driving the current, one integrates E_θ along a field line between the two cones, and H_φ along a circle around the axis of the cones:

$$v(r) = \int_\alpha^{\pi-\alpha} E_\theta r\, d\theta = E_0 e^{-j2\pi r/\lambda} \ln[\tan^{-2}(\alpha/2)] \tag{6}$$

$$i(r) = \int_0^{2\pi} H_\varphi r \sin\theta\, d\varphi = 2\pi H_0 e^{-j2\pi r/\lambda} \tag{7}$$

The ratio

$$Z_{00} = \frac{v(r)}{i(r)} = \frac{Z_0}{2\pi} \ln\left[\tan^{-2}\left(\frac{\alpha}{2}\right)\right] \tag{8}$$

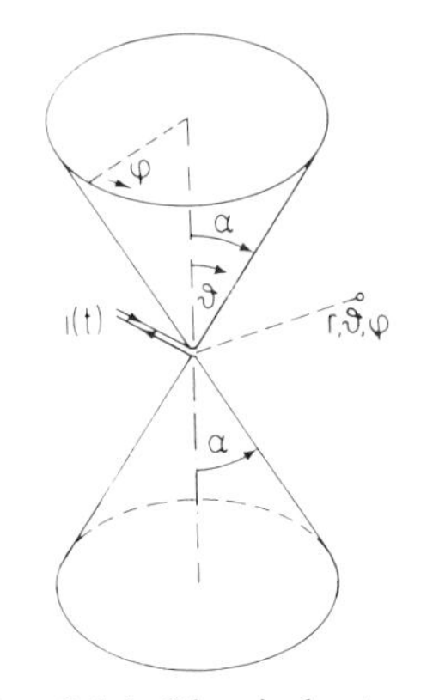

FIG. 3.1-1. Biconical antenna.

FIG. 3.1-2. Planar log-spiral antenna.

does not contain the wavelength, which makes Z_{00} independent of frequency. It also does not contain r; since $i(r)$ and $v(r)$ for $r = 0$ are the current flowing into the antenna and the voltage driving it, the quantity Z_{00} is the impedance presented to the power source. This impedance is a pure resistance, and it is frequency independent.

The biconical antenna is a three-dimensional structure. Our next example, the planar spiral or planar log-spiral antenna, is a two-dimensional structure. An example is shown in Fig. 3.1-2. One contour of the black, conducting area is defined by the equation:

$$r = r_0 e^{\varphi/\tan\alpha} \tag{9}$$

The other contours are obtained by rotation of this first contour. If the angles of this rotation are $\pi/2$, π, and $3\pi/2$ one obtains a so-called self-complementary pattern. The black and white areas of the antenna pattern of Fig. 3.1-2—with the spirals extended to infinity—are in this case equal except for a rotation by $\pi/2$. The impedance Z of all self-complementary antennas is given by Booker's relation (Booker, 1946):

$$Z = Z_0/2 \doteqdot 189\ \Omega \tag{10}$$

One of the best frequency-independent antennas from the practical point of view is the log-periodic dipole antenna shown in Fig. 3.1-3. The length L_n of the dipole n and its distance R_n from the apex are defined by the following equations:

$$L_n/R_n = 2\tan\alpha, \qquad R_{n+1}/R_n = L_{n+1}/L_n = \tau < 1 \tag{11}$$

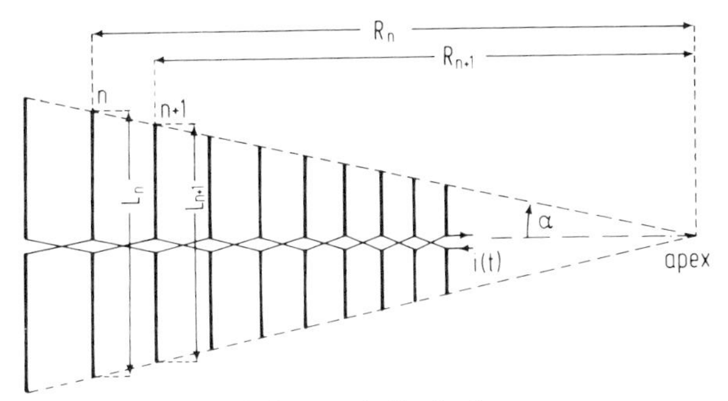

FIG. 3.1-3. Log-periodic dipole antenna.

This antenna can be implemented practically for the shortwave band, where the large size of the structures would make biconical and planar log-spiral antennas impractical. The design of such antennas is discussed in great detail in the literature (Smith 1966).

3.2 HERTZIAN DIPOLE

3.2.1 Frequency-Independent Operation

We had discussed the generation of electric and magnetic field strengths in the far zone with the time variation of Fig. 2.4-1d by means of a Hertzian dipole. We will now investigate the generation of the very same field strengths by means of frequency-independent antennas. To do so we plot the pulses of Fig. 2.4-1d again, but with a time shift to simplify the calculation, in Fig. 3.2-1a. A current $i(\theta) = IF(\theta)$ with this time variation fed into a frequency-independent antenna should produce an electric field strength in the far zone that has the same time variation $F(\theta - \theta_0)$ except for a time

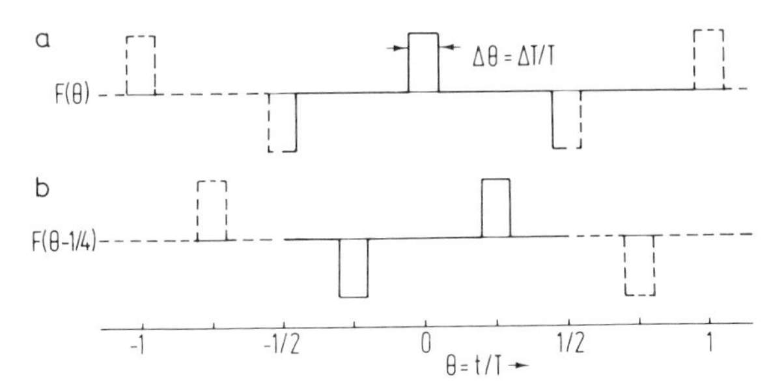

FIG. 3.2-1. Explanation of the operation of a frequency-independent antenna by means of a Fourier series of the driving current (a), which is differentiated and modified term by term to yield the electric and magnetic field strengths in the far zone (b).

shift; this time shift is not necessarily due to the propagation time of the wave only.

Let us represent $F(\theta)$ by its Fourier series:

$$\begin{aligned} a_c(i) &= \sqrt{2}\int_{-1/2}^{1/2} F(\theta)\cos 2\pi i\theta\, d\theta \\ &= 2\sqrt{2}\left\{\int_0^{\Delta\theta/2}\cos 2\pi i\theta\, d\theta - \int_{1/2-\Delta\theta/2}^{1/2}\cos 2\pi i\theta\, d\theta\right\} \\ &= \frac{\sqrt{2}}{\pi i}(1-\cos\pi i)\sin\pi i\,\Delta\theta \end{aligned} \quad (1)$$

$$\begin{aligned} F(\theta) &= \sqrt{2}\sum_{i=1}^{n} a_c(i)\cos 2\pi i\theta \\ &= \frac{2}{\pi}\sum_{i=1}^{n}\frac{1}{i}(1-\cos\pi i)\sin(\pi i\,\Delta\theta)\cos(2\pi i\theta), \qquad i = 1, 2, 3, \ldots \end{aligned} \quad (2)$$

Note that the upper limit of the sum is n rather than infinite. We can make n as large as we need to get a satisfactory representation of $F(\theta)$, but n must remain finite to avoid a problem with convergence in our next step.[1]

The electric and magnetic field strength in the far zone is proportionate to the first derivative of the antenna current for dipole radiation, and proportionate to higher derivatives for quadrupole, octupole, etc., radiation. The physical principle, that radiation is caused by the acceleration of charges, holds true regardless of the particular shape of an antenna. The time variation $dF/d\theta$ of the first derivative of the antenna current follows from Eq. (2):

$$\frac{dF}{d\theta} = -2\sqrt{2}\,\pi\sum_{i=1}^{n}[i(1-\cos\pi i)]\frac{\sqrt{2}}{\pi i}\sin(\pi i\,\Delta\theta)\sin(2\pi i\theta) \quad (3)$$

The time variation of $dF/d\theta$ looks totally different from that of $F(\theta)$, but the unique feature of frequency-independent antennas is to compensate this difference. To show how this is done, let us observe first that Eq. (3) is a Fourier sum with sinusoidal functions $\sin 2\pi i\theta$ only. The only nontrivial way to write the function $F(\theta)$ of Fig. 3.2-1a with sinusoidal functions only is to shift $F(\theta)$ by $\theta = -1/4$ as shown in line b. The Fourier series of this shifted function has the following form:

[1] We will differentiate the sum in Eq. (2). The differentiation of an infinite Fourier series is generally not permitted since the derivative of a convergent series is not necessarily convergent.

$$a_s(i) = \sqrt{2} \int_{-1/2}^{1/2} F\left(\theta - \frac{1}{4}\right) \sin 2\pi i\theta \, d\theta$$

$$= 2\sqrt{2} \int_{1/4-\Delta\theta/2}^{1/4+\Delta\theta/2} \sin 2\pi i\theta \, d\theta = 2 \frac{\sqrt{2}}{\pi i} \sin \frac{\pi i}{2} \sin \pi i \, \Delta\theta \quad (4)$$

$$F\left(\theta - \frac{1}{4}\right) = \sqrt{2} \sum_{i=1}^{n} a_s(i) \sin 2\pi i\theta$$

$$= \sqrt{2} \sum_{i=1}^{n} \left[2 \sin \frac{\pi i}{2}\right] \frac{\sqrt{2}}{\pi i} \sin(\pi i \, \Delta\theta) \sin(2\pi i\theta) \quad (5)$$

A comparison of Eqs. (3) and (5) shows that they differ in the terms shown in brackets only. Hence, the antenna has to multiply each term of the sum of Eq. (3) by

$$\frac{2 \sin \pi i/2}{i(1 - \cos \pi i)} = g(i) \quad (6)$$

to produce the sum of Eq. (5), except for the constant factor -2π. For a qualitative understanding of this process refer to Fig. 3.1-3. The shortest dipoles radiate the components in Eq. (3) with the largest value of the frequency i, or i/T in nonnormalized notation. Since the length of the dipoles decreases like $1/i$ the field strengths will be decreased by the same factor if the peak current is the same for any frequency i. This accounts for the factor $1/i$ in Eq. (6). The factor

$$\begin{aligned} \frac{2 \sin \pi i/2}{1 - \cos \pi i} &= 1 \qquad \text{for} \quad i = 1, 5, 9, \ldots, 4k+1, \ldots \\ &= -1 \qquad \text{for} \quad i = 3, 7, 11, \ldots, 4k+3, \ldots \\ & \qquad k = 0, 1, 2, \ldots \end{aligned} \quad (7)$$

shows, that the components of Eq. (3), which are not zero ($i = 1, 3, 5, \ldots$), should be left unchanged for $i = 1, 5, 9, \ldots$ and be amplitude-reversed for $i = 3, 7, 11, \ldots$. The antenna of Fig. 3.1-3 shows such an amplitude reversal for every other dipole. However, one should observe that an amplitude reversal is also caused by the propagation time along the antenna of the current fed to the antenna. This explains how an amplitude reversal can be achieved with the antennas of Figs. 3.1-1 and 3.1-2.

From this point of view, a frequency-independent antenna differentiates each term of the Fourier representation of the antenna current according to Fig. 3.2-1a, and modifies each term so that the electric and magnetic field strengths in the far zone vary according to Fig. 3.2-1b. A Hertzian dipole according to Eqs. (2.1-1) and (2.1-2), on the other hand, differentiates the antenna current without use of any series representation; it is made a "frequency-independent antenna" by feeding a current into it, that varies like the integral of the wanted variation of the field strengths in the far zone.

Let us generalize this principle. Consider an antenna that produces a field strength $EG(\theta)$ from a driving current $IF(\theta)$. In order to produce a field strength $H(\theta)$, one must then feed the current $IF(\theta)H(\theta)/G(\theta)$ into the antenna. One must be careful to note that nothing is said about the efficiency of radiation. It is inherently very attractive to replace the many dipoles in Fig. 3.1-3 by only one dipole and a modified antenna current, but there is so far no assurance that this possibility can be turned into a practical advantage.

3.2.2 Dipole Arrays

We calculate the power $P(1/r^2, t - r/c)$ flowing through the surface of a large sphere with radius r around a Hertzian dipole by integrating $\mathbf{P}(1/r^2)$ of Eq. (2.1-3) over the surface of the sphere. To this end the scalar product of $\mathbf{P}(1/r^2)$ with the vector $d\mathbf{O}$ of a surface element of the sphere is integrated. As shown in Fig. 3.2-2, the direction of $d\mathbf{O}$ is perpendicular to the sphere, pointing away from the center, while its magnitude equals $r^2 \sin\alpha \, d\alpha \, d\beta$.

$$
\begin{aligned}
P\left(\frac{1}{r^2}, t - \frac{r}{c}\right) &= \oiint \mathbf{P}\left(\frac{1}{r^2}\right) d\mathbf{O} \\
&= \int_{\alpha=0}^{\pi} \int_{\beta=0}^{2\pi} \left(\frac{\mathbf{P}(1/r^2)\mathbf{r}}{r}\right) r^2 \sin\alpha \, d\alpha \, d\beta \\
&= Z_0 \left(\frac{s}{4\pi c}\right)^2 \left(\frac{di}{dt}\right)^2 \int_0^{\pi} \int_0^{2\pi} \sin^3\alpha \, d\alpha \, d\beta \\
&= Z_0 \frac{s^2}{6\pi c^2} \left(\frac{di}{dt}\right)^2 \qquad (8)
\end{aligned}
$$

This is the power radiated at the time $t - r/c$ through the surface of the sphere. The average power radiated during a certain time interval of duration T through the surface of the sphere is denoted $P(1/r^2)$:

$$
\begin{aligned}
P\left(\frac{1}{r^2}\right) &= \frac{1}{T} \int_{+r/c}^{T+r/c} P\left(\frac{1}{r^2}, t - \frac{r}{c}\right) dt \\
&= \frac{Z_0 s^2}{6\pi c^2 T} \int_0^T \left(\frac{di}{dt'}\right)^2 dt', \qquad t' = t - r/c, \quad dt' = dt \qquad (9)
\end{aligned}
$$

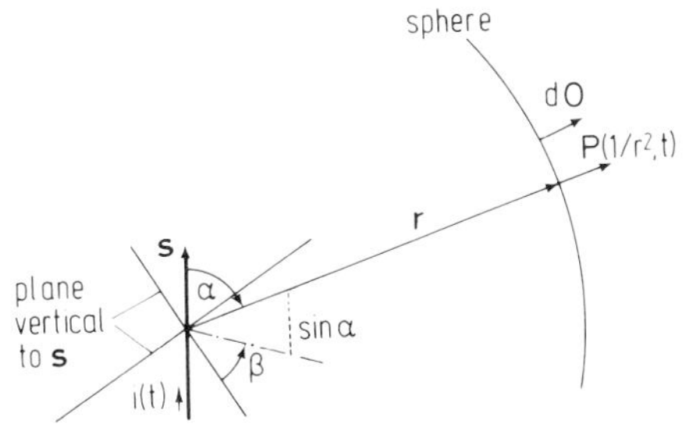

FIG. 3.2-2. A Hertzian dipole in polar coordinates r, α, β. The current $i(t)$ flows in the dipole with the dipole vector s. The location vector is denoted **r**, the surface element vector **dO**, and the component of Poynting's vector decreasing like $1/r^2$ by $\mathbf{P}(1/r^2, t)$.

Note that there is no assumption that the antenna current $i(t)$ and thus the power $P(1/r^2, t - r/c)$ are periodic with period T. Figure 3.2-3 shows the derivative di/dt' of an antenna current according to Fig. 2.4-1, the power $P(1/r^2, t')$ produced by it, and the average $P(1/r^2)$ taken over a time interval of duration T.

Let us now turn to the simple array of two Hertzian dipoles shown in Fig. 3.2-4, and calculate the field strengths at the point L in the far zone produced by the currents $i(t)$ in the two dipoles. The following relations are obtained from Fig. 3.2-4:

$$\mathbf{r}_1 = \mathbf{r} - \tfrac{1}{2}\gamma\mathbf{s}, \qquad \mathbf{r}_2 = \mathbf{r} + \tfrac{1}{2}\gamma\mathbf{s}$$

$$r_1 = (\mathbf{r}_1\mathbf{r}_1)^{1/2} \doteqdot r\left(1 - \frac{1}{2}\gamma\frac{s}{r}\cos\alpha\right), \qquad r \gg \gamma s$$

$$r_2 \doteqdot r\left(1 + \frac{1}{2}\gamma\frac{s}{r}\cos\alpha\right)$$

$$i\left(t - \frac{r_1}{c}\right) = i_{r1} = i\left[t - \frac{r}{c}\left(1 - \frac{1}{2}\gamma\frac{s}{r}\cos\alpha\right)\right]$$

$$= i\left(t' + \frac{1}{2}\gamma\frac{s}{c}\cos\alpha\right)$$

$$i\left(t - \frac{r_2}{c}\right) = i_{r2} = i\left[t - \frac{r}{c}\left(1 + \frac{1}{2}\gamma\frac{s}{r}\cos\alpha\right)\right]$$

$$= i\left(t' - \frac{1}{2}\gamma\frac{s}{c}\cos\alpha\right) \tag{10}$$

$$\mathbf{r}_1 \times \mathbf{s} = \mathbf{r}_2 \times \mathbf{s} = \mathbf{r} \times \mathbf{s}$$

The electric field strength $\mathbf{E}_r$ is the sum of the electric field strengths $\mathbf{E}_{r1}$ and $\mathbf{E}_{r2}$ produced by dipoles 1 and 2, which are given by Eq. (2.1-1). We

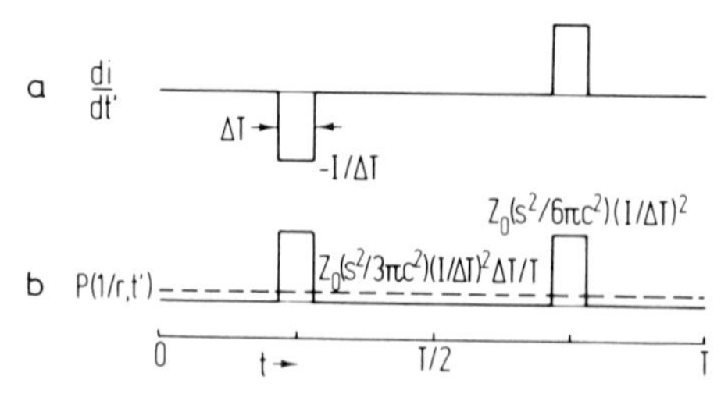

FIG. 3.2-3. Time derivative of the antenna current (a) in a Hertzian electric dipole, and the variation of the component of Poynting's vector decreasing like $1/r^2$ (b). The dashed line shows the average power in the interval $0 \leqq t \leqq T$.

use only the terms decreasing with $1/r$, and write $\mathbf{E}_r(1/r)$, $\mathbf{E}_{r1}(1/r)$ and $\mathbf{E}_{r2}(1/r)$ for these terms:

$$\mathbf{E}_r(1/r) = \mathbf{E}_{r1}(1/r) + \mathbf{E}_{r2}(1/r)$$

$$= Z_0 \frac{s}{4\pi c r}\left[\left(1 + \frac{1}{2}\gamma\frac{s}{r}\cos\alpha\right)\frac{di_{r1}}{dt}\frac{(\mathbf{r} - \frac{1}{2}\gamma\mathbf{s}) \times (\mathbf{r} \times \mathbf{s})}{sr^2}\right.$$

$$\times\left(1 + \gamma\frac{s}{r}\cos\alpha\right) + \left(1 - \frac{1}{2}\gamma\frac{s}{r}\cos\alpha\right)\frac{di_{r2}}{dt}$$

$$\left.\times\frac{(\mathbf{r} + \frac{1}{2}\gamma\mathbf{s}) \times (\mathbf{r} \times \mathbf{s})}{sr^2}\left(1 - \gamma\frac{s}{r}\cos\alpha\right)\right]$$

$$= Z_0 \frac{s}{4\pi c r}\left(\frac{di_{r1}}{dt} + \frac{di_{r2}}{dt}\right)\frac{\mathbf{r} \times (\mathbf{r} \times \mathbf{s})}{sr^2}, \qquad r \gg \gamma s \qquad (11)$$

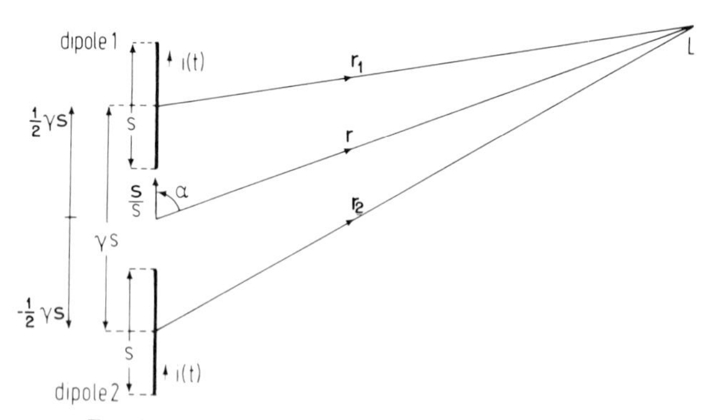

FIG. 3.2-4. Coaxial pair of Hertzian electric dipoles.

Note that terms of the form $(1/r^2)\,di/dt$ occur in Eq. (11), while there were no such terms in Eq. (2.1-1). Although we neglect these terms here to shorten the formulas, we will have to come back to them later on.

The magnetic field strength $\mathbf{H}_r(1/r)$ follows from Eq. (2.1-1) under the same condition $r \gg \gamma s$:

$$\mathbf{H}_r\left(\frac{1}{r}\right) = \mathbf{H}_{r1}\left(\frac{1}{r}\right) + \mathbf{H}_{r2}\left(\frac{1}{r}\right) = \frac{s}{4\pi cr}\left(\frac{di_{r1}}{dt} + \frac{di_{r2}}{dt}\right)\frac{\mathbf{s}\times\mathbf{r}}{sr} \tag{12}$$

Poynting's vector $\mathbf{P}_r(1/r^2)$ follows in analogy to Eq. (2.2-3):

$$\begin{aligned}\mathbf{P}_r\left(\frac{1}{r^2}\right) &= \mathbf{E}_r\left(\frac{1}{r}\right)\times\mathbf{H}\left(\frac{1}{r}\right)\\ &= Z_0\left(\frac{s}{4\pi cr}\right)^2\left(\frac{di_{r1}}{dt} + \frac{di_{r2}}{dt}\right)^2 \sin^2\alpha\,\frac{\mathbf{r}}{r}\end{aligned} \tag{13}$$

We have enough formulas to derive some radiation patterns. Consider first the power radiated by the single Hertzian dipole as given by Eq. (2.1-3). The only term depending on the angle α is the term $\sin^2\alpha$, where α is specified in Fig. 3.2-2. Hence, the relative power as function of α, or the power radiation pattern, is $\sin^2\alpha$. This pattern is shown in Fig. 3.2-5. It has rotational symmetry around the axis of the dipole, which means it has the same shape for all values of the angle β in Fig. 3.2-2.

The power radiation pattern of Eq. (13) has not the same shape, even though $\sin^2\alpha$ is the only explicit function of α. According to Eq. (10), the currents i_{r1} and i_{r2} also depend on the angle α. Hence, one must specify the time variation of the antenna current. We choose a square wave according to Fig. 2.4-1, but with linear transient of duration ΔT between positive and negative values of the current $i(t - r/c) = i(t')$ as shown in

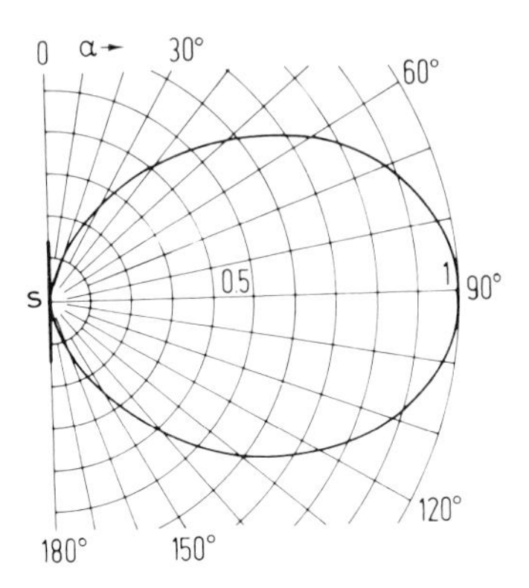

FIG. 3.2-5. Power radiation pattern of the Hertzian electric dipole of Fig. 3.2-2 according to Eq. (2.1-3). The pattern is rotationally symmetric for the angle β.

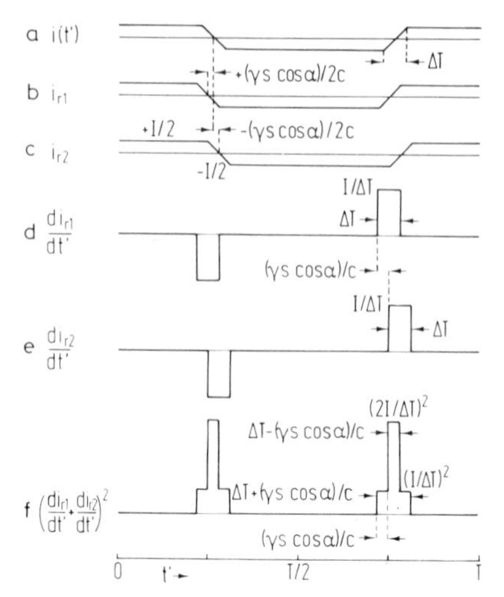

FIG. 3.2-6. Currents (a–c) according to Eq. (10) for the pair of coaxial Herzian electric dipoles of Fig. 3.2-4, their derivatives (d, e) and the square of their sum (f).

Fig. 3.2-6a. The current $i_{r1} = i_{r1}[t' + \frac{1}{2}\gamma(s/c)\cos\alpha]$ is advanced by $+(\gamma s\cos\alpha)/2c$, while the current $i_{r2} = i_{r2}[t' - \frac{1}{2}\gamma(s/c)\cos\alpha]$ is retarded by $-(\gamma s\cos\alpha)/2c$ relative to $i(t')$ according to Eq. (10). These currents, their derivatives, and the square of the sum of the derivatives are shown in Fig. 3.2-6b–f. For further clarification, the derivatives and the square of their sum are shown for typical values of α in Fig. 3.2-7.

Let us first study the *peak power radiation pattern* $P_{\text{peak}}(\alpha)$. In the sector $0 \leqq \cos\alpha \leqq c\,\Delta T/\gamma s$ the peak power is constant and the peak power radiation patterns equals $\sin^2\alpha$ according to Eq. (13). In the sector $c\,\Delta T/\gamma s \leqq \cos\alpha \leqq 1$ the peak power drops to 1/4 according to Fig. 3.2-7, and the peak radiation pattern equals $1/4\,\sin^2\alpha$ in this sector. Hence, there is a jump of 4:1 in the pattern for $\alpha = \alpha_0$,

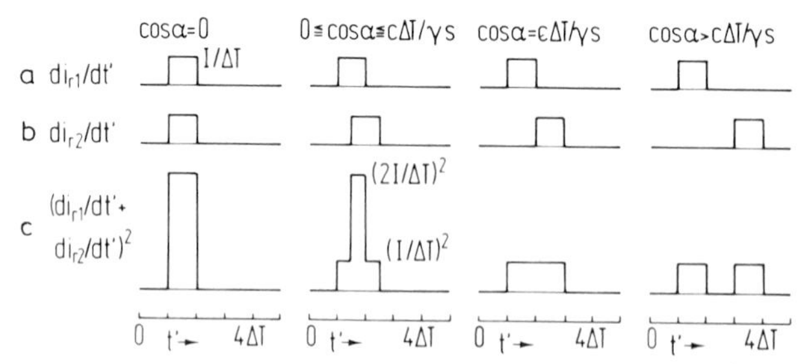

FIG. 3.2-7. Four typical time variations of $(di_{r1}/dt' + di_{r2}/dt')^2$ in Fig. 3.2-6f, and the values of the angle α for which they occur.

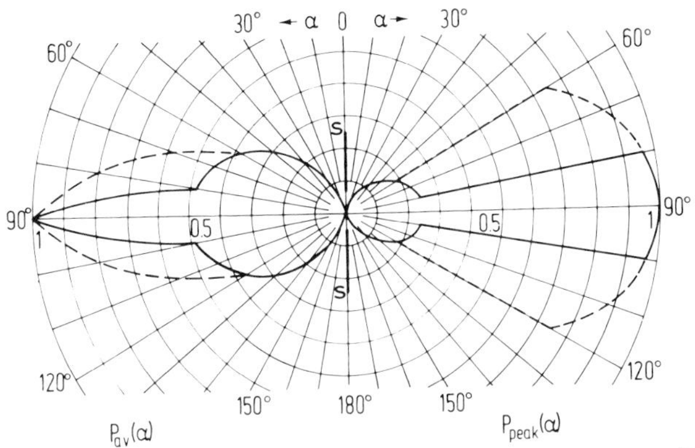

FIG. 3.2-8. Peak power radiation pattern $P_{peak}(\alpha)$ and average power radiation pattern $P_{av}(\alpha)$ for $\alpha_0 = 80°$ (solid lines) and $\alpha_0 = 60°$ (dashed lines) of the coaxial pair of dipoles in Fig. 3.2-4, for the antenna current $i(t')$ in Fig. 3.2-6a.

$$\begin{aligned} \cos\alpha_0 &= c\,\Delta T/\gamma s \qquad \text{for} \quad c\,\Delta T/\gamma s \leqq 1 \\ &= 1 \qquad\qquad\quad \text{for} \quad c\,\Delta T/\gamma s > 1 \end{aligned} \tag{14}$$

Respective patterns are shown[1] for $\alpha_0 = 60°$ and $\alpha_0 = 80°$ in Fig. 3.2-8, right half-plane. One may choose the angle α_0 by using the value

$$\gamma s = c\,\Delta T/\cos\alpha_0 \tag{15}$$

for the distance between the centers of the dipoles in Fig. 3.2-4.

The *average power radiation pattern* $P_{av}(\alpha)$ is obtained by averaging Poynting's vector of Eq. (13) over a time interval T:

$$\begin{aligned} &\frac{1}{T}\int_{r/c}^{r/c+T} \mathbf{P}_r\left(\frac{1}{r^2}\right) dt \\ &= Z_0\left(\frac{s}{4\pi rc}\right)^2 \frac{1}{T}\int_{r/c}^{r/c+T}\left(\frac{di_{r1}}{dt}+\frac{di_{r2}}{dt}\right)^2 dt\,\sin^2\alpha\,\frac{\mathbf{r}}{r} \end{aligned} \tag{16}$$

The integral

$$\begin{aligned} \frac{1}{T}\int_{r/c}^{r/c+T}\left(\frac{di_{r1}}{dt}+\frac{di_{r2}}{dt}\right)^2 dt &= \frac{1}{T}\int_0^T\left(\frac{di_{r1}}{dt}+\frac{di_{r2}}{dt}\right)^2 dt' \\ t' = t - r/c, \qquad dt' &= dt \end{aligned} \tag{17}$$

[1] The patterns for $0 \geqq \cos\alpha \geqq -1$ follow from the symmetry of Fig. 3.2-4 with respect to $\alpha = 90°$.

is a function of α, according to Figs. 3.2-6 and 3.2-7. One obtains from these illustrations the following relations:

$$\frac{1}{T}\int_0^T (di_{r1}/dt + di_{r2}/dt)^2\, dt' = 2(I/\Delta T)^2\,(2\,\Delta T/T) \qquad \text{for} \quad c\,\Delta T/\gamma s \leqq \cos\alpha \leqq 1 \tag{18}$$

$$\begin{aligned}\frac{1}{T}\int_0^T (di_{r1}/dt + di_{r2}/dt)^2\, dt' &= 2\frac{1}{T}(I/\Delta T)^2\,\{(\gamma s\cos\alpha)/c \\ &\quad + 4[\Delta T - (\gamma s\cos\alpha)/c] + (\gamma s\cos\alpha)/c\} \\ &= 2(I/\Delta T)^2[2 - (\gamma s\cos\alpha)/c\,\Delta T]\,(2\,\Delta T/T) \\ &\quad \text{for} \quad 0 \leqq \cos\alpha \leqq c\,\Delta T/\gamma s\end{aligned} \tag{19}$$

Equation (19) yields the largest value, $8(I/\Delta T)^2\,\Delta T/T$ for $\alpha = 90°$. The angle dependent part of Eq. (16), divided by the largest value of Eq. (19), is defined as the average power radiation pattern $P_{\text{av}}(\alpha)$:

$$\begin{aligned}P_{\text{av}}(\alpha) &= [1 - (\gamma s\cos\alpha)/2c\,\Delta T]\sin^2\alpha \qquad &&\text{for} \quad 0 \leqq \cos\alpha \leqq c\,\Delta T/\gamma s \\ &= \tfrac{1}{2}\sin^2\alpha &&\text{for} \quad c\,\Delta T/\gamma s \leqq \cos\alpha \leqq 1\end{aligned} \tag{20}$$

Substitution of Eq. (15) permits us to write Eq. (20) in terms of α_0:

$$\begin{aligned}P_{\text{av}}(\alpha) &= [1 - (\cos\alpha)/2\cos\alpha_0]\sin^2\alpha \qquad &&\text{for} \quad 0 \leqq \cos\alpha \leqq \cos\alpha_0 \\ &= \tfrac{1}{2}\sin^2\alpha &&\text{for} \quad \cos\alpha_0 \leqq \cos\alpha \leqq 1\end{aligned} \tag{21}$$

Plots of $P_{\text{av}}(\alpha)$ for $\alpha_0 = 60°$ and $80°$ are shown in Fig. 3.2-8, left half-plane.

The average power P_{av} radiated through the surface of a large sphere with radius r is obtained by integration of Eq. (16) over the surface of the sphere:

$$\begin{aligned}P_{\text{av}} &= \oiint \frac{1}{T}\int_{r/c}^{r/c+T} \mathbf{P}_r \left(\frac{1}{r^2}\right) dt\, d\mathbf{O} \\ &= Z_0\left(\frac{s}{4\pi rc}\right)^2 \\ &\quad \times \int_{\alpha=0}^{\pi}\int_{\beta=0}^{2\pi}\left[\frac{1}{T}\int_0^T\left(\frac{di_{r1}}{dt'} + \frac{di_{r2}}{dt'}\right)^2 dt'\right] r^2\sin^3\alpha\, d\alpha\, d\beta \\ &= Z_0\frac{s^2}{8\pi c^2}\int_0^{\pi}\left[\frac{1}{T}\int_0^T\left(\frac{di_{r1}}{dt'} + \frac{di_{r2}}{dt'}\right)^2 dt'\right]\sin^3\alpha\, d\alpha\end{aligned} \tag{22}$$

Substitution of Eqs. (18) and (19) yields,

$$P_{av} = Z_0 \frac{I^2 s^2}{\pi c^2} \frac{1}{T\,\Delta T} \left\{ \int_{\alpha=0}^{\alpha_0} \sin^3 \alpha \, d\alpha \right.$$

$$\left. + \int_{\alpha_0}^{\pi/2} [2 - (\gamma s \cos \alpha)/c\,\Delta T] \sin^3 \alpha \, d\alpha \right\}$$

$$= Z_0 \frac{I^2 s^2}{3\pi c^2} \frac{1}{T\,\Delta T} \left[2 - \cos^3 \alpha_0 + 3 \cos \alpha_0 - \frac{3\gamma s}{4c\,\Delta T} (1 - \sin^4 \alpha_0) \right]$$

$$\cos \alpha_0 = c\,\Delta T/\gamma s \qquad \text{for} \quad c\,\Delta T/\gamma s \leqq 1$$

$$= 1 \qquad \text{for} \quad c\,\Delta T/\gamma s > 1 \tag{23}$$

Consider a few particular cases. First, the two dipoles in Fig. 3.2-4 shall practically coincide, which means $\gamma s = 0$ and $\alpha_0 = 0$. One obtains from Eq. (23):

$$P_{av} = 4Z_0 \frac{I^2 s^2}{3\pi c^2} \frac{1}{T\,\Delta T}, \qquad \alpha_0 = 0 \tag{24}$$

Let us compare this with the average power radiated by a single dipole according to Eq. (9) with the current $i(t')$ of Fig. 3.2-6a:

$$P_{av} = \frac{Z_0 s^2}{6\pi c^2 T} \left[\left(\frac{I}{\Delta T} \right)^2 2\,\Delta T \right] = Z_0 \frac{I^2 s^2}{3\pi c^2} \frac{1}{T\,\Delta T} \tag{25}$$

Equation (25) equals Eq. (24) if I is replaced by $2I$. Hence, two dipoles close together radiate as much power as one dipole with twice the current. Putting it differently, the power radiated per dipole is increased by a factor 2 due to the interaction. This is the well-known result that the efficiency of power radiation of a radiator increases with n, if there are n interacting radiators.

Consider now the other extreme of two radiators with a very large distance $\gamma s \gg c\,\Delta T$ between them. From Eq. (23) follows with $\cos \alpha_0 = 0$ and $\sin \alpha_0 = 1$:

$$P_{av} = 2Z_0 \frac{I^2 s^2}{3\pi c^2} \frac{1}{T\,\Delta T}, \qquad \alpha_0 = 90^\circ \tag{26}$$

The radiated power is now twice the power radiated by one dipole. The power radiation pattern at a distance $r \gg \gamma s$ is shown in Fig. 3.2-9. The *rest lobes* have the same shape as in Fig. 3.2-8 but the *main lobe* is compressed so much that its beam width vanishes.

For the two intermediate cases $\alpha_0 = 60^\circ$ and 80°, for which radiation patterns are shown in Fig. 3.2-8, we obtain from Eq. (23):

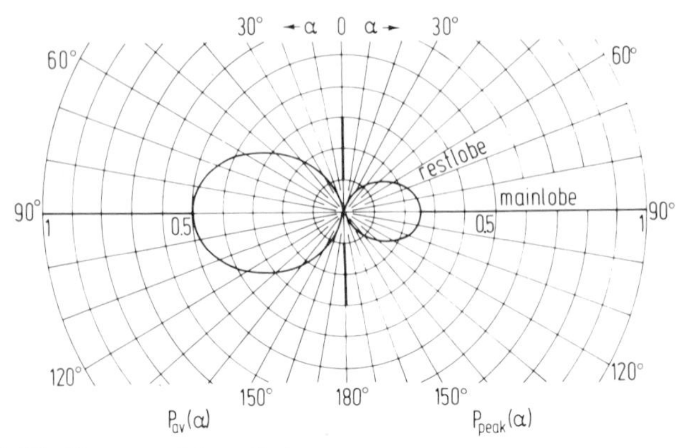

FIG. 3.2-9. Limiting case of the radiation patterns in Fig. 3.2-8 for $\alpha_0 = 90°$, when the distance between the two dipoles becomes very large, $\gamma s \gg c\,\Delta T$.

$$P_{av} = 2.72 Z_0 \frac{I^2 s^2}{3\pi c^2} \frac{1}{T\,\Delta T}, \qquad \alpha_0 = 60°, \quad c\,\Delta T/\gamma s = \cos\alpha_0$$

$$P_{av} = 2.26 Z_0 \frac{I^2 s^2}{3\pi c^2} \frac{1}{T\,\Delta T}, \qquad \alpha_0 = 80° \tag{27}$$

Hence, the average radiated power decreases monotonically with increasing values of α_0 from the limit of Eq. (24) to that of Eq. (26). Note that the power $P_{av}(\alpha)$ or $P_{peak}(\alpha)$ radiated in the direction $\alpha = 90°$ is always the same. The reduction of P_{av} is only due to the reduction of the beam width of the main lobe.

The increase of the power radiated by a dipole, that is part of an array of dipoles, is difficult to understand if one only considers the far-zone components in Eq. (11). However, it is quite evident from that equation that terms with the time variation di_{r1}/dt or di_{r2}/dt are not only multiplied by $1/r$ but also by $1/r^2$. Hence, there are near-zone as well as far-zone components with this time variation, while this is not so for the single dipole according to Eq. (2.1-1). These near-zone components have to be considered if one wants to study how the increase of radiated power due to interaction comes about. There is, in principle, no difficulty inserting $\mathbf{r}_1$, $\mathbf{r}_2$, i_{r1}, and i_{r2} from Eq. (10) into Eqs. (2.1-1) and (2.1-2) and to calculate Poynting's vector in Eq. (13) to a higher approximation in $1/r$, but the equations become very lengthy.

The main lobes in Figs. 3.2-8 and 3.2-9 are very satisfactory, but the rest lobes are not. In order to reduce them, one may use more than two dipoles. Figure 3.2-10 shows four dipoles. The distance between the centers of the dipoles D1 and D4 equals γs, just as in Fig. 3.2-4. Hence, the size of

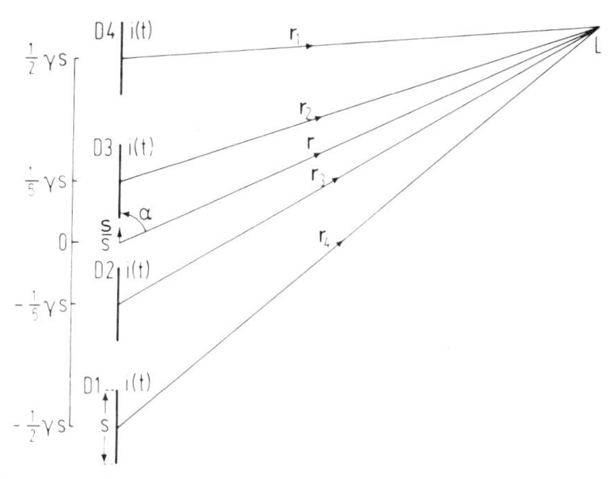

FIG. 3.2-10. Four coaxial Hertzian electric dipoles with equal distances between them.

the dipole array is the same as before. Equations (10)–(13) assume the following form:

$$\mathbf{r}_1 = \mathbf{r} - \tfrac{1}{2}\gamma\mathbf{s}, \quad \mathbf{r}_2 = \mathbf{r} - \tfrac{1}{6}\gamma\mathbf{s}, \quad \mathbf{r}_3 = \mathbf{r} + \tfrac{1}{6}\gamma\mathbf{s}, \quad \mathbf{r}_4 = \mathbf{r} + \tfrac{1}{2}\gamma\mathbf{s}$$

$$r_1 \doteq r\left(1 - \frac{1}{2}\gamma\frac{s}{r}\cos\alpha\right), \qquad r_2 \doteq r\left(1 - \frac{1}{6}\gamma\frac{s}{r}\cos\alpha\right)$$

$$r_3 \doteq r\left(1 + \frac{1}{6}\gamma\frac{s}{r}\cos\alpha\right), \qquad r_4 \doteq r\left(1 + \frac{1}{2}\gamma\frac{s}{r}\cos\alpha\right)$$

$$\begin{aligned}
i\left(t - \frac{r_1}{c}\right) &= i_{r1} = i\left[t - \frac{r}{c}\left(1 - \frac{1}{2}\gamma\frac{s}{r}\cos\alpha\right)\right] \\
&= i\left(t' + \frac{1}{2}\gamma\frac{s}{c}\cos\alpha\right) \\
i\left(t - \frac{r_2}{c}\right) &= i_{r2} = i\left(t' + \frac{1}{6}\gamma\frac{s}{c}\cos\alpha\right) \\
i\left(t - \frac{r_3}{c}\right) &= i_{r3} = i\left(t' - \frac{1}{6}\gamma\frac{s}{c}\cos\alpha\right) \\
i\left(t - \frac{r_4}{c}\right) &= i_{r4} = i\left(t' - \frac{1}{2}\gamma\frac{s}{c}\cos\alpha\right)
\end{aligned} \tag{28}$$

$$\begin{aligned}
\mathbf{E}_r\left(\frac{1}{r}\right) &= \mathbf{E}_{r1}\left(\frac{1}{r}\right) + \mathbf{E}_{r2}\left(\frac{1}{r}\right) + \mathbf{E}_{r3}\left(\frac{1}{r}\right) + \mathbf{E}_{r4}\left(\frac{1}{r}\right) \\
&= Z_0\frac{s}{4\pi cr}\left(\frac{di_{r1}}{dt} + \frac{di_{r2}}{dt} + \frac{di_{r3}}{dt} + \frac{di_{r4}}{dt}\right)\frac{\mathbf{r}\times(\mathbf{r}\times\mathbf{s})}{sr^2}
\end{aligned} \tag{29}$$

$$\mathbf{H}_r\left(\frac{1}{r}\right) = \frac{s}{4\pi c r}\left(\frac{di_{r1}}{dt} + \frac{di_{r2}}{dt} + \frac{di_{r3}}{dt} + \frac{di_{r4}}{dt}\right)\frac{\mathbf{s}\times\mathbf{r}}{sr} \tag{30}$$

$$\mathbf{P}_r\left(\frac{1}{r}\right) = Z_0\left(\frac{s}{4\pi c r}\right)^2\left(\frac{di_{r1}}{dt} + \frac{di_{r2}}{dt} + \frac{di_{r3}}{dt} + \frac{di_{r4}}{dt}\right)^2 \sin^2\alpha\,\frac{\mathbf{r}}{r} \tag{31}$$

Figure 3.2-11 shows on top a current $i(t')$ in analogy to Fig. 3.2-6a, but the time scale has been expanded by a factor 2, and only the transition from $-I/2$ to $+I/2$ is shown. The currents i_{r1} to i_{r4} shifted from $+(\gamma s \times \cos\alpha)/2c$ to $-(\gamma s\cos\alpha)/2c$ relative to $i(t')$ are shown in lines b to e, while lines f to i show the derivatives of these currents. The square of the sum of the derivatives according to Eq. (29) is shown in line j.

Figure 3.2-12 shows the square of the sum of the derivatives of the antenna currents for the regions where the peak values are $(4I/\Delta T)^2$, $(3I/\Delta T)^2$, $(2I/\Delta T)^2$, and $(I/\Delta T)^2$. The peak power radiation pattern $P_{\text{peak}}(\alpha)$ follows readily from Eq. (31):

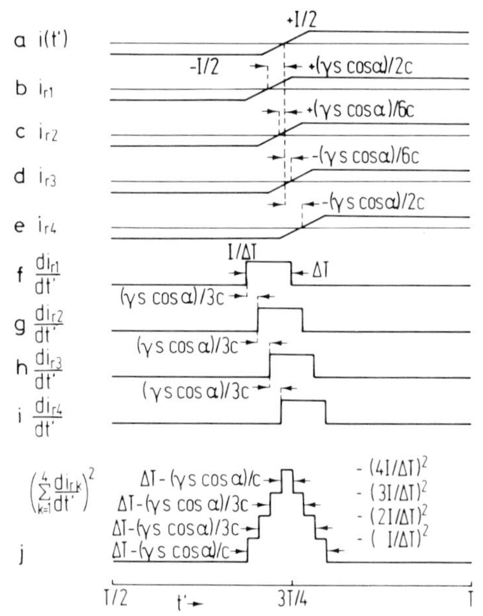

FIG. 3.2-11. Currents (a–e) according to Eq. (28) for the four Hertzian electric dipoles in Fig. 3.2-10, their derivatives (f–i), and the square of their sum (j).

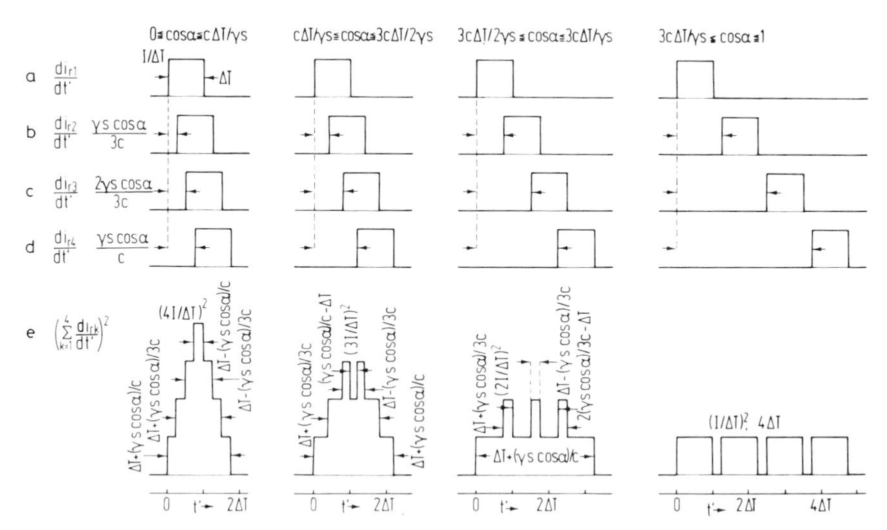

FIG. 3.2-12. Four distinct time variations of $(\Sigma\, di_{rk}/dt')^2$ in Fig. 3.2-11j, and the values of the angle α for which they occur.

$$
\begin{aligned}
P_{\text{peak}}(\alpha) &= \sin^2\alpha && \text{for} \quad 0 \leqq \cos\alpha \leqq \cos\alpha_0 \\
&= \tfrac{9}{16}\sin^2\alpha && \text{for} \quad \cos\alpha_0 \leqq \cos\alpha \leqq \tfrac{3}{2}\cos\alpha_0 \\
&= \tfrac{4}{16}\sin^2\alpha && \text{for} \quad \tfrac{3}{2}\cos\alpha_0 \leqq \cos\alpha \leqq 3\cos\alpha_0 \\
&= \tfrac{1}{16}\sin^2\alpha && \text{for} \quad 3\cos\alpha_0 \leqq \cos\alpha \leqq 1 \\
\cos\alpha_0 &= c\,\Delta T/\gamma s && \text{for} \quad c\,\Delta T/\gamma s \leqq 1 \\
&= 1 && \text{for} \quad c\,\Delta T/\gamma s > 1
\end{aligned} \tag{32}
$$

This pattern is shown for $\alpha_0 = 86°$ on the right half of Fig. 3.2-13. The angles where the pattern changes discontinuously are denoted α_0, α_1, and α_2:

$$
\begin{aligned}
\alpha_0 &= 86° \\
\alpha_1 &= \cos^{-1}(\cos\alpha_1) = \cos^{-1}(\tfrac{3}{2}\cos 86°) = 84° \\
\alpha_2 &= \cos^{-1}(\cos\alpha_2) = \cos^{-1}(3\cos 86°) = 78°
\end{aligned} \tag{33}
$$

The average power radiation pattern $P_{\text{av}}(\alpha)$ requires more work. We average Poynting's vector of Eq. (31) over the time interval T in analogy to Eq. (16):

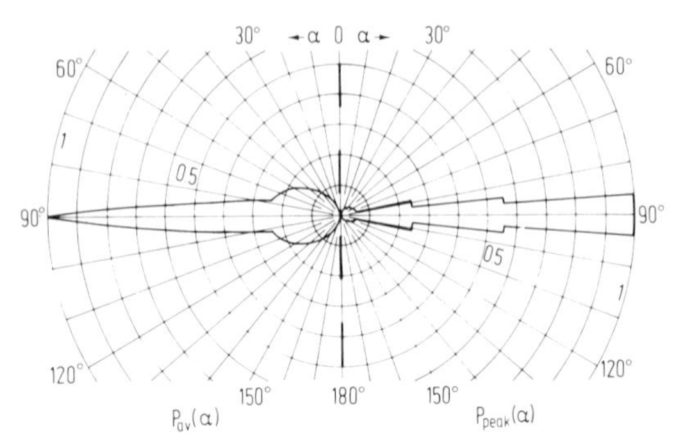

FIG. 3.2-13. Peak power radiation patterns $P_{\text{peak}}(\alpha)$ and average power radiation pattern $P_{\text{av}}(\alpha)$ for $\alpha_0 = 86°$ of the array of four dipoles in Fig. 3.2-10, for the antenna current $i(t')$ in Fig. 3.2-11a.

$$\frac{1}{T}\int_{r/c}^{r/c+T} \mathbf{P}_r\left(\frac{1}{r^2}\right) dt = Z_0\left(\frac{s}{4\pi cr}\right)^2 g(\alpha)\sin^2\alpha\,\frac{\mathbf{r}}{r}$$

$$g(\alpha) = \frac{1}{T}\int_{r/c}^{r/c+T}\left(\frac{di_{r1}}{dt} + \frac{di_{r2}}{dt} + \frac{di_{r3}}{dt} + \frac{di_{r4}}{dt}\right)^2 dt \tag{34}$$

The variable t in the integral is replaced by t':

$$g(\alpha) = \frac{1}{T}\int_0^T\left(\frac{di_{r1}}{dt'} + \frac{di_{r2}}{dt'} + \frac{di_{r3}}{dt'} + \frac{di_{r4}}{dt'}\right)^2 dt' \tag{35}$$

$$t' = t - r/c, \qquad dt' = dt$$

The function $g(\alpha)$ follows from Fig. 3.2-12 for the four sectors of α defined there:

$$\begin{aligned}
g(\alpha) &= 4(I/\Delta T)^2[8\,\Delta T - 10(\gamma s\cos\alpha)/3c]/T \\
&\quad \text{for} \quad 0 \leqq \cos\alpha \leqq c\,\Delta T/\gamma s \\
&= 4(I/\Delta T)^2[7\,\Delta T - 7(\gamma s\cos\alpha)/3c]/T \\
&\quad \text{for} \quad c\,\Delta T/\gamma s \leqq \cos\alpha \leqq 3c\,\Delta T/2\gamma s \\
&= 4(I/\Delta T)^2[5\,\Delta T - (\gamma s\cos\alpha)/c]/T \\
&\quad \text{for} \quad 3c\,\Delta T/2\gamma s \leqq \cos\alpha \leqq 3c\,\Delta T/\gamma s \\
&= 4(I/\Delta T)^2 2\,\Delta T/T \\
&\quad \text{for} \quad 3c\,\Delta T/\gamma s \leqq \cos\alpha \leqq 1
\end{aligned} \tag{36}$$

Equation (36) yields the largest value, $32(I/\Delta T)^2\,\Delta T/T$, for $\alpha = 90°$. The angle-dependent part of Eq. (34), divided by the largest value of Eq. (36), is the average power radiation pattern $P_{\mathrm{av}}(\alpha)$:

$$
\begin{aligned}
P_{\mathrm{av}}(\alpha) &= [1 - 5(\gamma s \cos\alpha)/12c\,\Delta T]\sin^2\alpha \\
&\qquad \text{for} \quad 0 \leqq \cos\alpha \leqq \cos\alpha_0 \\
&= \tfrac{1}{8}[7 - 7(\gamma s \cos\alpha)/3c\,\Delta T]\sin^2\alpha \\
&\qquad \text{for} \quad \cos\alpha_0 \leqq \cos\alpha \leqq \tfrac{3}{2}\cos\alpha_0 \\
&= \tfrac{1}{8}[5 - (\gamma s \cos\alpha)/c\,\Delta T]\sin^2\alpha \\
&\qquad \text{for} \quad \tfrac{3}{2}\cos\alpha_0 \leqq \cos\alpha \leqq 3\cos\alpha_0 \\
&= \tfrac{1}{4}\sin^2\alpha \\
&\qquad \text{for} \quad 3\cos\alpha_0 \leqq \cos\alpha \leqq 1 \qquad (37) \\
\cos\alpha_0 &= c\,\Delta T/\gamma s \qquad \text{for} \quad c\,\Delta T/\gamma s \leqq 1 \\
&= 1 \qquad\qquad\;\; \text{for} \quad c\,\Delta T/\gamma s > 1
\end{aligned}
$$

This pattern is shown for $\alpha_0 = 86°$ on the left half of Fig. 3.2-13.

For $\gamma s \gg c\,\Delta T$ we obtain again a main beam with vanishing beamwidth and a restlobe $\frac{1}{2}\sin^2\alpha$ for the peak power or $\frac{1}{4}\sin^2\alpha$ for the average power pattern in the sector $90 < \alpha \leqq 0$. These two patterns are shown in Fig. 3.2-14.

The comparison of the two limit patterns for $\gamma s \gg c\,\Delta T$ of Figs. 3.2-9 and 3.2-14 shows that the increase in the number of dipoles improves the restlobe. The beamwidth of the main lobe can be made vanishingly small for any number of dipoles. Equations (20) and (37) show that the largest value of the restlobe is never larger than 1/2 for 2 dipoles or 1/4 for 4 dipoles. A short reflection shows that this result can be generalized for n di-

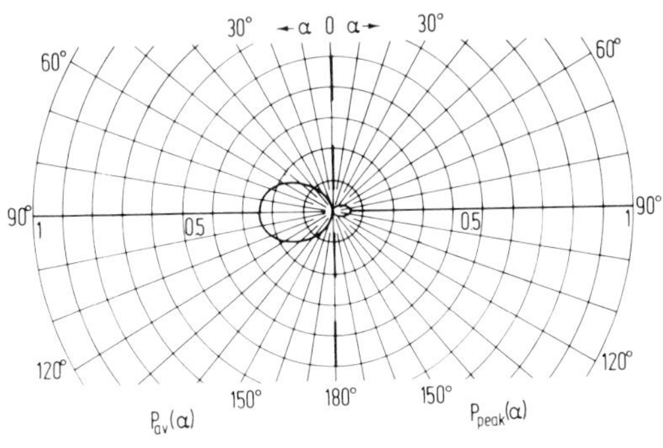

FIG. 3.2-14. Limiting case of the radiation patterns in Fig. 3.2-13 for $\alpha_0 = 90°$, when the distance between the dipoles becomes very large, $\gamma s \gg c\,\Delta T$.

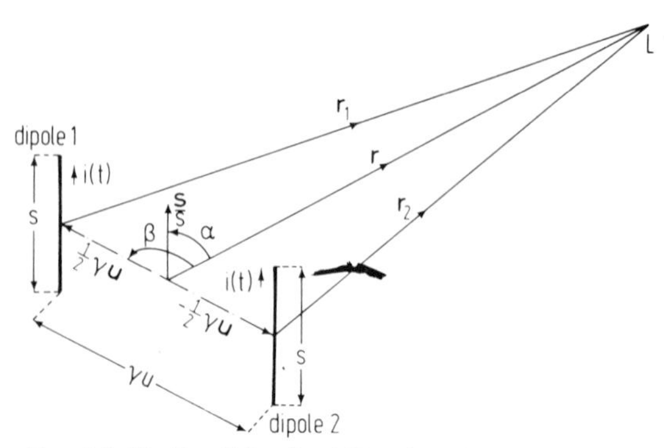

FIG. 3.2-15. Parallel pair of Hertzian electric dipoles.

poles. The restlobe of the peak power pattern is in this case not larger than $1/n^2$, and the restlobe of the average power pattern is not larger than $1/n$. Hence, a ratio $1/n = 1000$, or 30 dB, calls for 1000 dipoles, if the ratio refers to the average power pattern, but for only 32 dipoles, if it refers to the peak power pattern.

The dipole arrays of Figs. 3.2-4 and 3.2-10 are coaxial. As a result, the radiation patterns are circularly symmetric for the angle β defined in Fig. 3.2-2. In order to achieve a narrow beam for the angle β, one must use *parallel dipole arrays* rather than coaxial ones. Such an array with two dipoles is shown in Fig. 3.2-15. In analogy to Eq. (10) we find the following relations:

$$\mathbf{r}_1 = \mathbf{r} + \tfrac{1}{2}\gamma\mathbf{u}, \qquad \mathbf{r}_2 = \mathbf{r} - \tfrac{1}{2}\gamma\mathbf{u}$$

$$r_1 = (\mathbf{r}_1\mathbf{r}_1)^{1/2} \doteqdot r\left(1 - \frac{1}{2}\gamma\frac{u}{r}\cos\beta\right), \qquad r \gg \gamma u, \quad u^2 = \mathbf{uu}$$

$$r_2 \doteqdot r\left(1 + \frac{1}{2}\gamma\frac{u}{r}\cos\beta\right)$$

$$i(t - r_1/c) = i_{r1} = i\left[t - \frac{r}{c}\left(1 - \frac{1}{2}\gamma\frac{u}{r}\cos\beta\right)\right]$$

$$= i\left(t' + \frac{1}{2}\gamma\frac{u}{c}\cos\beta\right)$$

$$i(t - r_2/c) = i_{r2} = i\left(t' - \frac{1}{2}\gamma\frac{u}{c}\cos\beta\right) \tag{38}$$

The electric and magnetic field strengths as well as Poynting's vector in the far zone follow in analogy to Eqs. (11)–(13):

$$\mathbf{E}\left(\frac{1}{r}\right) = Z_0 \frac{s}{4\pi c r}\left(\frac{di_{r1}}{dt} + \frac{di_{r2}}{dt}\right)\frac{\mathbf{r} \times (\mathbf{r} \times \mathbf{s})}{sr^2} \tag{39}$$

$$\mathbf{H}\left(\frac{1}{r}\right) = \frac{s}{4\pi c r}\left(\frac{di_{r1}}{dt} + \frac{di_{r2}}{dt}\right)\frac{\mathbf{s} \times \mathbf{r}}{sr} \tag{40}$$

$$\mathbf{P}_r\left(\frac{1}{r^2}\right) = Z_0 \left(\frac{s}{4\pi c r}\right)^2 \left(\frac{di_{r1}}{dt} + \frac{di_{r2}}{dt}\right)^2 \sin^2\alpha\,\frac{\mathbf{r}}{r} \tag{41}$$

Note that the only difference between these three equations and Eqs. (11)–(13) is the different time variation of the currents i_{r1} and i_{r2}.

Let us again use a current $i(t')$ as shown in Fig. 3.2-16a. From line f one infers that the dipoles interact for

$$\Delta T - (\gamma u \cos\beta)/c > 0 \tag{42}$$

while there is no interaction for smaller values of $\beta \leqq \beta_0$:

$$\begin{aligned} \cos\beta_0 &= c\,\Delta T/\gamma u \qquad \text{for} \quad c\,\Delta T/\gamma u \leqq 1 \\ &= 1 \qquad\qquad\quad \text{for} \quad c\,\Delta T/\gamma u > 1 \end{aligned} \tag{43}$$

The peak value $P_{\mathrm{peak}}(1/r)$ of Poynting's vector for these two sectors of β follows from Fig. 3.2-16f:

$$\begin{aligned} P_{\mathrm{peak}}\left(\frac{1}{r^2}\right) &= Z_0\left(\frac{s}{4\pi c r}\right)^2\left(\frac{2I}{\Delta T}\right)^2 \sin^2\alpha \\ &\quad \text{for} \qquad 0 \leqq \cos\beta \leqq c\,\Delta T/\gamma u \\ &= Z_0\left(\frac{s}{4\pi c r}\right)^2\left(\frac{I}{\Delta T}\right)^2 \sin^2\alpha \\ &\quad \text{for} \quad c\,\Delta T/\gamma u \leqq \cos\beta \leqq 1 \end{aligned} \tag{44}$$

The peak power radiation pattern $P_{\mathrm{peak}}(\alpha, \beta)$ is $P_{\mathrm{peak}}(1/r^2)$ divided by its largest value $Z_0(s/4\pi c r)^2(2I/\Delta T)^2$:

$$\begin{aligned} P_{\mathrm{peak}}(\alpha, \beta) &= \sin^2\alpha \qquad \text{for} \qquad 0 \leqq \cos\beta \leqq c\,\Delta T/\gamma u \\ &= \tfrac{1}{4}\sin^2\alpha \quad\; \text{for} \quad c\,\Delta T/\gamma u \leqq \cos\beta \leqq 1 \end{aligned} \tag{45}$$

This pattern is shown[1] for

$$c\,\Delta T/\gamma u = \cos\beta_0 = \cos 76^\circ$$

and $\alpha = 90^\circ$ by the solid line in Fig. 3.2-17. The variation of $P_{\mathrm{peak}}(\alpha, \beta)$ with α is shown in Fig. 3.2-5.

---

[1] The patterns $P_{\mathrm{peak}}(\alpha, \beta)$ and $P_{\mathrm{av}}(\alpha, \beta)$ outside the sector $0 \leqq \cos\beta \leqq 1$ follow from the symmetry of Fig. 3.2-15 for $\beta = 0^\circ$ and 90°.

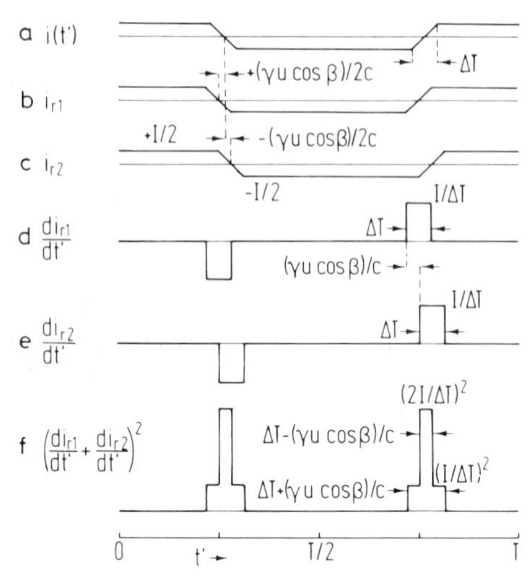

FIG. 3.2-16. Currents (a–c) according to Eq. (38) for the pair of parallel Hertzian electric dipoles of Fig. 3.2-15, their derivatives (d, e), and the square of their sum (f).

Let us turn to the average power radiation pattern $P_{av}(\alpha, \beta)$. The average value of Poynting's vector for the two sectors of β follows from Fig. 3.2-16f:

$$P_{av}(1/r^2) = Z_0 \left(\frac{s}{4\pi cr}\right)^2 \frac{8I^2}{T\,\Delta T}[1 - (\gamma u \cos\beta)/2c\,\Delta T] \sin^2\alpha \qquad \text{for} \quad 0 \leqq \cos\beta \leqq c\,\Delta T/\gamma u \tag{46}$$

$$= Z_0 \left(\frac{s}{4\pi cr}\right)^2 \frac{4I^2}{T\,\Delta T} \sin^2\alpha \qquad \text{for} \quad c\,\Delta T/\gamma u \leqq \cos\beta \leqq 1$$

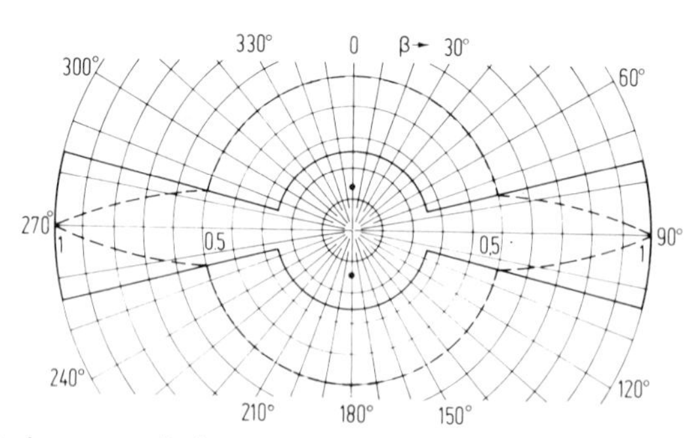

FIG. 3.2-17. Peak power radiation pattern $P_{peak}(90, \beta) = P_{peak}(\beta)$ (solid line), and average power radiation pattern $P_{av}(90, \beta) = P_{av}(\beta)$ (dashed line), for $\beta_0 = 76°$ of the array of two dipoles in Fig. 3.2-15, for the antenna current $i(t')$ in Fig. 3.2-16a.

The average power radiation pattern is $P_{av}(1/r^2)$ divided by its largest value $Z_0(s/4\pi cr)^2 8I^2/T\,\Delta T$:

$$
\begin{aligned}
P_{av}(\alpha, \beta) &= [1 - (\gamma u \cos\beta)/2c\,\Delta T]\sin^2\alpha \\
&\quad \text{for} \quad 0 \leqq \cos\beta \leqq c\,\Delta T/\gamma u \\
&= \tfrac{1}{2}\sin^2\alpha \\
&\quad \text{for} \quad c\,\Delta T/\gamma u \leqq \cos\beta \leqq 1
\end{aligned}
\tag{47}
$$

This pattern is shown for

$$c\,\Delta T/\gamma u = \cos\beta_0 = \cos 76°$$

and $\alpha = 90°$ by the dashed line in Fig. 3.2-17. The variation of $P_{av}(\alpha, \beta)$ with α is shown in Fig. 3.2-5. Average and peak radiation patterns in the sector $104° < \beta < 256°$ are repeated in the sector $284° < \beta < 76° + +360°$.

If the distance between the two dipoles in Fig. 3.2-15 becomes very large, $\gamma u \gg c\,\Delta T$, one obtains the limit of the peak and average power radiation pattern shown in Fig. 3.2-18. The restlobe is much larger than in Fig. 3.2-9, which holds for the two coaxial dipoles of Fig. 3.2-4. The reason is, of course, that the dipole in Fig. 3.2-2 has the directive radiation pattern for the variable α shown in Fig. 3.2-5, while its radiation pattern for the variable β equals 1 for the whole range $0 \leqq \beta \leqq 360°$.

Next we derive the radiation patterns for the array of four dipoles in Fig. 3.2-19. We do not need to actually do any calculations. A comparison of Eqs. (13) and (41) as well as (10) and (38) shows that we can derive the radiation patterns for parallel arrays from those of coaxial arrays by substituting u and $\cos\beta$ for s and $\cos\alpha$. One only has to be careful to preserve the term $\sin^2\alpha$ in Eq. (13) or (31) since it is due to the radiation pattern of the individual dipole rather than the array. The peak power radia-

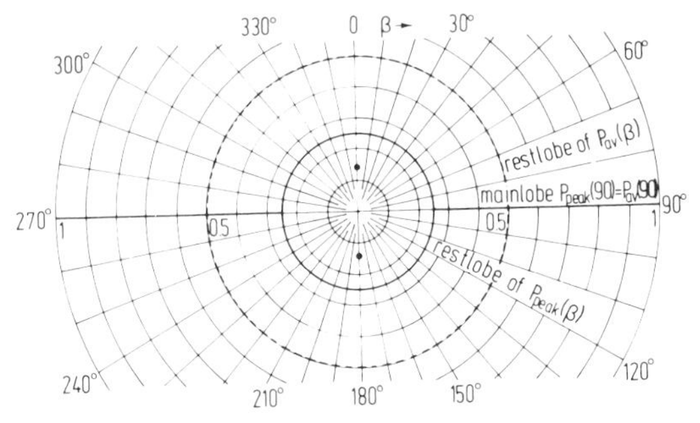

FIG. 3.2-18. Limiting case of the radiation patterns in Fig. 3.2-17, when the distance between the dipoles becomes very large, $\gamma u \gg c\,\Delta T$.

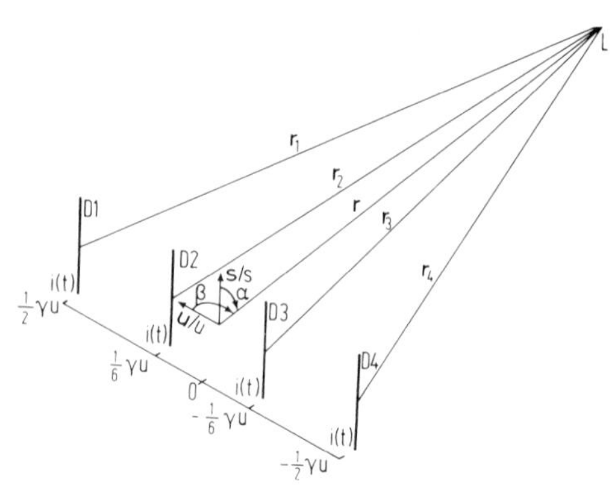

FIG. 3.2-19. Four parallel Hertzian electric dipoles with equal distances between them.

tion pattern $P_{\text{peak}}(\alpha, \beta)$ for the array of Fig. 3.2-19 follows thus from Eq. (32):

$$\begin{aligned}
P_{\text{peak}}(\alpha, \beta) &= \sin^2 \alpha && \text{for} \quad 0 \leqq \cos \beta \leqq \cos \beta_0 \\
&= \tfrac{9}{16} \sin^2 \alpha && \text{for} \quad \cos \beta_0 \leqq \cos \beta \leqq \tfrac{3}{2} \cos \beta_0 \\
&= \tfrac{4}{16} \sin^2 \alpha && \text{for} \quad \tfrac{3}{2} \cos \beta_0 \leqq \cos \beta \leqq 3 \cos \beta_0 \\
&= \tfrac{1}{16} \sin^2 \alpha && \text{for} \quad 3 \cos \beta_0 \leqq \cos \beta \leqq 1 \\
\cos \beta_0 &= c\, \Delta T/\gamma u && \text{for} \quad c\, \Delta T/\gamma u \leqq 1 \\
&= 1 && \text{for} \quad c\, \Delta T/\gamma u > 1 \\
& 0 \leqq \alpha \leqq 180^\circ
\end{aligned} \tag{48}$$

The average power radiation pattern $P_{\text{av}}(\alpha, \beta)$ follows from Eq. (37):

$$\begin{aligned}
P_{\text{av}}(\alpha, \beta) &= [1 - 5(\gamma u \cos \beta)/12c\, \Delta T] \sin^2 \alpha \\
& \text{for} \quad 0 \leqq \cos \beta \leqq \cos \beta_0 \\
&= \tfrac{1}{8}[7 - 7(\gamma u \cos \beta)/3c\, \Delta T] \sin^2 \alpha \\
& \text{for} \quad \cos \beta_0 \leqq \cos \beta \leqq \tfrac{3}{2} \cos \beta_0 \\
&= \tfrac{1}{8}[5 - (\gamma u \cos \beta)c\, \Delta T] \sin^2 \alpha \\
& \text{for} \quad \tfrac{3}{2} \cos \beta_0 \leqq \cos \beta \leqq 3 \cos \beta_0 \\
&= \tfrac{1}{4} \sin^2 \alpha \\
& \text{for} \quad 3 \cos \beta_0 \leqq \cos \beta \leqq 1
\end{aligned} \tag{49}$$

The definition of β_0 and the range of α are the same as for Eq. (48).

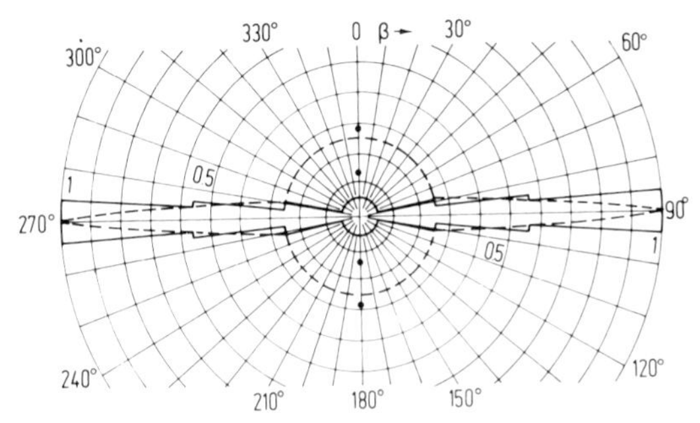

FIG. 3.2-20. Peak power radiation pattern $P_{peak}(90, \beta) = P_{peak}(\beta)$ (solid line), and average power radiation pattern $P_{av}(90, \beta) = P_{av}(\beta)$ (dashed line) for $\beta_0 = 86°$ of the array of four dipoles in Fig. 3.2-19, for the antenna current $i(t')$ in Fig. 3.2-11a with α replaced by β.

Plots of $P_{peak}(\alpha, \beta)$ and $P_{av}(\alpha, \beta)$ for $\alpha = 90°$ and $\beta_0 = 86°$ are shown in Fig. 3.2-20. This illustration should be compared with Fig. 3.2-13, which shows the radiation patterns of four coaxial dipoles for $\alpha_0 = 86°$. The main difference is the smaller restlobe of both $P_{peak}(\alpha)$ and $P_{av}(\alpha)$ in Fig. 3.2-13 caused by the factor $\sin^2\alpha$.

The limit of $P_{peak}(\alpha, \beta)$ and $P_{av}(\alpha, \beta)$ in Fig. 3.2-20 for $\gamma u \gg c\,\Delta T$ is shown in Fig. 3.2-21. Again, the difference with the comparable plots of Fig. 3.2-14 are the larger restlobes.

Consider the combination of the dipole arrays of Figs. 3.2-19 and 3.2-10 to an array with 4 parallel sets of 4 coaxial dipoles each. The radiation patterns as function of β will be those derived for Fig. 3.2-19, since the patterns obtained for Fig. 3.2-10 had the same value for any β. The variation of $P_{peak}(\alpha, \beta)$ and $P_{av}(\alpha, \beta)$ of Eqs. (48) and (49) as functions of α will no longer be $\sin^2\alpha$, but it will be replaced by Eqs. (32) and (37). Hence, $P_{peak}(\alpha, \beta)$ and $P_{av}(\alpha, \beta)$ would be represented by $P_{peak}(\alpha)P_{peak}(\beta)$ and

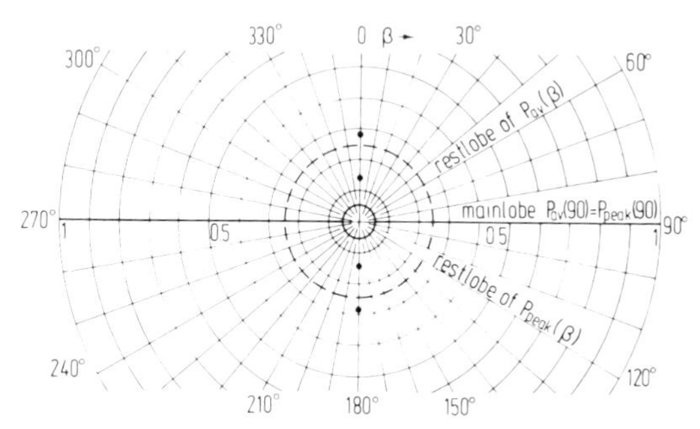

FIG. 3.2-21. Limiting case of the radiation patterns in Fig. 3.2-20, when the distance between the dipoles becomes very large, $\gamma u \gg c\,\Delta T$.

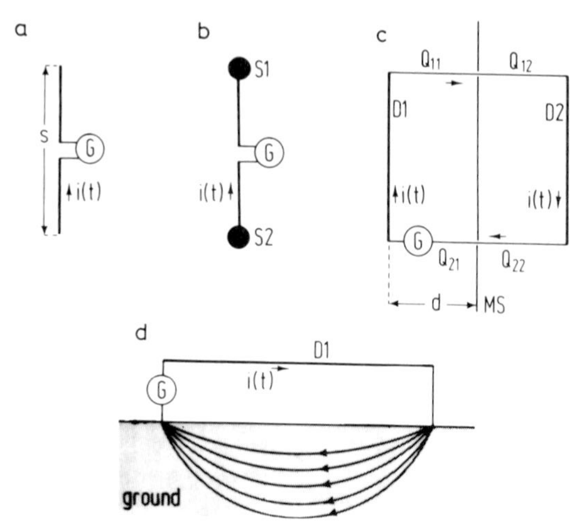

FIG. 3.2-22. Various forms of the Hertzian electric dipole. (a) Principle of dipole. (b) Spheres at the ends add capacity. (c) Current loop divided by a conducting plane MS. (d) Long-wave antenna with ground return.

$P_{av}(\alpha)P_{av}(\beta)$; examples of $P_{peak}(\alpha)$ and $P_{av}(\alpha)$ are shown by Figs. 3.2-13 and 3.2-14, while examples of $P_{peak}(\beta) = P_{peak}(90°, \beta)$ and $P_{av}(\beta) = P_{av}(90°, \beta)$ are shown by Figs. 3.2-20 and 3.2-21.

We have developed enough theory to permit the design of a large class of planar dipole arrays that are fed by currents $i(t)$. More complicated time variations of $i(t)$ than used here, and more complicated arrangements of dipoles may make the plotting of radiation patterns very tedious, but the principle should be quite clear. The main challenge left is to find a good implementation for the Hertzian dipole. We will now turn to this problem.

3.2.3 Large Current Dipoles

The typical Hertzian electric dipole shown in Fig. 3.2-22a consists of two rods driven by a generator G. The length s of the rods must be "sufficiently short." For a sinusoidal wave with wavelength λ this condition becomes specifically $s/\lambda \ll 1$, while it assumes the form $s/c\,\Delta T \ll 1$ for a wave with the time variation di_{r1}/dt' in Fig. 3.2-16d. Since the current along the rods is assumed to be equal everywhere at a certain time t, one obtains a better approximation to the theory, if two spheres are attached to the ends of the rods as shown in Fig. 3.2-22b. This added capactiy makes $i(t)$ fairly equal along the rods, but it also causes the power supplied by the generator G to be partly a reactive power. Since only the effective power is radiated, while the generator must be designed for the total power, the design of Fig. 3.2-22 is not desirable for high-power radiators.

Consider the design according to Fig. 3.2-22c. The current flows through the dipole D1 and returns through the dipole D2 to the generator. Since the currents in D1 and D2 flow in opposite directions, one does not obtain dipole-mode radiation but quadrupole radiation (Harmuth, 1977a, p. 250), which carries typically much less power. In essence, the dipole radiation of D1 is canceled by the dipole radiation of D2. In order to prevent this cancellation, one may insert a large conducting sheet of metal MS between D1 and D2 as shown. The reflection caused by this conducting plate creates the equivalent of the two dipoles in Fig. 3.2-15; it follows from Fig. 3.2-16 that one must choose the distance d in Fig. 3.2-22c small compared with $c\,\Delta T$ to obtain powerful radiation in the direction vertical to the conducting sheet ($\beta = 0$ in Fig. 3.2-5).

The currents in the conductors between D1 and D2 flow in opposite directions and thus produce low-power quadrupole radiation. The sections of these conductors are denoted Q_{11} to Q_{22} for this reason. One may, of course, shield these conductors and connect the shielding to the conducting sheet MS to suppress this unwanted radiation even more.

Another principle to turn the Hertzian dipole into a practical radiator is shown in Fig. 3.2-22d. The dipole D1 is again a rod, or rather a long wire in practical cases, but the dipole D2 is replaced by return paths through the ground. As a result, the physical structure of the "dipole" D2 becomes so different from that of dipole D1 that the dipole-mode radiation is not completely canceled[1]. This type of antenna is widely used for the radiation of long waves, and was extensively studied in connection with radio communication to deeply submerged submarines (Keiser, 1974; Chang and Wait, 1974; Davidson *et al.*, 1974; Burrows, 1974).

The principle of Fig. 3.2-22c can be used to radiate pulses with a duration of 1–0.1 ns duration, but the metallic sheet becomes rather large for 1 ns, since light travels 30 cm during this time. In order to prevent any interference by the radiation of dipole D2 at any point of interest in the far zone one may have to make the metallic shield MS several times as long as 30 cm. A simplification is achieved by shielding D2 by means of a coaxial cable whose shield is grounded as shown in Fig. 3.2-23a. Another method is to fill the space to the right of the metallic shield MS in Fig. 3.2-22c with an absorbing material ABS as shown in Fig. 3.2-23b. Dipole D2 will then still radiate, but the radiation will interfere very little with

[1] One may interpret the structures of Figs. 3.2-22c and d also as Hertzian magnetic dipoles.

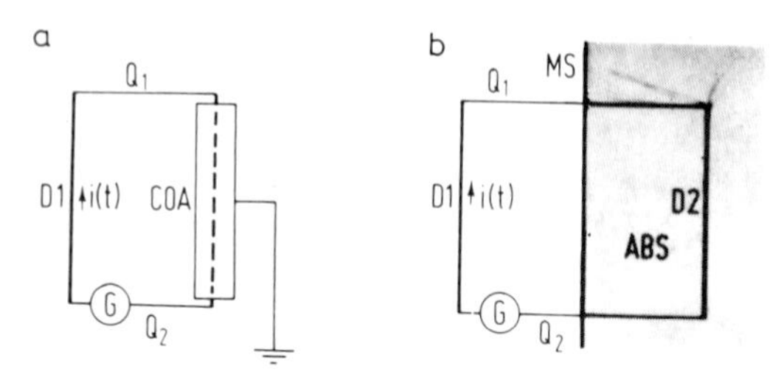

FIG. 3.2-23. Variation of the principle of the dipole of Fig. 3.2-22c by the use of a coaxial cable COA (a) and by the use of absorbing material ABS (b). MS is a metallic surface.

that of dipole D1. Note that dipole D1 radiates four times the power, if the metallic surface is properly positioned so that the mirror image of D1 doubles the electric and magnetic field strength produced by dipole D1. Dipole D2, on the other hand, does not interact with the metallic surface due to the absorbing material, and its radiated power is thus not increased. Hence, 4/5 of the power supplied to the dipoles is radiated, which implies a reduction of the power usefully radiated to the left of the metallic surface MS by only $10 \log 4/5 = -0.97$ dB. In Fig. 3.2-22c, on the other hand, half the power is radiated to the left and half to the right, implying a reduction of the useful radiation to the left by -3 dB.

The generator G in Fig. 3.2-22a and b must be able to operate with an extremely small load impedance. Ideally, the resistance of the dipoles and the connecting wires should be zero, leaving as load only the radiation impedance. The ideal generator should have an internal impedance equal to zero. Such a generator exists in the form of the thin-film *Josephson tunneling junction* (Anacker, 1979; Foner and Schwartz, 1974; Newhouse, 1975; Solymar, 1972).

For an explanation of this device refer to the current–voltage characteristic of Fig. 3.2-24. A current $i < i(0)$ can flow through the Josephson junction without any voltage drop across the junction. The junction is superconducting, and the current is called *dc Josephson current.* When the current reaches $i(0)$, a voltage of about 2.5 mV develops across the junction. The operating point for this condition is denoted a in Fig. 3.2-24. If the current is now reduced, the voltage does not drop to zero but follows branch 2 of the current–voltage characteristic to the point $v_{min}(0)$, i_{min}. For a still smaller current, the operating point jumps back to the superconducting branch 1 of the current–voltage characteristic.

If a magnetic flux Φ is applied to the Josephson junction, the maximum current on branch 1 is reduced from $i(0)$ to $i(\Phi)$, and the operating point b is reached for the current. Maintaining the flux and reducing the current

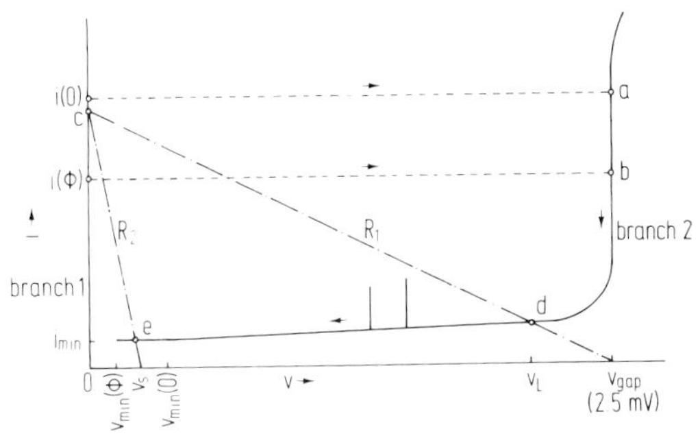

FIG. 3.2-24. Current–voltage characteristic of a thin-film Josephson tunneling junction with two load lines for resistances R_1 and R_2.

makes the voltage drop along branch 2 to the point $v_{min}(\Phi)$, i_{min}, before jumping back to branch 1.

Let a Josephson junction device be connected in a loop with a voltage source having the voltage v_{gap} and a resistor having the resistance R. This corresponds to the circuits of Fig. 3.2-23, with the Josephson junction as generator G, and the resistance R_1 due to the radiation resistance as well as the ohmic resistance of the dipoles and their connecting wires. The load line R_1 in Fig. 3.2-24 represents the operating condition. There is a stable operating point c at $v = 0$ and i somewhat smaller than $i(0)$, and another one denoted d at $v = v_L$ and i somewhat larger than i_{min}. Let the operating point be at c. Application of a magnetic flux Φ will terminate the superconductive state, and the operating point will jump to d. Turning off the magnetic flux will not have much effect, but a reduction of the current to i_{min} followed by an increase will get the operating point back to c.

Consider now a much smaller load resistance R_2 and a much smaller supply voltage v_s between $v_{min}(0)$ and $v_{min}(\Phi)$. For $v = 0$ one has the same operating point c as before, but operating point d is replaced by e. Application of the magnetic flux Φ now makes the operating point c jump to e but removal of the flux makes it jump back to c, since $v_{min}(\Phi)$ is smaller than $v_{min}(0)$.

Consider some numerical values. The largest value for the current $i(0)$ of existing Josephson junction devices is about 10 mA. The voltage v_{gap} is about 2.5 mV. This yields for R_1 a value of $2.5/10 = 0.25\ \Omega$. If one makes R_1 larger, one reduces the difference of current between the operating points c and d considerably. The smallest possible value for the load resistance is obtained for a supply voltage between $v_{min}(0)$ and $v_{min}(\Phi)$ and the same value of 10 mA for $i(0)$. This minimum resistance is about $R_1/10$ or

0.026 Ω. The switching time from operating point c to the points d or e is about 10 ps. The return time from d to c depends on the time required to reduce and increase the current, but we need only fast switching in one direction to produce a sawtooth current[1] as shown in Fig. 2.4-4a and field strengths according to Fig. 2.4-4b.

Thin-film Josephson junction devices appear ideal, but there are other good switching devices such as Trapatt diodes or tunnel diodes (Chow, 1964) that switch fast, draw much larger currents and do not require operation at the temperatures of liquid helium. However, they require supply voltages in the order of hundreds of millivolts rather than millivolts, and they do not have zero internal resistance.

Josephson junction devices have switching times ΔT of about 10 ps while we are interested in pulses with a duration from about 100 ps to 1 ns. One obtains the longer pulses by using dipoles D1 in Fig. 3.2-23 that have a length

$$L = c\,\Delta T' \tag{50}$$

where c is the velocity with which a transient propagates along the dipole, and $\Delta T'$ is the wanted duration of the rectangular pulses according to Fig. 2.4-4b for the time variation of the electric and magnetic field strengths in the far zone.

For an explanation of Eq. (50) consider a current step, that rises proportionately with t from $i = 0$ to $i = I$ during the time ΔT, applied to a dipole of length L. The current $i(x, t)$ for any point x along the dipole for any time is shown by Fig. 3.2-25a. At a certain location $x = x_0$ the current i is zero from $t = 0$ to $t = t_0$, rises to the value $i = I$ at the time $t = t_0 + \Delta T$, and remains constant from then on. Similarly, at the time $t = t_0$ the current i equals I from $x = 0$ to $x = x_0 - c\,\Delta T$, drops to zero at the location $x = x_0$, and equals zero from $x = x_0$ to $x = L$. The section $x_0 - c\,\Delta T \leqq x \leqq x_0$ of the dipole radiates at the time t. The length s of the radiator and the derivative di/dt have the values,

$$s = c\,\Delta T, \qquad di/dt = I/\Delta T \tag{51}$$

and the product $s\,di/dt$ in Eqs. (2.1-1)–(2.1-3) of the Hertzian dipole becomes

[1] Figure 2.4-4 shows a fast increase and a slow decrease of the current, while the Josephson junction device produces a fast decrease and a slower increase. This reverses the amplitude of the derivative in Fig. 2.4-4, which is of no censequence since one can drive the current $i(t)$ in either direction around the loops in Fig. 3.2-23.

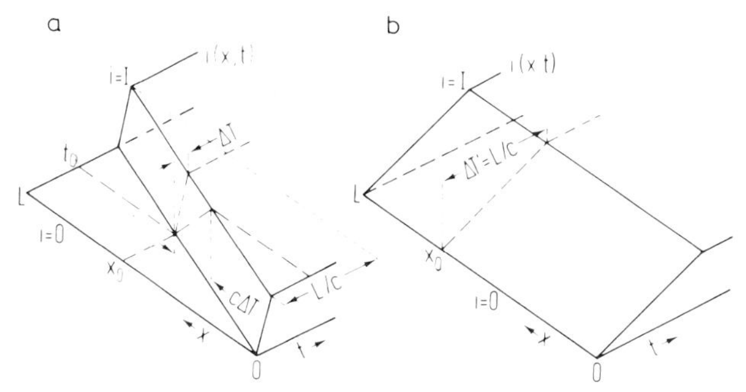

FIG. 3.2-25. (a) The current $i(x, t)$ flowing at the point x at the time t in a conductor of length L, if a unit step current pulse with a short transient time ΔT is applied at $x = 0$ at the time $t = 0$. (b) The current $i(x, t)$ with transient time $\Delta T' = L/c$ applied simultaneously at all points $0 \leqq x \leqq L$ at the time $t = 0$.

$$s \frac{di}{dt} = Ic \tag{52}$$

during the time interval

$$0 \leqq t \leqq L/c, \qquad \Delta T \ll L/c \tag{53}$$

Consider now the current $i(x, t)$ shown in Fig. 3.2-25b. It rises at any location x along the dipole of length L proportionate with t from $i = 0$ at $t = 0$ to $i = I$ at $t = \Delta T' = L/c$. Hence, the length s of the radiator and the derivative di/dt have the values,

$$s = L, \qquad di/dt = I/\Delta T' = Ic/L \tag{54}$$

the product $s\, di/dt$ has the same value Ic as in Eq. (52), and this product exists during the time interval defined by Eq. (53). Hence, the current with short transient time ΔT in Fig. 3.2-25a produces the same electric and magnetic field strengths as the current with long transient time $\Delta T' = L/c$ in Fig. 3.2-25b, provided ΔT is much smaller than $\Delta T'$.

3.3 Resonating Radiators

Consider a rod of length L. The propagation of an electromagnetic wave along this rod is described by the partial differential equations of the transmission line, Eqs. (1.3-1) and (1.3-2), discussed in Section 1.3. For distortion-free transmission, we obtained Eq. (1.3-17) for the current $i(x, t)$:

$$i(x, t) = \frac{1}{Z} e^{-at}[f(x - ct) - g(x + ct)] \tag{1}$$

Consider now a particular solution that is suggested by Bernoulli's product method of Eq. (1.3-21). The current $i(x, t)$ shall be expressed in the form $I\varphi(x)\psi(t)$, where $\psi(x)$ is a function of x only, and $\psi(t)$ a function of t only. There are two choices,

$$f_1(x - ct) = V \cos[2\pi(x - ct)/\lambda]$$
$$g_1(x + ct) = V \cos[2\pi(x + ct)/\lambda] \tag{2}$$
$$f_2(x - ct) = V \sin[2\pi(x - ct)/\lambda]$$
$$g_2(x + ct) = V \sin[2\pi(x + ct)/\lambda] \tag{3}$$

where λ is a constant with the dimension of a length, and V a constant with the dimension of a voltage:

$$i_1(x, t) = (2V/Z)e^{-at} \sin(2\pi x/\lambda) \sin(2\pi ct/\lambda) \tag{4}$$
$$i_2(x, t) = (2V/Z)e^{-at} \cos(2\pi x/\lambda) \sin(2\pi ct/\lambda) \tag{5}$$

Consider now Fig. 3.3-1a. The current $(2V/Z)e^{-at} \sin(2\pi ct/\lambda)$ shall flow at $x = 0$, and the current must be zero for $x = L$. This condition can be satisfied by Eq. (5) for

$$2\pi L/\lambda = (2k + 1)\pi/2 \quad \text{or} \quad \lambda = 4L/(2k + 1), \quad k = 0, 1, 2, \ldots \tag{6}$$

The current $i_2(x, t)$ for $k = 0$ is shown in Fig. 3.3-1a.

Let us turn to the dipole of Fig. 3.3-1b. The current must be zero for

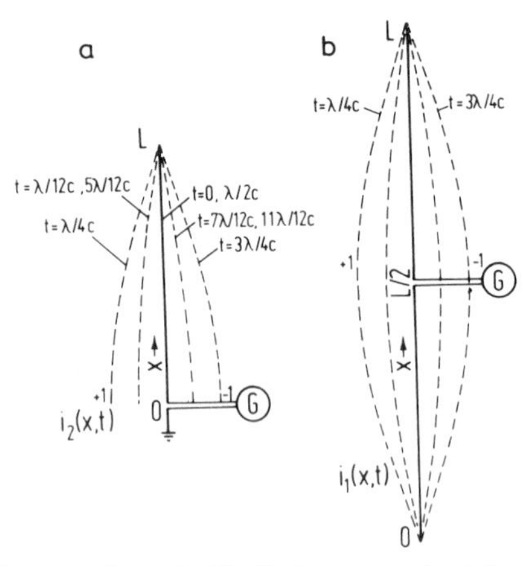

FIG. 3.3-1. Resonant monopole excited in the lowest mode at the point $x = 0$ (a), and resonant dipole excited in the lowest mode at point $x = L/2$ (b). The currents shown hold for $a = 0$ in Eqs. (4) and (5).

$x = 0$, which eliminates $i_2(x, t)$ of Eq. (5), and at $x = L$, which yields for Eq. (4) the condition

$$2\pi L/\lambda = (k + 1)\pi \quad \text{or} \quad \lambda = 2L/(k + 1), \quad k = 0, 1, 2, \ldots \quad (7)$$

For $x = L/2$ one obtains

$$\sin(2\pi x/\lambda) = \sin[(k + 1)\pi/2]$$

For $k = 1, 3, 5, \ldots$ one obtains $i_1(L/2, t) = 0$. Only the values $k = 0, 2, 4, \ldots$ in Eq. (7) yield useful solutions. The current $i_1(x, t)$ for $k = 0$ is shown in Fig. 3.3-1b.

The factor e^{-at} in Eqs. (4) and (5) implies that the current decays with time. This lost power must be supplied by the generator G in Fig. 3.3-1 to produce nondecaying currents $i_1(x, t)$ and $i_2(x, t)$.

Let us generalize the concept of the resonant antenna from sinusoidal functions to more general functions $f(x - ct)$ and $g(x + ct)$. Let $f(x - ct)$ in Eq. (1) stand for a periodic function with period T,

$$f(x - ct) = f[x - c(t + T)] \quad (8)$$

while $g(x + ct)$ shall equal zero. A current $i(t)$ is fed by the generator G via the hybrid coupler HC to the radiator in Fig. 3.3-2. The current flows through the coaxial cable to the second input terminal of the hybrid coupler. The length of the loop shall be L, the total delay time $L/c = T$. At a certain time $t = 0$ the current $i(x, 0)$ at any point x of the loop is defined by

$$i(x, 0) = (V/Z)f(x) \quad (9)$$

while the current at the time $t = T$ is defined by

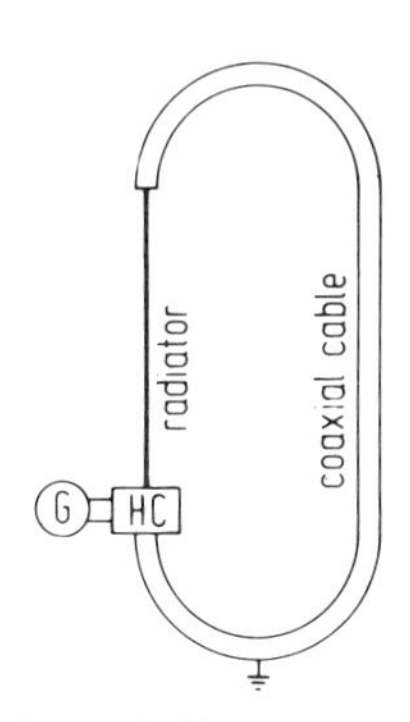

FIG. 3.3-2. Resonating radiator for periodic waves with the general time variation $f(t) = f(t - T)$. The delay time T of the feedback loop equals the period of the driving current. G, Generator producing the driving current; HC, hybrid coupler.

$$i(x, T) = (V/Z)e^{-aT}f(x - cT) = (V/Z)e^{-aT}f(x) \qquad (10)$$

Except for the attenuation e^{-aT} the current for any value of x is the same as in Eq. (9). Hence, by adding at the time $t = T$ the current $(V/Z)(1 - e^{-aT})f(x)$, one brings Eq. (10) back to the form of Eq. (9). In Fig. 3.3-2 this current is added by the generator G via the hybrid coupler. Any periodic current with period T can be maintained indefinitely. The power lost due to radiation and dissipation is resupplied by the generator. Figure 3.3-2 is thus a generalization of the resonant antennas of Fig. 3.3-1 from periodic sinusoidal to general periodic functions.

3.4 Receptors

The electric and magnetic field strengths produced by a Hertzian dipole in the far zone are proportionate to the first derivative of the current. This has occasionally created the belief that a Hertzian dipole used as receptor will yield a current that is proportionate to the integral of the field strength, but this is not so. The field strength in the far zone is by definition at a great distance from the radiator, but it is right at the location of the receptor. Hence, the physical configuration of the receptor is not the reverse of that of the radiator. Common observation shows that we receive the field strengths and not their integral. An oscilloscope in a laboratory for digital circuits will shown huge spikes, if the probe is held into the air as a receptor. The spikes are produced by the fast switching of currents in the digital circuits, which results in large values of the derivative di/dt, even though the currents or the supply voltages are small. The oscilloscope shows these spikes rather than their insignificant integral, and the probe must thus respond to the field strength rather than to its integral.

Figure 3.4-1 shows the reception of a wave by a Hertzian dipole. The dipole vector **s** is in the same plane as the vector **E** of the electric field strength. The angle between **s** and Poynting's vector **P** is denoted α, which makes the angle between **s** and **E** equal to $\alpha - \pi/2$. The voltage induced in the dipole equals

$$v(t) = q\mathbf{Es} = qEs\cos(\alpha - \pi/2) = qEs\sin\alpha \qquad (1)$$

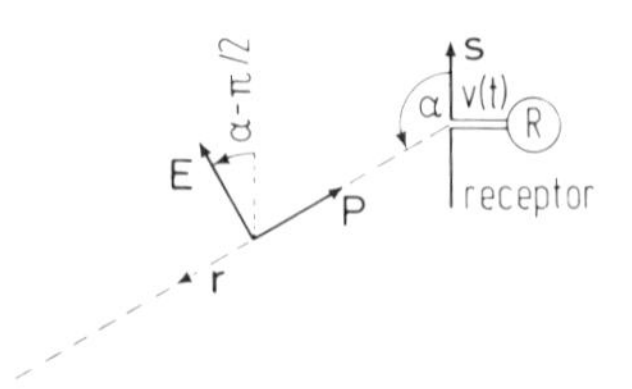

Fig. 3.4-1. Hertzian electric dipole as receptor feeding into a receiver with resistance R.

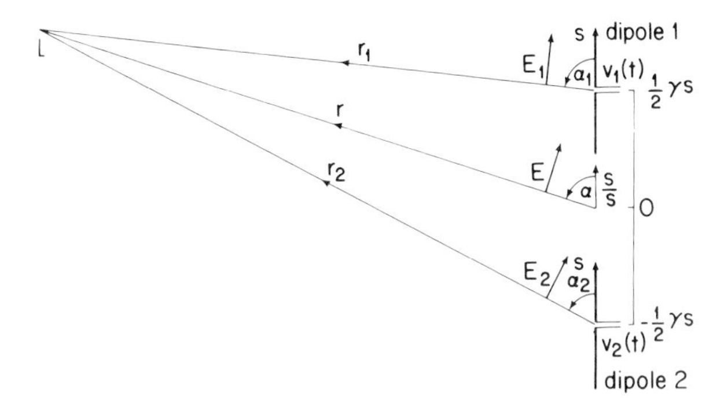

FIG. 3.4-2. Coaxial pair of Hertzian electric dipole receptors.

where q is a factor that depends on a more detailed specification of the receptor. The power delivered to the receiver with input resistance R becomes

$$P_{\text{in}} = v^2(t)/R = q^2E^2s^2 \sin^2 \alpha \tag{2}$$

The power *reception* pattern defined by Eq. (2) equals $\sin^2 \alpha$, and is thus equal to the power radiation pattern defined by Eq. (2.1-3). The equality of radiation and reception patterns holds for dipole arrays as well as for the single dipole. We discuss this with the help of Fig. 3.4-2, which is the receptor array corresponding to the radiator array of Fig. 3.2-4. The following relations hold for a wave arriving from a point L at a great distance:

$$\mathbf{r}_1 = \mathbf{r} - \tfrac{1}{2}\gamma\mathbf{s}, \qquad \mathbf{r}_2 = \mathbf{r} + \tfrac{1}{2}\gamma\mathbf{s}$$

$$r_1 \doteqdot r\left(1 - \frac{1}{2}\gamma\frac{s}{r}\cos\alpha\right), \qquad r_2 \doteqdot r\left(1 + \frac{1}{2}\gamma\frac{s}{r}\cos\alpha\right)$$

$$r \sin \alpha = r_1 \sin \alpha_1 = r_2 \sin \alpha_2$$

$$\mathbf{E}_1\mathbf{s} = E_1 s \sin \alpha_1 \doteqdot E_1 s\left(1 + \frac{1}{2}\gamma\frac{s}{r}\cos\alpha\right)\sin\alpha$$

$$\mathbf{E}_2\mathbf{s} = E_2 s \sin \alpha_2 \doteqdot E_2 s\left(1 - \frac{1}{2}\gamma\frac{s}{r}\cos\alpha\right)\sin\alpha$$

$$E_1 = E\left(t - \frac{r_1}{c}\right) = E\left[t - \frac{r}{c}\left(1 - \frac{1}{2}\gamma\frac{s}{r}\cos\alpha\right)\right] = E\left(t' + \frac{1}{2}\gamma\frac{s}{r}\cos\alpha\right)$$

$$E_2 = E\left(t - \frac{r_2}{c}\right) = E\left(t - \frac{r}{c}\left(1 + \frac{1}{2}\gamma\frac{s}{r}\cos\alpha\right)\right] = E\left(t' - \frac{1}{2}\gamma\frac{s}{r}\cos\alpha\right) \tag{3}$$

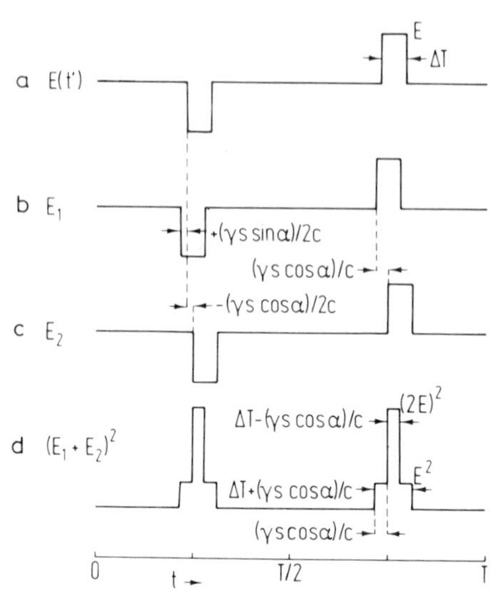

FIG. 3.4-3. Time variation (a–c) of the magnitudes $E(t') = E$, E_1, and E_2 of the field strengths in Fig. 3.4-2, and the square of the sum (d) of E_1 and E_2.

For large values of r, one obtains:

$$v_1(t) + v_2(t) = q(\mathbf{E}_1\mathbf{s} + \mathbf{E}_2\mathbf{s}) = q(E_1 + E_2)s \sin\alpha, \qquad r \gg \gamma s \tag{4}$$

The power delivered to a receiver with input resistance R becomes

$$P_{\text{in}} = [v_1(t) + v_2(t)]^2/R = (q^2s^2/R)(E_1 + E_2)^2 \sin^2\alpha \tag{5}$$

Let us compare this formula with Eq. (3.2-13). The angular dependence of Poynting's vector there is contained in the factor

$$(di_{r1}/dt + di_{r2}/dt)^2 \sin^2\alpha$$

The time variation of the electric and the magnetic field strength in the far zone equals that of the derivative of the antenna current. Hence, E_1 and E_2 have the time variation of di_{r1}/dt and di_{r2}/dt, except for a time shift. The magnitudes $E(t') = E$, E_1, and E_2 of the field strengths in Fig. 3.4-2 are plotted in Fig. 3.4-3 for the current $i(t')$ of Fig. 3.2-6. The function $(E_1 + E_2)^2$ equals, except for a constant factor, the function $(di_{r1}/dt + dr_{r2}/dt)^2$ in Fig. 3.2-6. Hence, the peak power as well as the average power *reception* pattern for the dipoles in Fig. 3.4-2 equal the peak power and the average power *radiation* pattern of the dipoles in Fig. 3.2-4.

4 Selective Receivers

4.1 Receiver for General Periodic Waves

Figure 4.1-1 shows on top a block diagram of the usual receiver for periodic sinusoidal waves or for signals with a small relative frequency bandwidth. The input signal from the antenna passes through a radio-frequency filter RFF that reduces the noise, rejects any mirror image signal, attenuates signals with wrong frequency, and amplifies. Next comes the frequency converter FC that shifts the signal to a lower fixed frequency. The following intermediate-frequency filter IFF may thus be built for a fixed frequency band and does not have to be tunable. The intermediate-frequency filter usually performs the functions of amplification and automatic volume control in addition to the filtering. The demodulator DEM following the intermediate-frequency filter removes the frequency-shifted sinusoidal carrier either by rectification or by synchronous demodulation.

Let us now turn to the receiver for general periodic waves in Fig. 4.1-1.

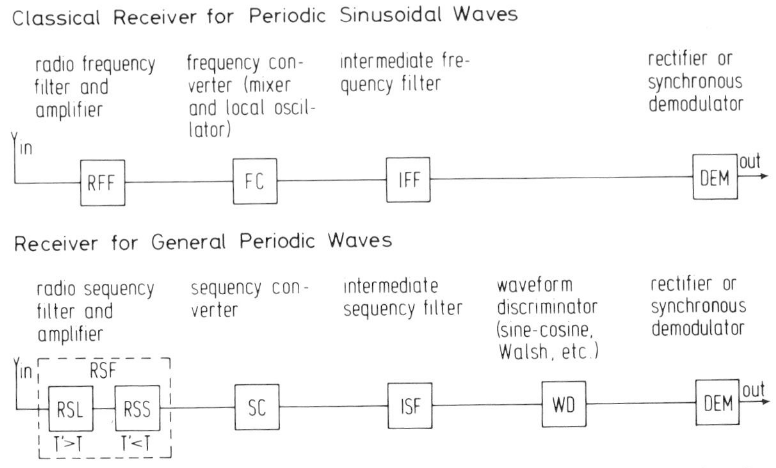

FIG. 4.1-1. Stages of classical receiver for periodic sinusoidal waves and receiver for general periodic waves. RFF, radio frequency filter; FC, frequency converter; IFF, intermediate frequency filter; DEM, demodulator; RSF, radio sequency filter; RSL, radio sequency filter for long periods; RSS, radio sequency filter for short periods; SC, sequency converter; ISF, intermediate sequency filter.

It is designed to receive any wave of period T with sufficiently many periods, regardless of the time variation of the wave during one period. In particular, the receiver is capable of receiving a sinusoidal wave with the frequency $f = 1/T$. The discrimination between the wave of period T having the selected waveform, from all other waves of period T but wrong waveform, is done in a special stage of the receiver.

The general receiver starts with a radio-sequency filter and amplifier RSF. This filter contains two separate stages. The radio-sequency filter for long periods RSL attenuates periodic signals with a period T' longer than the desired period T. The radio-sequency filter for short periods RSS, suppresses periodic signals with a period T' shorter than the desired period T or—more generally—shorter than a fraction of the desired period T. Nonperiodic signals are attenuated by both filters.

The radio-sequency filter is followed by the sequency converter SC, which converts the period T into a much longer period $2^n T$. The purpose of the sequency converter is similar to the purpose of the frequency converter. The longer period, like a lower frequency, simplifies the construction of the circuits following the converter. However, there is no desire to make $2^n T$ a fixed period independent of the period T, while the frequency converter produces a fixed output frequency regardless of the input frequency. The main role of the sequency converter is to make $2^n T$ so long that sampled signal circuits can be used easily. Specifically, one wants to use digital circuits after the sequency converter while analog circuits have to be used for the radio-sequency filter.

The intermediate-sequency filter ISF further attenuates signals with too long or too short a period.

The waveform discriminator WD follows the intermediate sequency filter. It distinguishes between signals having the same period T, but different time variation during this time, e.g., $\sin 2\pi kt/T$ and $\sin 2\pi mt/T$ or $\mathrm{sal}(k, t/T)$ and $\mathrm{sal}(m, t/T)$ for $k \neq m$. There is, of course, little interest in using this receiver for the reception of sinusoidal waves, but it is important to recognize that this receiver works for general periodic waves, not just for Walsh or some other special wave.

Following the waveform discriminator is a demodulator DEM, which may be either a rectifier or a synchronous demodulator, just as in the case of the sinusoidal receiver.

The classical receiver in Fig. 4.1-1 can often be simplified. A very strong signal can be received with an antenna and the rectifier DEM, while a somewhat weaker signal will in addition require the radio frequency filter. On the other hand, a sophisticated receiver may use more than one frequency converter, and more than one intermediate-frequency filter. Corresponding remarks hold for the receiver for general period waves.

4.2 Radio-Sequency Filter

4.2.1 Sequency Low-Pass Filter

A radio-sequency filter for long periods having three stages is shown in Fig. 4.2-1. Each stage contains two hybrid couplers HC, a nonreversing amplifier AM, an adjustable attenuator AT, and a delay circuit D. The delay in the feedback loop of each stage equals the period T of the signal one wants to receive.

Let a periodic signal with period T and amplitude A arrive at the input terminal of the first stage. The attenuator shall be set so that the feedback loop has exactly the gain 1. The delayed first period of the signal will be added by the hybrid coupler HC11 to the second arriving period of the signal. A signal with amplitude $2A$ will then circulate in the feedback loop. Generally, the circulating signal will have the amplitude nA after n periods have been received. Signals with a wrong period or no periodicity will add more slowly. For instance, the mean-square deviation of the distribution of the amplitudes of thermal noise will increase in proportion to only the square root of n. This difference in summation provides a filtering effect.

Let there be some attenuation in the feedback loop so that an input amplitude A becomes qA after one circulation, where $q < 1$. After n periods one obtains the amplitude

$$A(1 + q + q^2 + \cdots + q^n) = A(1 - q^n)/(1 - q) \tag{1}$$

For most practical applications, one may assume that n is so large that q^n can be neglected and the sum becomes $A/(1 - q)$. For $q = 0.9$ one obtains the sum $10A$, while $q = 0.99$ yields the sum $100A$. The second case provides a much better selectivity and amplification than the first, but the stability of the circuit must be very good to prevent $q = 0.99$ from becoming $q > 1$, which would make the circuit oscillate.

The second stage in Fig. 4.2-1 receives from the first stage the input signal $A/(1 - q)$. Since the second stage is equal to the first it yields an

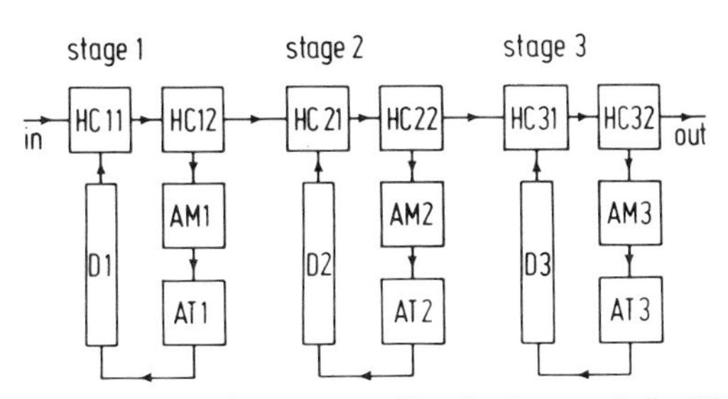

Fig. 4.2-1. Three stages of a radio-sequency filter for long periods. HC, hybrid coupler; D, delay circuit (typically a coaxial cable); AM, amplifier; AT, adjustable attenuator.

output signal $[A/(1 - q)]/(1 - q) = A(1 - q)^2$. Generally, m stages will produce the amplitude $A/(1 - q)^m$ of the signal with the correct period T.

Consider now a signal with the period $2T$ arriving at the input terminal of stage 1 in Fig. 4.2-1. A representative signal is shown in line a of Fig. 4.2-2. Let the pulses have the amplitude A. The first pulse will be attenuated to qA after one circulation through the delay loop of stage 1 in Fig. 4.2-1, and to q^2A after two circulations. The next pulse is then added to yield $A + q^2A$. After two circulations this becomes $q^2(A + q^2A)$, and another pulse with amplitude A is added. The following geometric series is obtained

$$A + q^2A + q^4A + q^6A + \cdots \doteqdot A/(1 - q^2) \tag{2}$$

The amplitude $A/(1 - q^2)^m$ will be produced by m stages, as compared with $A/(1 - q)^m$ for the signal with the correct period T.

Let now the period of the signal in line a of Fig. 4.2-2 have the general value rT, with $r = 2, 3, 4, \ldots$. After r circulations through the delay loop, the amplitude A will be reduced to q^rA, and the next pulse of the signal will be added to yield $A + q^rA$. The terms of the geometric series are now multiplied with powers of q^r:

$$A + q^rA + q^{2r}A + \cdots \doteqdot A/(1 - q^r) \tag{3}$$

The amplitude $A/(1 - q^r)^m$ is produced by m stages. A signal with period T and amplitude A will have the amplitude $A/(1 - q)^m$ after passing through m stages. The ratio of $A/(1 - q^r)^m$ and $A/(1 - q)^m$ is the relative attenuation Q:

$$Q = \frac{A/(1 - q^r)^m}{A/(1 - q)^m} = \left(\frac{1 - q}{1 - q^r}\right)^m \tag{4}$$

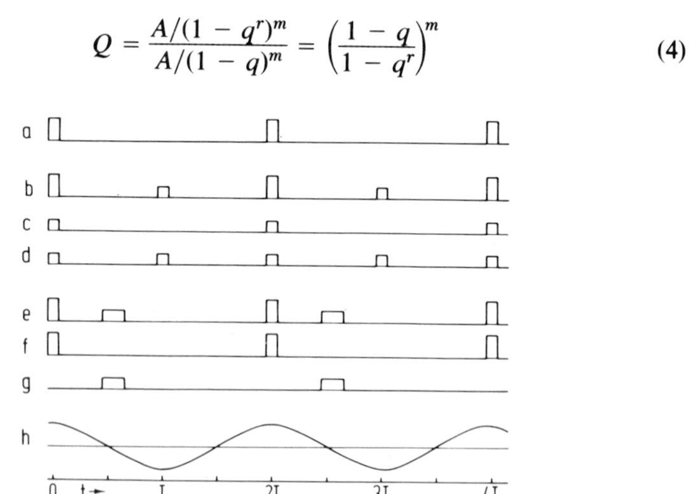

FIG. 4.2-2. Typical signals with period $2T$.

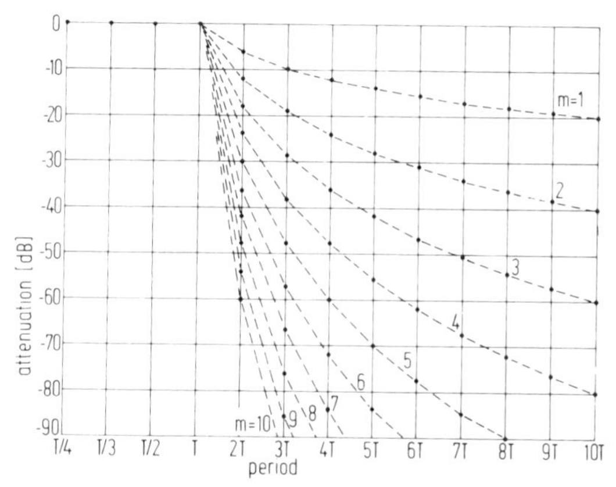

FIG. 4.2-3. Relative attenuation $-20m \log r$ of signals with period rT by $m = 1, 2, \ldots, 10$ filter stages tuned to the period T. The plots hold for the limit $q = 1$ of lossless feedback loops. Only the values plotted for $T/4$, $T/2$, ..., $10T$ have a physical meaning; the dashed lines only indicate which plotted points belong to the same number m of tuned stages.

If q is close to 1, we can write $q = 1 - \epsilon$ and use the approximation $q^r = 1 - r\epsilon$ to obtain

$$Q = \left(\frac{1 - q}{1 - q^r}\right)^m = \left(\frac{1}{r}\right)^m \tag{5}$$

where $r = 2, 3, \ldots$ is the multiple of the period T, and m the number of stages tuned to the period T.

Figure 4.2-3 shows $-20m \log r$, which is the relative attenuation according to Eq. (5) expressed in decibels, for periods $rT = 2T, 3T, \ldots, 10T$ and $m = 1, 2, \ldots, 10$ tuned stages. The feedback loops have unit gain, $q = 1$.

Figure 4.2-4 shows the relative attenuation $-20m \log (1 - q)/(1 - q^r)$ according to Eq. (4) for $q = 0.9$ (points connected by dashed lines) and for $q = 1$ (points connected by solid lines). Practical feedback loops will have gains between 0.9 and 1; these practical ranges are indicated by heavy vertical bars.

The plots of Figs. 4.2-3 and 4.2-4 permit one to determine the number of tuned stages required for a certain relative attenuation of the signals with period rT. Let us assume we want a relative attenuation of at least -50 dB. According to Fig. 4.2-4, we need six stages to get more than -50 dB for $rT = 3T, 4T, \ldots$. A signal with period $2T$ will be attenuated only -35 dB. However, we will see later on that signals with period $2T$ can be attenuated in a special circuit by an additional 20–30 dB.

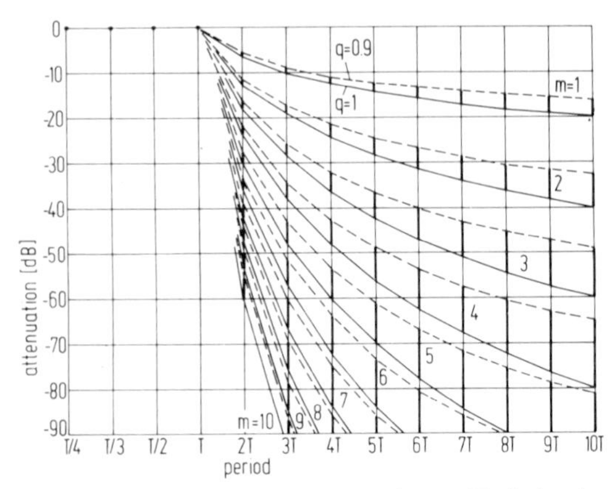

FIG. 4.2-4. Relative attenuation $-20m \log(1-q)/(1-q)^r$ of signals with period rT by $m = 1, 2, \ldots, 10$ filter stages tuned to the period T. The plots hold for the range $0.9 \leqq q \leqq 1$.

Consider an input signal with an amplitude of about 1 mV. The amplification in m stages equals $1/(1-q)^m$. To reach an amplitude of about 1 V with a gain $q = 0.9$ we need $m = 3$ stages, $1/(1-0.9)^3 = 1000$. Hence, we decide to divide the six stages required for selectivity into three radio sequency stages and three intermediate sequency stages.

Let us return to Fig. 4.2-2. We have so far only discussed the signal of line a with periods $2T, 3T, \ldots$. The signal in line b has also the period $2T$. It can be decomposed into the signals of lines c and d. Our calculation of relative attenuation is applicable to the signal of line c; the one in line d is not attenuated at all since it has the period T.

A further case is shown by line e. This signal can be decomposed into the signals of lines f and g. Our calculation applies to each one of them, and one has to use the sum of the two signals after attenuation to obtain the filtered signal.

Of considerable practical interest is the reception of a sinusoidal wave $A \sin 2\pi t/2T$ as in line h of Fig. 4.2-2. We will calculate its relative attenuation in some detail. The instantaneous amplitude $A \sin 2\pi t/2T$ for a certain time t becomes $qA \sin 2\pi t/2T$ after one circulation in the delay loop. The instantaneous amplitude $A \sin 2\pi(t-T)/2T = -A \sin 2\pi t/2T$ is added to it to yield $-A(1-q) \sin 2\pi t/2T$. This amplitude becomes $-qA(1-q) \sin 2\pi t/2T$ after a second circulation, and the amplitude $A \sin 2\pi(t-2T)/2T = +A \sin 2\pi t/2T$ is added. One obtains in successive steps the following instantaneous amplitudes:

(a) $$A \sin 2\pi t/2T$$

(b) $$-A(1-q) \sin 2\pi t/2T$$

$$\text{(c)}\quad +A[1 - q(1 - q)] \sin 2\pi t/2T$$
$$= A(1 - q + q^2) \sin 2\pi t/2T$$

$$\text{(d)}\quad -A\{1 - q[1 - q(1 - q)]\} \sin 2\pi t/2T$$
$$= -A(1 - q + q^2 - q^3) \sin 2\pi t/2T \quad (6)$$

The series $1 - q + q^2 - q^3 + \cdots$ can be rewritten in the following form:

$$1 - q + q^2 - q^3 + \cdots = (1 - q)(1 + q^2 + q^4 + \cdots)$$
$$\doteq (1 - q)/(1 - q^2) \quad (7)$$

The output voltage $v(2n, t)$ after $2n$ circulations thus becomes

$$v(2n, t) = A \frac{1 - q}{1 - q^2} \sin \frac{2\pi t}{2T} \quad (8)$$

while $v(2n + 1, t)$ after $2n + 1$ circulations becomes

$$v(2n + 1, t) = -A \frac{1 - q}{1 - q^2} \sin \frac{2\pi t}{2T} = +A \frac{1 - q}{1 - q^2} \sin \frac{2\pi(t - T)}{2T} \quad (9)$$

The comparision of Eqs. (8) and (9) with Eq. (2) shows that the amplitude of the sinusoidal function of Fig. 4.2-2h is reduced by an additional factor $1 - q$ compared with the function of line a. For $q = 0.9$, this implies an additional attenuation of $20 \log(1 - q) = -20$ dB. In m tuned stages the sinusoidal function will be attenuated by an additional factor $(1 - q)^m$ or $20m \log (1 - q)$.

The filter of Fig. 4.2-1 will only attenuate waves with a multiple period rT; $r = 2, 3, \ldots$. The waves with a fractional period T/r are not attenuated. This is shown in Figs. 4.2-3 and 4.2-4 for T/r equal to $T/2$, $T/3$, and $T/4$. The circuit of Fig. 4.2-1 is therefore called the radio sequency filter for long periods.

Let us note that our results about attenuation apply only to periodic signals with periods rT or T/r, where r is an integer. If one wants the attenuation of signals for fractional values of r, one must specify their time variation in more detail. The derivation of the attenuation of the sinusoidal signal in Eqs. (6)–(9) shows how one must proceed. Analytically, this is very cumbersome, but there is no difficulty in determining the attenuation numerically with a computer. The difference between Eqs. (1) and (2) for $0.9 \leqq q < 1$ equals less than 1% for $n = 44$. Hence, a given input signal $f(t)$ with period T_0 has to shifted by $+T$, $+2T$, ..., $+nT$ and the sum $f(t) + qf(t + T) + q^2f(t + 2T) + \cdots + q^nf(t + nT)$ has to be formed, where n does not have to be larger than 44, for most practical purposes, to obtain the output signal. It is evident, that the resulting signal will not have the time variation of $f(t)$ unless T_0 equals either rT or T/r.

Let us now generalize the sequency low-pass filter to a sequency band-pass filter. Two methods are known to accomplish this. They will be discussed in the following two sections.

4.2.2 Sequency Band-Pass Filter

In order to obtain a band-pass filter one may use a low-pass filter that attenuates signals with a period larger than T, a second low-pass filter that attenuates signals with a period, e.g., larger than $T/2$, and subtract the output signal of this second low-pass filter from that of the first. Figure 4.2-5 shows such a circuit. The input signal is fed by the hybrid coupler HC10 to the low-pass filter with delay T at the bottom, and the second with delay $T/2$ at the top. The amplitude reverser AR1 reverses the amplitude, and the hybrid coupler HC15 produces the difference of the two signals. A second stage can be connected to the output terminal of the hybrid coupler HC15 to increase the filtering effect.

Refer now to Fig. 4.2-6, which shows the attenuation of the two low-pass filters of stage 1 in Fig. 4.2-5 according to the formula $-20 \log(rT/T) = -20 \log r$ for the delay time T (points connected by the dash–dot line), and $-20 \log[rT/(T/2)] = -20 \log 2r$ for the delay time $T/2$ (points connected by the dashed line). The difference between the output signals of the two low-pass filters at the output terminal of the hybrid coupler HC15 in Fig. 4.2-5 will be nominally zero for signals with the periods $T/8$, $T/6$, $T/4$, and $T/2$. For the periods T, $2T$, $3T$, ... we have to

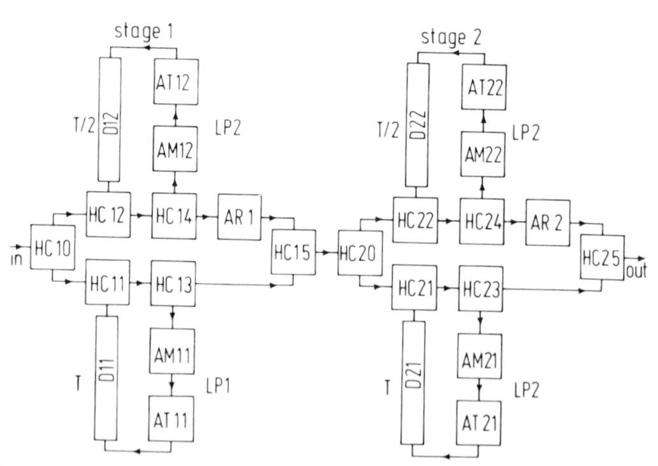

FIG. 4.2-5. Sequency band-pass filter using two sequency low-pass filters LP1 and LP2 with delays T and $T/2$, and producing the difference of the output signals. HC, hybrid coupler; D, delay line; AM, amplifier; AT, adjustable attenuator; AR, amplitude reverser.

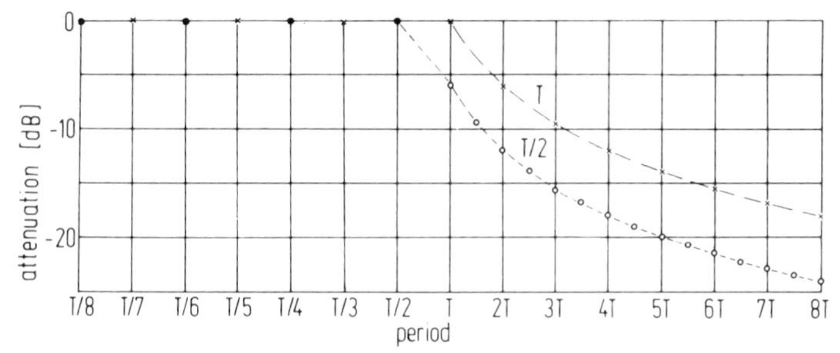

FIG. 4.2-6. Relative attenuation $-20 \log r$ (points connected by dash–dot line) and $-20 \log 2r$ (points connected by dashed line) of signals with period rT or $r(T/2)$ by the sequency low-pass filters LP1 and LP2 with delay T and $T/2$ in Fig. 4.2-5.

calculate the attenuation, while nothing can be said about signals with the periods $T/7$, $T/5$, $T/3$, $3T/2$, $5T/2$, $7T/2$, ... without specifying their time variation.

The amplitude of a pulse at the output of the low-pass filter LP1 in Fig. 4.2-5 is given by Eq. (3); the same formula with r replaced by $2r$ holds for the low-pass filter LP2. The difference between the two output amplitudes becomes

$$v(r) = A/(1 - q^r) - A/(1 - q^{2r}) \tag{10}$$

In analogy to Eq. (4) we use the output $v(1)$ for a signal with period $rT = T$ as reference:

$$v(1) = A/(1 - q) - A/(1 - q^2) \tag{11}$$

The relative attenuation for $r = 2, 3, \ldots$ is the ratio of Eq. (10) and Eq. (11). If one uses m rather than one stage we must take the mth power of this ratio:

$$Q = \left(\frac{v(r)}{v(1)}\right)^m = \left(q^{r-1}\frac{1 - q^2}{1 - q^{2r}}\right)^m \tag{12}$$

For small values of q one may again write $q = 1 - \epsilon$, which yields

$$Q = (1/r)^m \tag{13}$$

for $\epsilon \to 0$, which is the same value at that of Eq. (5).

Figure 4.2-7 shows the relative attenuation $-20m \log r$ according to Eq. (13) for the periods T, $2T$, $3T$, ... For $T/2$, $T/4$, $T/6$, ... one obtains nominally no output amplitude and thus infinite attenuation. Practically, one can expect a subtracting circuit to yield an error of about 10%, which means that the amplitude is reduced by a factor 0.1 and the power by a

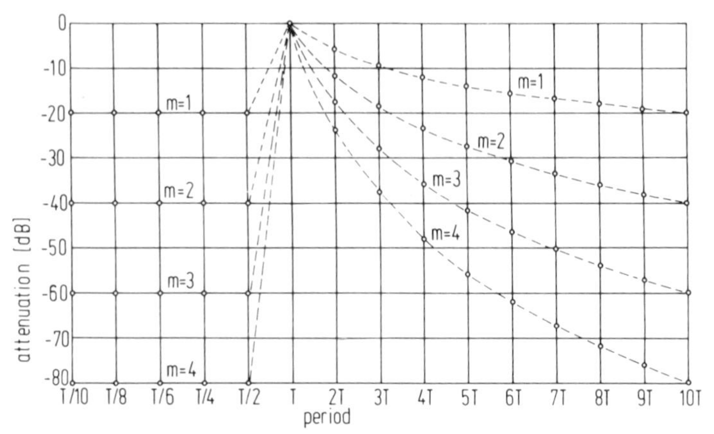

FIG. 4.2-7. Relative attenuation of sequency band-pass filters according to Fig. 4.2-5 with $m = 1 \cdots 4$ stages.

factor $0.1^2 = 0.01$ or -20 dB. Hence, a relative attenuation of -20 dB per stage is shown for $T/2$, $T/4$, ... in Fig. 4.2-7.

In Fig. 4.2-3 it was possible to state the relative attenuation for any signal with a period rT or T/r, without need to specify its time variation any closer. In Fig. 4.2-7 one can still state the relative attenuation for all multiples rT of the period T, but only for the fractions $T/2r$ rather than T/r, where $r = 1, 2, 3, \ldots$. If the time variation is specified we can proceed as at the end of Section 4.2.1. The input signal $f(t)$ produces an output signal of the low-pass filter LP1 in Fig. 4.2-5 that is obtained from the sum $f(t) + qf(t + T) + q^2f(t + 2T) + \cdots + q^nf(t + nT)$, where n does not have to be larger than about 44, as discussed previously. The output signal of the low-pass filter LP2 in Fig. 4.2-5 is the sum $f(t) + qf(t + T/2) + q^2f(t + 2T/2) + \cdots + q^nf(t + nT/2)$. The difference of the two sums is the output signal of the bandpass filter.

We turn now to a second, quite different method for turning the sequency low-pass filter of Fig. 4.2-1 into a bandpass filter.

4.2.3. *Short-Period Eliminator*

Let the output terminal of the third stage of the sequency low-pass filter in Fig. 4.2-1 be connected to the input terminal of the radio-sequency filter for short periods in Fig. 4.2-8. The operation of this circuit is as follows. The switch s1 connects the input terminal during a certain interval $0 < t < \tau$ to the integrator INT1. The output voltage of INT1 is sampled by the switch s4 at $t = \tau$, and the sampled voltage is held in the

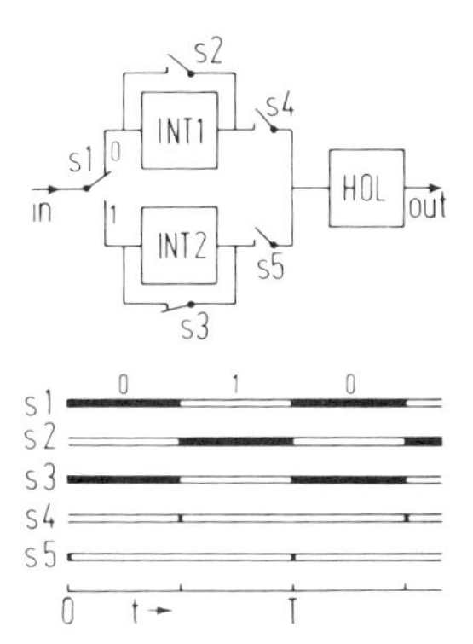

FIG. 4.2-8. Radio-sequency filter for short periods. INT, integrator; HOL, hold circuit; s1, routing switch; s2, s3, resetting switches; s4, s5, sampling switches. Time diagram: black indicates switch closed or in position 0, white indicates switch open or in position 1.

hold circuit HOL. During the time $\tau < t < 2\tau$ the integrator INT1 is reset by switch s2. The input terminal is connected during this time via s1 to the integrator INT2, the output voltage of which is sampled at $t = 2\tau$ by switch s5 and held in the hold circuit HOL. Hence, the integrators INT1 and INT2 integrate alternately; this allows the time τ to reset them. The value of τ depends on the waveform one wants to receive. We restrict ourselves to a simple case here and choose $\tau = T/2$. This value is used in the time diagram of Fig. 4.2-8.

The switch s1 must switch rapidly between the positions 0 and 1. The sampling switches s4 and s5 must be able to close and transfer the sampled voltage in a fraction of the time τ. The reset switches s2 and s3 can operate more slowly.

The sampling switches s4 and s5 produce samples at the times 0, $T/2$, T, $3T/2$, ... Hence, only signals with a period $T/2$, T, $3T/2$, ... can occur at the output of the circuit of Fig. 4.2-8. We say that signals with periods shorter than $T/2$ have been eliminated.[1] One should keep in mind that part of the energy of these eliminated signals may have been transferred to

---

[1] The integration over a period τ may be interpreted as the multiplication of the signal in this interval with a constant, followed by an integration. Since the constant is also the first function wal(0, t/τ) in the system of Walsh functions, the system of sine–cosine functions in Fig. 1.2-2, or the system of Legendre polynomials in Fig. 1.2-6, the multiplication and integration implies that all other functions are suppressed due to their orthogonality to the constant function, and no aliasing can occur. One should be careful to observe that the statements about radio-sequency filters apply only to the integer values of r or the later-used letter s. For noninteger values, one cannot make such general statements, but one must specify the time variation of the function to be filtered.

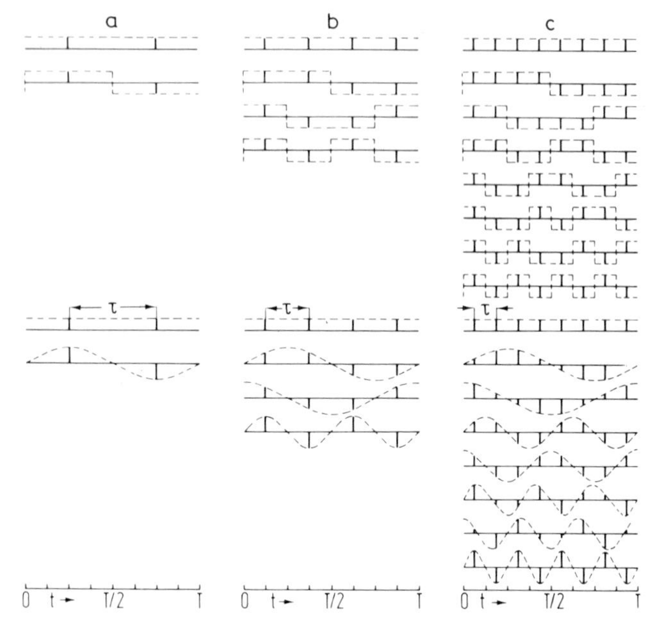

FIG. 4.2-9. First 2, 4 or 8 functions of a Walsh or Fourier representation of the output voltages of a radio-sequency filter for short periods in Fig. 4.2-8. The integration times τ equal $T/2$ (a), $T/4$ (b), and $T/8$ (c).

signals with a period equal to $T/2$ or longer. This is one reason why the attenuation of signals with period $2T$, $3T$, ... should not be done exclusively in the filter RSL in Fig. 4.1-1, which is ahead of the filter RSS, but also in the intermediate sequency filter ISF, which follows RSS.

For a better understanding of the role of the radio sequency filter for short periods, refer to Fig. 4.2-9. We had chosen the integration time τ equal to $T/2$, which means two samples are produced per period T. This case is shown in column a of Fig. 4.2-9. Two samples can be represented by the sample functions[1] Wal(0, t/T)—which consists of two positive samples—and Wal(1, t/T)—which consists of a positive sample followed by a negative sample. The dashed lines in the upper part of column a connect the samples so that the more familiar Walsh functions wal(0, t/T) and wal(1, t/T) = sal(1, t/T) can be seen. Farther down in column a we show the samples connected by dashed lines according to the first two functions of a Fourier series; this interpretation is as valid as the one by Walsh functions. One could also use the first two functions of the system

[1] The notation Wal(k, t/T) is used for sampled Walsh functions, and the notation wal(k, t/T) for the continuous Walsh functions (Ahmed *et al.*, 1973).

of Legendre polynomials in Fig. 1.2-6. We prefer the Walsh functions, because they are the easiest ones to plot, and the sine–cosine functions, because they are the most familiar ones.

Let the radio-sequency filter for short periods use the integration time $\tau = T/4$. There will now be four samples per period T. The resulting four orthogonal sample functions Wal(0, t/T), Wal(1, t/T), Wal(2, t/T), and Wal(3, t/T) are shown in the upper part of column b in Fig. 4.2-9. The dashed lines connecting them show the more familiar functions wal(0, t/T), wal(1, t/T), wal(2, t/T), and wal(3, t/T). Furthermore, the first four functions of the Fourier series are shown in column b in sampled and continuous form.

Column c of Fig. 4.2-9 shows the functions of the Walsh and Fourier series required if the integration time τ of the radio sequency filter for short periods equals $T/8$ and one obtains eight samples per interval T.

In general, an integration time $\tau = T/s$ with $s = 2, 3, \ldots$ produces s samples per interval T, and a sum of s orthogonal functions with proper coefficients will represent these s samples. Which functions we choose for the set is still open. Walsh and sine–cosine functions are two possible choices, but there are infinitely many others. A final decision concerning which set to use need not be made until the waveform discriminator WD in Fig. 4.1-1 is reached. However, the choice of τ narrows the choice of functions. We have used $T/2$, $T/4$, and $T/8$ in Fig. 4.2-9, and this favors Walsh functions. The choice of $\tau = T/12$ would not permit the use of Walsh functions but still would allow the very similar two-valued Paley functions (Harmuth, 1977a, p. 50), while the choice $\tau = T/3$ or $\tau = T/13$ would make it difficult to find a practically useful system of functions.

One may object to the use of the constant functions with positive samples only in Fig. 4.2-9. Since the electric and magnetic field strengths in the far zone vary like the first derivative of the antenna current $i(t)$, one would need a current $i(t) = It/T$ that increases proportionate to time in order to produce a constant field strength. This makes it practically impossible to produce a dc wave, while the constant functions in Fig. 4.2-9 seem to imply such a wave. This is not so. Even though one cannot produce dc waves, one can radiate and receive waves that are essentially like the sampled constant functions in Fig. 4.2-9. Such waves are of considerable practical interest, since they are difficult to detect unless one knows how to do it.

4.3 Sequency Converter

The purpose of the sequency converter is to change the period T of the received signal to a multiple of T. This multiple will generally be a power

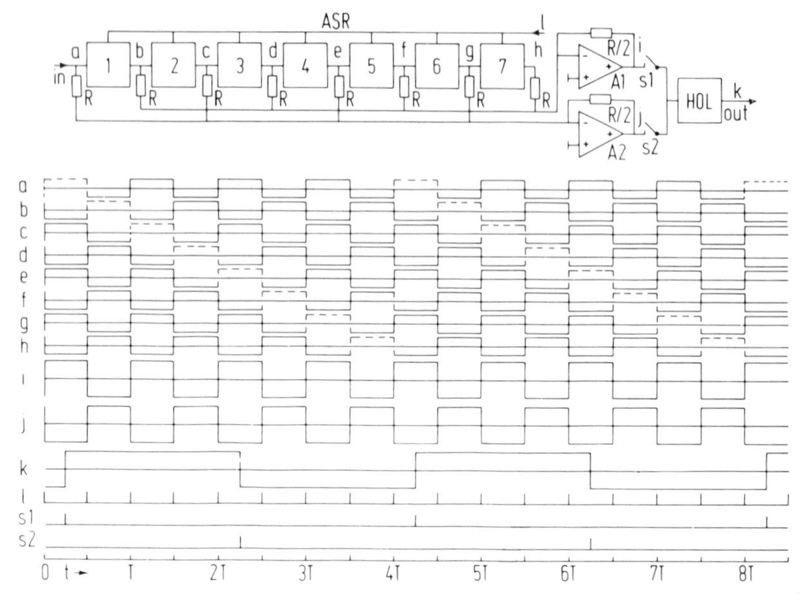

FIG. 4.3-1. Sequency converter implemented by an analog shift register. The period T of the input signal at point (a) is converted to the period $4T$ at the output terminal (k).

of 2 due to practical advantages of such a choice. The reason for wanting to increase the period T to 2^iT is the simplification of the processing equipment. The ultimate goal is to use purely digital circuits beyond the sequency converter.

The output voltage of the radio-sequency filter for short periods in Fig. 4.2-8 is fed to the input terminal of the sequency converter of Fig. 4.3-1. This circuit contains an analog shift register ASR with seven stages, two summing amplifiers A1 and A2, sampling switches s1, s2, and a hold circuit HOL. Such a circuit can be implemented at the present for periods $T = 100$ ns or longer. For shorter periods, one has to use coaxial cables instead of the analog shift register, and hybrid couplers for the summing circuit; the principle of operation remains unchanged.

The sampled voltages are shifted through the shift register ASR by trigger pulses on the line l. The shifting of a square wave is shown by the time diagram for the points a to h. The voltages at the points a, c, e, and g are summed at the point j, while the voltages at the points b, d, f, and h are summed at the point i. Operation of the sampling switches s1 and s2 as shown in the time diagram produces a square wave with period $4T$ at the output terminal k. The period of the output signal is four times as long as the period of the input signal. Furthermore, the summing process has pro-

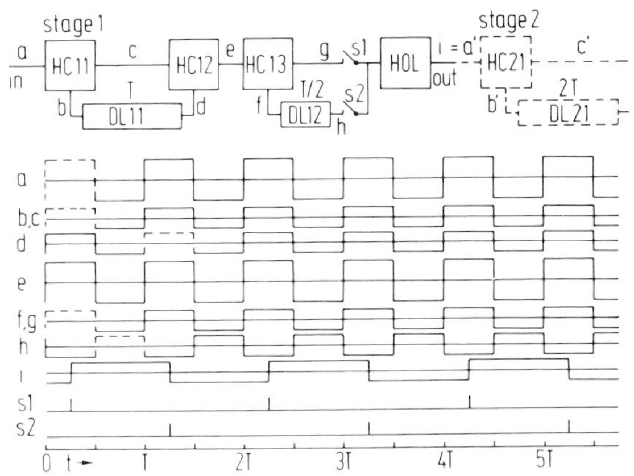

FIG. 4.3-2. High-speed sequency converter implemented by coaxial cables as delay lines (DL) and hybrid couplers (HC). Each stage doubles the period. HOL, hold circuit.

vided a filter effect against noise as well as against a square wave with period $2T$. This is one method to increase the attenuation of signals with period $2T$, which was pointed out to be insufficient when Figs. 4.2-3 and 4.2-4 were discussed.

Using two circuits according to Fig. 4.3-1 in series increases the period of the square wave from $4T$ to $16T$, etc. The most economical circuit is obtained if one doubles the period in each stage. Such a circuit, using coaxial cables and hybrid couplers for fast operation, is shown in Fig. 4.3-2. There is, however, a price for the simplicity of this circuit. It can handle only signals with two samples per period T, which means signals composed of the functions of column a in Fig. 4.2-9. The circuit of Fig. 4.3-1, on the other hand, can handle signals with either two or four samples per period T, which means signals composed of the functions in column a or b in Fig. 4.2-9; the unit T has to be replaced by $T/2$ on the time scale of Fig. 4.3-1 for the case of four samples per period T. One may easily infer from this remark how sequency converters for signals with 8, 16, ... samples per period T can be built.

4.4 INTERMEDIATE-SEQUENCY FILTER

The intermediate-sequency filter ISF in Fig. 4.1-1 is in principle equal to the radio-sequency filter for long periods RSL, but the implementation is completely different due to the longer signal periods produced by the sequency converter.

An intermediate-sequency filter with three stages is shown in Fig. 4.4-1. The attenuation provided by this filter was already discussed in connec-

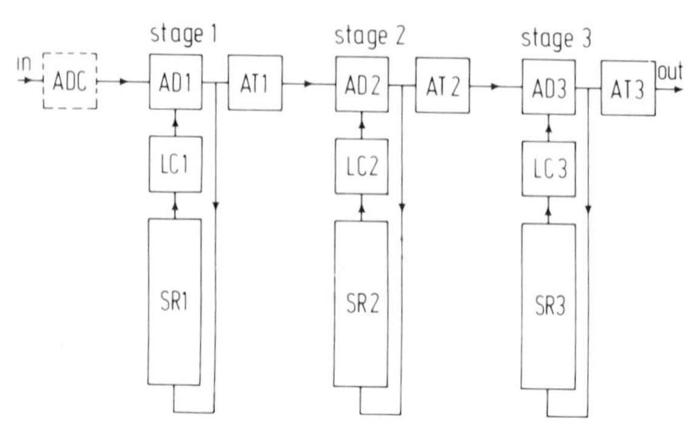

FIG. 4.4-1. Intermediate-sequency filter with three stages. ADC, analog-to-digital converter; AD, digital adder or analog summer; SR, shift register; LC, loss controller; AT, attenuator.

tion with Figs. 4.2-3 and 4.2-4. Hence, we need to describe the circuit only. First, we consider analog circuits for sampled signals. The analog-to-digital converter ADC is not needed in this case. Each feedback loop consists of a summing amplifier AD, an analog shift register SR, and a loss controller LC, which is simply an amplifier with adjustable gain. The attenuators AT are fixed voltage dividers that reduce the output voltage of each stage to about the level of the input voltage. The intermediate-sequency filter is supposed to filter, not to amplify.

Consider now the implementation of the intermediate-sequency filter by digital circuits. The analog-to-digital converter ADC in Fig. 4.4-1 converts the incoming voltage samples into binary characters. A binary adder AD1 adds each character to the binary number circulating in the feedback loop. The sum is fed to the attenuator AT1 as well as to the binary shift register SR1. We assume that the operation of all digital circuits is in parallel, not in series. Hence, each digit of a character is shifted simultaneously through the shift register. The shift register requires one stage per sample per period T. This means that signals according to column a of Fig. 4.2-9 require a parallel shift register with two stages, signals according to column b one with four stages, etc.

Parallel adders AD and parallel shift registers SR are known circuits. The loss controller LC is a new circuit. Without it, the number circulating in the feedback loop would be increased every time a new character is delivered from the analog-to-digital converter to the adder AD. A digital feedback loop operates inherently with a gain $q = 1$, while any analog feedback loop has usually more losses than one wants, and requires amplification to partially compensate them. In essence, the loss controller is a multiplier that multiplies with $q = 0.9$ or some other value of q close to

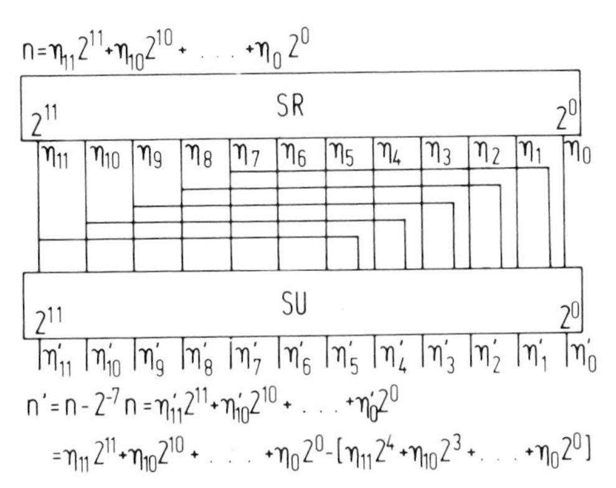

FIG. 4.4-2. Implementation of a digital loss controller for numbers q slightly smaller than 1. SR, parallel shift register; SU, parallel subtractor.

1. However, a real multiplier should not be used due to its cost and slow speed.

A more practical loss controller is shown in Fig. 4.4-2. On top is the parallel shift register SR delivering the number n as a binary character with 12 digits. This number is fed to a parallel subtractor SU. Furthermore, the number n is multiplied by 2^{-7}, and the resulting number $2^{-7}n$ is subtracted in SU from the number n. The multiplication by 2^{-7} is accomplished by wiring SR to SU as shown. The number $n' = n - 2^{-7}n$ rather than n is fed to the adder AD1. The ratio $q = n'/n = 1 - 2^{-7} = \frac{127}{128} = 0.9922$ is the gain of the feedback loop. Let us observe that the gain is actually $q = 1$ as long as the circulating number n is smaller than 2^7 and decreases with increasing n toward the limit $q = \frac{127}{128}$.

A digital version of the attenuators AT in Fig. 4.4-1 is simpler than an analog version. If we want to attenuate the character at the output terminal of a digital adder AD in Fig. 4.4-1 by a factor 2^{-7}, we simply have to

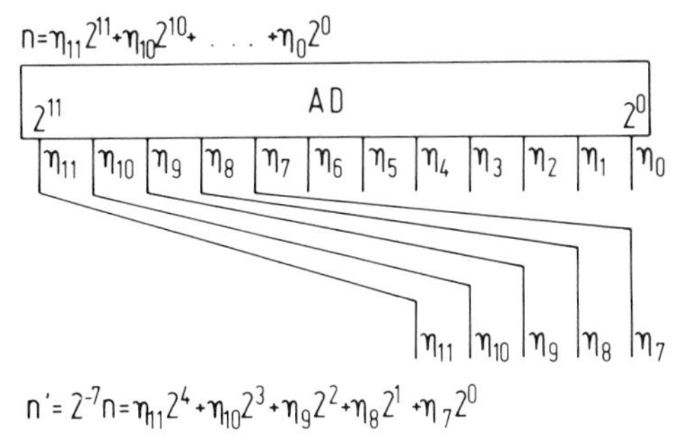

FIG. 4.4-3. Implementation of a digital attenuator for binary ratios $n'/n = 2^{-s} = 2^{-7}$ by wiring. AD, digital adder with parallel output.

wire the terminals for the five most significant digits to the input terminals for the five least significant digits of the adder AD in the following stage as shown in Fig. 4.4-3. The attenuators AT in Fig. 4.4-1 can be implemented by this type of wiring.

4.5 Waveform Discriminator

Following the intermediate-sequency filter ISF in Fig. 4.1-1 is the waveform discriminator WD. We discuss its principle with the help of the fast circuit shown in Fig. 4.5-1. A circuit of this type is not applicable if shift registers are used in the sequency converter or the intermediate-sequency filter. The circuit can be used if the sequency converter is built according to Fig. 4.3-2 and no intermediate-sequency filter is used, or if it is built according to Fig. 4.2-1. Alternately, the circuit can be connected directly to the radio-sequency filter RSF. A slow waveform discriminator for use with shift registers will be discussed later on.

Let one of the two sampled signals Wal(0, t/T) of Wal(1, t/T) of Fig. 4.5-2a be applied to the input terminal of the waveform discriminator in Fig. 4.5-1. The four switches s1 to s4 shall be in the position 0 as shown. The voltages at the points b1 to b4 are all equal; they are shown delayed in lines b to g of Fig. 4.5-2a. One may readily see from this time diagram that the input signal Wal(0, t/T) passes through the circuit and reaches the output terminal h, while the signal Wal(1, t/T) is suppressed.

Let now the switches s1 to s4 in Fig. 4.5-1 be in the position 1. The time diagram of Fig. 4.5-2b applies. The signal Wal(1, t/T) passes now through the circuit while the signal Wal(0, t/T) is suppressed.

The signals in Fig. 4.5-2 have the period T. Figure 4.5-3 shows the four sampled signals Wal(0, $t/2T$), Wal(1, $t/2T$), Wal(2, $t/2T$), and Wal(3, $t/2T$) with the period $2T$, and the effect of the circuit of Fig. 4.5-1 on them. The four switches s1 to s4 in Fig. 4.5-1 are in the position 0. The signal Wal(0, $t/2T$) passes the circuit, since it is equal to Wal(0, t/T) except for the notation, while the three other signals are suppressed. We have discussed in connection with Figs. 4.2-3 and 4.2-4 that signals with

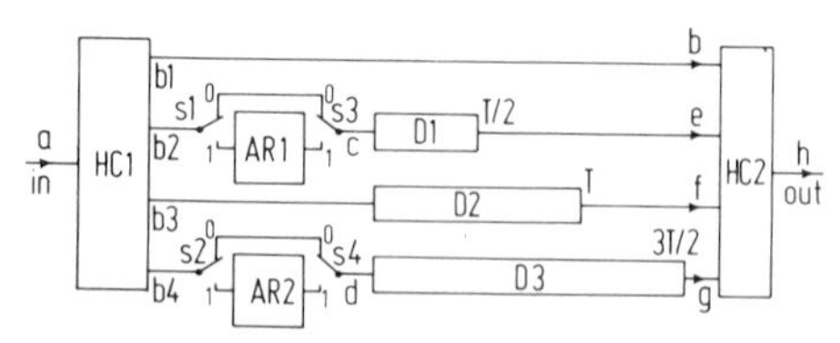

Fig. 4.5-1. Principle of the discrimination of waveforms with four samples per period T according to column (b) of Fig. 4.2-9. HC, hybrid coupler; AR, amplitude-reversing amplifier with unit gain; D, delay line.

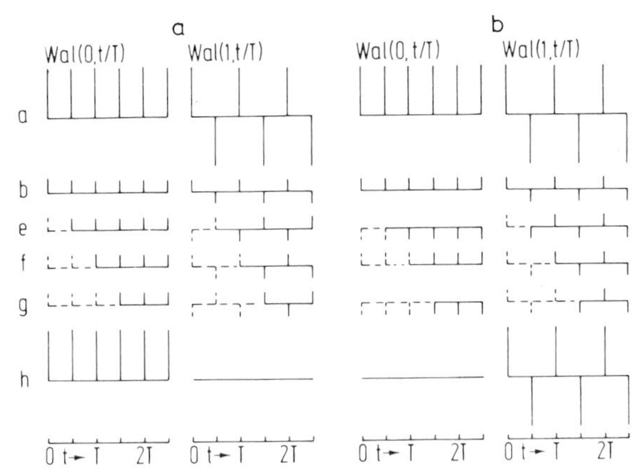

FIG. 4.5-2. Time diagram for the circuit of Fig. 4.5-1 if signals Wal(0, t/T) and Wal(1, t/T) are applied.

period $2T$ are only marginally attenuated by the sequency filters of Figs. 4.2-1 and 4.4-1. The waveform discriminator, like the sequency converter, improves this attenuation by some 20–40 dB.

Figure 4.5-4 shows again the signals Wal(0, $t/2T$) to Wal(3, $t/2T$). The switches s1 to s4 in Fig. 4.5-1 are now in the position 1. The signal Wal(3, $t/2T$) passes through the circuit since it is equal to the signal Wal(1, t/T), but the other three signals are suppressed.

The circuit of Fig. 4.5-1, modified for slow operation, is shown in Fig. 4.5-5. The delay lines are replaced by shift registers SR, the hybrid couplers and amplitude-reversing amplifiers by adding circuits AD and subtracting circuits SU. The time diagram is drawn for the use of

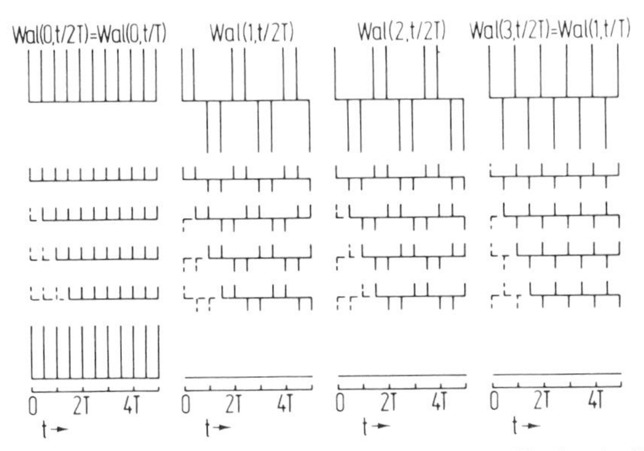

FIG. 4.5-3. Time diagram for the circuit of Fig. 4.5-1 if signals Wal(0, $t/2T$) to Wal(3, $t/2T$) are applied and the circuit is set for the detection of Wal(0, t/T).

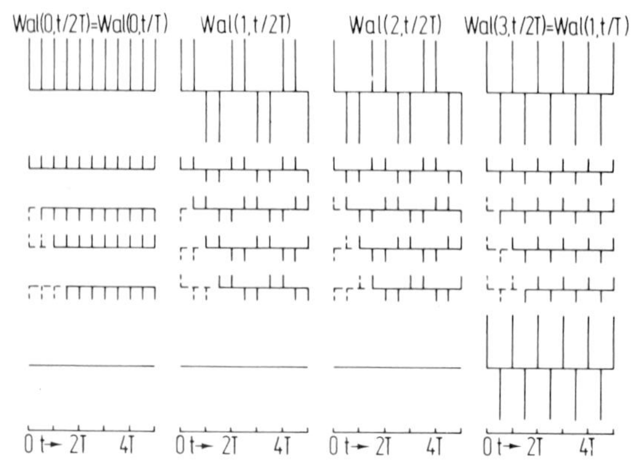

FIG. 4.5-4. Time diagram for the circuit of Fig. 4.5-1 if signals Wal(0, $t/2T$) to Wal(3, $t/2T$) are applied and the circuit is set for the detection of Wal(1, t/T).

analog-sampled signals, but there is no problem in modifying it for digital signals. The circuit diagram holds for either case, since we have not specified how the shift register, the subtractor, and the adders are to be built.

The circuits of Figs. 4.5-1 and 4.5-5 have been designed for Walsh functions, which means that the additions and subtractions are done without any weighting factors. These same circuits can be used for sampled sinusoidal functions if weighting factors according to the sampled amplitudes in Fig. 4.2-9 are used. The extension to other systems of sampled orthogonal functions is straightforward. One may readily recognize that the

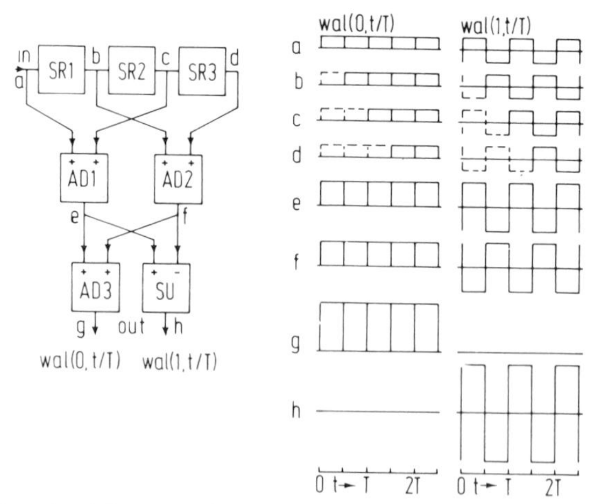

FIG. 4.5-5. Slow circuit for a waveform discriminator based on Walsh functions. SR, shift register stage; AD, adder or summer; SU, subtractor.

waveform discriminator is essentially a sliding correlator, and use this knowledge to design more sophisticated waveform discriminators.

4.6 Receiver for Nonperiodic Waves

Periodic signals are only a small subset of all possible signals. It will be seen later on that many applications require nonperiodic signals. We will give here a brief introduction to the design of selective receivers for nonperiodic radio signals. The topic will be taken up again in Chapter 6, since the field of nonperiodic radio signals is so large that one has to narrow down the class of desirable signals before one can discuss their selective reception in any detail.

Consider a conventional radar that sends out pulses at the times 0, T_D, $2T_D$, ... as shown in Fig. 4.6-1, line a. Each pulse has a fine structure, e.g., that of a Barker code shown in line b with an expanded time scale (Barker, 1953; Lange, 1971, Vol. 3, p. 315). The hyperfine structure, which is the sinusoidal carrier, is shown in line c on a still more expanded time scale.[1]

Let us now turn to a class of nonsinusoidal signals for which the fine structure of Fig. 4.6-1 is used as hyperfine structure. Figure 4.6-2 shows on top a pulsed radar signal that is radiated at the times 0, T_D, $2T_D$, ... Line b shows a fine structure that is clearly very different from the one in Fig. 4.6-1. A sequence of 15 pulses at the locations $T_D - 219T$, $T_D -$

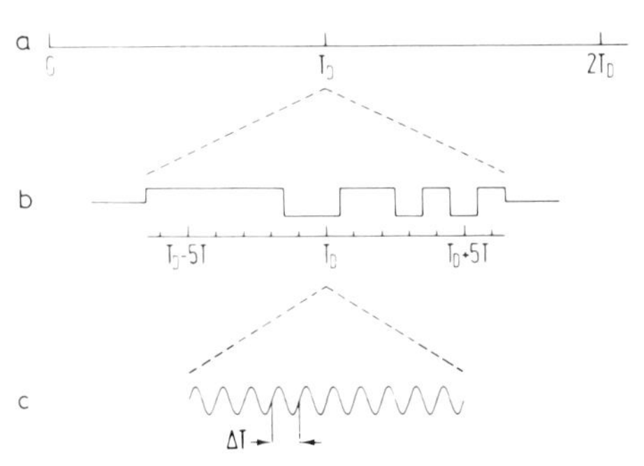

Fig. 4.6-1. Signals of a pulsed radar (a), with its fine structure (b) shown on an expanded time scale, and its sinusoidal hyperfine structure (c) shown on a still more expanded time scale. The fine structure is the 13-digit Barker code. $T_D \gg T \gg \Delta T$.

[1] This separation into fine and hyperfine structure is also applicable to radar signals produced by frequency modulation; instead of frequency modulation of a sinusoidal carrier by a simple baseband signal one can amplitude-modulate a carrier with a more complicated baseband signal.

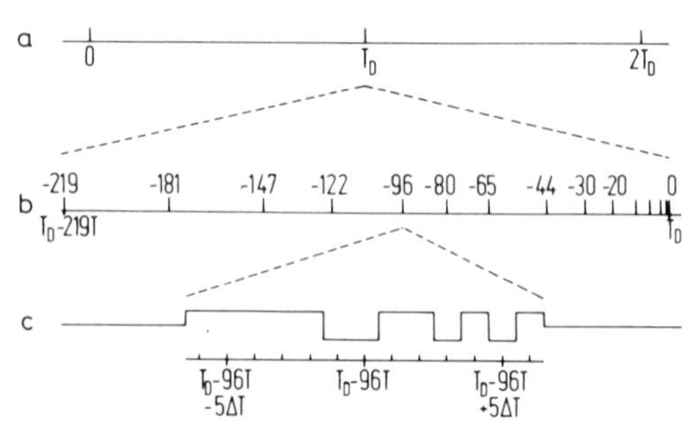

FIG. 4.6-2. Signals of a pulsed radar (a), with its fine structure (b) shown on an expanded time scale, and its nonsinusoidal hyperfine structure (c) on a still more expanded time scale. The hyperfine structure is the 13-digit Barker code. $T_D \gg T \gg \Delta T$.

$181T$, $T_D - 147T$, ..., T_D is used. We will discuss later on why these particular pulse locations were chosen and how they were arrived at. Line c in Fig. 4.6-2 shows the hyperfine structure, which is the structure of each one of the 15 pulses of line b. The hyperfine structure in Fig. 4.6-2 is equal to the fine structure in Fig. 4.6-1 except for the changed time scale.

Consider first the generation of radio signals according to Fig. 4.6-2. The sequence of pulses in line b is shown again in line b of Fig. 4.6-3. In order to produce field strengths with this time variation in the far zone, one must feed the sawtooth current of line a into a radiator operating in the dipole mode. Note that only the short positive jumps are important. The long sections of the sawtooth current with negative slope must only decrease slowly compared with the duration of the jumps, but this decrease does not have to be linear as shown.

The addition of a hyperfine structure to the pulses in line b of Fig. 4.6-2 is shown in Fig. 4.6-4. The jump in the sawtooth current in line a of Fig. 4.6-3 at the time $T_D - 96T$ is shown on an enlarged time scale in line a, and the resulting time variation of the field strengths in the far zone in line b. The short positive pulse in line b is now shown to have the duration $13\,\Delta T$, which is the duration of the hyperfine structure signal in line c of Fig. 4.6-2. The positive jump of the sawtooth current is shown in more detail in line a′, and the field strengths in the far zone in line b′.

A practical way to produce a current according to line a′ of Fig. 4.6-4 is to connect five current drivers to a dipole radiator. The first is turned on at the time $T_D - 96T - 6\,\Delta T - \Delta T/2$ and reaches saturation at the time $T_D - 96T - 6\,\Delta T + 3\,\Delta T/2$, etc. Hence, in order to produce the electromagnetic wave with the time variation of Fig. 4.6-2 we need timing circuits that trigger current drivers at the times $lT_D - mT - (n + 1/2)\Delta T$, where $l = 0, 1, 2, \ldots$, $m = 219, 181, \ldots, 1, 0$, and $n = 6, 5, \ldots, -5, -6$.

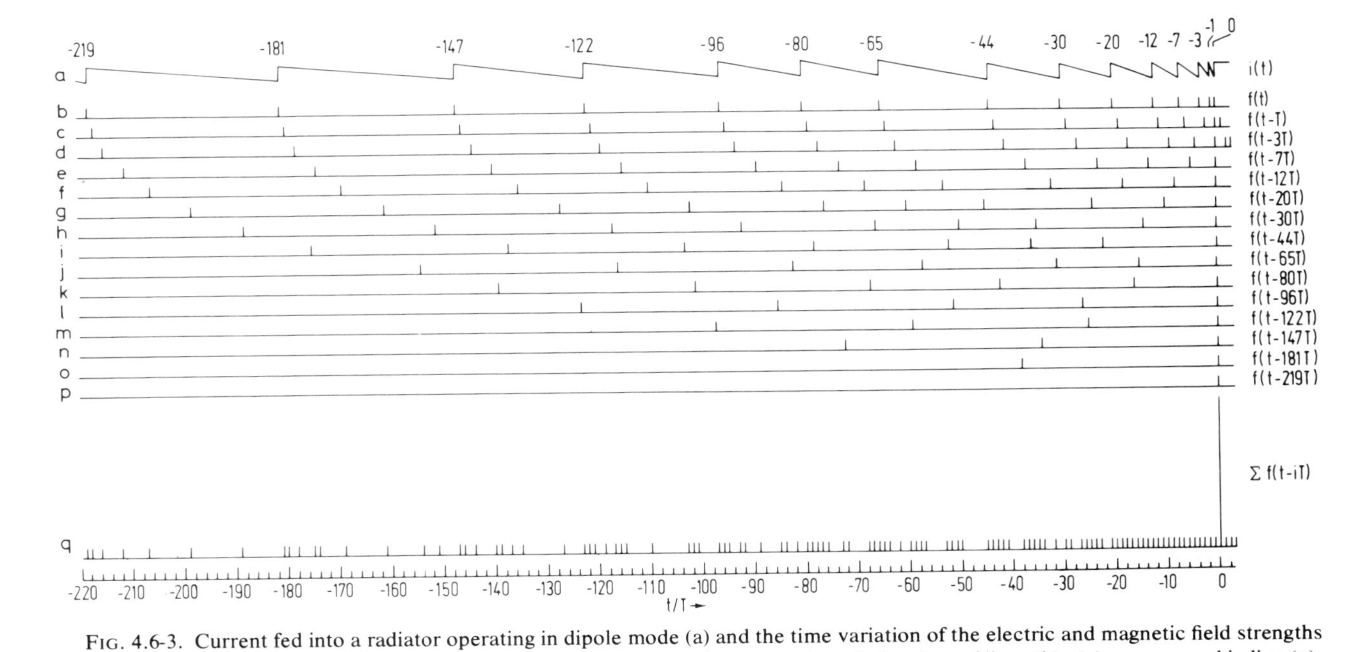

FIG. 4.6-3. Current fed into a radiator operating in dipole mode (a) and the time variation of the electric and magnetic field strengths in the far zone (b). The pulses of line (b) are delayed by T, $3T$, ... in lines (c)–(p), and all pulses of lines (b)–(p) are summed in line (q). Line (q) is symmetric for $t/T > 0$.

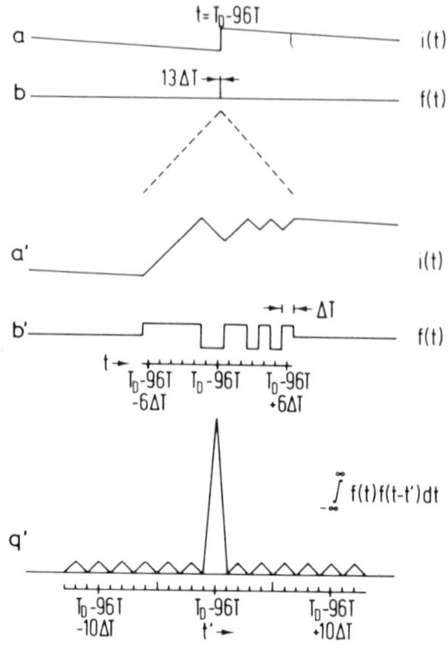

FIG. 4.6-4. Hyperfine structure of the pulses in line (a) of Fig. 4.6-2. The sawtooth current fed into the radiator according to line (a) of Fig. 4.6-3 at the time $T_D - 96T$ (a), and the field strengths produced in the far zone (b). The same current and field strengths with a larger time scale (a′, b′). Line (q′) shows the auto correlation function of the Barker code of line (b′).

For very fast switching current drivers according to Fig. 3.2-25 one must replace ΔT by L/c.

We turn to the problem of reception. The one pulse at time T_D in line a of Fig. 4.6-2 consists of the 15 pulses in line b, each of which consists of the 13 pulses of line c with a duration ΔT each. This yields a total of $15 \times 13 = 195$ pulses of duration ΔT each. If we want to measure the actual length of the nominal time interval $0 \leqq T \leqq T_D$ with an error on the order of ΔT or less, we must compress these 195 pulses into a single pulse with an approximate duration of ΔT. This pulse compression is the primary difference between the receiver required here and the general wave receiver discussed in the preceding sections.

Let us discuss the compression of the 15 pulses of line b in Fig. 4.6-3 into one. In order to do so we need the same sequence of pulses delayed by T, $3T$, $7T$, ..., $219T$ as shown in lines c–p of Fig. 4.6-3. The sum of the pulses in lines b–p is shown in line q. Fifteen pulses are summed to produce the large pulse at $t/T = 0$. At all other times one gets either nothing or a pulse with the original amplitude. In readily understandable terminology we have an enhancement ratio 15:1 for the pulse at the time $t/T = 0$, and a mainlobe-to-sidelobe ratio of 15:1.

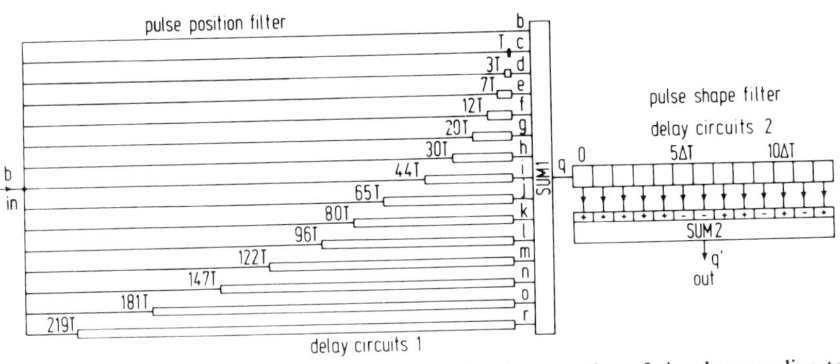

FIG. 4.6-5. Principle of the pulse compression for the reception of signals according to Fig. 4.6-2. The pulse position filter compresses the 15 pulses of line (b), Fig. 4.6-2, into 1, the pulse shape filter compresses the 13 pulses of line (c) by means of cross correlation.

The rule by which the sequence of pulses in line b, Fig. 4.6-3, was obtained, will be discussed in Chapter 6.

Figure 4.6-5 shows a circuit that implements the pulse compression according to Fig. 4.6-3. The received signal is fed directly via terminal b to the summer SUM1 or via the delay circuits with delays T, $3T$, ..., $219T$. The signals at the input terminals b–p of the summer are the same as the ones shown in lines b–p of Fig. 4.6-3. The summer SUM1 delivers the signal of line q to its output terminal q. This circuit is called the *pulse position filter*, since signals with pulses at the positions shown in lines a and b in Fig. 4.6-3 will be enhanced relative to other signals and noise. It is a generalization of filters for periodic pulse sequences.

Consider next the compression of the hyperfine structure signal in line c of Fig. 4.6-2 or line b′ in Fig. 4.6-4. Since this is a Barker code, it can be compressed by means of a sliding correlator. Such a correlator consists of delay circuits that have a delay ΔT each. Twelve[1] are needed as shown in Fig. 4.6-5 by the delay circuits 2. The signals at the 13 output terminals of the delay circuits 2 are multiplied by $+1$ or -1, according to the polarity of the 13 pulses of the signal in line b′ in Fig. 4.6-4, and summed. If the signal of line b′ was received, one will obtain the autocorrelation function of line q′, in Fig. 4.6-5. The rise time of the mainlobe of this function is ΔT. Hence, we have succeeded in compressing $15 \times 13 = 195$ pulses of nominal duration ΔT into one large pulse with rise and fall times of duration ΔT. In addition to this one large pulse there are 12 pulses with amplitudes reduced to $\frac{1}{13}$, according to line q′ of Fig. 4.6-4, and many more pulses with amplitudes reduced to $\frac{1}{15}$ or $\frac{1}{195}$ according to Fig. 4.6-3, line q.

[1] The stage 0 of the delay circuits 2 does not need to provide any delay.

5 Applications in Radar

5.1 Low-Angle Tracking

5.1.1 Principle of Amplitude Reversal

One of the most important problems in low-angle radar tracking was stated by Barton (1976) as follows:

> Accurate tracking of low-elevation targets is an important objective in many radar and navigation systems. It has remained, for decades, an almost unattainable goal for designers of practical radars, and the many recent programs for development of compensation techniques attest to the continuing importance of the problem.
>
> In radar, there are two separate problem areas: surface clutter, or back-scatter, and multipath or forward scatter of target energy to the radar via the surface. Navigation or guidance systems using cooperative transponders with frequency shift can eliminate clutter, leaving multipath as the only problem. Doppler and MTI techniques are used in radar to minimize clutter, again leaving multipath as the most intractable problem in low-angle tracking.

The multipath problem is demonstrated by Fig. 5.1-1. The signal can propagate directly from the radar to the target and return directly; this is the desired *direct signal.* It may also propagate in one direction on the direct path but take a detour via the ground or sea surface in the other direction; we call this the first-order multipath signal. A second-order multipath signal is obtained if the signal propagates in both directions via the ground or the sea surface. The second-order multipath signal is meaningless if a transponder is used by the target. In the case of a radar with a passive target, the second-order multipath signal is weaker and more delayed than the first-order multipath signal. Hence, we will restrict ourselves to the problem of the first-order multipath signal.

There are a number of known methods to combat multipath errors. A high carrier frequency combined with an antenna of large aperture yields a very narrow beam that avoids multipath signals. Such equipment, with antenna dimensions of about 1 m operating at frequencies around 16 and 34 GHz (K_u and K_a bands), have been built. According to Fig. 1.5-1, a wave with a frequency of 34 GHz will suffer an attenuation of 30 dB in fog with 30-m visibility and of 60 dB in medium rain, if the distance to the target is 10 km. Hence, such a radar is a typical fair-weather radar. At 16 GHz, the attenuation is reduced to about 6 and 18 dB, but another

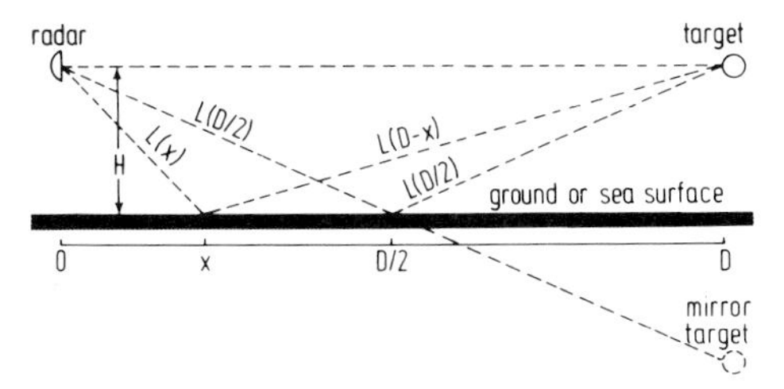

FIG. 5.1-1. Direct path and two indirect paths between a radar and a target illustrate the multipath problem.

problem is introduced. We will see later on that a pulse width of about 0.1 ns is required for a low-angle radar operating over a distance of 10 km. The bandwidth of such a pulse is essentially 10 GHz, and the resulting relative bandwidth $\eta = \Delta f/f_c = 10/16 = 0.625$ is far too large to yield any resonance effect according to Fig. 1.4-1, or permit amplitude modulation in the usual form. A low-angle tracking radar for all-weather use cannot be built with a small relative bandwidth, unless one is satisfied with a distance to the tracked target of 1 km or less. At the high frequencies, one also encounters difficulties with clutter due to frequency instability and clutter motion. Among the methods using signal processing to overcome these problems are *off-axis monopulse tracking* (Kirkpatrick, 1974), *double-null tracking* (White, 1974, 1976), and *maximum likelihood aperture sampling* (Willwerth and Kupiec, 1976).

We will use a method based on the use of certain nonsinusoidal waves that are changed in a recognizable way by reflection or scattering on the ground or the surface of the sea. For an explanation of the principle refer to Fig. 5.1-2.

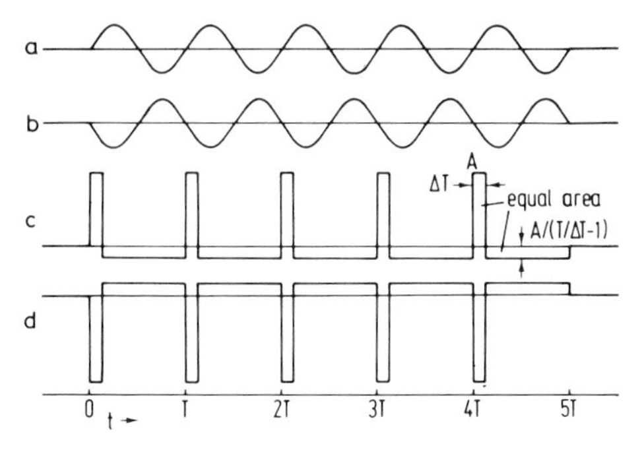

FIG. 5.1-2. Electric field strengths with sinusoidal (a) and nonsinusoidal (c) time variation, and the same field strengths amplitude reversed (b, d). The nonsinusoidal function (c) is intended for illustration only. A more practical signal would be the one in Figs. 4.6-2–4.6-4. The design of good signals is discussed in Chapter 6.

Line a shows a sinusoidal pulse with five cycles. This is a typical time variation of the electric field strength of a radar pulse. Such a pulse contains typically 100 and more sinusoidal cycles, in order to achieve a small relative bandwidth, but 5 are adequate for illustration. A target that is a good conductor—e.g., the aluminum surface of an airplane—will return an electric field strength with the time variation of line b in Fig. 5.1-2, which differs by an amplitude reversal from line a. The pulses of lines a and b are noticeably different at their beginning and end, but differ everywhere else by a shift of half a period or half a wavelength only. This difference of radiated and returned sinusoidal waves is well known. It is usual to say that the electric field strength is phase-shifted by 180°, but the phase shift is actually an amplitude reversal. The effect cannot be exploited in radar using sinusoidal carriers since the difference between an amplitude reversal and a phase shift of 180° shows up at the beginning and end of a pulse only. A radar pulse consisting of 100 sinusoidal cycles will have only about 1% of its energy in its "beginning" and "end," which implies a very poor signal-to-noise ratio for the effect. The phase of the cycles *not* at the beginning or end of the pulse, on the other hand, is not only due to the amplitude reversal at the target, but also due to the distance of the target, its shape, and its surface properties.

Consider now the electric field strength of a wave with the time variation of line c in Fig. 5.1-2. It contains five periods of short positive pulses with large amplitude, followed by long pulses with small negative amplitude. The time average of the field strength must be zero, which implies that the area of the positive and the negative pulses must be equal.

Let the wave be reflected or scattered by a conducting target, such as the metallic surface of an airplane or the surface of the sea. The electric field strength will now vary as shown in line d. The amplitude reversal between lines c and d is recognizable everywhere, not just at the beginning and end of the pulse. A different distance to the target will produce time-shifted pulses, the shape and surface properties of a scattering target will produce distorted pulses, but the amplitude reversal will not be affected.

The amplitude reversal of pulses like those in Fig. 5.1-2c has been verified experimentally (Harmuth, 1977a, oscillogram,[1] p. 319). Hence, we are not talking about a theoretically possible effect of nonsinusoidal waves but an experimentally proven fact. Let us apply this amplitude reversal effect to the problem of low-angle radar tracking.

---

[1] The illustrations over the captions Fig. 352-4 and 352-6 have been interchanged by mistake.

5.1.2 *Reflection of Linearly Polarized Nonsinusoidal Waves*

Figures 5.1-3a and b show the reflection and transmission of a planar electromagnetic wave, polarized either in the plane of incidence or vertical to it, at a planar boundary between medium 1 (air) and medium 2 (ground or seawater). The general form of the incident wave is determined by d'Alembert's solution of the wave equation in one-dimensional space:

$$\begin{aligned} e_i(x, z, t) &= E_i f(r_i - ct) = E_i f[(x \sin \varphi - z \cos \varphi) - ct] \\ h_i(x, z, t) &= H_i f(r_i - ct) \end{aligned} \tag{1}$$

To avoid unnecessary generality, we think of $f(r_i - ct)$ as a periodic function with period T, similar to one obtained from Fig. 5.1-2c by periodic continuation. Since most readers feel more comfortable with a Fourier expansion of $f(r_i - ct)$, rather than the function itself, we will use a Fourier series of $f(r_i - ct)$. To circumvent the problems of convergence introduced by a series expansion, we assume that the series defines $f(r_i - ct)$ rather than the other way round. It is well known that a Fourier expansion of pulses as shown in Fig. 5.1-2c does not converge at the jumps because of the Gibbs phenomenon, but our investigation is not affected if we add the little overshoots of the Gibbs phenomenon to the pulses of Fig. 5.1-2c. The definition of $f(r_i - ct)$ by a Fourier series is written as follows:

$$\begin{aligned} f(r_i - ct) &= \sum_{j=-\infty}^{\infty} c_i(j) \exp(-ik_{1,j} r_i) \exp(i2\pi jt/T) \\ k_{1,j} &= [-i\omega\mu_1(i\omega\epsilon_1 + \sigma_1)]^{1/2}, \qquad \omega = 2\pi j/T, \\ j &= 0, \pm 1, \pm 2, \ldots \end{aligned} \tag{2}$$

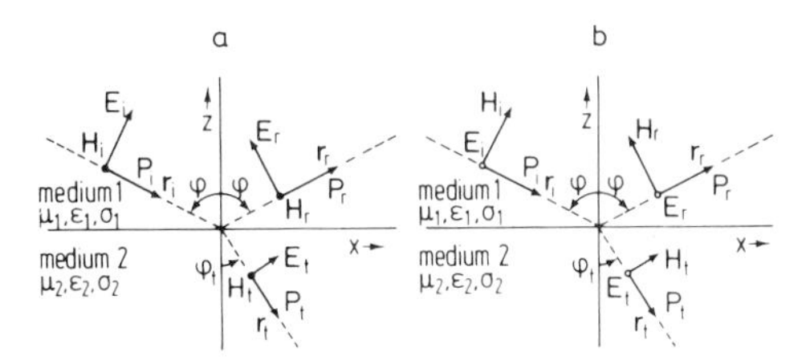

FIG. 5.1-3. Reflection of a planar wave polarized in the plane of incidence (a) and perpendicular to the plane of incidence (b) at the boundary of two media. The vectors $\mathbf{H}_i$, $\mathbf{H}_r$, and $\mathbf{H}_t$ of the magnetic field strengths in (a) are pointing toward the viewer, while the vectors $\mathbf{E}_i$, $\mathbf{E}_r$, and $\mathbf{E}_t$ in (b) point away from the viewer. μ, permeability constant; ϵ, dielectric constant; σ, conductivity constant.

The Fourier components of the incident wave have the form:

$$\begin{aligned} e_{i,j}(x, z, t) &= E_i c_i(j) \exp(-ik_{1,j} r_i) e^{i\omega t} \\ h_{i,j}(x, z, t) &= H_i c_i(j) \exp(-ik_{1,j} r_i) e^{i\omega t} \end{aligned} \tag{3}$$

The space–time variation of the reflected wave will be described by the function $g(r_r - ct)$. One obtains the following set of equations in analogy to Eqs. (1)–(3):

$$\begin{aligned} e_r(x, z, t) &= E_r g(r_r - ct) = E_r g[(x \sin \varphi + z \cos \varphi) - ct] \\ h_r(x, z, t) &= H_r g(r_r - ct) \end{aligned} \tag{4}$$

$$g(r_r - ct) = \sum_{j=-\infty}^{\infty} c_r(j) \exp(-ik_{1,j} r_r) \exp(i2\pi jt/T) \tag{5}$$

$$\begin{aligned} e_{r,j}(x, z, t) &= E_r c_r(j) \exp(-ik_{1,j} r_r) e^{i\omega t} \\ h_{r,j}(x, z, t) &= H_r c_r(j) \exp(-ik_{1,j} r_r) e^{i\omega t} \end{aligned} \tag{6}$$

The reflected wave uses the same wavenumbers $k_{1,j}$ as the incident wave, since both propagate in medium 1. The transmitted wave propagates in medium 2 and thus requires wavenumbers $k_{2,j}$. In addition to this change, one has to replace the index r by t and the function $g(r_r - ct)$ by $h(r_t - ct)$ in Eqs. (4)–(6) to obtain the respective relations for the wave transmitted into medium 2. We shall not need the transmitted wave.

Figure 5.1-3 applies to the wave $e_i(x, z, t)$, $h_i(x, z, t)$ as well as to its Fourier components $e_{i,j}(x, z, t)$, $h_{i,j}(x, z, t)$. The reflection of sinusoidal waves is discussed by means of equal or similar illustrations in many textbooks (Jordan, 1950, p. 13; Wagner, 1953, pp. 29, 32). We may use the ratios $E_r c_r(j)/E_i c_i(j) = H_r c_r(j)/H_i c_i(j)$ derived there.

(a) Polarization parallel to plane of incidence (Fig. 5.1-3a):

$$q_p = \frac{E_r c_r(j)}{E_i c_i(j)} = \frac{H_r c_r(j)}{H_i c_i(j)} = \frac{n^2 \cos \varphi - (\mu_2/\mu_1)(n^2 - \sin^2 \varphi)^{1/2}}{n^2 \cos \varphi + (\mu_2/\mu_1)(n^2 - \sin^2 \varphi)^{1/2}} \tag{7}$$

(b) Polarization vertical to plane of incidence (Fig. 5.1-3b):

$$q_v = \frac{E_r c_r(j)}{E_i c_i(j)} = \frac{H_r c_r(j)}{H_i c_i(j)} = \frac{\cos \varphi - (\mu_1/\mu_2)(n^2 - \sin^2 \varphi)^{1/2}}{\cos \varphi + (\mu_1/\mu_2)(n^2 - \sin^2 \varphi)^{1/2}} \tag{8}$$

The refraction index is defined as follows:

$$n = k_{2,j}/k_{1,j} \tag{9}$$

The value of $k_{1,j}$ is shown in Eq. (2), whereas $k_{2,j}$ is obtained by the substitution of μ_2, ϵ_2, σ_2 for μ_1, ϵ_1, σ_1:

$$k_{2,j} = [-i\omega\mu_2(i\omega\epsilon_2 + \sigma_2)]^{1/2} \tag{10}$$

Consider the reflection at the boundary of air and seawater. The following value for n is obtained:

$$\begin{gathered}\mu_1 = \mu_2 = \mu, \qquad \sigma_1 = 0, \qquad \sigma_2 \gg \omega\epsilon_2 \\ k_{1,j} = \omega(\mu\epsilon_1)^{1/2}, \qquad k_{2,j} = (-i\omega\mu\sigma_2)^{1/2} \\ n = \sqrt{-i}(\sigma_2/\epsilon_1\omega)^{1/2} = \sqrt{-i}(Z\lambda\sigma_2/2\pi)^{1/2}\end{gathered} \tag{11}$$

With the wave impedance for air $Z = 377\ \Omega$ and a typical conductivity of seawater $\sigma_2 = 4$ A/Vm, one obtains for the magnitude of n^2 the value 7.2 at 10 GHz. The magnitude of $(n^2 - \sin^2 \varphi)^{1/2}$ must thus be between 2.9 and 2.5. The "low-elevation angles" we are interested in are about 1°, which means that the angle of incidence φ is about 89°. The cosine of 89° equals 0.017, and is thus small compared with $|(n^2 - \sin^2 \varphi)^{1/2}|$. Hence, we obtain from Eq. (8):

$$q_{\mathrm{v}} = \frac{\cos \varphi - (n^2 - \sin^2 \varphi)^{1/2}}{\cos \varphi + (n^2 - \sin^2 \varphi)^{1/2}} = -1, \quad \cos \varphi \ll |(n^2 - \sin^2 \varphi)^{1/2}| \tag{12}$$

We infer from this result that $\mathbf{H}_{\mathrm{r}}$ and $\mathbf{E}_{\mathrm{r}}$ in Fig. 5.1-3b should point in the opposite direction. They are clearly amplitude-reversed compared with $\mathbf{H}_{\mathrm{i}}$ and $\mathbf{E}_{\mathrm{i}}$ for φ close to 90°. Since q_{v} in Eq. (12) is independent of j, which means it is frequency independent, we infer further that the reflected wave is not distorted, $f(r - ct) = g(r - ct)$, and that the magnitude of $\mathbf{E}_{\mathrm{r}}$ and $\mathbf{H}_{\mathrm{r}}$ is the same as that of $\mathbf{E}_{\mathrm{i}}$ and $\mathbf{H}_{\mathrm{i}}$.

Consider now the reflection by metal, such as the surface of an airplane. The main change is that the value of σ_2 is increased from 4 to about 10^7 A/Vm. Hence, n increases by more than a factor of 1000 and Eq. (12) is extremely well satisfied.

If a plane wave polarized parallel to the plane of incidence is reflected by seawater, we obtain the following value for q_{p} from Eq. (7):

$$q_{\mathrm{p}} = \frac{n^2 \cos \varphi - (n^2 - \sin^2 \varphi)^{1/2}}{n^2 \cos \varphi + (n^2 - \sin^2 \varphi)^{1/2}} \tag{13}$$

The real part of the numerator vanishes for the Brewster angle φ_{B}:

$$(n^2 \cos \varphi_{\mathrm{B}})(n^2 \cos \varphi_{\mathrm{B}})^* = (n^2 - \sin^2 \varphi_{\mathrm{B}})^{1/2}[(n^2 - \sin^2 \varphi_{\mathrm{B}})^{1/2}]^* \tag{14}$$

If φ is much larger than the Brewster angle, in particular, if it is close to 90°, we obtain $q_{\mathrm{p}} = -1$, and a reversal of the direction of $\mathbf{E}_{\mathrm{r}}$ and $\mathbf{H}_{\mathrm{r}}$ in Fig. 5.1-3a. If φ is much smaller than the Brewster angle, which means that φ is close to 0°, we obtain $q_{\mathrm{p}} = +1$; the direction of $\mathbf{E}_{\mathrm{r}}$ and $\mathbf{H}_{\mathrm{r}}$ is then as shown in Fig. 5.1-3a.

For a frequency of 10 GHz, one obtains from Eq. (11) the value $|n^2| = 7.2$. Neglecting $\sin^2 \varphi_B$ compared with n^2 in Eq. (14) yields about 68° for the Brewster angle; at 1 GHz, one obtains $|n^2| = 72$ and 83°, and at 100 MHz one obtains $|n^2| = 720$ and 88°. Hence, at angles of incidence around 89°, one obtains an amplitude reversal of the electric and magnetic field strengths just like for waves polarized perpendicular to the plane of incidence, but the conditions for distortion-free reflection are less well satisfied.

Let the boundary between air and seawater by replaced by an air–metal boundary. The value of $|n^2|$ increases by about a factor of 10^6, and the Brewster angle becomes essentially 90°. Hence, $\mathbf{E}_r$ and $\mathbf{H}_r$ have the direction shown in Fig. 5.1-3a for essentially any angle φ.

Consider now the reflection at the boundary of air and ground. We base our numerical values on a relative dielectric constant $\epsilon_r = 16$ and a conductivity $\sigma_2 = 10^{-2}$ A/Vm, which represents typical conditions (Jordan, 1950, pp. 612–617). Using these values, we may ignore σ_2 for frequencies above about 100 MHz. Hence, we obtain

$$k_{1,j} = \omega(\mu\epsilon)^{1/2}, \qquad k_{2,j} = \omega(16\mu\epsilon)^{1/2}, \qquad n^2 = 16 \tag{15}$$

Equation (8) yields again $q_v = -1$, just like for reflection by water according to Eq. (12). For waves polarized in the plane of incidence, we again obtain Eq. (13) from Eq. (7), but the interpretation is somewhat different since n^2 is now a real constant, while it was a complex function of frequency in Eq. (11). The Brewster angle follows for all frequencies above about 100 MHz from the relation $4 \cos \varphi_B = 1$, and it equals about 75°. The conditions for distortion-free reflection for angles of incidence close to 90° are now well satisfied.

The angle of incidence for which most of the power is returned by the target to the radar in Fig. 5.1-1 is close to 0°; this holds true for the direct as well as the indirect path. If the target is made of metal, such as the aluminum or steel surface of an airplane, the electric field strength of the wave returned to the radar will be amplitude-reversed. It does not matter whether the returned wave is reflected or scattered; the amplitude reversal applies in both cases, but the scattered wave will also be distorted. Figure 5.1-4 shows the directions of the vectors **E**, **H**, and **P** for propagation along the direct or the indirect path from the radar to the target in Fig. 5.1-1 and for the return along the direct path. The vectors are shown as an observer standing behind the radar looking to the target would "see" them. One may readily infer that the same illustrations, with some changes in the wording, apply for a wave propagation on the direct path to the target, but returning via the direct and the indirect path.

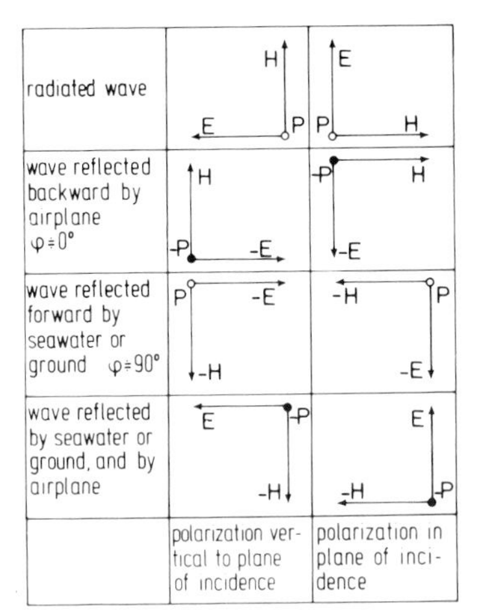

FIG. 5.1-4. The vectors **E**, **H**, and **P** along the direct and the indirect path from the radar to the target in Fig. 5.1-1, and for the return along the direct path. The vector **P** represented by a circle points away from the viewer, while a black dot indicates that **P** points toward the viewer.

Unless the planes of reflection in Fig. 5.1-3 are extremely smooth, one will not get a wave reflected at one point but the sum of waves reflected at many points, i.e., a scattered wave. The amplitude reversals discussed will remain in effect. We can make use of these amplitude reversals only if there is a sufficient time difference between the arrival of the signal that propagated on the direct path in both directions, and any signal that propagated via the reflection plane on the way to or from the target. Hence, we have to investigate the required time or range resolution. But first, some comments on the use of Fourier analysis seem to be in order.

5.1.3 On the Use of Fourier Analysis

The mathematical analysis in the preceding section was presented in terms of Fourier analysis. As a result, we could use the known formulas for the reflection of waves, and we could compute the refraction indices in Eqs. (11) and (15). Furthermore, it is sufficiently evident how the results derived for periodic functions by means of a Fourier series can be applied to nonperiodic functions by means of a Fourier transform. In addition, the analysis of the various values of the Brewster angle—and the amplitude reversal or nonreversal determined by it—for metal, seawater, and the ground may show a way for a look-down radar to discriminate an airplane from its background, which is not based on the movement but on the material of the airplane. The one thing one could *not* obtain from Fourier analysis was the concept of using the amplitude reversal of the polarity-unsymmetric function[1] of Fig. 5.1-2c. The shape of the function defined by the Fourier series of Eqs. (2) and (5) is hard to recognize because of all the terms in the series, just as one sometimes cannot see the forest because of all the trees. Hence, it is not enough for a mathematical description to be correct, it must also be sufficiently lucid to permit easy recognition of the physical processes described.

Let us briefly discuss when we can use a Fourier series or transform. The general solution of the wave equation in one dimension, $f(x - ct) + g(x + ct)$, is more general than the class of functions that can be represented by convergent Fourier series or Fourier transforms. Hence, a rigorous and general mathematical investigation will often have to avoid Fourier analysis. In electrical communications, however, the signals always have a finite energy, which is another way of saying that they are quadratically integrable. This smaller class of functions can be approximated by a Fourier series or transform with a vanishing mean square error. If a better type of convergence is needed, such as uniform convergence, one can generally not use Fourier representations, or Fourier-type representations that are based on complete systems of orthogonal functions other than sine–cosine functions.

Another difficulty is created if one wants to differentiate a series expansion or a transform. This holds true for Fourier series and transforms as well as for series and transforms based on general systems of orthogonal functions. If a convergent series is differentiated term by term, the resulting series is not necessarily convergent. This mathematical fact is of great practical importance, since the field strengths in the far zone due to

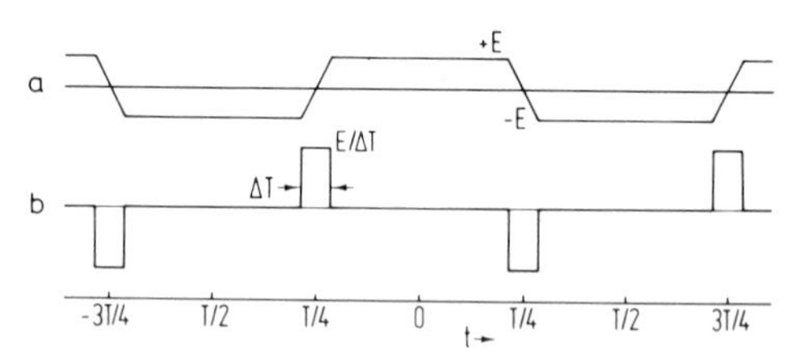

FIG. 5.1-5. Trapezoidal antenna current (a) and its derivative (b).

[1] A polarity symmetric function is defined as one for which an amplitude reversal has the same effect as a time shift. Sinusoidal functions are the best known example, but there are many others.

dipole radiation are proportionate to the derivative of the antenna current. We will investigate this in some more detail.

Consider an antenna current $i(t)$ with the trapezoidal time variation of Fig. 5.1-5a. Its Fourier series can be written with only cosinusoidal terms $a_c(i)$ by choosing the interval $-T/2 \leqq t \leqq T/2$ for the expansion:

$$a_c(i) = 2\sqrt{2}\frac{E}{T}\int_0^{T/2-\Delta T/2} \cos\left(\frac{2\pi it}{T}\right) dt$$
$$+ \int_{T/4-\Delta T/2}^{T/4+\Delta T/2} \left(\frac{T}{2\,\Delta T} - \frac{2}{\Delta T}t\right) \cos\left(\frac{2\pi it}{T}\right) dt$$
$$- \int_{T/4+\Delta T/2}^{T/2} \cos\left(\frac{2\pi it}{T}\right) dt \tag{16}$$
$$a_c(i) = \sqrt{2}E\frac{\sin(\pi i/2)}{\pi i/2}\frac{\sin(\pi i\,\Delta T/T)}{\pi i\,\Delta T/T}, \qquad i = 1, 2, \ldots$$

The Fourier series of $i(t)$ becomes

$$i(t) = \sum_{i=1}^{\infty} a_c(i)\sqrt{2}\cos(2\pi it/T)$$
$$= 2E\sum_{i=1}^{\infty}\frac{\sin(\pi i/2)}{\pi i/2}\frac{\sin(\pi i\,\Delta T/T)}{\pi i\,\Delta T/T}\cos\left(\frac{2\pi it}{T}\right) \tag{17}$$

The derivative di/dt is obtained formally by differentiating the terms of the series, but the convergence of the resulting series still has to be determined:

$$\frac{di}{dt} = \frac{8E}{T}\sum_{i=1}^{\infty}\sin\left(\frac{\pi i}{2}\right)\frac{\sin(\pi i\,\Delta T/T)}{\pi i\,\Delta T/T}\sin\left(\frac{2\pi it}{T}\right) \tag{18}$$

The coefficients in Eq. (18) are by a factor $\pi i/2$ larger than in Eq. (17). The series still converges thanks to the term $(\pi i\,\Delta T/T)^{-1}$, but the convergence is obviously very poor for small values of $\Delta T/T$. For $\Delta T/T \to 0$ the series diverges, even though the series of Eq. (17) remains convergent when the trapezoidal function of Fig. 5.1-5a becomes a square wave. One may readily verify that Eq. (18) is also obtained when the function of Fig. 5.1-5b is expanded into a Fourier series, which contains now sinusoidal functions only:

$$a_s(i) = -2\sqrt{2}\frac{E}{T\,\Delta T}\int_{T/4-\Delta T/2}^{T/4+\Delta T/2}\sin\left(\frac{2\pi it}{T}\right) dt$$
$$= 4\sqrt{2}\frac{E}{T}\sin\left(\frac{\pi i}{2}\right)\frac{\sin(\pi i\,\Delta T/T)}{\pi i\,\Delta T/T} \tag{19}$$

$$di/dt = \sum_{i=1}^{\infty} a_s(i)\sqrt{2}\,\sin(2\pi it/T) \tag{20}$$

The use of the functions $i(t)$ and di/dt of Fig. 5.1-5a and b is much simpler and more lucid than the use of the series expansions of Eqs. (17), (18), or (20). Furthermore, it is evident from Fig. 5.1-5b that di/dt consists of Dirac pulses for $\Delta T/T \to 0$, which one cannot learn[1] from Eq. (18). On the other hand, if one wants to learn which frequency bands are occupied by a signal, one has no choice but to calculate the coefficients of the Fourier series or transform. The problem of divergence due to differentiation is avoided by transforming the field strengths rather than the antenna current. Nothing can be done about a poor or slow convergence of a Fourier expansion, since this is merely a sign that a signal occupies a large frequency band.

5.1.4 Required Range Resolution

Figure 5.1-6 shows a radar at the height H_R and a target at the height H_T. The length of the direct path is denoted D',

$$D' = [D^2 + (H_T - H_R)^2]^{1/2} \tag{21}$$

while the length of the indirect path is denoted $L(x) + L(D - x)$:

$$L(x) + L(D - x) = (x^2 + H_R^2)^{1/2} + [(D - x)^2 + H_T^2]^{1/2} \tag{22}$$

The length of the indirect path varies with x. The shortest indirect path is obtained for $x = x_{min}$:

$$x_{min} = DH_R(H_R + H_T)^{-1}, \qquad D, x \gg H_R, H_T \tag{23}$$

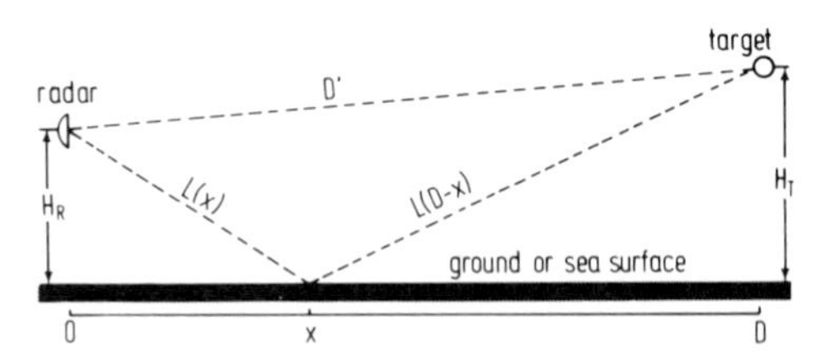

FIG. 5.1-6. Calculation of the propagation time difference between the direct path D' and the shortest indirect path $L(x_{min}) + L(D - x_{min})$.

[1] For mathematical correctness, one must point out that the series of Eq. (18) can be made convergent for $\Delta T/T = 0$ by means of the theory of summable series. The problem seems to be never "what can be done," but what is simpler and more lucid.

The difference between the shortest indirect path $L(x_{\min}) + L(D - x_{\min})$ and the direct path D' follows from Eqs. (22) and (23):

$$L(x_{\min}) + L(D - x_{\min}) - D' = 2H_R H_T/D \tag{24}$$

The difference of propagation time along these two paths is denoted Δt:

$$\Delta t = 2H_R H_T/cD \tag{25}$$

Figures 5.1-7 and 5.1-8 show Δt as a function of the distance D for the two radar heights $H_R = 5$ and 10 m, and target heights H_T from 5 to 50 m. It is apparent that Δt is of the order of 1 to 0.1 ns for target heights and distances of practical interest. Hence, one must use pulses with a duration between 1 and 0.1 ns. This is in line with the pulse widths required by the existing low-elevation-angle radars, but there is an important difference. A pulse with a duration of 1 ns occupies essentially the band $0 \leqq f \leqq 1$ GHz, while one with a duration of 0.1 ns occupies the band $0 \leqq f \leqq 10$ GHz. If we cut off the lowest 10% of these frequency bands, we obtain the slightly distorted pulses of Figs. 2.2-2b or 2.2-3b. Absorption by rain or fog would be not much of a problem for these pulses. However, let

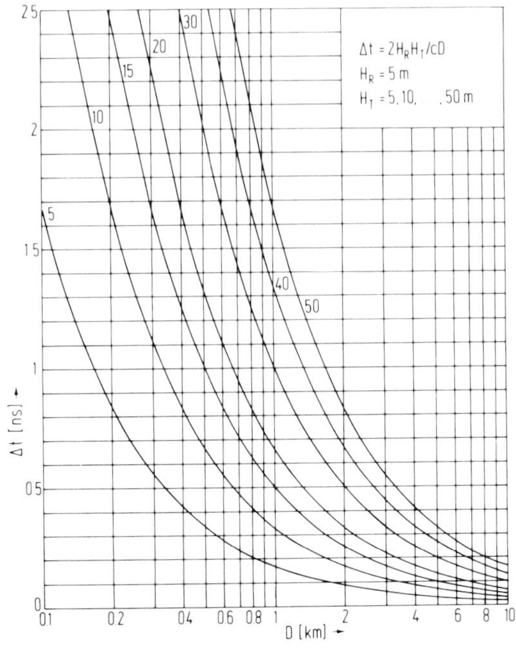

FIG. 5.1-7. The minimum propagation time difference Δt as function of the target distance D for a radar height $H_R = 5$ m and target heights H_T from 5 to 50 m.

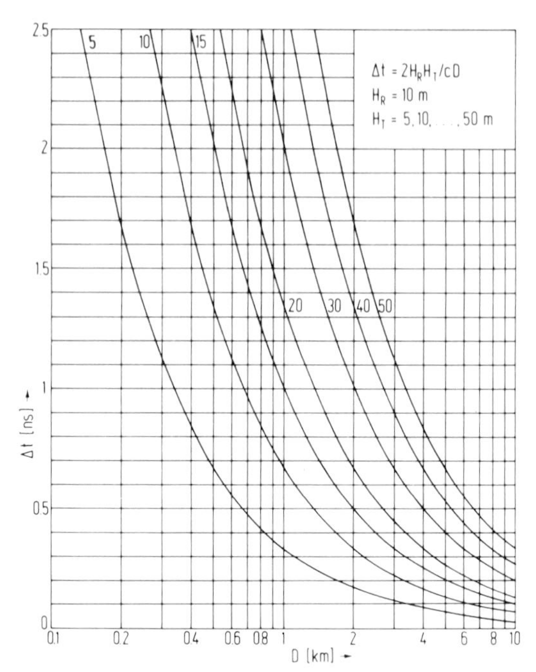

FIG. 5.1-8. The minimum propagation time difference Δt as function of the target distance D for a radar height $H_R = 10$ m and target heights H_T from 5 to 50 m.

these short pulses be amplitude-modulated onto a sinusoidal carrier and let the relative bandwidth be small. The carrier frequency would have to be at least 30 GHz, in order to avoid the resonance of the H_2O molecule at 22.2 GHz, and would have to be on the order of 100 GHz and greater for the pulse with a duration of 0.1 ns. The absorption losses would now be prohibitive according to Fig. 1.5-1.

By not using a small relative bandwidth we can not only avoid the absorption losses but we can also make use of the amplitude reversal of the electric or magnetic field strength. This works best with carrier-free signals as shown in Fig. 5.1-2c, but a signal with a sinusoidal carrier and a large relative bandwidth will also show the amplitude reversal, as may readily be seen from Figs. 2.3-1c and g or 2.3-2c and g.

Figure 5.1-9 shows representative pulse shapes of a low-elevation-angle carrier-free radar. A sawtooth-shaped current (a) is fed to the radiator. The slow rising section of the sawtooth is of no importance; it may rise proportionate with time or otherwise as long as the current increases slowly. The important part is the fast dropping edge of the sawtooth. A radiator operating in the dipole mode will produce in the far zone electric

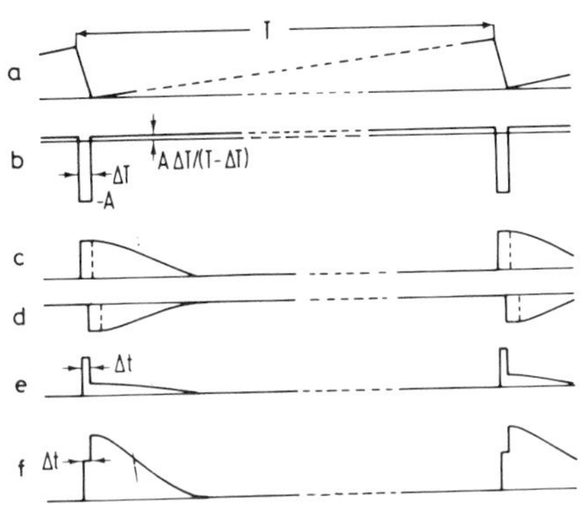

FIG. 5.1-9. Typical pulse shapes of low-elevation-angle carrier-free radar: (a) antenna current; (b) electric and magnetic field strengths in the wave zone; (c) electric field strength of the wave returned directly from a scattering airplane with metallic surface; (d) the same, but one way was traveled on an indirect path; (e) sum of the direct return (c) and indirect return (d); (f) sum of the absolute values of the direct return (c) and the indirect return (d).

and magnetic field strengths that vary like the first derivative of the radiator current. This time variation is shown in line b. Short negative pulses with amplitude $-A$ and duration Δt, and long positive pulses with amplitude $A\,\Delta T/(T - \Delta T)$ and duration $T - \Delta T$ are produced. To obtain some idea of numerical values, let us assume that we want to track the target ambiguity-free over a distance of 1.5 km. The round-trip distance equals 3 km and the time required by the signal to travel 3 km is 10 μs. Hence, T in Fig. 5.1-9 is on the order of 10 μs, while ΔT is on the order of 1–0.1 ns. The ratio of the magnitude of the negative and positive amplitudes in line b is thus at least 10,000 : 1, and the positive pulses will generally not be visible.

Let the pulses in line b be returned by a metallic reflector, such as a radar reflector, with an angle of incidence close to 0°. The amplitude of the electric field strength will be reversed, that of the magnetic field strength will not be reversed, but the time variation will otherwise remain unchanged. This situation is not so if the pulses are returned by a metallic scatterer, such as an airplane. The nose of the airplane will be first to return a pulse, but the more distant parts of the airplane will return pulses with a time delay. Since a pulse duration of 1–0.1 ns implies a propagation distance of 30–3 cm, which is small compared with the dimensions of an airplane, the delayed pulses will sum up to produce a distorted pulse that is upward of 10 times as long as the original pulse. However, the amplitude reversal of the electric field strength will not be affected. Hence, the electric field strength of the returned wave will have the time variation shown in line c of Fig. 5.1-9. A more detailed discussion of the distortions caused by scatterers may be found in the literature (Harmuth, 1977a, p. 278).

Line d in Fig. 5.1-9 shows the time variation of the electric field strength of the wave that propagated in one direction between radar and target on the direct path, and in the other direction on an indirect path. It is delayed by the minimum propagation time difference Δt and amplitude-reversed. The magnitude of the amplitude will generally be smaller than for the direct signal of line c, but the pulse will also be generally longer since it was scattered twice.

Line e shows the sum of lines c and d. This is the signal actually received; it may be displayed on a scope or fed to automatic detection equipment. It is characterized by a large positive pulse of duration Δt. It does not matter much whether the duration ΔT of the radiated pulse is shorter or longer than Δt, since the scattered pulses in lines c and d will always be longer than Δt. The use of a pulsewidth ΔT shorter than Δt would make sense if one wants to discriminate a radar reflector, since the pulse in line c would not be widened in this case.

Line f shows the signal obtained by summing the absolute value of the pulses in lines c and d. A radar not exploiting amplitude reversal, such as any radar using a sinusoidal carrier, would display a signal fluctuating between the two limits of lines e and f. Line e shows clearly a better signal than line f. This is the improvement over radar with a sinusoidal carrier and a small relative bandwidth that one can expect for the multipath problem.

5.1.5 *Comparison with Conventional Radar*

We have already discussed the importance of noise, rain, and fog for high-resolution radar in connection with Figs. 1.5-1 and 1.5-2, and thus need to add here only a few comments.

For a pulse duration of 1 ns the band with the lowest noise temperature according to Fig. 1.5-2 is between 1 and 2 GHz for a low-elevation-angle radar. The noise temperature varies there from 60 K at 2 GHz and an elevation angle of 1°, to 90 K at 1 GHz and an elevation angle of 0°. The relative bandwidth equals $\eta = 1/(1 + 2) = 0.33$. The Fourier transform of a typical signal with this relative bandwidth is shown in Fig. 2.3-3f, while the corresponding time signals are shown in Fig. 2.3-1h and 2.3-2h. It is evident that the peak of duration Δt shown in Fig. 5.1-9e would not show up for signals with such a small relative bandwidth, unless one uses demodulation and thus doubles the required bandwidth. Accepting the doubled bandwidth makes it easier to compare the large-relative-bandwidth system with a conventional one operating with a carrier frequency of 34 GHz, and also requiring twice the baseband bandwidth. According to Fig. 1.5-2 the noise temperature at 34 GHz equals somewhat less than 300 K for elevation angles between 0° and 1°, which is about three times the temperature in the band 1 GHz $\leqq f \leqq$ 3 GHz.

The attenuation due to rain and fog is insignificant in the band from 1 to 3 GHz according to Fig. 1.5-1. At 34 GHz we have 3 dB/km for fog with 30-m visibility, 7 dB/km for medium rain, and some 20 dB/km for heavy rain. These numbers are not too bad, since a pulse with a duration of 1 ns cannot be used for distances much beyond 1 km according to Figs. 5.1-7 and 5.1-8. Of course, this is not a useful range for tracking an aircraft, but it is good enough for tracking armored vehicles on the ground, or for short-distance homing guidance.

Consider now a pulse of about 0.1 ns duration. If the band 1 GHz $\leqq f \leqq$ 10 GHz is used for its transmission one obtains the pulse of Fig. 2.2-2b. This pulse fits the idealization of Fig. 5.1-9 quite well. The average noise temperatures for angles between 0° and 1° is about 100 K according to Fig. 1.5-2. Let this pulse be modulated onto a carrier. The bandwidth would be doubled to 20 GHz. This is beyond what can be currently done. A bandwidth of about 10 GHz with a carrier frequency of 94 GHz is the current state of the art. Using such a radar for comparison, we see from Fig. 1.5-2 that the noise temperature is about three times as high as for the carrier-free radar, but in addition the resolution is only 0.2 ns rather than 0.1 ns, which implies that the useful range is only half of what it is for the carrier-free radar.

The attenuation in decibels per kilometer due to fog increases by about a factor 100 when the frequency is increased from 10 to 100 GHz according to Fig. 1.5-1. The fact that the increase is so huge makes it unnecessary to discuss it in any detail. One cannot use a small relative bandwidth for tracking an aircraft at a distance of several kilometers if the visibility is 100 m or less. For light and medium rain one has similar results. Heavy rain reduces the difference between the carrier-free radar and the radar with small relative bandwidth substantially; the attenuation increases only somewhat more than by a factor 10 between 10 and 100 GHz.

5.1.6 Armored Vehicle Tracking Radar

The recognizability of an amplitude reversal, the different noise temperature, and the different attenuation by moisture are not the only effects one can exploit by not using a sinusoidal carrier for short radar pulses. As a more subtle effect, consider the use of absorbing materials to reduce the returned energy of a radar signal. The wavelength at 100 GHz is 3 mm, and a layer of absorbing material would have to be about that thick to be effective. A pulse covering the frequency band from 1 to 10 GHz contains wavelengths from 30 to 300 mm. Absorbing layers would become too unwieldy at those wavelengths.

The use of absorbing materials is of great significance for a low-elevation-angle radar, if the target is not an airplane but an armored vehicle on the ground (Wiltse, 1978). A range of 1 km is in this case acceptable, but the altitude of the target is very low, and this means that pulses of about 0.1 ns duration must be used to overcome the multipath problem. A layer 3 mm thick is not much of a burden for a tank, if this neutralizes a tracking radar operating in the band from 90 to 100 GHz.

In addition to absorbing materials, there is another problem with armored vehicle tracking that is not encountered with airplanes. The airplane must move, and it is usually not surrounded by targets that scatter electromagnetic waves like an airplane but are otherwise of no interest. A vehicle on the ground is either stationary or it moves slowly compared with an aircraft. There are usually many other scatterers in its vicinity that must be discriminated. Our eyes perform this discrimination by forming an image from the backscattered light. The resolution of a radar is usually much too poor to form an image, the exception being the side-looking radar, or synthetic aperture radar, that produces images resembling aerial photographs.

Armored vehicle tracking radars operating at a frequency of 94 GHz achieve a nominal beamwidth of about 0.5° with dishes of the approximate size of tank searchlights (Wiltse, 1978). This implies a resolution of about 9 m at a distance of 1 km, which is not good enough to produce an image of vehicles of usual size. Using sophisticated signal processing and somewhat higher frequencies—which implies higher absorption losses due to rain and fog—one may perhaps achieve a resolution of about 1 m at a distance of 1 km, but the whole approach is obviously marginal.

Using synthetic aperture techniques seems to offer a better solution, but the usual synthetic aperture techniques are based on the Doppler shift of the frequency of a sinusoidal wave, and this requires a fast relative motion between radar and target. Let us consider in very general terms whether there is hope to do better by using nonsinusoidal waves.

It is a fundamental rule of communications that a pure sinusoidal (time) function transmits information at the rate zero. In passive beam forming, we receive information about the location of a source that emits a wave. In the one case, we deal with the time variable t and in the other with the space variable x, but there must be a fundamental rule for space that is equivalent to the rule "the transmission rate of information is zero for sinusoidal functions."

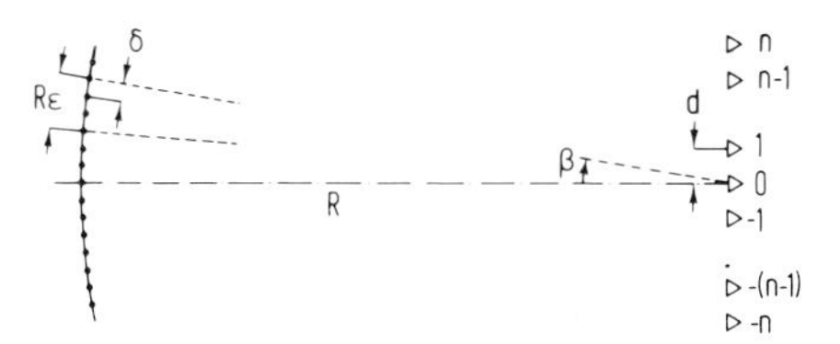

FIG. 5.1-10. The relative number of points resolved at a distance R by an array of $2n + 1$ sensors using sinusoidal waves.

For the derivation of this rule consider Fig. 5.1-10. A line array with $2n + 1$ equally spaced sensors is shown on the right. This array can resolve $2n + 1$ plane waves with different angles of incidence, or $2n + 1$ "points." The resolution angle equals $\epsilon = \lambda/(2n + 1)d$. The resolvable points at the distance R have the distance $\Delta x = R\epsilon$ from each other. It is usual to give them the locations $0, \pm\epsilon R, \pm 2\epsilon R, \ldots, \pm n\epsilon R$. Note that we do not have to assume that R is infinite. Our investigation is thus valid for focused beams that resolve points at a finite distance.

How many points are there on a circle of radius R in the sector $-n\epsilon < \beta < +n\epsilon$? Figure 5.1-10 shows on the left points with a distance δ, while only points with distance $R\epsilon$ can be resolved by the array. Since we are deriving a theoretical limit, we have three different choices for the distance δ, or the number N of points in the sector $-n\epsilon < \beta < +n\epsilon$:

(a) The distance δ is zero in such a way that N is nondenumerably infinite. This corresponds to our usual assumption of a continuous space.
(b) The distance δ is zero in such a way that N is denumerably infinite.
(c) The distance δ is larger than zero and finite.

For the cases a and b the relative number of resolved points equals $(2n + 1)/N = 0$, regardless of the distance R. However, one is not easily impressed by the practical significance of such a result, since it is based on a mathematical abstraction. The case c with a finite value of δ is the only one with physical significance. At a distance R there will be $(2n + 1)\epsilon R/\delta$ points[1] with distance δ in the sector $-n\epsilon < \beta < +n\epsilon$. The relative number of resolvable points becomes $(2n + 1)/[(2n + 1)\epsilon R/\delta] = \delta/\epsilon R$. For $R \to \infty$ the ratio $\delta/\epsilon R$ converges to zero. This is a physically significant result. Note that the physically nonsignificant cases a and b yielded a relative number of resolvable points equal to zero for any value of R, while the significant case c only yielded convergence to zero for increasing values of R. Hence, for planar wavefronts, coming from points at infinity, we have the following theorem:

[1] The number is actually either $2n\epsilon R/\delta$ or $(2n + 1)\epsilon R/\delta$, but this is of no consequence.

The relative number of resolved points is zero for sinusoidal waves if the wavefronts are planar and if a line array with a finite number of equally spaced sensors is used.

The obvious way to get around this limitation is to use signals with a bandwidth to resolve points in space, just as we use signals with a bandwidth to resolve points in time. The radiation patterns for dipole arrays derived in Section 3.2.2 showed that these patterns are totally different from the ones for sinusoidal waves. One needs only two dipoles to get an arbitrarily narrow mainlobe, which is not so for sinusoidal waves, while the number of dipoles affects the restlobe-to-mainlobe ratio only. We will thus investigate in the following sections synthetic aperture techniques based on nonsinusoidal signals, with the goal of achieving a resolution sufficient for imaging without a need for relative motion between radar and target.

5.2 Synthetic Aperture Radar

5.2.1 Resolution of Two Points

The principle of synthetic aperture radar based on sinusoidal functions was introduced by C. Wiley in 1951. A book by Harger (1970), a collection of papers by Kovaly (1976), and a translation of a Russian book edited by Reutov and Mikhaylov (1970) provide a survey of the field. We will investigate here the use of signals with large relative bandwidth—or nonsinusoidal signals for short—for synthetic aperture radar. Analysis and results will look totally different from the conventional synthetic aperture radar, but eventually we will recognize that the sinusoidal as well as the nonsinusoidal synthetic aperture radar use signals with a bandwidth to determine the direction as well as the distance to a point in space.[1] The conventional synthetic aperture radar produces a signal with bandwidth from a pure sinusoidal signal by means of the Doppler shift caused by the relative motion between a radar and a target, which means the bandwidth is produced mechanically. The bandwidth of nonsinusoidal signals, on the other hand, is produced either mechanically or electronically. The electronic generation of signals makes it possible to use a much larger variety of signals than Doppler-shifted sinusoids, and this makes everything ap-

[1] The importance of a bandwidth for the conventional sidelooking radar was clearly stated by Kovaly (1976; paragraph following Fig. 9 on the fifth page of the Introduction; the page is not numbered): "Each resolution function is inversely proportional to a bandwidth—in the range dimension, the RF bandwidth; in the azimuth dimension, the Doppler bandwidth. Fundamentally, then, resolution is obtained in both dimensions by generating a bandwidth."

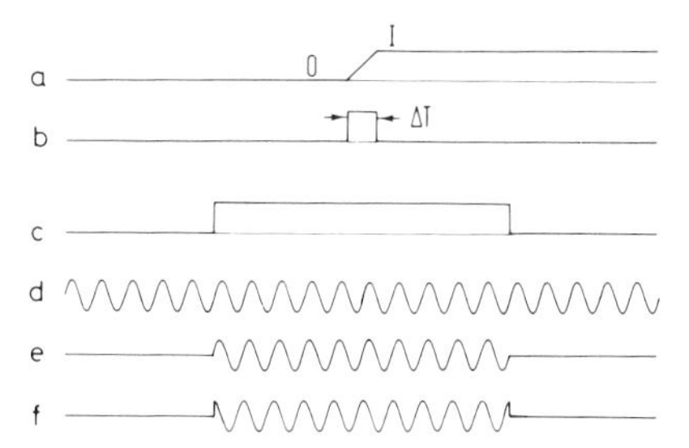

FIG. 5.2-1. Current step function supplied to a radiator (a) and the time variation of the electric and magnetic field strengths of the radiated wave in the far zone (b) rectangular pulse (c), periodic sinusoidal carrier (d), antenna current of a typical radar pulse (e), and its first derivative (f).

pear so different. Let us emphasize once more, that the basic principle is the same for the conventional and the nonsinusoidal synthetic aperture radar: The transmission of information requires bandwidth, regardless of whether this information is about points in time or points in space; the larger the bandwidth, the more information can be transmitted, and the better can be the resolution of a synthetic aperture radar.

Some of the results obtained can be applied to the conventional radar using a pulsed sinusoidal carrier. For an explanation of this point refer to Fig. 5.2-1. A current step function with a linear transient of duration ΔT from 0 to I is shown in line a. If this current is fed to a typical radiator, one obtains electric and magnetic field strengths in the far zone that have the time variation of line b. Consider now a simple conventional radar. It uses

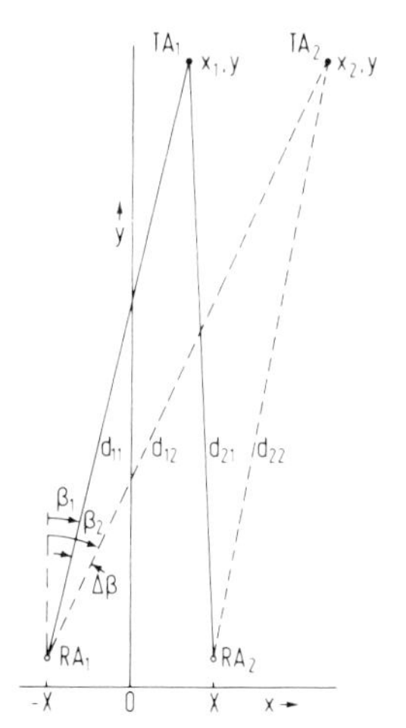

FIG. 5.2-2. Angular or polar coordinate resolution and Cartesian coordinate resolution achievable with two radars RA_1 and RA_2 located at $-X$ and $+X$.

the rectangular pulse of line c as baseband signal, the periodic sinusoidal function of line d as carrier, and produces the antenna current of line e by multiplication of the baseband signal with the carrier; the typical time variation of electric and magnetic field strengths in the far zone depends upon the first derivative of the antenna current as shown by line f. Beam forming is always based on the sinusoidal carrier, even though the baseband signal and the carrier are contained in the radiated signal as equally important factors of a product. Since the baseband signal of line c differs only in the time scale from the rectangular pulse of line b, an investigation of synthetic-aperture techniques for the rectangular pulse also applies to beam forming based on the baseband signal in conventional radar.

Figure 5.2-2 shows two radars RA_1 and RA_2, located at $-X$ and $+X$, measure the two distances d_{11} and d_{21} to the target TA_1 with the coordinates x_1, y. These measurements require only the rectangular pulses of lines b or c in Fig. 5.2-1, since the sinusoidal carrier can be—and usually is—removed by demodulation. The distances d_{11} and d_{21} are expressed in Cartesian coordinates:

$$d_{11} = [(x_1 + X)^2 + y^2]^{1/2} \tag{1}$$

$$\begin{aligned} d_{21} &= [(x_1 - X)^2 + y^2]^{1/2} = [(x_1 + X - 2X)^2 + y^2]^{1/2} \\ &= d_{11}(1 - 4Xx_1/d_{11}^2)^{1/2} \\ &\doteq d_{11}(1 - 2Xx_1/d_{11}^2) \qquad \text{for} \quad Xx_1 \ll d_{11}^2 \end{aligned} \tag{2}$$

Consider the *linear* difference between the two distances d_{11} and d_{21}:

$$d_{11} - d_{21} \doteq 2xX_1/d_{11} \tag{3}$$

It vanishes for large values of d_{11}:

$$\lim_{d_{11}\to\infty} (d_{11} - d_{21}) = 0 \tag{4}$$

Hence, for any finite value of X and x_1, the linear difference $d_{11} - d_{21}$ will have the same value 0, if d_{11} is large enough, and we will not be able to calculate x_1 from the linear difference $d_{11} - d_{21}$, the distance d_{11} of the target, and the distance $2X$ between the two radars.

Let a sinusoidal wave be radiated, reflected or backscattered by the target TA_1. The propagation times of a wavefront to the two radars equal d_{11}/c and d_{21}/c. We cannot measure these times, we can only measure phase differences. However, if the distance $2X$ between the two radars is no larger than half the wavelength λ, the measurement of the phase difference becomes equivalent to the measurement of the relative arrival time $t_{r,11}$ of the wavefront at radar RA_1:

$$t_{r,11} = d_{11}/c - d_{21}/c, \qquad 2X \le \lambda/2 \tag{5}$$

The right-hand side of Eq. (5) equals the left-hand side of Eq. (3), except for the constant $1/c$, and the relative time of arrival $t_{r,11}$ must thus vanish for large values of d_{11} like the right-hand side of Eq. (3). Hence, basing the resolution on the linear difference $d_{11} - d_{21}$ obtained from the absolute propagation time of two pulses, or on the phase difference of two sinusoidal oscillations, yields a very similar result. The two targets TA_1 and TA_2 in Fig. 5.2-2 must have an infinite distance $x_2 - x_1$ for $y \to \infty$, but the ratio $(x_2 - x_1)/y$ can have a finite value $\Delta\beta$:

$$\Delta\beta \doteqdot (x_2 - x_1)/y, \qquad x_2 - x_1, \quad y \to \infty, \quad \Delta\beta \ll 1 \tag{6}$$

Let us emphasize that Eqs. (3) and (5) do *not* yield the same resolution for a finite distance. The spacing $2X$ of the two receivers in Fig. 5.2-2 for a sinusoidal wave must not be larger than $\lambda/2$ according to Eq. (5), while there is no such restriction in Eq. (3). This is one reason why one should not ignore the baseband signal for beam forming.

The sinusoidal wave yields only the difference $d_{11} - d_{21}$, but the rectangular pulse yields the two distances d_{11} and d_{12} separately. Hence, we can square d_{11} and d_{21} to obtain the *quadratic* difference from Eqs. (1) and (2), which is exact for any value of xX_1:

$$d_{11}^2 - d_{21}^2 = 4Xx_1 \tag{7}$$

The quadratic difference does not vanish for large values of d_{11} as does the linear difference of Eq. (3). Note that Eq. (7) is exact, while Eq. (3) is an approximation.

Let us calculate the minimum distance $x_2 - x_1$ between the targets TA_1 and TA_2 in Fig. 5.2-2, based on the quadratic difference of distances. In analogy to Eqs. (1), (2), and (7), one obtains:

$$\begin{gathered} d_{12} = [(x_2 + X)^2 + y^2]^{1/2}, \qquad d_{22} = [(x_2 - X)^2 + y^2]^{1/2} \\ d_{12}^2 - d_{22}^2 = 4Xx_2 \end{gathered} \tag{8}$$

The resolvable distance $x_2 - x_1$ follows from Eqs. (7) and (8):

$$x_2 - x_1 = [(d_{12}^2 - d_{22}^2) - (d_{11}^2 - d_{21}^2)]/4X \tag{9}$$

The smallest resolvable distance depends on the accuracy with which the distances d_{11} to d_{22} can be measured, i.e., on the time resolution of the radar, and it depends on the distance $2X$ between the two radars. The coordinate y is not explicitly contained in Eq. (9), and the resolvable distance $x_2 - x_1$ thus remains constant as y increases toward infinity.[1] Such a result is too good to be true. Let us see where the limitation for the resolvable distance comes from.

[1] This is the same idealized result as known for the focused conventional synthetic aperture radar (Cutrona and Hall, 1962). The resolution angle is zero and the minimum separation of two resolvable points does not depend on their distance from the radar.

5.2.2 *Distance and Location Error*

Equation (3) is rewritten to yield x_1 as function of d_{11}, d_{12}, and X:

$$x_1 \doteqdot d_{11}(d_{11} - d_{21})/2X \tag{10}$$

A radar cannot measure a distance d_{ij} to a target with unlimited accuracy, but only with a certain error $\Delta d_{ij} \ll d_{ij}$. This error is partly due to features of the equipment, such as jitter, and partly due to the features of the transmission path. We do not need to make any assumption about the error Δd_{ij} at this time beyond the condition that Δd_{ij} is small compared with d_{ij}.

The error of x_1 due to inaccurate measurements of distance is denoted Δx_1. Introduction of the errors Δd_{ij} into Eq. (10) yields the following result:

$$x_1 + \Delta x_1 \doteqdot (d_{11} + \Delta d_{11})(d_{11} + \Delta d_{11} - d_{21} - \Delta d_{21})/2X \tag{11}$$

Ignoring the error terms of higher than first order, we obtain

$$\Delta x_1 = [d_{11}(2\,\Delta d_{11} - \Delta d_{21}) - d_{21}\,\Delta d_{11}]/2X \tag{12}$$

The corresponding error for the quadratic difference is calculated from Eq. (7):

$$x_1 + \Delta x_1 = [(d_{11} + \Delta d_{11})^2 - (d_{21} + \Delta d_{21})^2]/4X \tag{13}$$

$$\Delta x_1 \doteqdot (d_{11}\,\Delta d_{11} - d_{21}\,\Delta d_{21})/2X \tag{14}$$

The comparison of Eqs. (12) and (14) shows that the error Δx_1 increases in both cases proportionate to the distances d_{11} and d_{21}. The factors of d_{21} are essentially the same in both equations, but the factors of d_{11} are not. Since the errors Δd_{ij} are not constants but random variables, the factor $2\,\Delta d_{11} - \Delta d_{21}$ in Eq. (12) is less desirable than the factor Δd_{11} in Eq. (14). Hence, even though the quadratic difference $d_{11}^2 - d_{21}^2$ is not as good as it appears to be from Eq. (9), it is better than the linear difference $d_{11} - d_{21}$. Beyond the better resolution, the quadratic difference has a second advantage. The linear difference of Eq. (3) holds only for large values of d_{11}. For smaller values, one must use more terms of the series expansion in Eq. (2). This process is called focusing. The quadratic difference of Eq. (7) is correct for any distance, and the problem of focusing does not arise. The importance of focusing for the conventional synthetic-aperture radar is well understood (Cutrona and Hall, 1962).

Let us now investigate the effect of an error ΔX in the distance $2X$ between the radars in Fig. 5.2-2. For the linear difference, we obtain from Eq. (3):

$$\begin{aligned} x_1 + \Delta x_1 &= d_{11}(d_{11} - d_{21})/(2X + \Delta X) \\ \Delta x_1 &\doteqdot -\Delta X d_{11}(d_{11} - d_{21})/4X^2 \end{aligned} \tag{15}$$

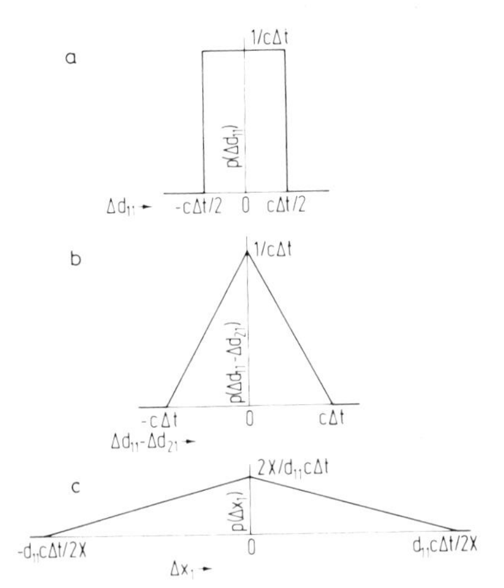

FIG. 5.2-3. Constant density function of the distance error Δd_{11}, Δd_{21} (a), the density function of $\Delta d_{11} - \Delta d_{21}$ (b), and the density function of the location error x_1 (c).

For the quadratic difference we obtain a corresponding formula from Eq. (7):

$$\Delta x_1 = -\Delta X(d_{11}^2 - d_{21}^2)/8X^2 \tag{16}$$

The error Δx_1 of Eq. (16) seems to be only half as large as the error Δx_1 in Eq. (15). However, one may readily verify that both equations yield the same error for a small difference between d_{11} and d_{21}. Note that the error ΔX is multiplied in Eqs. (15) and (16) by the squares of the distances to the target, while the errors Δd_{ij} in Eqs. (12) and (14) are multiplied by the distances only.

In order to obtain some idea about the statistical properties of the error Δx_1 in Eq. (14), let us consider the case of a target with the coordinate $x = 0$ in Fig. 5.2-2, which implies $d_{11} = d_{21}$. One obtains

$$\Delta x_1 = (\Delta d_{11} - \Delta d_{21})d_{11}/2X$$

from Eq. (14). If the shortest time resolved by the radar equals Δt one obtains a distance error Δd_{11} or Δd_{21} between $-c\,\Delta t/2$ and $+c\,\Delta t/2$. Assume that both Δd_{11} and Δd_{21} have the constant probability density function shown in Fig. 5.2-3a. The density function of $\Delta d_{11} - \Delta d_{21}$ is then the triangular function shown in Fig. 5.2-3b. The density function for Δx_1 follows by the change of scale shown in Fig. 5.2-3c.

5.2.3 Sidelooking Radar

We will now try to transform the distance measurement by pulses and the quadratic difference into workable concepts. Since the sidelooking

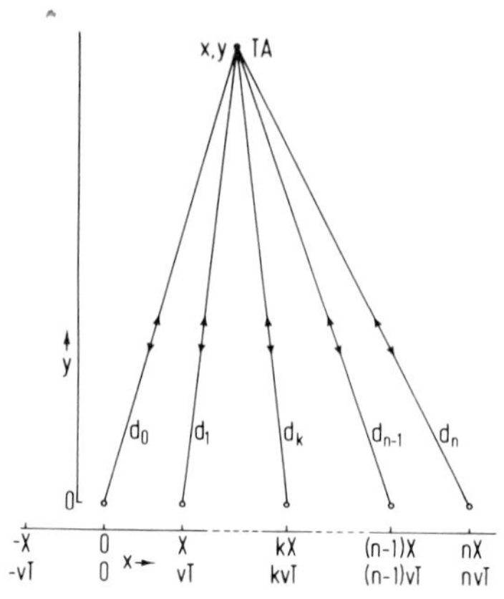

FIG. 5.2-4. Geometric relations for a sidelooking radar moving along the x axis with velocity v, and a target TA.

radar is a known workable concept for synthetic aperture radar based on sinusoidal functions, we will first investigate sidelooking radar for nonsinusoidal functions.

Consider an airplane moving with velocity v in the direction of the x axis in Fig. 5.2-4. A radar carried by it radiates pulses into a certain large sector of the half-plane $y > 0$. The pulse radiated at the location $x = 0$ shall be *reflected* from a target with the coordinates x, y. The distance to the target is denoted d_0:

$$d_0 = (x^2 + y^2)^{1/2} \tag{17}$$

Further pulses are radiated when the airplane is at the locations $X, \ldots, kX, \ldots, nX$. The general distance d_k is expressed with the help of the reference distance d_0:

$$d_k = [(x - kX)^2 + y^2]^{1/2} = d_0[1 - (2kXx - k^2X^2)/d_0^2]^{1/2} \tag{18}$$

The round-trip travel times of the two pulses radiated at $x = 0$ and $x = kX$ are t_0 and t_k:

$$t_0 = 2d_0/c, \qquad t_k = 2d_k/c \tag{19}$$

Since the radar measures the times t_0 and t_k rather than the distances d_0 and d_k, we will use the times. The square difference $t_0^2 - t_k^2$ follows from Eqs. (17) and (18):

$$t_0^2 - t_k^2 = 4(2kXx - k^2X^2)/c^2 \tag{20}$$

For the explanation of the required data processing, it is useful to rewrite this equation in two steps:

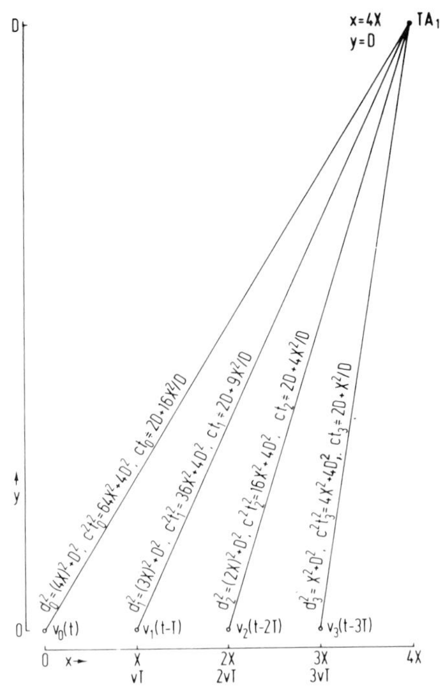

FIG. 5.2-5. Example of a sidelooking radar measuring the distance of the target TA_1 from the locations $x = 0, X, 2X, 3X$ at the time $t = 0, T, 2T, 3T$. The radar moves with velocity $v \ll c$ in the direction of the x axis. The round-trip travel time of a signal to the target TA_1 is negligibly small compared with T.

$$t_0^2 - [t_k^2 - 4(kX/c)^2] = 8kXx/c^2 \tag{21}$$

$$t_0^2 - [t_k^2 - 4(kX/c)^2 + 8(x/c^2)kX] = 0 \tag{22}$$

In order to explain how the coordinate x of the target TA in Fig. 5.2-4 is obtained from these equations, we will consider a specific numerical case. Figure 5.2-5 shows a sidelooking radar moving along the x axis and measuring the distance to the target TA_1 from the four locations $0, X, 2X, 3X$. The coordinates of the target are $x = 4X$, $y = D$. The four squared distances $d_0^2 \dots d_3^2$ according to Eqs. (17) and (18) are shown. Also shown are the round-trip travel times $c^2t_0^2 \dots c^2t_3^2$ according to Eq. (19), and the first-order approximations of $ct_0 \dots ct_3$. The output voltages of the radar are denoted $v_0(t)$, $v_1(t - T)$, $v_2(t - 2T)$, and $v_3(t - 3T)$.

Let us now turn to Fig. 5.2-6. On top (a) are shown the four voltages $v_0(t)$ to $v_3(t)$ due to the target TA_1 in Fig. 5.2-5; the time shifts $-T$, $-2T$, and $-3T$ have been removed. The same voltages as functions of t^2 rather

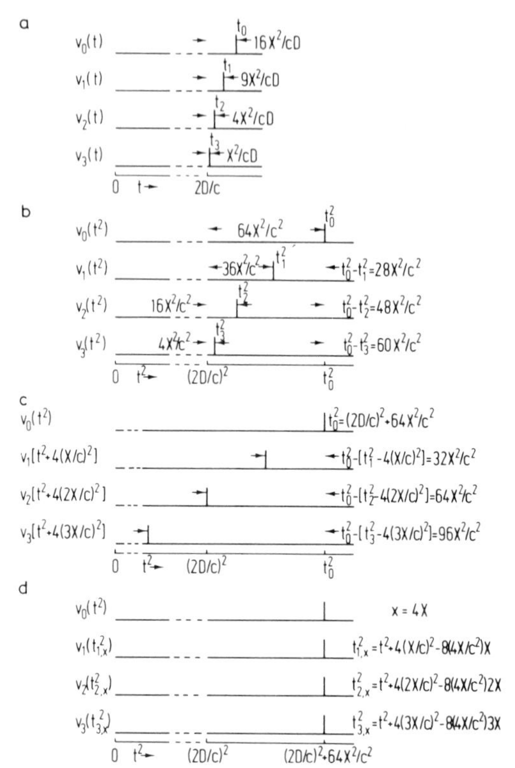

FIG. 5.2-6. Time diagram for the processing of the four voltages $v_0(t)$ to $v_3(t - 3T)$ of Fig. 5.2-5 received by the sidelooking radar.

than t are shown below (b). These voltages would be seen on a cathode ray tube if the deflection voltage for the time axis increased proportionate to t^2 rather than to t. In digital processing equipment, the transition from t to t^2 means a relabeling of storage addresses from t to t^2.

The square differences $t_0^2 - t_k^2$ of Eq. (20) are shown on the right-hand side of Fig. 5.2-6b.

Figure 5.2-6c shows the transition from Eq. (20) to Eq. (21). A voltage $v_k(t^2)$ is time-shifted to become $v_k[t^2 + 4(kX/c)^2]$. There is no difficulty in performing these shifts in digital processing equipment, while analog equipment would call for a sufficiently large delay to make sure that shifted voltages always are delayed and not advanced in time.

The final step from Eq. (21) to Eq. (22) requires the time shift $8kXx/c^2$ for a particular value of x; the value $x = 4X$ is chosen in Fig. 5.2-6d. The output voltages $v_0(t)$ to $v_3(t - 3T)$ are now properly lined up. The sum of the voltages in Fig. 5.2-6d yields four times—or generally n times—the amplitude of the individual voltages. This increase in amplitude would not be achieved for a value $x \neq 4X$. We recognize here that the use of n radars in Fig. 5.2-4 instead of the two radars in Fig. 5.2-2 increases the dynamic range or the contrast ratio.

Figure 5.2-6d yields the x coordinate of the target TA_1 in Fig. 5.2-5. The y coordinate follows from the time t_0^2 in Fig. 5.2-6d and Eq. (17) or Eqs. (18) and (19):

$$
\begin{aligned}
y^2 &= d_0^2 - x^2 = c^2 t_0^2/4 - x^2 \\
y^2 &= d_k^2 - (x - kX)^2 = c^2 t_k^2/4 - (x - kX)^2
\end{aligned}
\tag{23}
$$

The sign of y remains undetermined, since the geometric relations in Fig. 5.2-5 would be the same for a target located at $y = -D$ rather than $y = +D$. There are a variety of ways to resolve this ambiguity. One is to use a radar that does not radiate omnidirectionally but only into a sector that is no larger than the half-plane $y > 0$.

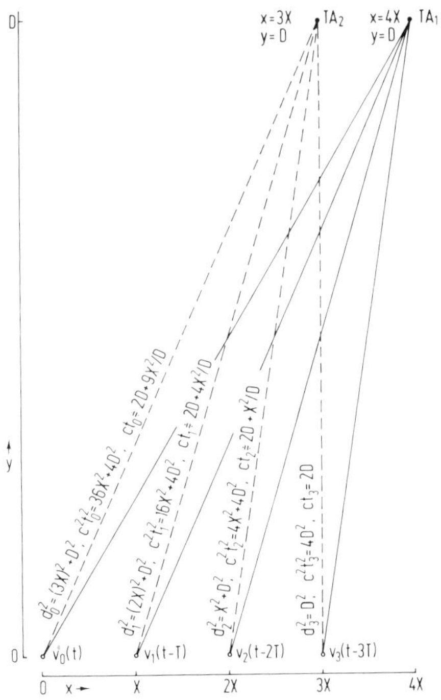

FIG. 5.2-7. Example of a sidelooking radar measuring the distance to the targets TA_1 and TA_2 from the locations $x = 0, X, 2X, 3X$ at the times $t = 0, T, 2T, 3T$. The radar moves with velocity $v \ll c$ in the direction of the x axis. The round-trip travel time of a signal to either target is negligibly small compared with T. The values of d_i^2, $c^2t_i^2$, and ct_i, $i = 0 \ldots 3$, for target TA_1 are shown in Fig. 5.2-5.

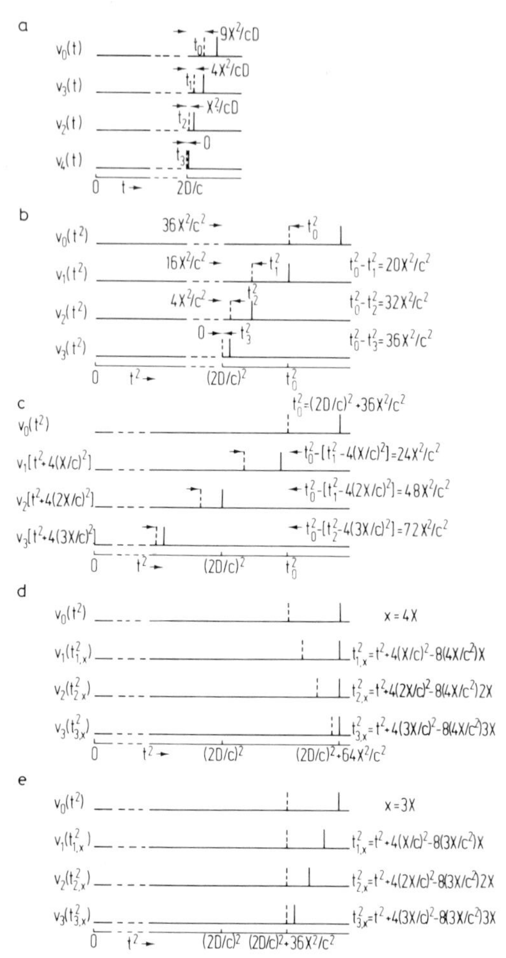

FIG. 5.2-8. Time diagram for the processing of the four voltages $v_0(t)$ to $v_3(t - 3T)$ of Fig. 5.2-7 received by the sidelooking radar. The dashed pulses hold for target TA_2, the solid pulses for target TA_1; the numerical values for the solid pulses are shown in Fig. 5.2-6.

A sidelooking radar is not normally used to locate a single target, but to make a map that contains many targets or points. We will show next how two targets—and by implication many targets with sufficient separation—are resolved, but we will have to return to the resolution of many points again at a later time.

Figure 5.2-7 shows the target TA_1 of Fig. 5.2-5 and, in addition, a second target TA_2 with the coordinates $x = 3X$, $y = D$. The values of d_i^2, $c^2t_i^2$, and ct_i, for $i = 0 \ldots 3$ are shown for this second target, while the respective values for the first target TA_1 are the same as shown in Fig. 5.2-5. The four voltages $v_0(t)$, $v_1(t - T)$, $v_2(t - 2T)$, and $v_3(t - 3T)$—corrected for the delays T, $2T$, and $3T$—are shown in Fig. 5.2-8a. The pulses shown by solid lines are due to the target TA_1; they are the same pulses as shown in Fig. 5.2-6. The pulses shown by dashed lines are due to target TA_2. The round-trip delay times $t_0 = 2D/c + 9X^2/cD$ to $t_3 = 2D/c$ are only shown for the dashed pulses; the values shown in Fig. 5.2-6 hold for the solid pulses.

The voltages $v_0(t^2)-v_3(t^2)$ of the squared time are shown in Fig. 5.2-8b, and the shifted voltages $v_k[t^2 + 4(kX/c)^2]$, $k = 0 \ldots 3$ in Fig. 5.2-8c. A further time shift corresponding to $x = 4X$ yields the voltages of Fig. 5.2-8d, while shifts corresponding to $x = 3X$ yield the voltages of Fig. 5.2-8e. Summation of the voltages of Fig. 5.2-8d produces one solid pulse with four times the amplitude of the original pulses, and four dashed pulses with the old amplitude; the contrast or the dynamic range is thus 4:1. An equivalent statement holds for Fig. 5.2-8e, except that the role of the solid and the dashed pulses is interchanged.

Equation (14) shows that the resolution error Δx for the x coordinate depends on the accuracy of the distance measurements and the spacing $2X$ of the two radars in Fig. 5.2-2. This spacing $2X$ is replaced by nX or nvT in Fig. 5.2-4, and by $3X$ in Figs. 5.2-5 and 5.2-7. The dynamic range or contrast is determined by the number of receivers, which is 2 in Fig. 5.2-2, $n + 1$ in Fig. 5.2-4, and 4 in Figs. 5.2-5 and 5.2-7. Of course, the signal-to-noise ratio also increases with the number of receivers.

Let us note that the Doppler effect has not yet been used, even though it is considered to be essential for the conventional sidelooking radar. There are two readily recognizable uses one can make of this fact. If the pulses that we have been using are transmitted with a sinusoidal carrier, one may use the Doppler shift of the carrier to distinguish between moving and nonmoving targets. On the other hand, since we do not need the Doppler shift, we can transmit the pulses without a carrier. This makes it possible to use pulses with a duration of 1 to 0.1 ns, while avoiding the high losses due to rain and fog that a carrier frequency commensurate with such short pulses would imply. A very different use of the Doppler shift is the elimination of ghost targets, which works both with the conventional and the nonsinusoidal sidelooking radar. This topic must be deferred until later.

5.2.4 Stationary Radar with Synthetic Aperture

Since we have not used the Doppler shift for the sidelooking radar, one must be able to achieve the same resolution by using stationary radar transmitters at the locations $0, X, \ldots, nX$ in Fig. 5.2-4, or $0, X, 2X, 3X$ in Figs. 5.2-5 and 5.2-7. This is not the same as using a phased array for beamforming based on sinusoidal waves. For instance, if we want to form a beam using 100 sensors for a wave with a frequency of 10 MHz, we must space the sensors at a distance of no more than $\lambda/2 = 15$ m according to Eq. (5); the length of the phased array is thus limited to $L =$ 1.5 km, and the resolution angle to about $\epsilon = \lambda/L = 0.02$. The only way to increase the aperture is to increase the number of sensors.

No restrictions on the spacing of the sensors is implied by Eqs. (3) or (7). One may space the sensors over an aperture of megameters rather than kilometers. The number of sensors will determine the dynamic range and the signal-to-noise ratio, but there will be no ambiguities due to grating lobes.

There are some advantages of a stationary radar over one mounted on an aircraft or a satellite. One may operate at the low frequencies used for over-the-horizon radar, which require large radiators and sensors. Furthermore, the errors due to an inaccurate course or altitude, as well as errors due to pitch, yaw, and roll of an aircraft, are avoided; positioning stationary sensors very accurately over distances of megameters is a more practical proposition than to do so with an airborne sensor. Another application for a stationary high-resolution radar is the taxi radar at an airport; if the resolution is good enough to show the shape of an aircraft, one can instantly recognize in which direction it is heading, while, at a lower resolution, one must derive this information from the movement of the aircraft. This application is very similar to that of the armored vehicle tracking radar with imaging capability called for in Section 5.1.6.

In the case of the sidelooking radar according to Fig. 5.2-4, one must radiate a signal at each position $0, X, \ldots, nX$. For a stationary radar, we need to radiate only one signal rather than n, since all receivers are permanently at the positions $0, X, \ldots, nX$. This fact implies an increase of the signal-to-noise ratio by a factor n, a reduction of the time required to produce an image by a factor $1/n$, and a reduction of the required radiators also by $1/n$.

Figure 5.2-9 shows a stationary radar with a radiator at $x = 0$ and n sensors at $x = 0, X, \ldots, nX$. The distances d_0 and d_k are the same as given by Eqs. (17) and (18). The round-trip travel time t_0 is the same as defined by Eq. (19), but t_k has a different value, since the signal travels now from the radiator at $x = 0$ to the target TA and a sensor at $x = kX$:

$$t_0 = 2d_0/c, \qquad t_k = (d_0 + d_k)/c \tag{24}$$

With the help of Eq. (18) we obtain the following relation to replace Eq. (20):

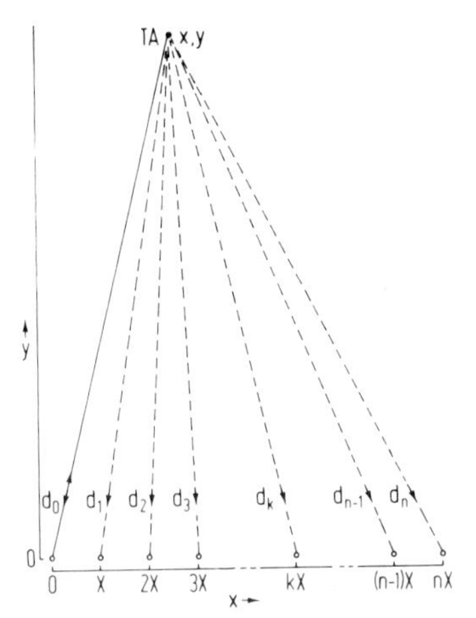

FIG. 5.2-9. Geometric relations for a stationary radar with no restriction on the distance between the sensors at $x = 0, X, \ldots, nX$. The only radiator is located at $x = 0$, the target TA at x, y.

$$t_0^2 - t_k^2 = (3d_0^2 - 2d_0 d_k + d_k^2)/c^2$$
$$= [2d_0^2 + 2kXx - k^2X^2$$
$$- 2(d_0 - 2kXx + k^2X^2)^{1/2}]/c^2 \qquad (25)$$

In contrast to Eq. (20), we have now a square root. For large values of d_0 we can use a series expansion for the square root. The unfocused case is obtained if only the first two terms of the series expansion are used:

$$t_0^2 - t_k^2 \doteq 2(2kXx - k^2X^2)/c^2, \qquad d_0 \gg kX \qquad (26)$$

The factor 2 occurs on the right-hand side instead of the factor 4 in Eq. (20). This is in analogy to the well-known fact that the synthetic aperture of a sidelooking radar needs to be only half as large as the real aperture of a stationary radar to obtain the same resolution. In Fig. 5.2-4, each signal travels twice over the respective distance d_k, $k = 0 \ldots n$, but, in Fig. 5.2-9, each signal travels only once over the distance d_k on the way from the target TA to the sensor at kX, while the distance d_0 from the radiator to the target is the same for all signals. Equation (26) is rewritten in analogy to Eqs. (21) and (22):

$$t_0^2 - [t_k^2 - 2(kX/c)^2] \doteq 4kXx/c^2 \qquad (27)$$

$$t_0^2 - [t_k^2 - 2(kX/c)^2 + 4(x/c^2)kX] \doteq 0 \qquad (28)$$

In order to explain the data processing required to obtain the coordinate

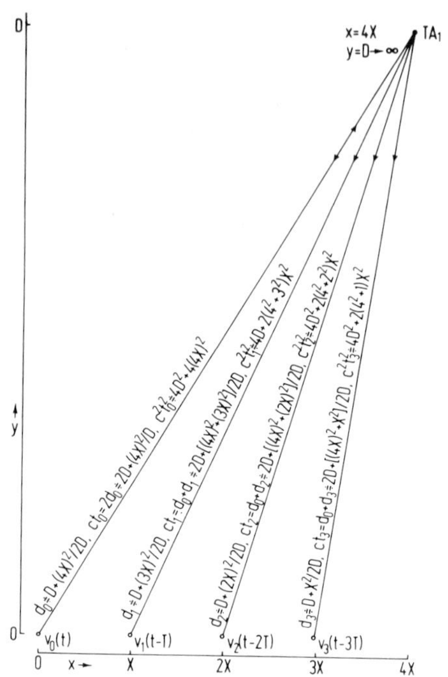

FIG. 5.2-10. Example of a stationary synthetic aperture radar measuring the distances $2d_0$, $d_0 + d_1$, $d_0 + d_2$, and $d_0 + d_3$ to the target TA_1. One radiator at $x = 0$ and four sensors at $x = 0, X, 2X, 3X$ are used. The distances d_0 to d_3 are all large compared with the synthetic aperture $3X$.

x of the target TA, we consider again a specific numerical case. Figure 5.2-10 shows four sensors at the locations $x = 0, X, 2X, 3X$, and one radiator at $x = 0$. The distances d_k and $d_0 + d_k$, as well as the squares $(d_0 + d_k)^2$ are shown for $k = 0 \ldots 3$.

Figure 5.2-11 shows on top the received voltages $v_0(t)$ to $v_3(t)$ as functions of the time t, and below as functions of t^2. The implementation of Eq. (27) is shown by Fig. 5.2-11c, and the implementation of Eq. (28) by Fig. 5.2-11d.

Let us turn to the focused case. We start from Eq. (25), but do not use a series expansion for the square root. The distance d_0^2 is replaced by $x^2 + y^2$ according to Eq. (17):

$$t_0^2 - \left[t_k^2 - \left(\frac{kX}{c}\right)^2\right]$$

$$= \frac{2kXx}{c^2} + \frac{2(x^2 + y^2)}{c^2}\left[1 - 2\left(1 - \frac{2kXx - (kX)^2}{x^2 + y^2}\right)^{1/2}\right] \qquad (29)$$

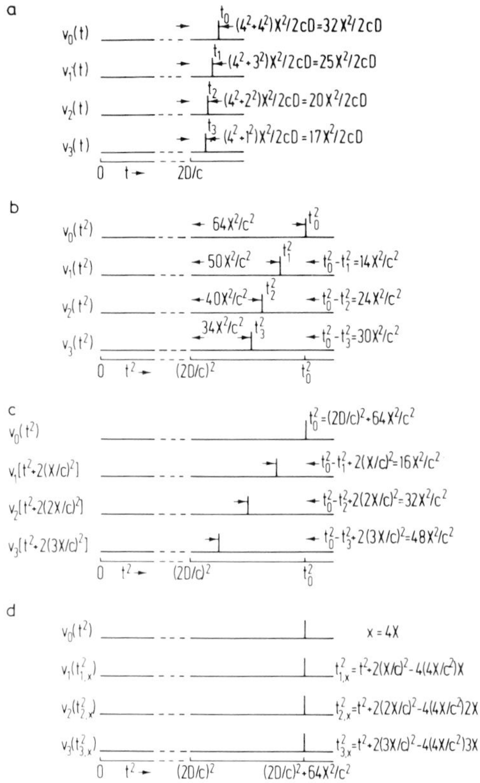

FIG. 5.2-11. Time diagram for the processing of the four voltages $v_0(t)$ to $v_3(t)$ of Fig. 5.2-10 received by the stationary synthetic aperture radar.

$$t_0^2 - \left[t_k^2 - \left(\frac{kX}{c}\right)^2 + 2\left(\frac{x}{c^2}\right) kX\right] = \frac{2(x^2 + y^2)}{c^2}\left[1 - 2\left(1 - \frac{2kXx - (kX)^2}{x^2 + y^2}\right)^{1/2}\right] \tag{30}$$

$$t_0^2 - \left\{t_k^2 - \left(\frac{kX}{c}\right)^2 + 2\left(\frac{x}{c^2}\right) kX + \frac{2(x^2 + y^2)}{c^2}\left[1 - 2\left(1 - \frac{2kXx - (kX)^2}{x^2 + y^2}\right)^{1/2}\right]\right\} = 0 \tag{31}$$

Equations (29) and (30) are implemented like Eqs. (27) and (28), the only difference being the reduction of the factors in front of the terms

$(kX/c)^2$ and $(x/c^2)kX$ to one-half. Equation (31) calls for a different time shift for every value of y. Hence, this *focusing term* calls for a lot of computing, but creates no further problems; a set of voltages $v_k(t)$, $k = 0 \ldots n$, permits the production of images of points at all coordinates y. This situation is in contrast to the usual focusing based on sinusoidal functions, which permits a focused image only of points on a circle with a *certain distance* from the center of the sensor array, but not for points on circles with *any distance* from the center of the sensor array.

5.2.5 Two-Dimensional Synthetic Aperture

The conventional synthetic aperture radar implements a one-dimensional array or a *line array*. In principle, one could implement two-dimensional or *area arrays* too, but doing so would call for rather skillful navigation of aircrafts or satellites. No such problem occurs with stationary arrays. Before investigating the two-dimensional synthetic aperture array, let us first see what it could be used for. A square array of 10×10 sensors spread over an area of 1000×1000 km², using only one radiator, could produce very good maps of the moon or Venus. Since the Doppler effect due to the rotation of a planet is not needed as in conventional planetary mapping (Rodgers and Ingalls, 1970; Kovaly, 1972), it is still available as an unused parameter. However, it would be useless to devote one's effort to advocating such a purely scientific application. Science has been advanced by using military applications as a vehicle at least as far back as Archimedes, and the world has not changed all that much since.[1] There are plenty of military applications for two-dimensional arrays. Every radar using a parabolic dish is, in essence, using a two-dimensional array. Since we would not want to replace a parabolic dish by a larger two-dimensional array, it is not evident what could be gained. We will show later on that a gain is indeed possible theoretically, but one requires digital circuits operating in subnanosecond time.

Figure 5.2-12 shows a two-dimensional array of nm sensors, located at the intersection of the n columns and the m rows. A radiator is located at $x = y = 0$. The distance between a sensor located at $x = kX$, $y = lY$ and the target TA is denoted d_{kl}, where $k = 0 \ldots n$ and $l = 0 \ldots m$. Equations (17) and (18) are replaced by the following generalized equations:

[1] Much of Archimedes' work was funded by the military-industrial complex of Syracuse in Sicily; it yielded a high benefit/cost ratio for the sponsors (Livius 17, Book XXIV, Section 34). For instance, his work on applied optics and solar energy resulted in the burning of ships of the hostile Roman Navy during Punic War II (Durant, 1939, Chapter XXVIII, Section II).

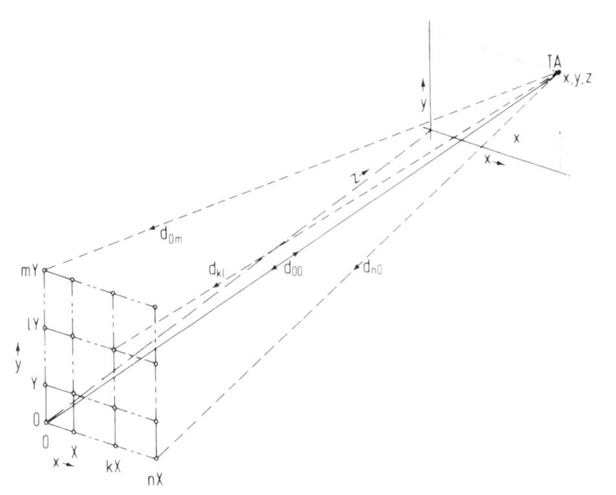

FIG. 5.2-12. Geometric relations for a stationary radar with a two-dimensional sensor array. The sensors are located at the points kX, lY with $k = 0 \cdots n$ and $l = 0 \cdots m$. The only radiator is located at $x = y = 0$, the target TA is located at x, y, z.

$$d_{00} = (x^2 + y^2 + z^2)^{1/2} \tag{32}$$

$$\begin{aligned} d_{kl} &= [(x - kX)^2 + (y - lY)^2 + z^2]^{1/2} \\ &= d_{00}[1 - (2kXx + k^2X^2 + 2lYy - l^2Y^2)/d_{00}^2]^{1/2} \end{aligned} \tag{33}$$

The generalization of the travel times given in Eq. (24) yields

$$t_{00} = 2d_{00}/c, \qquad t_{kl} = (d_{00} + d_{kl})/c \tag{34}$$

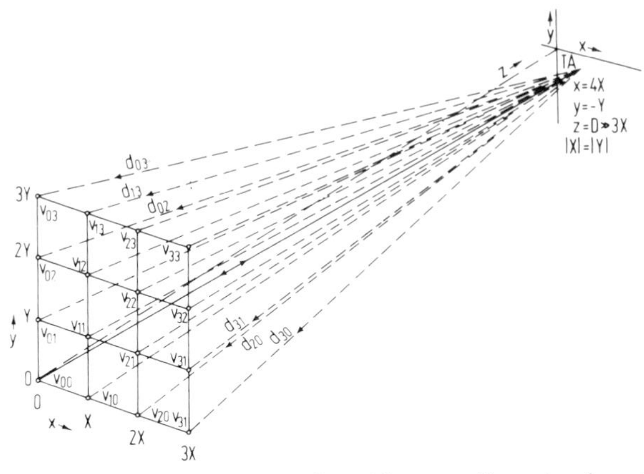

FIG. 5.2-13. Example of a stationary radar with a two-dimensional synthetic aperture measuring the distances $2d_{00} \cdots d_{00} + d_3$ to a target TA. One radiator at $x = y = 0$ and 16 sensors at $x = kX$, $y = lY$ are used. The distances d_{00} to d_{33} are all large compared with the synthetic apertures $3X$ and $3Y$ in the x and y directions.

TABLE 5.2-1

THE VOLTAGES $v_{kl} = v_{kl}(t)$ OF FIG. 5.2-13 AND THE DISTANCES d_{kl}, TOGETHER WITH THE QUANTITIES ct_{kl}, $c^2t_{kl}^2$, $t_{00}^2 - t_{kl}^2$, $t_{c,kl}^2$, AND $t_{00}^2 - t_{c,kl}^2$ DERIVED FROM THEM

$v_{03}(t)$ $d_{03} = D + [(4X)^2 + (4Y)^2]/2D$ $ct_{03} = d_{00} + d_{03} = 2D + (32X^2 + 17Y^2)/2D$ $c^2t_{03}^2 = 4D^2 + 2(32X^2 + 17Y^2)$ $t_{00}^2 - t_{03}^2 = 2(0X^2 - 15Y^2)/c^2$ $t_{c,03}^2 = t_{03}^2 - 2(0X/c)^2 - 2(3Y/c)^2$ $t_{00}^2 - t_{c,03}^2 = 0(X/c)^2 - 12(Y/c)^2$	$v_{13}(t)$ $d_{13} = D + [(3X)^2 + (4Y)^2]/2D$ $ct_{13} = d_{00} + d_{13} = 2D + (25X^2 + 17Y^2)/2D$ $c^2t^2 = 4D^2 + 2(25X^2 + 17Y^2)$ $t_{00}^2 - t_{13}^2 = 2(7X^2 - 15Y^2)/c^2$ $t_{c,13}^2 = t_{13}^2 - 2(X/c)^2 - 2(3Y/c)^2$ $t_{00}^2 - t_{c,03}^2 = 16(X/c)^2 - 12(Y/c)^2$
$v_{02}(t)$ $d_{02} = D + [(4X)^2 + (3Y)^2]/2D$ $ct_{02} = d_{00} + d_{03} = 2D + (32X^2 + 10Y^2)/2D$ $c^2t_{02}^2 = 4D^2 + 2(32X^2 + 10Y^2)$ $t_{00}^2 - t_{02}^2 = 2(0X^2 - 8Y^2)/c^2$ $t_{c,02}^2 = t_{02}^2 - 2(0X/c)^2 - 2(2Y/c)^2$ $t_{00}^2 - t_{c,02}^2 = 0(X/c)^2 - 8(Y/c)^2$	$v_{12}(t)$ $d_{12} = D + [(3X)^2 + (3Y)^2]/2D$ $ct_{12} = d_{00} + d_{12} = 2D + (25X^2 + 10Y^2)/2D$ $c^2t_{12}^2 = 4D^2 + 2(25X^2 + 10Y^2)$ $t_{00}^2 - t_{12}^2 = 2(7X^2 - 8Y^2)/c^2$ $t_{c,12}^2 = t_{12}^2 - 2(X/c)^2 - 2(2Y/c)^2$ $t_{00}^2 - t_{c,12}^2 = 16(X/c)^2 - 8(Y/c)^2$
$v_{01}(t)$ $d_{01} = D + [(4X)^2 + (2Y)^2]/2D$ $ct_{01} = d_{00} + d_{01} = 2D + (32X^2 + 5Y^2)/2D$ $c^2t_{01}^2 = 4D^2 + 2(32X^2 + 5Y^2)$ $t_{00}^2 - t_{01}^2 = 2(0X^2 - 3Y^2)/c^2$ $t_{c,01}^2 = t_{01}^2 - 2(0X/c)^2 - 2(Y/c)^2$ $t_{00}^2 - t_{c,01}^2 = 0(X/c)^2 - 4(Y/c)^2$	$v_{11}(t)$ $d_{11} = D + [(3X)^2 + (2Y)^2]/2D$ $ct_{11} = d_{00} + d_{12} = 2D + (25X^2 + 5Y^2)/2D$ $c^2t_{11}^2 = 4D^2 + 2(25X^2 + 5Y^2)$ $t_{00}^2 - t_{11}^2 = 2(7X^2 - 3Y^2)/c^2$ $t_{c,11}^2 = t_{11}^2 - 2(X/c)^2 - 2(Y/c)^2$ $t_{00}^2 - t_{c,11}^2 = 16(X/c)^2 - 4(Y/c)^2$
$v_{00}(t)$ $d_{00} = D + [(4X)^2 + Y^2]/2D$ $ct_{00} = 2d_{00} = 2D + (32X^2 + 2Y^2)/2D$ $c^2t_{00}^2 = 4D^2 + 2(32X^2 + 2Y^2)$	$v_{10}(t)$ $d_{10} = D + [(3X)^2 + Y^2]/2D$ $ct_{10} = d_{00} + d_{01} = 2D + (25X^2 + 2Y^2)/2D$ $c^2t_{10}^2 = 4D^2 + 2(25X^2 + 2Y^2)$ $t_{00}^2 - t_{10}^2 = 2(7X^2 + 0Y^2)/c^2$ $t_{c,10}^2 = t_{10}^2 - 2(X/c)^2 - 2(0Y/c)^2$ $t_{00}^2 - t_{c,10}^2 = 16(X/c)^2 - 0(Y/c)^2$

(Continued)

One obtains the generalization of Eq. (25):

$$t_{00}^2 - t_{kl}^2 = (3d_{00}^2 - 2d_{00}d_{kl} - d_{kl}^2)/c^2$$
$$= \{2d_{00}^2 + 2kXx - k^2X^2 + 2lYy - l^2Y^2$$
$$- 2[d_{00} - 2kXx + k^2X^2 - 2lYy + l^2Y^2]^{1/2}\}/c^2 \qquad (35)$$

For large values of d_{00}, we use again a series expansion for the square root:

$$t_{00}^2 - t_{kl}^2 \doteq 2(2kXx - k^2X^2 + 2lYy - l^2Y^2)/c^2, \qquad d_{00} \gg kX, lY \quad (36)$$

The two Eqs. (27) and (28) are replaced by four equations:

TABLE 5.2-1 *(Continued)*

$v_{23}(t)$ $d_{23} = D + [(2X)^2 + (4Y)^2]/2D$ $ct_{23} = d_{00} + d_{23} = 2D + (20X^2 + 17Y^2)/2D$ $c^2t_{23}^2 = 4D^2 + 2(20X^2 + 17Y^2)$ $t_{00}^2 - t_{23}^2 = 2(12X^2 - 15Y^2)/c^2$ $t_{c,23}^2 = t_{23}^2 - 2(2X/c)^2 - 2(3Y/c)^2$ $t_{00}^2 - t_{c,23}^2 = 32(X/c)^2 - 12(Y/c)^2$	$v_{33}(t)$ $d_{33} = D + [X^2 + (4Y)^2]/2D$ $ct_{33} = d_{00} + d_{33} = 2D + (17X^2 + 17Y^2)/2D$ $c^2t_{33}^2 = 4D^2 + 2(17X^2 - 17Y^2)$ $t_{00}^2 - t_{33}^2 = 2(15X^2 - 15Y^2)/c^2$ $t_{c,33}^2 = t_{33}^2 - 2(3X/c)^2 - 2(3Y/c)^2$ $t_{00}^2 - t_{c,33}^2 = 48(X/c)^2 - 12(Y/c)^2$
$v_{22}(t)$ $d_{22} = D + [(2X)^2 + (3Y)^2]/2D$ $ct_{22} = d_{00} + d_{22} = 2D + (20X^2 + 10Y^2)/2D$ $c^2t_{22}^2 = 4D^2 + 2(20X^2 + 10Y^2)$ $t_{00}^2 - t_{22}^2 = 2(12X^2 - 8Y^2)/c^2$ $t_{c,22}^2 = t_{22}^2 - 2(2X/c)^2 - 2(2Y/c)^2$ $t_{00}^2 - t_{c,22}^2 = 32(X/c)^2 - 8(Y/c)^2$	$v_{32}(t)$ $d_{32} = D + [X^2 + (3Y)^2]/2D$ $ct_{32} = d_{00} + d_{32} = 2D + (17X^2 + 10Y^2)/2D$ $c^2t_{32}^2 = 4D^2 + 2(17X^2 + 10Y^2)$ $t_{00}^2 - t_{32}^2 = 2(15X^2 - 8Y^2)$ $t_{c,32}^2 = t_{32}^2 - 2(3X/c)^2 - 2(2Y/c)^2$ $t_{00}^2 - t_{c,32}^2 = 48(X/c)^2 - 8(Y/c)^2$
$v_{21}(t)$ $d_{21} = D + [(2X)^2 + (2Y)^2]/2D$ $ct_{21} = d_{00} + d_{21} = 2D + (20X^2 + 5Y^2)/2D$ $c^2t_{21}^2 = 4D^2 + 2(20X^2 + 5Y^2)$ $t_{00}^2 - t_{21}^2 = 2(12X^2 - 3Y^2)/c^2$ $t_{c,21}^2 = t_{21}^2 - 2(2X/c)^2 - 2(Y/c)^2$ $t_{00}^2 - t_{c,21}^2 = 32(X/c)^2 - 4(Y/c)^2$	$v_{31}(t)$ $d_{31} = D + [X^2 + (2Y)^2]/2D$ $ct_{31} = d_{00} + d_{31} = 2D + (17X^2 + 5Y^2)/2D$ $c^2t_{31}^2 = 4D^2 + 2(17X^2 + 5Y^2)$ $t_{00}^2 - t_{31}^2 = 2(15X^2 - 3Y^2)/c^2$ $t_{c,31}^2 = t_{31}^2 - 2(3X/c)^2 - 2(Y/c)^2$ $t_{00}^2 - t_{c,31}^2 = 48(X/c)^2 - 4(Y/c)^2$
$v_{20}(t)$ $d_{20} = D + [(2X)^2 + Y^2]/2D$ $ct_{20} = d_{00} + d_{20} = 2D + (20X^2 + 2Y^2)/2D$ $c^2t_{20}^2 = 4D^2 + 2(20X^2 + 2Y^2)$ $t_{00}^2 - t_{20}^2 = 2(12X^2 + 0Y^2)/c^2$ $t_{c,20}^2 = t_{20}^2 - 2(2X/c)^2 - 2(0Y/c)^2$ $t_{00}^2 - t_{c,20}^2 = 32(X/c)^2 - 0(Y/c)^2$	$v_{30}(t)$ $d_{30} = D + (X^2 + Y^2)/2D$ $ct_{30} = d_{00} + d_{30} = 2D + (17X^2 + 2Y^2)/2D$ $c^2t_{30}^2 = 4D^2 + 2(17X^2 + 2Y^2)$ $t_{00}^2 - t_{30}^2 = 2(15X^2 + 0Y^2)$ $t_{c,30}^2 = t_{30}^2 - 2(3X/c)^2 - 2(0Y)^2$ $t_{00}^2 - t_{c,30}^2 = 48(X/c)^2 - 0(Y/c)^2$

$$t_{00}^2 - [t_{kl}^2 - 2(kX/c)^2] \doteqdot 4kXx/c^2 + 2(2lY - l^2Y^2)/c^2 \tag{37}$$

$$t_{00}^2 - [t_{kl}^2 - 2(kX/c)^2 - 2(lY/c)^2] \doteqdot 4kXx/c^2 + 4lYy/c^2 \tag{38}$$

$$t_{00}^2 - [t_{kl}^2 - 2(kX/c)^2 - 2(lY/c)^2 + 4(x/c^2)kX] \doteqdot 4lYy/c^2 \tag{39}$$

$$t_{00}^2 - [t_{kl}^2 - 2(kX/c)^2 - 2(lY/c)^2 + 4(x/c^2)kX + 4(y/c^2)lY] \doteqdot 0 \tag{40}$$

Equation (37) corrects t_{kl}^2 for the column kX in which the sensor is positioned, while Eq. (38) corrects for the row lY. For easier writing, we define the corrected time $t_{c,kl}$:

$$t_{c,kl}^2 = t_{kl}^2 - 2(kX/c)^2 - 2(lY/c)^2 \tag{41}$$

a

	$\times X^2/2cD$	
$v_{00}(t)$	34	t_{00}
$v_{10}(t)$	27	t_{10}
$v_{20}(t)$	22	t_{20}
$v_{30}(t)$	19	t_{30}
$v_{01}(t)$	37	t_{01}
$v_{11}(t)$	30	t_{11}
$v_{21}(t)$	25	t_{21}
$v_{31}(t)$	22	t_{31}
$v_{02}(t)$	42	t_{02}
$v_{12}(t)$	35	t_{12}
$v_{22}(t)$	30	t_{22}
$v_{32}(t)$	27	t_{32}
$v_{03}(t)$	49	t_{03}
$v_{13}(t)$	42	t_{13}
$v_{23}(t)$	37	t_{23}
$v_{33}(t)$	34	t_{33}

0 $t \rightarrow$ $2D/c$ t_{00}

c

	$t_{0l}^2 - t_{kl}^2 + 2(kX/c)^2 = (X/c)^2 \times$
$v_{00}(t^2)$	t_{00}^2
$v_{10}[t^2+2(X/c)^2]$	16
$v_{20}[t^2+2(2X/c)^2]$	32
$v_{30}[t^2+2(3X/c)^2]$	48
$v_{01}(t^2)$	t_{01}^2
$v_{11}[t^2+2(X/c)^2]$	16
$v_{21}[t^2+2(2X/c)^2]$	32
$v_{31}[t^2+2(3X/c)^2]$	48
$v_{02}(t^2)$	t_{02}^2
$v_{12}[t^2+2(X/c)^2]$	16
$v_{22}[t^2+2(2X/c)^2]$	32
$v_{32}[t^2+2(3X/c)^2]$	48
$v_{03}(t^2)$	t_{03}^2
$v_{13}[t^2+2(X/c)^2]$	16
$v_{23}[t^2+2(2X/c)^2]$	32
$v_{33}[t^2+2(3X/c)^2]$	48

0 $t^2 \rightarrow$ $(2D/c)^2$ t_{00}^2

b

	$\times X^2/c^2$	$t_{00}^2 - t_{kl}^2 = (X/c)^2 \times$
$v_{00}(t^2)$	68	0
$v_{10}(t^2)$	54	14
$v_{20}(t^2)$	44	24
$v_{30}(t^2)$	38	30
$v_{01}(t^2)$	74	-6
$v_{11}(t^2)$	60	8
$v_{21}(t^2)$	50	18
$v_{31}(t^2)$	44	24
$v_{02}(t^2)$	84	-16
$v_{12}(t^2)$	70	-2
$v_{22}(t^2)$	60	8
$v_{32}(t^2)$	54	14
$v_{03}(t^2)$	98	-30
$v_{13}(t^2)$	84	-16
$v_{23}(t^2)$	74	-6
$v_{33}(t^2)$	68	0

0 $(2D/c)^2$ $t^2 \rightarrow$ t_{00}^2

d

	$t_{00}^2 - t_{c,kl}^2$
$v_{00}(t^2)$	t_{00}^2
$v_{10}[t^2+2(X/c)^2]$	$16(X/c)^2$
$v_{20}[t^2+2(2X/c)^2]$	$32(X/c)^2$
$v_{30}[t^2+2(3X/c)^2]$	$48(X/c)^2$
$v_{01}[t^2+2(Y/c)^2]$	$4(Y/c)^2$
$v_{11}[t^2+2(X/c)^2+2(Y/c)^2]$	$16(X/c)^2$
$v_{21}[t^2+2(2X/c)^2+2(Y/c)^2]$	$32(X/c)^2$
$v_{31}[t^2+2(3X/c)^2+2(Y/c)^2]$	$48(X/c)^2$
$v_{02}[t^2+2(2Y/c)^2]$	$8(Y/c)^2$
$v_{12}[t^2+2(X/c)^2+2(2Y/c)^2]$	$16(X/c)^2$
$v_{22}[t^2+2(2X/c)^2+2(2Y/c)^2]$	$32(X/c)^2$
$v_{32}[t^2+2(3X/c)^2+2(2Y/c)^2]$	$48(X/c)^2$
$v_{03}[t^2+2(3Y/c)^2]$	$12(Y/c)^2$
$v_{13}[t^2+2(X/c)^2+2(3Y/c)^2]$	$16(X/c)^2$
$v_{23}[t^2+2(2X/c)^2+2(3Y/c)^2]$	$32(X/c)^2$
$v_{33}[t^2+2(3X/c)^2+2(3Y/c)^2]$	$48(X/c)^2$

0 $(2D/c)^2$ $t^2 \rightarrow$ t_{00}^2

FIG. 5.2-14. Time diagram of the processing of the 16 voltages $v_{00}(t)$ to $v_{33}(t)$ of Fig. 5.2-13 received by the stationary radar with two-dimensional synthetic aperture.

Equation (39) shifts the received voltages according to the x coordinate of the target, and Eq. (40) shifts them according to the y coordinate.

Let us again show the data processing by means of an example. Figure 5.2-13 shows an array with 4×4 sensors and a target located at $x = 4X$, $y = -Y$, $z = D$. For simplicity, we assume that X and Y have the same magnitude. Since there is not enough space in Fig. 5.2-13 to show d_{kl}, ct_{kl}, and $c^2t_{kl}^2$ in analogy to Fig. 5.2-10, we list these quantities in Table 5.2-1, ordered by the voltages $v_{00}(t)$ to $v_{33}(t)$. Also shown are $t_{00}^2 - t_{kl}^2$, $t_{c,kl}^2$, and $t_{00}^2 - t_{c,kl}^2$.

Figure 5.2-14 is the time diagram for the two-dimensional array. It starts with the 16 voltages $v_{00}(t)$ to $v_{33}(t)$ of Fig. 5.2-14a. The pulses produced by the target TA in Fig. 5.2-13 and the respective travel times t_{00} to t_{33} are shown; also shown are the times $t_{00} - 2D/c = 34X^2/2cD$, $t_{10} - 2D/c = 27X^2/2cD$, ...

The voltages $v_{00}(t^2)$ to $v_{33}(t^2)$ are shown in Fig. 5.2-14b. The differences $t_{00}^2 - (2D/c)^2 = 68(X/c)^2$, $t_{10}^2 - (2D/c)^2 = 54(X/c)^2$, ... are listed as well as the differences $t_{00}^2 - t_{10}^2 = 14(X/c)^2$, $t_{00} - t_{20}^2 = 24(X/c)^2$, ... Figs. 5.2-14a and b are in complete analogy to Figs. 5.2-11a and b.

Figure 5.2-14c shows the correction of the x-shift $2(kX/c)^2$ according to Eq. (37), which is analogous to Fig. 5.2-11c. The differences $t_{00}^2 - t_{10}^2 + 2(kX/c)^2 = 16(X/c)^2$, $t_{00}^2 - t_{20}^2 + 2(kX/c)^2 = 32(X/c)^2$, ... are shown. Note that the first four voltages $v_{00}(t^2)$ to $v_{30}[t^2 + 2(3X/c)^2]$ are corrected relative to $v_{00}(t^2)$; the next four voltages $v_{01}(t^2)$ to $v_{31}[t^2 + 2(3X/c)^2]$ are corrected relative to $v_{01}(t^2)$; etc.

The correction of the y-shift according to Eq. (38) is shown in Fig. 5.2-14d. There is no analog for this diagram in Fig. 5.2-11. The differences $t_{00}^2 - t_{c,kl}^2$ are shown; for $l = 0$ they are identical to the shifts in Fig. 5.2-14c, but this is not so for $l = 1, 2, 3$. The additional shifts $4(Y/c)^2$ for $l = 1$, $8(Y/c)^2$ for $l = 2$, and $12(Y/c)^2$ for $l = 3$ are introduced.

The time diagrams for the implementation of Eqs. (39) and (40) are not shown. In analogy to Fig. 5.2-11d, one must introduce the shift $-4(x/c^2)kX$ for each coordinate x for which one wants to produce an image, to implement Eq. (39). Then one has to introduce the additional shift $-4(y/c^2)lY$ for each coordinate y for which one wants to produce an image, to implement Eq. (40).

Let us look at the effort required to produce images with a two-dimensional synthetic aperture radar. If we want to resolve n values of the x coordinate and m values of the y coordinate, we have to produce a number of time shifts that is proportionate to nm. Hence, the data-processing effort increases like n^2 for $n = m$, or quadratically. One would not expect to be able to do better than this. The advance in data-processing equipment is one of the most conspicuous developments of our time, and a quadratic increase in the data-processing effort causes no concern. The situation is very different for the number of sensors. Fig. 5.2-13 shows 4^2 sensors while Fig. 5.2-10 shows only 4 sensors. However, a comparison of Fig. 5.2-14d with Fig. 5.2-11c shows that the dynamic range has increased from 4:1 to 16:1. A short reflection reveals that we need the same number of sensors for a one-dimensional and a two-dimensional array, if we want the same dynamic range in both cases. A one-dimensional array with 100 sensors does not become a two-dimensional array with 10,000 sensors, but an array with 100 sensors positioned in a square pattern.

5.2.6 *Resolution of Pulses*

Equation (16) gives the error Δx_1 of the x coordinate of a point x_1, y_1 as a function of the error ΔX of the x coordinate of the two radars, while Eq. (14) gives the error Δx_1 as a function of the errors Δd_{11} and Δd_{21} of the measured distances d_{11} and d_{21}. The error ΔX can be made very small for a stationary radar. The errors Δd_{11} and Δd_{21} are the more interesting limitations on the achievable error Δx_1.

Let us assume that a rectangular pulse is used for the distance measurement; this may be either an inherently single pulse or a single pulse produced by pulse compression. The edges will not be infinitely steep since the frequency bandwidth of the equipment will be finite. Let us assume that the frequency response of the radar equipment can be represented by an idealized low-pass filter with bandwidth Δf. A step function with amplitude A and infinitesimally short transient time from 0 to A will have the time variation of the sine integral[1] shown in Fig. 5.2-15 after passing

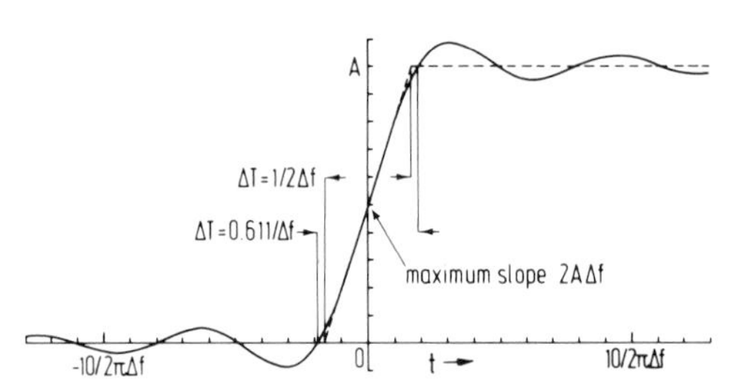

FIG. 5.2-15. Transient of a step function with amplitude A after passing through a low-pass filter with nominal bandwidth Δf, approximated by the sine integral (solid line) and by a linear transient with transient time $\Delta T = 1/2\,\Delta f$ (dashed line).

[1] The theoretical difficulties of using the sine integral and the practical way to overcome them are well known to any student of communications engineering (Close, 1966, p. 479). It is usual to say that a function cannot be time- and frequency-limited; the assumption of a finite bandwidth Δf thus creates the unrealistic infinitely long transition from 0 to A. Another point of view is that one needs a voltage source, a switch, and a filter to produce a transient like the one in Fig. 5.2-15; the switch makes this a linear, time-variable circuit while a description in terms of sinusoidal functions calls for a linear, time-invariant circuit. Either way, the use of the sine integral calls mathematically for an infinite delay that can be approximated practically quite well by a finite delay.

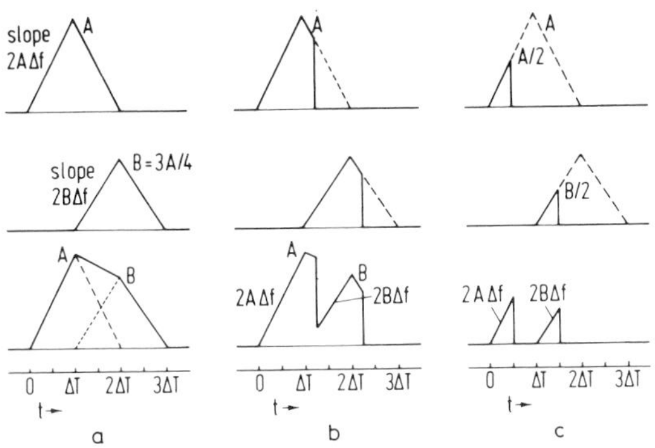

FIG. 5.2-16. (a) Rectangular pulses after passing through an idealized low-pass filter with transient time $\Delta T = 1/2\ \Delta f$ according to Fig. 5.2-15. The columns b and c show the suppression of the pulses after determination of their amplitude.

through such a filter. Following common practice, we use the simpler linear transient with transition time $\Delta T = 1/2\ \Delta f$ shown by the dashed line. The slope of the linear transient and the maximum slope of the sine integral have the same value $2A\ \Delta f$.

Let us use the approximation of the linear transient for the rectangular pulse. If the pulse has the duration ΔT and the amplitude A, it will be transformed by an idealized low-pass filter with bandwidth $\Delta f = 1/2\ \Delta T$ into the triangular pulse with amplitude A shown in the top part of Fig. 5.2-16a. A second pulse with amplitude $3A/4$, delayed by ΔT, is shown below. The sum of both pulses is shown in the third row. Sampling at the times ΔT and $2\ \Delta T$ yields the amplitudes A and $3A/4$. This is the basis for the usual statement that a bandwidth Δf permits a time resolution $\Delta T = 1/2\ \Delta f$ and a distance resolution $\Delta d = c\ \Delta T/2 = c/4\ \Delta f$; the factor 1/2 in the last equation is due to the fact that the radar signal has to travel twice the distance to the target.

Let us look at column b of Fig. 5.2-16. Once we know the position $t = \Delta T$ and the amplitude A of the triangular pulse in the first row, we do not need the rest or the trailing part of the pulse. We can calculate that part of the pulse and subtract it from the received pulse. The resulting truncated pulse is shown by the solid line. The pulse in the second row with amplitude $3A/4$ is truncated in the same way. It would be difficult to do this truncation with analog circuits. However, synthetic aperture radar images are routinely produced by computer processing, usually not directly but from data stored on magnetic tape, and the pulse truncation discussed here is not a difficult process for a computer.

The sum of the two truncated pulses is shown in the third row of Fig. 5.2-16b. This sum looks better than the one in column a, but one actually gains nothing. The amplitude measured at the time $t = \Delta T$ would be the amplitude A of the first pulse plus a certain fraction of the amplitude $3A/4$ of the second pulse, if the pulses were not at least the time ΔT apart. Hence, a wrong amplitude would be sampled for a spacing between the pulses shorter than ΔT, and the use of this wrong amplitude for the truncation of the first pulse would also modify the amplitude of the second pulse.

In order to achieve a better resolution than $\Delta T = 1/2\,\Delta f$, we must obtain the amplitudes A and $3A/4$ of the pulses before they have reached their peak values. Since the slope of a pulse with amplitude A equals $2A\,\Delta f$, and the bandwidth Δf is known, we can derive A from the slope. In the absence of noise, one obtains the slope from the difference of two samples taken closely together. The knowledge of the slope $2A\,\Delta f$ gives us the amplitude A of the pulse, while the knowledge of the time and the amplitudes of the samples taken gives us the time $t = \Delta T$ where the peak of the triangular pulse is located. With this knowledge we can calculate the instantaneous amplitudes of the triangular pulse at any time t, and subtract them from the received instantaneous amplitudes. The results are the truncated pulses in the first two rows of Fig. 5.2-16c. Their sum in the third row shows that a resolution time $\Delta T'$ smaller than $\Delta T = 1/2\,\Delta f$ can be achieved. The minimum value of $\Delta T'$ is no longer determined by the bandwidth Δf, but by statistical disturbances of the signals and the imperfections inherent in practical equipment.

Let us assume a sample with amplitude A_1 is obtained at the time t_1 and a sample with amplitude $A_1 + \Delta A$ at the time $t_1 + \Delta T'$. The slope has the value

$$[(A_1 + \Delta A) - A_1]/\Delta T' = \Delta A/\Delta T' = 2A\,\Delta f \tag{42}$$

and the amplitude A of the received pulse equals

$$A = \Delta A/2\,\Delta T'\,\Delta f \tag{43}$$

The values of $\Delta T'$ and Δf are fixed and are not subject to statistical disturbances. Hence, the error of A caused by statistical disturbances is strictly due to ΔA. Let the statistical disturbances be additive thermal noise. The signal-to-noise ratio of ΔA is $20 \log(\Delta A/A)$ dB lower than that of A. In order to obtain A with the same statistical error from Eq. (43) as from sampling the peak values of the triangular pulses in Fig. 5.2-16a, one must increase the signal power by $20 \log(A/\Delta A)$ dB.

Let us next investigate the precision with which the time of arrival of a pulse can be determined. Figure 5.2-16a only shows that we can measure the amplitudes A and $3A/4$ independently, if we know that A occurs at $t = \Delta T$ and $3A/4$ at $t = 2\,\Delta T$, but nothing is said about where this knowledge of the sampling times came from.

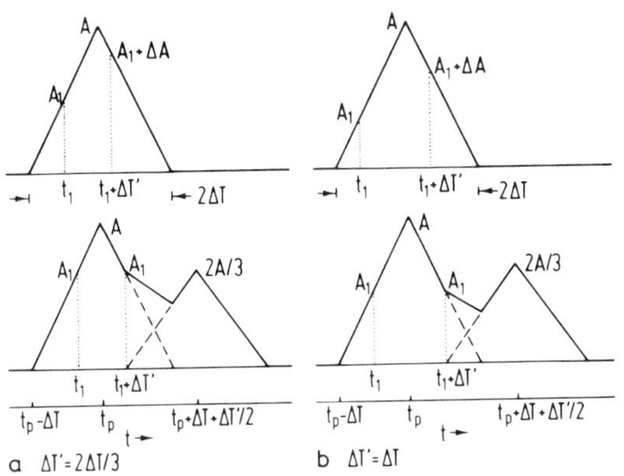

FIG. 5.2-17. Determination of the time t_p of the peak amplitude of a triangular pulse by a vanishing difference ΔA of two sampled amplitudes A_1 and $A_1 + \Delta A$.

The top of Fig. 5.2-17a shows how the time t_p of the peak of a triangular pulse can be determined. A sample A_1 is taken at the time t_1 and another sample $A_1 + \Delta A$ at the time $t_1 + \Delta T'$. We obtain the arrival time t_p of the peak when ΔA equals zero:

$$t_p = t_1 + \Delta T'/2 \qquad \text{for} \quad (A_1 + \Delta A) - A_1 = \Delta A = 0 \tag{44}$$

Instead of using the difference $(A_1 + \Delta A) - A_1$, one may also use the quotient $(A_1 + \Delta A)/A_1$:

$$(A_1 + \Delta A)/A_1 = 1 \qquad \text{or} \qquad \Delta A/A_1 = 0 \tag{45}$$

In either case, the minimum distance between two adjacent pulses is no longer ΔT, as in Fig. 5.2-16a, but $\Delta T + \Delta T'/2$, as one may readily see from the bottom of Fig. 5.2-17a. In the absence of noise, one may make $\Delta T'$ very small, but one will hesitate to do so when the signals are corrupted by noise. To see what determines the choice of $\Delta T'$, we plot ΔA as a function of the difference between sampling time t_1 and pulse peak time t_p for various values of $\Delta T'$ in Fig. 5.2-18; the plotting is done for $(t_1 - \Delta T') - (t_p - \Delta T')$ rather than to $t_1 - t_p$ in order to let all plots start at zero.

The rising slope of all four plots shown equals $A/\Delta T$; it becomes zero at $\Delta A = A\,\Delta T'/\Delta T$ and stays so till the time ΔT is reached. Then it follows a falling slope $-2A/\Delta T$, equal for all plots with $\Delta T' \leqq \Delta T$; the more complicated plot for $\Delta T' > \Delta T$ is not of interest here.

In the presence of additive thermal noise, we are interested in a long and steep negative slope. For instance, a noise amplitude within the limits $+A/3$ and $-A/3$ superimposed on the sampled amplitude $A_1 + \Delta A$ can make the solid line $(\Delta T' = \Delta T)$ in Fig. 5.2-18 zero anywhere in the interval

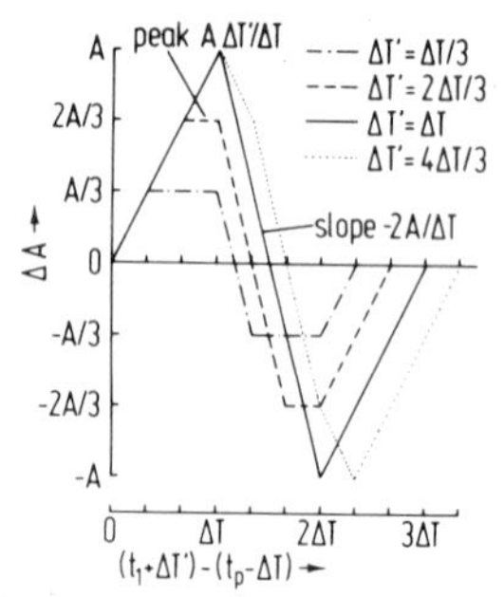

FIG. 5.2-18. Variation of ΔA in Fig. 5.2-17 as a function of the sampling times t_1 and $t_1 + \Delta T$ relative to the time $t_p - \Delta T$ of the beginning of the pulse for various values of ΔT. (dot–dash line) $\Delta T' = \Delta T/3$; (dashed line) $\Delta T' = 2\,\Delta T/3$; (solid line) $\Delta T' = \Delta T$; (dotted line) $\Delta T' = 4\,\Delta T/3$.

$$4\,\Delta T/3 \le (t_1 + \Delta T') - (t_p - \Delta T) \le 5\,\Delta T/3$$

but the dash–dot line ($\Delta T' = T/3$) can be made zero anywhere in the much larger interval

$$0 \le (t_1 + \Delta T') - (t_p - \Delta T) \le 7\,\Delta T/3$$

A value $\Delta T' > \Delta T$ does not yield either a steeper slope or a longer one in Fig. 5.2-18; it increases the required distance between two resolvable pulses in Fig. 5.2-17 without giving any advantage in return. Hence, $\Delta T'$ should vary from almost zero at large signal-to-noise ratios to ΔT at small signal-to-noise ratios.

In a practical case, one generally tries to use $\Delta T' = \Delta T$. This choice implies that the middle of the rising and falling edges of a pulse are used to measure its time of arrival. According to Fig. 5.2-15, this point ($t = 0$) is least affected by the ringing at the beginning and end of the transient; experience shows that this situation holds generally, not only for the idealized sine integral transient.

Let us turn to the determination of the arrival time t_p of the pulse peak if the pulse is truncated according to Fig. 5.2-16. From Fig. 5.2-19 and Eq. (42), one may readily recognize the following relation:

$$(A - A_1)/(t_p - t_1) = 2A\,\Delta f = \Delta A/\Delta T' \tag{46}$$

Reordering of the terms yields t_p:

$$t_p = t_1 + 1/2\Delta f - A_1/(\Delta A/\Delta T') \tag{47}$$

The values of t_1, Δf, and $\Delta T'$ are not subject to statistical disturbances, but the amplitude A_1 and the slope $\Delta A/\Delta T'$ are. With the notation

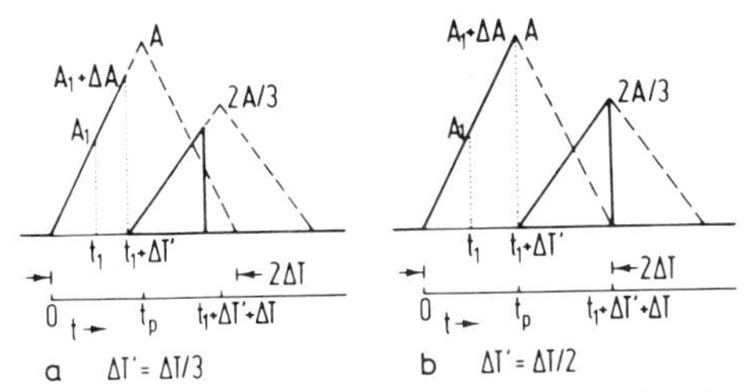

FIG. 5.2-19. Determination of the time t_p of the peak amplitude of a triangular pulse by a vanishing difference $\Delta A - A_1$ of two sampled amplitudes A_1 and $A_1 + \Delta A$.

$A_1/\Delta A = q$, we may rewrite Eq. (47) in the form of Eqs. (44) and (45):

$$t_p = t_1 + 1/2\Delta f - q\,\Delta T' \quad \text{for} \quad A_1/\Delta A = q \quad \text{or} \quad q\,\Delta A - A_1 = 0 \quad (48)$$

We chose $q = 1$ to simplify the investigation. The calculation of the best value of q for a certain model of noise is rather difficult.[1]

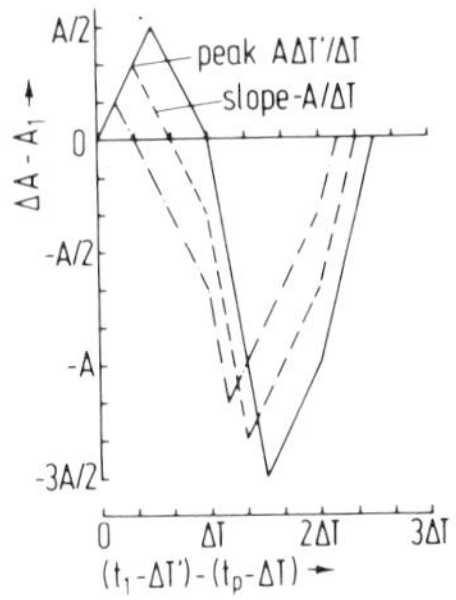

FIG. 5.2-20. Variation of $\Delta A - A_1$ in Fig. 5.2-19 as a function of the sampling times t_1 and $t_1 + \Delta T'$ relative to the time $t_p - \Delta T$ of the beginning of the pulse for various values of $\Delta T'$. (solid line) $\Delta T' = \Delta T/2$; (dashed line) $\Delta T' = \Delta T/3$; (dash–dot line) $\Delta T' = \Delta T/6$.

[1] Let the noise samples a_1 and a_2 be added to the sampled signal amplitudes A_1 and $A_1 + \Delta A$ to yield $A_1 + a_1$ and $A_1 + \Delta A + a_2$. The difference $q\,\Delta A - A_1$ in Eq. (48) yields $q(\Delta A - a_1 + a_2) - (A_1 + a_1)$, which equals $\Delta A - A_1 + a_2 - 2a_1$ for $q = 1$. The difference ΔA in Eq. (44) becomes $(A_1 + \Delta A + a_2) - (A_1 + a_1) = \Delta A + a_2 - a_1$. If a_1 and a_2 are statistically independent, a decision based on the equation $\Delta A + a_2 - a_1 = 0$ is less affected by noise than a decision based on the equation $\Delta A - A_1 + a_2 - 2a_1 = 0$. However, the assumption for Eq. (48) is that $\Delta T'$ is smaller than ΔT, and this implies that a_1 and a_2 are not statistically independent. Hence, the best choice of q and the effect of noise on the decision based on Eqs. (44) and (48) depend not only on the noise model and filter but also on $\Delta T'$.

The minimum distance between two adjacent pulses equals $t_1 + \Delta T'$ according to Fig. 5.2-19. For $q = 1$, one obtains $t_1 = \Delta T'$, and the minimum distance between two pulses becomes $2\,\Delta T'$. Since $2\,\Delta T'$ can be made arbitrarily small in the absence of noise, we must determine how noise influences the choice of $\Delta T'$. We plot, in Fig. 5.2-20, the difference $\Delta A - A_1$ as function of the difference between sampling time t_1 and pulse peak time t_p for various values of $\Delta T'$; as before, the plotting is done for $(t_1 - \Delta T') - (t_p - \Delta T)$ rather than for $t_1 - t_p$ in order to let all plots start at zero.

The initial rising slope of all plots in Fig. 5.2-20 equals $A/\Delta T$; the plots reach the peak $A\,\Delta T'/\Delta T$, and drop then with slope $-A/\Delta T$ to zero. For negative values of $\Delta A - A_1$, the slope is either $-A/\Delta T$ or steeper. The negative peak value is also absolutely larger than the positive one.

Let us compare the plots of Fig. 5.2-20 with the plots for $\Delta T' = \Delta T$ in Fig. 5.2-18, which had been found to be the best in the presence of thermal noise. The minimum time between two resolvable pulses in Fig. 5.2-18—determined by $\Delta A = 0$—equals $\Delta T + \Delta T'/2 = 3\,\Delta T/2$ for $\Delta T' = \Delta T$, while this time in Fig. 5.2-20—determined by $\Delta A - A_1 = 0$—equals $2\,\Delta T' \leqq \Delta T$. The ratio of the minimum distance of resolvable pulses is denoted ξ:

$$\xi = 2\,\Delta T'/(3\,\Delta T/2) = 4\,\Delta T'/3\,\Delta T \tag{49}$$

To determine the price to be paid in terms of signal power for this improved time resolution, we consider two cases. For a large signal-to-noise ratio, the slope around $\Delta A = 0$ in Fig. 5.2-18 and around $\Delta A - A_1 = 0$ in Fig. 5.2-20 will be the important quantity; note that we do not assume more about the noise than that the signal-to-noise ratio will be large. The slope in Fig. 5.2-20 equals $-A/\Delta T$ while the slope in Fig. 5.2-18 equals $-2A/\Delta T$. Doubling the pulse amplitude A and thus increasing the signal power by 3 dB makes the slope in Fig. 5.2-20 as large[1] as in Fig. 5.2-18.

If the signal-to-noise ratio is not large, we will want to make the peak $A\,\Delta T'/\Delta T$ in Fig. 5.2-20 equal to the peak A of the plots for $\Delta T' = \Delta T$ in Fig. 5.2-18. This requires that the amplitude A of the pulses in Fig. 5.2-19 be increased to $A\,\Delta T/\Delta T'$. The slope $-A/\Delta T$ in Fig. 5.2-20 becomes in this case $-A/\Delta T'$, which means it is steeper than in Fig. 5.2-18 for $\Delta T' <$

[1] The equality of the slope only implies that the influence of noise is approximately the same. The noise samples a_1 and a_2 in the equation $\Delta A + a_2 - a_1 = 0$ in the previous footnote are statistically independent, if we assume thermal noise, due to the assumption $\Delta T' = \Delta T$ for this case. On the other hand, the samples a_1 and a_2 neither occur in the form $a_2 - a_1$ in the equation $\Delta A - A_1 + a_2 - 2a_1 = 0$ nor are they statistically independent due to the relation $\Delta T' \leqq \Delta T/2$.

$\Delta T/2$; hence, the timing error will actually be smaller for the curves of Fig. 5.2-20 than for the curve $\Delta T' = \Delta T$ in Fig. 5.2-18. We introduce the amplitude ratio η,

$$\eta = (A\,\Delta T/\Delta T')/A = \Delta T/\Delta T' \tag{50}$$

and the power ratio η^2,

$$\eta^2 = (\Delta T/\Delta T')^2 \tag{51}$$

The product of the ratio ξ of the minimum distance of resolvable pulses and the amplitude ratio η is constant:

$$\xi\eta = (4\,\Delta T'/3\,\Delta T)(\Delta T/\Delta T') = \tfrac{4}{3} \tag{52}$$

In readily understandable terminology, we have a linear interchange between resolution time and pulse amplitude, and a quadratic interchange between resolution time and "pulse power," where pulse power stands either for peak power or average power of the pulse.

Let us return to Eq. (14) which gives the error Δx_1 of the x coordinate as a function of the distance errors Δd_{11} and Δd_{21}. The minimum time distance $2\,\Delta T'$ of two resolved pulses yields the bounds $\pm\Delta d = \pm c\,\Delta T'/2$ of the radar range error. The bounds for the error of the x coordinate follow from Eq. (14):

$$-c\,\Delta T'(d_{11} + d_{21})/4X \leq \Delta x_1 \leq c\,\Delta T'(d_{11} + d_{21})/4X \tag{53}$$

If one makes many measurements, one obtains an average error that is zero, and an rms error that is generally substantially smaller than the bounds of Eq. (53). We do not discuss this improvement of the resolution by statistical methods since a very sophisticated model of the causes of the errors is required to make such an investigation worthwhile. We may recognize this fact from the following example.

Let the distance[1] $2X$ between two radar receivers in Eq. (53) be 2500 km, while the two distances d_{11} and d_{21} shall be 5000 km each. The factor $(d_{11} + d_{21})/4X$ then equals 2. A bandwidth of 2.5 MHz can be achieved for an over-the-horizon radar.[2] For $\Delta f = 2.5$ MHz, we obtain $\Delta T = 200$ ns from Fig. 5.2-15. A realistic value of $\Delta T' = T/5 = 40$ ns follows from Fig. 5.2-20. The bounds for the error Δx_1 of Eq. (53) yield

$$-24\text{ m} \leqq \Delta x_1 \leqq +24\text{ m}$$

[1] This is approximately the distance between the DEW-line stations at Ft. Yukon, Alaska, and Thule, Greenland.

[2] This bandwidth cannot be achieved with a technology based on a small relative bandwidth, since a baseband bandwidth of 2.5 MHz and a carrier frequency of 25 MHz yield a relative bandwidth 0.1. However, over-the-horizon radar using the concepts of Section 2.3 can use such a large bandwidth.

Statistical averaging would reduce this worst case error significantly. Hence, the limitation for the resolution is no longer the aperture of the sensor array; it is replaced by statistical phenomena, particularly fluctuations of the ionosphere.

5.2.7 Angular Resolution with Finite Bandwidth

The beam pattern or diffraction pattern of a line array of receptors with length[1] L receiving a planar wavefront of sinusoidal waves with wavelength λ is given by the equation

$$v(\beta) = \frac{\sin \pi L\beta/\lambda}{\pi L\beta/\lambda} \tag{54}$$

The first zero crossing of this function is at

$$\beta = \epsilon = \lambda/L = cT/L = c/fL, \qquad \lambda = cT = c/f \tag{55}$$

This is the classical resolution angle for the line array. A circular array or a lens with diameter L yields the following diffraction pattern $v_c(\beta)$ instead of Eq. (54),

$$v_c(\beta) = \frac{2J_1(\pi L\beta/\lambda)}{\pi L\beta/\lambda} \tag{56}$$

where J_1 is a Bessel function. The first zero crossing of $v_c(\beta)$ occurs at

$$\beta = \epsilon \doteqdot 1.22\lambda/L \tag{57}$$

which is a slightly larger value than for the line array.

The classical resolution angle has caused a great deal of controversy in the past, since basic principles of information theory demand that the resolution should be determined by the average power P and the bandwidth Δf of the received wave in such a way that P and Δf could be traded according to the relation:

$$\Delta f \log P = \text{constant} \tag{58}$$

Neither a bandwidth, in the form of a bandwidth $\Delta\lambda$ of wavelengths rather than a bandwidth Δf of frequencies, nor a signal power or a signal-to-noise power ratio occurs in Eqs. (55) and (57). By using pulses rather than sinusoids we have succeeded in introducing bandwidth as well as power into

---

[1] We denote the length by L and call it the aperture. Originally, an area of a lens rather than a length was called aperture but there is a trend to use this term for a length (e.g., Cutrona, 1970).

the formulas for angular resolution. We will investigate this result in more detail to derive from Eq. (52) a formula that resembles Eq. (58) more closely. Furthermore, we want to show that signals with a bandwidth resolve more points than signals with bandwidth zero, as discussed in connection with Fig. 5.1-10.

Equation (52) may be rewritten in terms of bandwidth Δf and average signal power P. Let $\Delta T = 1/2\,\Delta f$ be substituted from Fig. 5.2-15 into Eq. (49):

$$\xi = 4\,\Delta T'/3(1/2\,\Delta f) = (8/3)\Delta T'\,\Delta f \tag{59}$$

Let P_0 denote the average power of a rectangular pulse of duration T, which produces the triangular pulses with amplitude A and duration $2\,\Delta T$ in the top rows of Figs. 5.2-16 and 5.2-17; the average power of a pulse of duration $\Delta T'$ is denoted P. The amplitudes of these two rectangular pulses are $A = K\sqrt{P_0}$ and $A\,\Delta T/\Delta T' = K\sqrt{P}$, where K is a constant. We obtain from Eq. (50)

$$\eta = (P/P_0)^{1/2} \tag{60}$$

Substitution of Eqs. (59) and (60) into Eq. (52) yields

$$\Delta f\sqrt{P} = \tfrac{3}{8}\sqrt{P_0}/\Delta T' = \text{constant} \tag{61}$$

This equation states that a certain resolution time $\Delta T'$ and a certain signal-to-noise ratio represented by the reference signal power[1] P_0 can be achieved for various values of Δf and P, as long as the product $\Delta f\sqrt{P}$ remains unchanged.

Let us compare Eqs. (58) and (61). The structure is the same, but $\log P$ is replaced by $\sqrt{P}$. Since $\sqrt{P}$ increases faster than $\log P$, one will expect that Eq. (61) can still be improved, but $\Delta f\sqrt{P}$ = constant is a remarkable good exchange of power for bandwidth.

In connection with Fig. 5.1-10, we had derived the distance Δx of two resolvable points at a distance R with $\Delta x = R\epsilon = R\lambda/A$. Neither signal power nor bandwidth enter this equation. The signal power is not in it, since the resolution is limited by the diffraction pattern rather than by noise, and an increase of the signal power does not change the diffraction pattern. The bandwidth is not in the equation because it is zero.

Let $\Delta T'$ be expressed from Eq. (61),

$$\Delta T' = \tfrac{3}{8}\sqrt{P_0}/\Delta f\sqrt{P} \tag{62}$$

and inserted into Eq. (53):

[1] We have chosen P_0 arbitrarily, but it would actually be determined by the signal-to-noise ratio one wants to achieve.

$$-\tfrac{3}{8}c(\sqrt{P}_0/\Delta f\sqrt{P})(d_{11}+d_{12})/4X \leqq \Delta x_1 \leqq \tfrac{3}{8}c(\sqrt{P}_0/\Delta f\sqrt{P})(d_{11}+d_{12})/4X \qquad (63)$$

The error Δx_1 now takes the place of the distance Δx of two resolvable points. The error Δx_1 can be reduced not only by using a large "aperture" X, but also by increasing the bandwidth Δf or the signal power P. The frequency bandwidth Δf, or rather the wavelength bandwidth $c/\Delta f$, may readily be viewed as a generalization of the wavelength λ in the classical formula $\Delta x = R\epsilon = R\lambda/L$; furthermore, d_{11} and d_{12} are obvious generalizations of R. The most important term in Eq. (63) is $\sqrt{P}$ since it has no equivalent in the classical formula.

Let us return to the theorem derived at the end of Section 5.1.6, which said that the relative number of resolved points is zero for sinusoidal waves, if the wavefronts are planar and if a line array with a finite number of equally spaced sensors is used. The basis for this theorem was that one could produce only $2n + 1$ beams with $2n + 1$ sensors. The situation is now completely different. If Δx_1 is as large as δ in Fig. 5.1-10, we can maintain the ratio $\Delta x_1/\delta = 1$ for any distance R by keeping the ratio $(d_{11} + d_{21})/\sqrt{P}$ in Eq. (63) constant. This completes the analogy for imaging or beam forming of the theorem in communications, which says that information can only be transmitted at a rate larger than zero with signals that have a bandwidth larger than zero.

Let us end with a short proof that the determination of the amplitude of a sample from its slope does not lead to an ever-increasing error, if the processing is not done with unlimited precision. Assume that an amplitude determined to have the value A_0 actually has the value $(1 + q)A_0$, where $1 > q > -1$. The error qA_0 is thus introduced. At the next determination of the amplitude, which should have the value A_1, we will determine[1] $A_1 + qA_0$, and this value should actually be $(1 + q)(A_1 + qA_0)$. Hence the error is now $qA_1 + q^2A_0$. After n samples we will have the total error $qA_n + q^2A_{n-1} + \cdots + q^{n+1}A_0$. Let all amplitudes $A_0 \cdots A_n$ have the same value. We may then use the sum of the geometric series, $qA_0/(1 - q)$, for the total error. For an error of $\pm 10\%$ or less, $q = \pm 0.1$, we obtain for the total error $qA_0(1 + q)$. The single error qA_0 is about 10% and the total error $qA_0(1 + q)$ about 11%. Hence, imperfect processing does not lead to a significant error accumulation.

Let us observe that the breaking of the limit of the classical resolution of Eqs. (55) and (57) required nothing more than the application of the

[1] Generally, the error qA_0 will be somewhat smaller when the second, third, ... sample is determined. We are calculating here a worse than typical case.

principle, that the transmission of information requires signals with a bandwidth. We must terminate the topic here, even though many scientists will not be satisfied with such a brief discussion of such an important problem. However, the brevity permits the principle to remain clear, while even the most thorough analysis would fail to satisfy everybody. We will return to this topic once more in Section 5.2.15, where a formula will be obtained that is more similar to the classical limit of Eq. (55) than Eq. (63).

5.2.8 Beam Forming by Means of the Doppler Effect

The methods discussed so far permit obtainment of a resolution[1] much beyond that of the classical formula $\epsilon = \lambda/L$, provided not too many

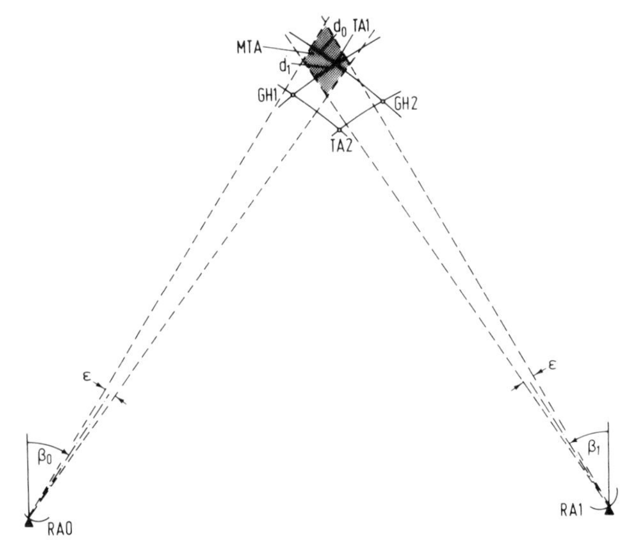

FIG. 5.2-21. Principle of ghosts in synthetic aperture radar. RA0, RA1 are radars; TA1, TA2 are targets; GH1, GH2 are ghost targets; MTA is multiple target area; d_0, d_1 are point-like targets as seen by radars RA0 and RA1.

[1] Some authors define an *effective aperture* for the synthetic aperture radar in such a way that the aperture L in the classical formula can be replaced by the effective aperture. An unfortunate by-product of this approach is that the word *effective* is sometimes ignored, which makes synthetic aperture techniques appear to yield the classical limit of resolution. The fact remains that the classical formula assumes the frequency bandwidth $\Delta f = 0$, while synthetic aperture techniques should use bandwidths $\Delta f > 0$. Any attempt to make the classical aperture and the synthetic aperture appear to be equal must thus be reducible to the definition $0 = 1$.

targets are present. Otherwise, the problem of *ghosts* is introduced. Refer to Fig. 5.2-21 for an explanation. A point-like target TA1 is shown. It will be seen by radar RA0 as a line d_0, the length of which is determined by the angular resolution ϵ, while its thickness is determined by the time resolution of the radar. The time resolution is generally much better than the angular resolution, and the ratio of width-to-length of the line d_0 in Fig. 5.2-21 reflects this fact.

In order to improve the resolution, one must primarily decrease the length of the line d_0. This can be done by means of a larger dish or antenna array for the radar RA0 or by synthetic aperture techniques. We started our investigation with the two radars in Fig. 5.2-2, and we add thus a second radar RA1 in Fig. 5.2-21. This radar sees the target TA1 as the line d_1 in Fig. 5.2-21. Since the target must be at the intersection of the lines d_0 and d_1, its location is now known with the precision of the range resolution rather than the angular resolution.

If there were only one target, one would not need any angular resolution in Fig. 5.2-21. But assume that there is a second target TA2. Without the angular resolution one would also obtain two ghosts GH1 and GH2, which are indistinguishable from the real targets. Using radar beams with a resolution angle eliminates all possible ghosts, except those located within the multiple target area MTA common to both beams. A sidelooking radar on a moving platform such as an airplane or a satellite can make measurements at the points 0, X, $2X$, ... along its course as shown in Fig. 5.2-4. In addition, one can use the Doppler effect instead of a dish or an antenna array to produce a beam with an angle ϵ. The ability to produce narrow beams without an antenna array or dish is one of the major advantages of a moving sidelooking radar over a stationary synthetic aperture radar.

Refer to Fig. 5.2-22 for the principle of beam forming by means of the Doppler effect. The radar RA emits a sinusoidal wave of frequency f. The wave returned from the target TA has the frequency f',

$$f' = f(1 + 2v_r/c) \tag{64}$$

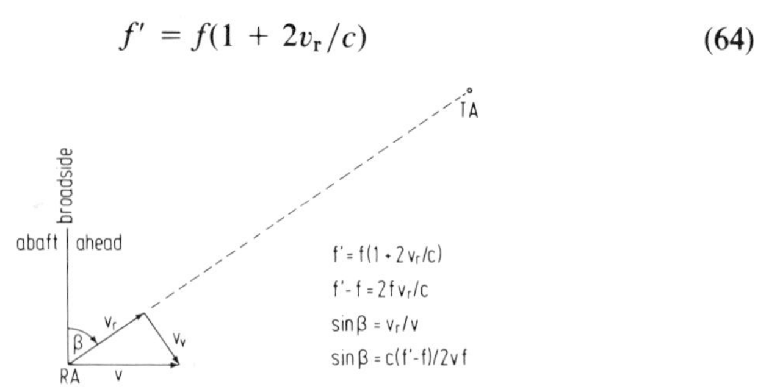

FIG. 5.2-22. Principle of beam forming by means of the frequency difference $f'-f$ due to a relative velocity v_r between the radar RA and a target TA with the azimuth β.

where v_r is the relative velocity between the radar and the target; the usual assumption is made that v_r is small compared with the velocity of light c. The frequency shift $f' - f$,

$$f' - f = 2fv_r/c \tag{65}$$

can be measured.

The angle β in Fig. 5.2-22 follows from the velocity v of the radar platform and the relative velocity v_r between radar and target:

$$\sin \beta = v_r/v \tag{66}$$

Insertion into Eq. (65) shows that the angle β can be determined from the frequency shift $f' - f$

$$\sin \beta = c(f' - f)/2vf \tag{67}$$

Hence, for every value of the Doppler shift, we obtain a different angle. For $f' - f > 0$, the target is ahead (of beam) of the radar; for $f' - f = 0$, the target is abeam or broadside to the radar; and, for $f' - f < 0$, the target is abaft (of beam).

For a more accurate description of beam forming by the Doppler shift, one must take into consideration that a radar does not radiate a periodic sinusoidal function but a pulse with m sinusoidal cycles of period T each, as shown in Fig. 5.2-1e. The pulse $\sqrt{2}\,\sin 2\pi t/T$, $-mT/2 \leqq t \leqq mT/2$, is radiated, and it comes back as the pulse $\sqrt{2}\,\sin[2\pi(t - t_0)/T(1 - 2v_r/c)]$, $-mT(1 - 2v_r/c)/2 \leqq t - t_0 \leqq mT(1 - 2v_r/c)/2$. In order to determine the relative velocity v_r, we multiply the returned pulse with periodic sine and cosine functions having the period $\tau = T$ and a phase shift α; the products are integrated[1]

$$K_c\left(\frac{v_r}{c}\right) = \frac{2}{mT(1 - 2v_r/c)} \int_{-\infty}^{\infty} \sin \frac{2\pi t'}{T(1 - 2v_r/c)} \sin\left(\frac{2\pi t'}{T} - \alpha\right) dt' \tag{68}$$

$$K_s\left(\frac{v_r}{c}\right) = \frac{2}{mT(1 - 2v_r/c)} \int_{-\infty}^{\infty} \sin \frac{2\pi t'}{T(1 - 2v_r/c)} \cos\left(\frac{2\pi t'}{T} - \alpha\right) dt' \tag{69}$$

$$t' = t - t_0$$

[1] This is in essence the cross-correlation function in the range-Doppler domain for a known distance $ct_0/2$ to the target (Harmuth, 1977a, pp. 332–340). For a discussion of optimal detection based on correlation of the in-phase and quadrature components, see, e.g., Sakrison (1970, Chapter 3).

The pulse $\sqrt{2}\,\sin[2\pi t'/T(1 - 2v_r/c)]$ is zero outside the interval $-mT(1 - 2v_r/c)/2 \leqq t' \leqq +mT(1 - 2v_r/c)/2$. The integrals have to be taken only over this interval. The terms of higher than first order in v_r/c may be ignored, since the Doppler effect in the form of Eq. (64) is only correct in first order of v_r/c. The following highly accurate approximations are obtained for $K_c(v_r/c)$ and $K_s(v_r/c)$:

$$K_c\left(\frac{v_r}{c}\right) = \frac{\sin 2\pi m v_r/c}{2\pi m v_r/c}\left(1 + \frac{v_r}{c}\right)\cos\alpha \tag{70}$$

$$K_s\left(\frac{v_r}{c}\right) = \frac{\sin 2\pi m v_r/c}{2\pi m v_r/c}\left(1 + \frac{v_r}{c}\right)\sin\alpha \tag{71}$$

For $m \gg 1$, i.e., for a signal with a small relative bandwidth, one may replace $1 + v_r/c$ by 1, even though mv_r/c is of the order of 1. With this approximation, one obtains from Eqs. (70) and (71) a quantity $v(0, \beta)$, that will turn out to be a beam pattern,

$$v(0, \beta) = \left[K_c^2\left(\frac{v_r}{c}\right) + K_s^2\left(\frac{v_r}{c}\right)\right]^{1/2} = \frac{\sin 2\pi m v_r/c}{2\pi m v_r/c} \tag{72}$$

With the help of Eq. (66) we may transform Eq. (72) into the following form:

$$v(0, \beta) = \frac{\sin[\pi(2mv/c)\sin\beta]}{\pi(2mv/c)\sin\beta} \tag{73}$$

Let us derive next the beam pattern of a line array of m sensors as shown in Fig. 5.2-23. The planar wavefront $F(\beta)$ with sinusoidal time vari-

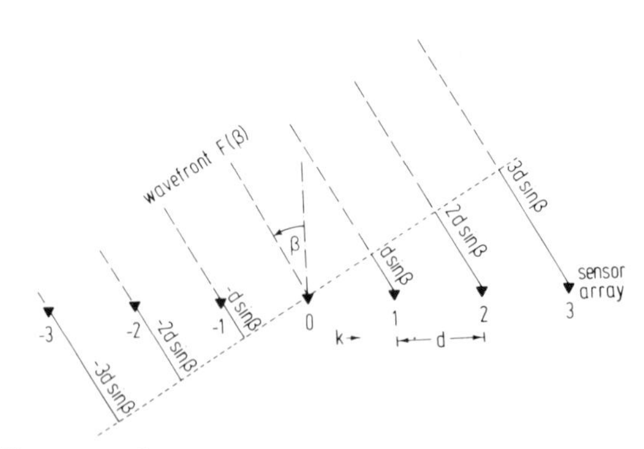

FIG. 5.2-23. Planar wavefront $F(\beta)$ arriving at a line array of $m = 2n + 1$ sensors. The sensors are equally spaced so that sensor k has the distance kd from the sensor $k = 0$ at the center of the array.

ation and frequency f is received. The output voltage produced at the sensor $k = 0$ shall be

$$v_0(t) = V \cos 2\pi f t$$

The general sensor k then produces the output voltage

$$v_k(t) = V \cos \left[2\pi f \left(t - k\frac{d}{c} \sin \beta\right)\right]$$

$$= V \cos \left[2\pi \left(ft - k\frac{d}{\lambda} \sin \beta\right)\right] \tag{74}$$

where c is the phase velocity of the wavefront and λ the wavelength. A beam is produced by summing the m voltages $v_k(t)$:

$$v(0, \beta) = V \sum_{k=-n}^{n} \cos \left[2\pi f t - k\frac{d}{\lambda} \sin \beta\right], \qquad m = 2n + 1 \tag{75}$$

One obtains from Eq. (75):

$$v(0, \beta) = V \cos 2\pi f t \sum_{k=-n}^{n} \cos(2\pi k d \lambda^{-1} \sin \beta) \tag{76}$$

The sum of the terms $\sin 2\pi f t \sin(2\pi k d \lambda^{-1} \sin \beta)$ is zero. One may compute the sum of Eq. (76) exactly (Urick, 1967, pp. 41–42), but we approximate it by an integral which yields an excellent approximation[1] for large values of n. For this approximation we multiply and divide Eq. (76) by n:

$$v(0, \beta) = nV \cos 2\pi f t \sum_{k=-n}^{n} n^{-1} \cos(2\pi k d \lambda^{-1} \sin \beta) \tag{77}$$

Each term $n^{-1} \cos(2\pi k d \lambda^{-1} \sin \beta)$ of the sum represents the area of a rectangle with one side equal to $1/n$ and the other equal to $\cos(2\pi k d \lambda^{-1} \sin \beta)$. The sum itself represents the area under a step function. For large values of n one may approximate this area by an integral,

$$v(0, \beta) \doteq nV \cos 2\pi f t \int_{-1}^{+1} \cos \left(\frac{2\pi n d}{\lambda} \frac{k}{n}\right) d\left(\frac{k}{n}\right)$$

$$\doteq mV \frac{\sin[\pi(md/\lambda) \sin \beta]}{\pi(md/\lambda) \sin \beta} \cos 2\pi f t \tag{78}$$

where $m \doteq 2n$ has been substituted from Eq. (75). The amplitude pattern is defined as the factor

---

[1] The requirement of a small relative bandwidth for the conventional sidelooking radar means that $m = 2n + 1$ must be about 100 or more. The integral yields an excellent approximation already for $m = 16$.

$$v(0, \beta) = \frac{\sin[\pi(md/\lambda) \sin \beta]}{\pi(md/\lambda) \sin \beta} \tag{79}$$

Equations (79) and (73) are identical for

$$2v/c = d/\lambda \tag{80}$$

The function $v(0, \beta)$ is plotted with a solid line in Fig. 5.2-24 for $\sin \beta \doteq \beta$.

In order to generalize the beam pattern of Eq. (73) from the broadside direction to the general direction γ, we change the period $\tau = T$ of the functions $\sin(2\pi t'/T - \alpha)$ and $\cos(2\pi t'/T - \alpha)$ in Eqs. (68) and (69) to

$$\tau = T(1 - l/m) \tag{81}$$

where $l = 0, \pm 1, \pm 2, \ldots$ and $l/m \ll 1$. Instead of Eq. (73), one obtains now:

$$v(\gamma, \beta) = \frac{\sin[\pi(2mv/c)(\sin \beta - \sin \gamma)]}{\pi(2mv/c)(\sin \beta - \sin \gamma)} \tag{82}$$

$$\sin \gamma = lc/2mv \tag{83}$$

In the case of the line array one obtains a beam in the general direction γ rather than the special direction $\gamma = 0$, for which Eq. (79) holds, by delaying the output voltages of the sensors before summation. Let the output of the sensor k be delayed[1] by

$$t_k = -kdc^{-1} \sin \gamma \tag{84}$$

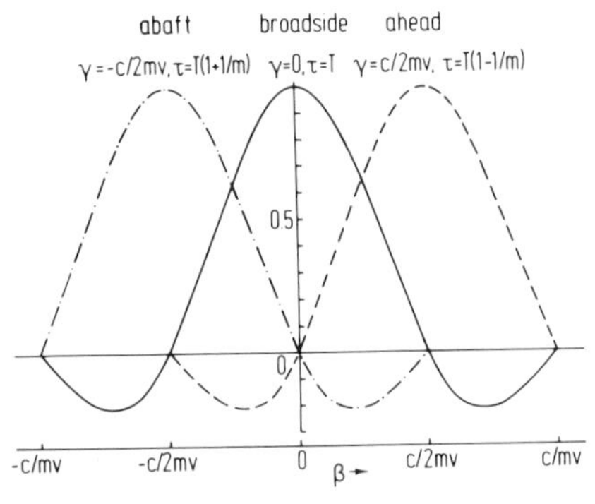

FIG. 5.2-24. Beam patterns produced by means of the Doppler effect; $\sin \beta \doteq \beta$, $\sin \gamma \doteq \gamma$.

---

[1] For positive values of k the delay becomes an advancement. This problem can be overcome in practical arrays by adding a sufficiently large delay t_0 so that $t_0 + t_k$ is always positive. Alternately, one may use phase-shift circuits instead of delay circuits, since the phase can be advanced as well as retarded.

Equation (74) becomes:

$$\begin{aligned} v_k(t) &= V \cos[2\pi f/t - t_k - kdc^{-1} \sin \beta)] \\ &= V \cos\{2\pi[ft - kd\lambda^{-1}(\sin \beta - \sin \gamma)]\} \end{aligned} \tag{85}$$

The summation according to Eq. (75) yields the general pattern $v(\gamma, \beta)$ instead of the pattern $v(0, \beta)$ of Eq. (79):

$$v(\gamma, \beta) = \frac{\sin[\pi(md/\lambda)(\sin \beta - \sin \gamma)]}{\pi(md/\lambda)(\sin \beta - \sin \gamma)} \tag{86}$$

From Eqs. (83) and (84) follows

$$t_k = -kld/2mv \tag{87}$$

as a condition that the beams of the array have the same direction as those formed by means of the Doppler effect. For a small relative bandwidth, $m \gg 1$, and first-order approximation in v/c, as well as v_r/c, the beam forming by means of the Doppler shift and the line array with delay or phase shifting are equivalent. Beam patterns for $l = +1$ (dashed line) and $l = -1$ (dash–dot line) according to Eqs. (82) and (83) are shown in Fig. 5.2-24.

We will now turn to beam forming by means of the Doppler effect for simple, nonsinusoidal waves.

5.2.9 Doppler Effect of a Simple Nonsinusoidal Carrier

The nonsinusoidal signal of Fig. 2.4-1d periodically repeated is shown again in Fig. 5.2-25a. We call this periodic nonsinusoidal signal a carrier. By multiplying it—or amplitude-modulating it—with a rectangular pulse of duration mT we obtain m periods of the carrier. The duration of a

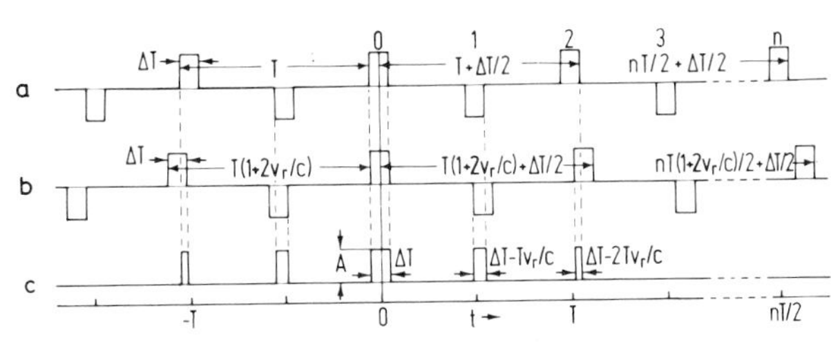

FIG. 5.2-25. Simple nonsinusoidal signal (a); the same signal returned by a reflector moving away with velocity v (b); the product of the two signals for a vanishing time shift $t_p = 0$ (c).

period is T, and the duration of the positive and negative pulses[1] is ΔT. The center of the pulse n has the distance $nT/2$ from $t = 0$, while the edges of the pulse have the distances $nT/2 + \Delta T/2$ or $nT/2 - \Delta T/2$ from $t = 0$.

This nonsinusoidal carrier returned by a reflector moving away with relative velocity v_r will have the period $T(1 + 2v_r/c)$ instead of T, as shown in Fig. 2.5-25b. The center of the pulse n has the distance $nT(1 + 2v_r/c)/2$ from $t = 0$, while the edges of this pulse have the distances $nT(1 + 2v_r/c)/2 + \Delta T(1 + 2v_r/c)$ or $nT(1 + 2v_r/c) - \Delta T(1 + 2v_r/c)/2$ from $t = 0$. We ignore $\Delta T v_r/c$ compared with ΔT, but we do not ignore nTv_r/c due to the factor n. This approach yields for the edges of the pulse n the distances $nT(1 + 2v_r/c)/2 + \Delta T/2$ and $nT(1 + 2v_r/c)/2 - \Delta T/2$ used in Fig. 2.5-25b.

The product of the original carrier of Fig. 2.5-25a and the returned carrier[2] with Doppler effect of Fig. 2.5-25b is shown in Fig. 2.5-25c. The time shift t_p between the radiated and the returned carrier is assumed to be zero, which implies that the Doppler effect is investigated for a known range.

Consider the integral $K(v_r/c)$ of the product of Fig. 2.5-25c. The area of the center pulse equals $A\,\Delta T$, that of the first pulse to the right or left $A(\Delta T - Tv_r/c)$, etc. Hence, one obtains

$$\begin{aligned} K(v_r/c) &= A\,\Delta T\left[1 + 2\sum_{n=1}^{n'}(1 - nTv_r/\Delta Tc)\right] \\ &= A\,\Delta T\left[1 + 2n' - 2(Tv_r/\Delta Tc)\sum_{n=1}^{n'} n\right] \\ &= A\,\Delta T[1 + 2n' - n'(n' + 1)Tv_r/\Delta Tc] \end{aligned} \tag{88}$$

The quantity n' denotes the largest value of n for which $1 - nTv_r/\Delta Tc$ is not yet zero or negative, but n' cannot be larger than the number of periods m. One must distinguish two cases:

$$n' \doteqdot \frac{\Delta T/T}{v_r/c} < m, \qquad n' = m \leq \frac{\Delta T/T}{v_r/c} \tag{89}$$

[1] For an all-weather, line-of-sight radar, the duration of ΔT should be between about 1 and 0.1 ns to keep most of the energy in the band from 500 MHz to 10 GHz.

[2] We are producing the product that corresponds to the product under the integral in Eq. (68). The product corresponding to the one under the integral in Eq. (69) is obtained by shifting the function in Fig. 2.5-25a by $+T/4$ to the left; this product is zero for small values of t, and we do not use it here for this reason. However, one could develop the theory in closer similarity to Eq. (69). This has not been done yet.

For $v_r = 0$, one obtains $K(0)$:

$$K(0) = A\,\Delta T(1 + 2m) \tag{90}$$

Equations (88) and (90) yield the following values for $K(v_r/c)/K(0)$ with the values of n' taken from Eq. (89):

$$\frac{K(v_r/c)}{K(0)} = \begin{cases} \dfrac{\Delta T/T}{2mv_r/c}, & \text{for } n' \doteq \dfrac{\Delta T/T}{v_r/c} < m, \quad m \gg 1 \\[2ex] 1 - \dfrac{mv_r/c}{2\,\Delta T/T}, & \text{for } n' = m < \dfrac{\Delta T/T}{v_r/c}, \quad m \gg 1 \\[2ex] \dfrac{1}{2}, & \text{for } m \doteq \dfrac{\Delta T/T}{v_r/c} \gg 1 \end{cases} \tag{91}$$

From Fig. 5.2-22, we introduce $\sin\beta = v_r/v$

$$v(0, \beta) = \frac{K(v_r/c)}{K(0)} = \begin{cases} \dfrac{\Delta T/T}{(2mv/c)\sin\beta}, \\ \qquad \text{for } n' \doteq \dfrac{\Delta T/T}{(v/c)\sin\beta} < m, \quad m \gg 1 \\[2ex] 1 - \dfrac{(mv/c)\sin\beta}{2\,\Delta T/T}, \\ \qquad \text{for } n' = m < \dfrac{\Delta T/T}{(v/c)\sin\beta}, \quad m \gg 1 \\[2ex] \dfrac{1}{2}, \quad \text{for } m \doteq \dfrac{\Delta T/T}{(v/c)\sin\beta} \gg 1 \end{cases} \tag{92}$$

Beam patterns $v(0, \beta)$ according to this formula are plotted in Fig. 5.2-26 for $\Delta T/T = 1/5$ and $\Delta T/T = 1/10$. For comparison, the beam pattern for a sinusoidal carrier defined by Eq. (73) is also shown. The beam patterns for the nonsinusoidal carriers drop faster than the one for a sinusoidal carrier in the neighborhood of $\beta = 0$, and the magnitude of their "sidelobes" is smaller. These are the two advantages one obtains for the larger frequency bandwidth occupied by a signal according to Fig. 5.2-25a compared with the pulsed sinusoid of Fig. 5.2-1e.

The main drawback of a nonsinusoidal signal according to Fig. 5.2-25a with m periods, or of a pulsed sinusoid also with m periods, is the poor range resolution, which is determined by the duration mT of these signals. Three methods have been developed for sinusoidal signals to overcome this drawback, and all can be applied to nonsinusoidal signals. The first uses coded baseband signals for amplitude modulation of the carrier (Rihaczek, 1969; Deley, 1970); the nonsinusoidal signal of Fig. 5.2-25 is in

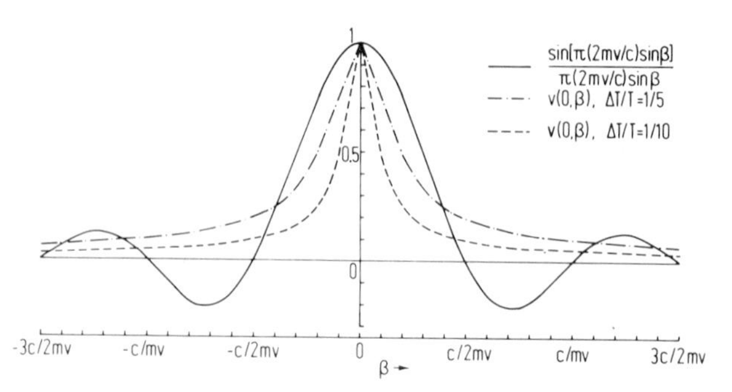

FIG. 5.2-26. Beam pattern $v(0, \beta)$ according to Eq. (92) produced by the Doppler effect of the nonsinusoidal signal of Fig. 5.2-25*a* with the ratios $\Delta T/T = 1/5$ (dash–dot line) and $\Delta T/T = 1/10$ (dashed line). The beam pattern due to the Doppler effect of a sinusoidal carrier according to Eq. (73) is shown for comparison: (solid line) $\sin[\pi(2mv/c)\sin\beta]/\pi(2mv/c)\sin\beta$.

this case treated as a carrier according to Section 2.4 and amplitude-modulated, e.g., with a Barker code (Harmuth, 1977a, p. 337). The second method uses pulse compression based on frequency modulation of the carrier (Farnett *et al.*, 1970). The third uses the pulse-Doppler principle (Mooney and Skillman, 1970). We will discuss in the next section the use of carrier coding—as opposed to baseband signal coding—to produce a good range resolution with a long signal required for a good Doppler resolution. This method requires signals with a large relative bandwidth, which explains why there is no analog to carrier coding for signals transmitted in the usual way with a sinusoidal carrier.

5.2.10 *Doppler Effect of a Coded Carrier*

Figure 5.2-27a shows a sawtooth-shaped antenna current. The time variation of the electric and magnetic field strengths in the far zone, produced by a radiator operating in the dipole mode, is shown in line b. There are short positive pulses with a large amplitude where the sawtooth current rises, and long negative pulses with a very small amplitude where the sawtooth current drops; the long negative pulses are not shown since their amplitudes are generally too small to be visible.

The signal $f(t)$ of Fig. 5.2-27b is a straightforward generalization of the signal of Fig. 5.2-25a. For this reason, we use the term *coded carrier* for it. However, one must point out that the distinction between baseband signal and carrier becomes difficult when one permits the carrier to be coded. We will see later on that a distinction between pulse position and pulse shape becomes more useful than a distinction between baseband

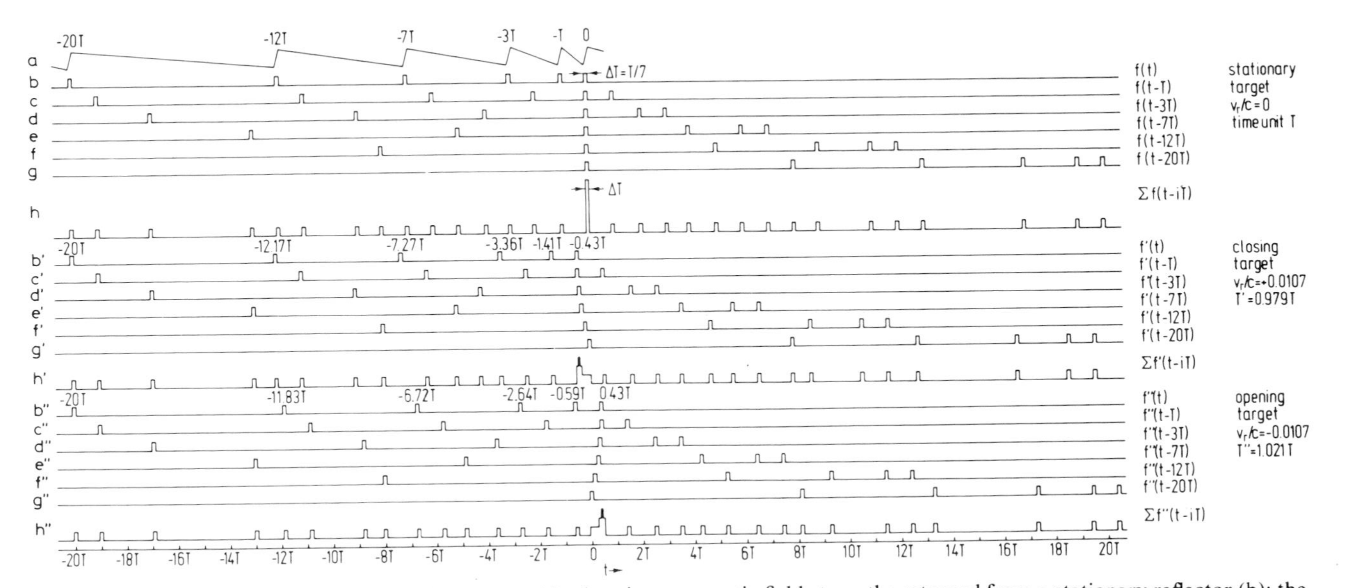

FIG. 5.2-27. Sawtooth current fed into the antenna (a); electric or magnetic field strengths returned from a stationary reflector (b); the same field strengths with delays T to $20T$ (c–g); sum of the field strengths (h); the same field strengths and their sum for a reflector approaching with velocity $v_r/c = -0.0107$ (b'–h'), and for a reflector moving away with velocity $v_r/c = +0.0107$ (b''–h'').

signal and carrier. We continue to use the term carrier primarily because the Doppler effect is generally associated with a carrier.

Consider the pulses of line b in Fig. 5.2-27 to be a signal returned to the radar by a reflector. The lines c–g show the same pulses delayed by T, $3T$, $7T$, $12T$, and $20T$, respectively. Line h shows the sum of the pulses in lines b–g. At $t = 0$, the pulses in lines b–g coincide and yield a pulse with large amplitude in line h. At all other times, the pulses in lines b–g do not coincide, and the summation yields no increase in the pulse amplitude. In readily understandable terminology, the pulse compression has yielded a mainlobe with an enhancement[1] ratio of 6:1 and a sidelobe-to-mainlobe ratio of 1:6.

A practical circuit for this compression is shown in the center of Fig. 5.2-28. The incoming signal has the form shown in line b of Fig. 5.2-27. The delay circuits denoting T, $3T$, $7T$, $12T$, and $20T$ produce the delayed signals of lines c–g in Fig. 5.2-27 and the summing circuit SUM0 produces the signal of line h.

Consider now the pulses of line b in Fig. 5.2-27 returned by a reflector that approaches with a velocity $v_r/c = -0.0107$. The resulting pulses are compressed so that a time interval sT becomes sT':

$$sT' = sT(1 - 2v_r/c) \tag{93}$$

For instance, the interval $8T$ between the pulse at $t = -20T$ and the pulse at $t = -12T$ in line b is reduced to

$$8T' = 8T(1 - 2 \times 0.0107) = 8 \times 0.979T = 7.83T \tag{94}$$

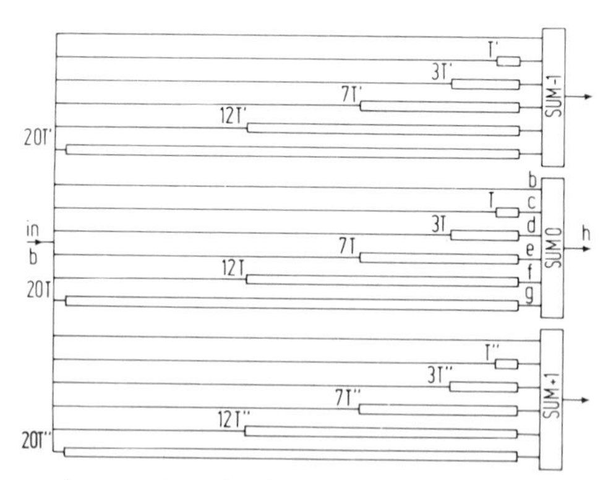

FIG. 5.2-28. Delay and summation circuits (SUM) for the pulse compression according to Fig. 5.2-27.

---

[1] The enhancement or compression ratio of a chirp radar is orders of magnitude larger than 6:1 (Haggarty *et al.*, 1975). We will derive nonsinusoidal signals with arbitrarily large enhancement and arbitrarily small sidelobe-to-mainlobe ratios in Chapter 6.

and the second pulse in line b′ in Fig. 5.2-27 is thus located at

$$-20T + 7.83T = -12.17T$$

The position of the other pulses in line b′, relative to the first pulse at $t = -20T$, may be obtained in this way by substituting $s = 13, 17, 19, 20$ into Eq. (93).

Let the pulses of line b′ be fed to the delay and summing circuit in the center of Fig. 5.2-28, with the delay times $T, 3T, \ldots, 20T$. The pulses of lines c′–g′ are obtained at the output terminals of the delay circuits, and their summation yields line h′ in Fig. 5.2-27. The big pulse at $t = 0$ in line h is replaced by a widened pulse with half the peak amplitude in line h′.

Let the pulses of line b′ now be applied to the delay and summing circuit on top of Fig. 5.2-28 with the delay times $T', 3T', \ldots, 20T'$, where $T' = 0.979T$ according to Eq. (94). The delayed pulses will now line up like in lines b–g, with T replaced by T' on the time scale, and the sum of line h will be obtained. The big pulse will not occur at $t = 0$, but at $t = -0.43T$, since this is the location of the sixth pulse in line b′. The measured range of the target will be the range at the time the sixth pulse is reflected.

Consider next a reflector that moves away with velocity $v_r/c = +0.0107$. The pulses of line b in Fig. 5.2-27 are now expanded so that a time interval sT becomes sT'':

$$sT'' = sT(1 + 2v_r/c)$$

The interval between the pulse at $t = -20T$ and the pulse at $t = -12T$ in line b is increased to

$$8T'' = 8T(1 + 2 \times 0.0107) = 8 \times 1.021T = 8.17T$$

and the second pulse in line b″ is thus located at $-20T + 8.17T = -11.83T$.

If the pulses of line b″ are fed to the delay and summing circuit in the center of Fig. 5.2-28, with the delay times $T, 3T, \ldots, 20T$, one obtains the pulses of lines c″–g″, and the sum of line h″. The delay and summing circuit at the bottom of Fig. 5.2-28 with the delay times $T'', 3T'', \ldots, 20T''$, where $T'' = 1.021T$, will produce the pulses of lines b–h, with T replaced by T'' on the time scale. The big pulse will now occur at $t = +0.43T$.

Let us summarize what the coded antenna current of line a or the coded signal of line b in Fig. 5.2-27 accomplish. The range resolution is determined by the pulse width ΔT. The Doppler resolution is determined by the duration of the pulse sequence, which is $20T$ in our example, and generally nT. Powerful pulses with a duration $\Delta T = 1$ ns can currently be produced, which implies a range resolution of 15 cm. Nature permits us to reduce the pulse duration to about 0.1 ns, before the attenuation by rain and fog becomes a problem; hence, the limit for the range resolution is about 1.5 cm in bad weather. This is about two orders of magnitude better than a radar with a sinusoidal carrier can achieve, since the carrier would have to have about 100 cycles during the duration of a pulse in order to permit resonating circuits and structures designed for sinusoidal functions to actually resonate. The pulses of a radar with sinusoidal carrier must thus be about 100 times as long as pulses without a sinusoidal carrier, in order to have the same highest significant frequency in their Fourier transform.

The practically achievable Doppler resolution depends on the length of the time interval nT over which a synchronization error smaller than ΔT can be maintained. This synchronization error is due to inaccurate timing of the positive ramps of the sawtooth function in line a of Fig. 5.2-27, and due to tolerances of the delay circuits in Fig. 5.2-28. Let the time interval nT have an error $\pm\Delta T$. If the time interval nT' due to a Doppler shift has the same value $nT \pm \Delta T$ as the inaccurate time interval nT without a Doppler shift, we cannot decide whether the changed time interval is due to instability or an actual Doppler shift:

$$nT \pm \Delta T = nT' = nT(1 - 2v_r/c), \qquad \Delta T/nT = 2v_r/c \tag{95}$$

For a relative velocity $v_r = 300$ m/s, one obtains $\Delta T/nT = 2 \times 10^{-6}$. Hence, a short-term deviation of about one part in 10^{-6} or better is required to detect a relative velocity of 300 m/s. This is in line with the required stability of the usual Doppler radar based on the frequency shift of the sinusoidal carrier.

5.2.11 Beam Forming with a Coded Carrier

The beam patterns shown by the dashed and dash–dot lines in Fig. 5.2-26 were derived from the carrier in Fig. 5.2-25. We now have to derive corresponding beam patterns for the carrier in line b in Fig. 5.2-27. The straightforward way to do so is to use the sum signal of line h in Fig. 5.2-27 as a sample function with which the output voltages of the three summers SUM−1 to SUM+1 in Fig. 5.2-28 are correlated. This means producing the autocorrelation function of the signal in line h for $v_r/c = 0$, the cross-correlation functions of the signal in line h with the signals in lines h′ and h″ for $v_r/c = -0.0107$ and $v_r/c = +0.0107$, and more such cross-correlation functions for other values of v_r/c. The irregular spacing of the pulses in line b of Fig. 5.2-27 has so far made it too difficult to derive the correlation function analytically. A purely numerical computation is not difficult, but we are more interested here in elucidating principles than in deriving accurate numbers, and the numerical computation would be of

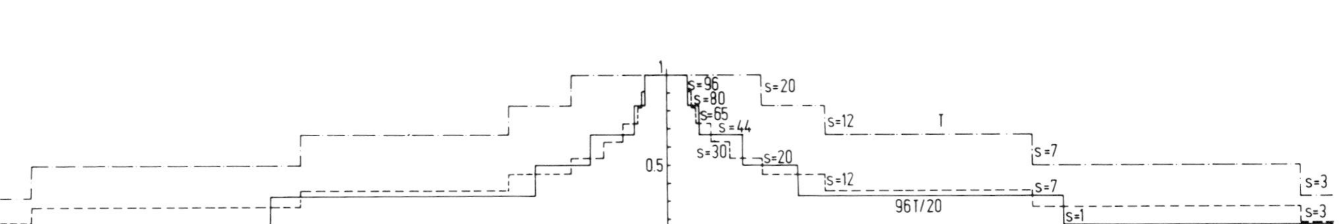

FIG. 5.2-29. Beam patterns for the carrier of Fig. 5.2-27 (dash–dot line), of Fig. 5.2-30 (dashed line), and of Fig. 5.2-27 with T replaced by $96T/20$ (solid line).

little help for this purpose. To avoid the numerical computation, we will investigate the peak amplitudes obtained at the terminal h in Fig. 5.2-28 for various values of v_r/c. The use of the cross-correlation functions would imply that a sliding correlator is connected to the same terminal h, and that the peak of the cross-correlation function of the output at terminal h and the sample function in line h of Fig. 5.2-27 is investigated.

Let sT' in Eq. (93) for $s = 20$ have the value $sT - \Delta T$:

$$sT - \Delta T = sT\left(1 - \frac{2v_r}{c}\right), \qquad \frac{v_r}{c} = \frac{1}{2s}\frac{\Delta T}{T} = \frac{1}{40}\frac{\Delta T}{T} \tag{96}$$

For this value of v_r/c, the five pulses at $t = 0$ in lines b–f in Fig. 5.2-27 will still overlap to produce the peak amplitude 5/6, but the pulse in line g will be too far to the left to overlap with the pulse in line b. Hence, for v_r/c in the interval $0 \leqq v_r/c \leqq \Delta T/40T$, we will have the relative peak amplitude $6/6 = 1$, but for $v_r/c = \Delta T/40T$ the relative peak amplitude will drop to $5/6 = 0.83$, as shown by the dash–dot line in Fig. 5.2-29.

Let s in Eq. (96) have the value 12 now. The pulses at $t = 0$ in lines b–e in Fig. 5.2-27 will still overlap, but the pulses in lines f and g will have shifted too much, and the relative peak amplitude $4/6 = 0.67$ is obtained for $v_r/c = \Delta T/24T = 1.67(\Delta T/40T)$. Progressing in this fashion to $s = 7$ and $s = 3$, one obtains the whole dash–dot plot of Fig. 5.2-29 for $v_r/c > 0$; the jump for $s = 1$ occurs at $v_r/c = \Delta T/2T = 20(\Delta T/40T)$, and is thus too far to the right to show. For $v_r/c < 0$, one recognizes from Eq. (96) that the plot is symmetric about $v_r/c = 0$.

The dash–dot line in Fig. 5.2-29 is so far plotted for v_r/c rather than for an angle β as required for a beam pattern. The relation $v_r = v \sin\beta$ of Fig. 5.2-22 introduces the variable $(v \sin\beta)/c$ for v_r/c in Fig. 5.2-29 or the variable $\sin\beta$ if we multiply the scale with c/v.

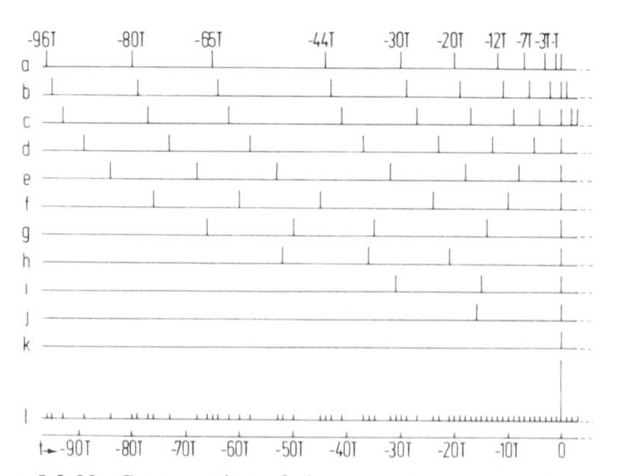

FIG. 5.2-30. Compression of eleven positive pulses into one.

The main use of a plot as shown in Fig. 5.2-29 is to compare beam patterns due to carriers with different coding. Consider the pulse diagram of Fig. 5.2-30. The lines a–k correspond to the lines b–g in Fig. 5.2-27, while line l corresponds to line h in Fig. 5.2-27. The pulses are represented by lines, and only the part of the diagram for $t \leqq 0$ is shown. This is done, of course, to reduce the width of the illustration. One may readily see that the pulses at $t = 0, -3T, -7T, -12T, -20T$ have the same location as those in line b of Fig. 5.2-27, and that the coded carrier of Fig. 5.2-30 is an extension of the one of Fig. 5.2-27 from 6 to 11 pulses. The beam pattern, shown by the dashed line in Fig. 5.2-29 is obtained by inserting $s = 96, 80, \ldots, 1$ in Eq. (96), and observing that the relative peak amplitude drops to 10/11, 9/11, ..., 1/11 at these values of v_r/c.

The beam pattern shown by the dashed line in Fig. 5.2-29 is substantially better than the one shown by the dash–dot line. Part of this improvement is due to the coding, and part is due to the increase of the duration of the signal from $20T$ in Fig. 5.2-27 to $96T$ in Fig. 5.2-30. To compensate for this increased duration, we substitute $96T/20$ for T in Fig. 5.2-27, so that the duration of the signal is increased from $20T$ to $96T$. The resulting beam pattern is shown by the solid line in Fig. 5.2-29; it is obtained by compressing the pattern shown by the dash–dot line by 20/96.

The beam patterns shown by the dashed and the solid lines in Fig. 5.2-29 look quite similar, but there are still two important differences. For large values of v_r/c, the dashed line will drop to 1/11 while the solid line will never drop below 1/6. Furthermore, the signal in Fig. 5.2-30 contains 11 pulses while the signal in Fig. 5.2-27—stretched over a time interval of $96T$—contains only 6 pulses, which implies a reduction of the signal-to-noise ratio by 6/11, if each pulse has the same energy. Hence, the coded carrier of Fig. 5.2-30 is better than the one of Fig. 5.2-27.

5.2.12 Pulse Position and Pulse Shape

We have seen in the preceding section that coding of the carrier affects the beam pattern as well as the signal-to-noise ratio. In the present section, we explain briefly how carrier coding differs from the usual coding of baseband signals. The codes discussed here are intended to serve tutorial purposes only. They are not good for practical use, but they show the principles quite well. The subject of pulse position and pulse shape coding will be taken up again in Chapter 6 on a more advanced level.

Error detecting and correcting codes for data transmission with binary signals can be written as binary numbers, e.g., 1001101, and coding means finding sets of such numbers with desirable features. The coding of the spreading function or spreading carrier in spread-spectrum com-

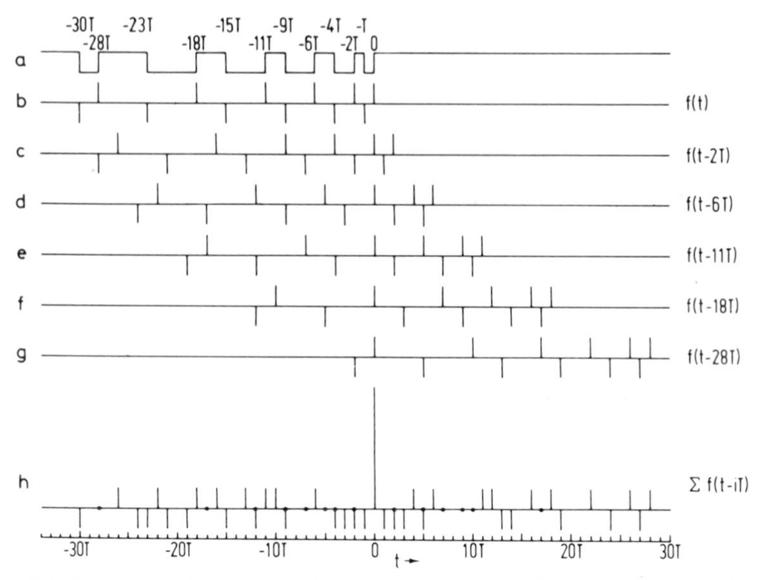

FIG. 5.2-31. Compression of six positive and six negative pulses into one pulse with amplitude enhancement 6/1 and a sidelobe-to-mainlobe ratio 1/6. The advantage of this code is that the sidelobes have positive or negative polarity, in contrast to Fig. 5.2-30.

munications is also defined by such a binary number. Baseband signals in radar can be written in the same way, but it is often preferable to write -1 instead of 0; e.g., the notation $+1\ +1\ -1\ +1$ for a Barker code is more lucid than 1101. Coding of ternary, quarternary, etc. signals calls for the use of ternary, quarternary, etc. numbers, but nothing is changed otherwise.

The carriers in line b of Fig. 5.2-27 and in line a of Fig. 5.2-30 can also be written as sequences of six or eleven digits "1" with many digits "0" between them. We are still dealing with a binary code, although one will want to use a notation different from that of binary numbers in order to avoid writing millions of digits "0." Putting it more technically, because of the sparse nonzero values, the usual 0, 1 strings are very cumbersome.

Consider now the two-valued antenna current in line a of Fig. 5.2-31. An antenna operating in the dipole mode produces electric and magnetic field strengths with the time variation of line b in the far zone. This carrier can be written with six digits "$+1$," six digits "-1," and many zeros between them. This is a ternary number and the carrier uses a trilevel waveform.

The binary and ternary numbers can describe the location of the pulses in Fig. 5.2-30, line a and in Fig. 5.2-31, line b, as well as their polarity. The

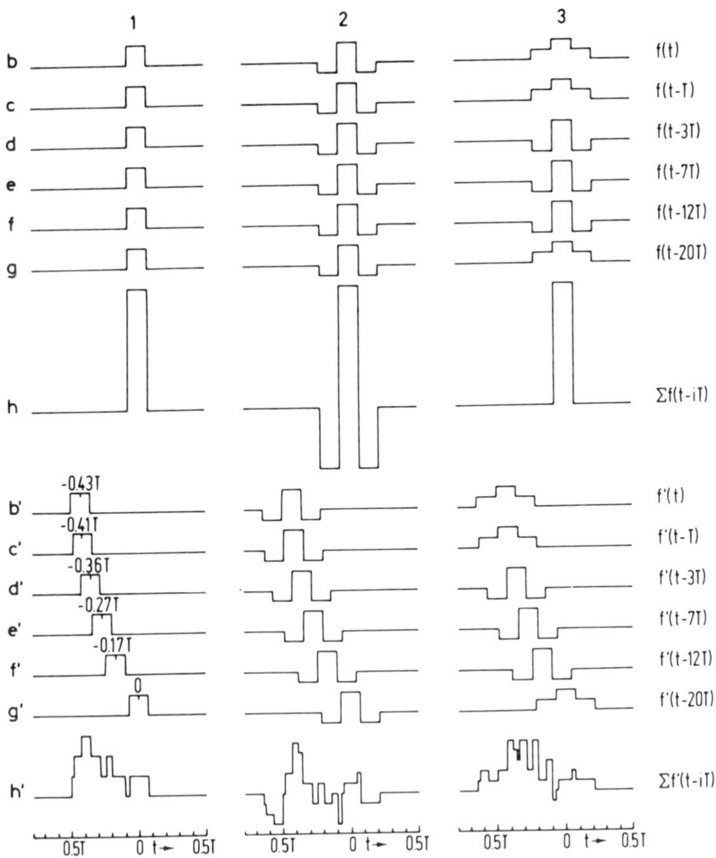

FIG. 5.2-32. The pulse diagrams of lines b–h′ of Fig. 5.2-27 in the neighborhood of $t = 0$ (column 1), and the same diagrams but with different pulse shapes (columns 2 and 3).

shape of the pulses is not specified by these numbers, and we have so far only assumed that the pulses are short compared with the time unit T.

For some examples of pulse shapes, let us turn to Fig. 5.2-32. Column 1 shows the pulses of lines b–h′ of Fig. 5.2-27 in the neighborhood of $t = 0$; the function in line h′ of Fig. 5.2-32 is somewhat different from the same function in Fig. 5.2-27, since the larger scale permits us to show finer details than Fig. 5.2-27.

Let the rectangular pulses of lines b–g, column 1, be replaced by the more complicated pulses of column 2. The sum in line h has now two large sidelobes, which is undesirable. However, the peak of the sum of the Doppler-shifted pulses in line h′ is smaller than in column 1, which is desirable.

The pulses in column 3 avoid the sidelobes in line h, and the sum in line h′ has the same peak value as the sum in column 2. The price for this very desirable result is that the pulses in lines b–g or b′–g′ in column 3 are no longer all equal. This implies that each pulse of the carrier in line b of Fig. 5.2-27 has its individual shape.

The shape of the pulses in lines b–g of column 3 may be represented by the numbers 1, 2, 1, and −1, 2, −1. In this simple case we can incorporate the code of the pulse shape into the code of the location of the pulses by substituting 1, 2, 1, or −1, 2, −1, for a digit "+1" of the pulse-location code, and −1, −2, −1, or 1, −2, 1, for a digit "−1." In general, the pulses in lines b–g of Fig. 5.2-32 can have more complicated shapes than the shown rectangular ones, and the simple description by numbers like 1, 2, 1, or −1, 2, −1 is not possible.

5.2.13 Ghost Suppression

Let us return to Fig. 5.2-21. We had seen there that the presence of more than one target creates the problem of ghost targets, which can be combated by beam forming. The goal of the investigation since Fig. 5.2-21 was to exploit the Doppler effect for this beam forming, if the radar is mounted on a moving platform. For a stationary radar, the beam forming can be accomplished by a phased array. A radar dish will not do the same, even though dishes are shown in Fig. 5.2-21, since a dish produces only one beam at a time while the Doppler effect as well as phased arrays simultaneously produce beams in different directions.

A second effect that helps suppress ghost targets is due to the multiple measurements made. A total of $n + 1$ measurements are made in Figs. 5.2-4 and 5.2-9, while $(n + 1)(m + 1)$ are made in Fig. 5.2-12. Refer to Fig. 5.2-33 to see what a large number of measurements does to the ghost targets.

There are four radars, RA0 to RA3, and two targets, TA1 and TA2. All four radars receive returns from these two targets. This is shown by the fact that four circles indicating the distances from the four radars intersect at the targets TA1 and TA2.

Only two circles intersect at the ghost targets GH0,1 to GH2,3 in Fig. 5.2-33, which means that only two radars receive returns from any one ghost target. According to Figs. 5.2-6d and 5.2-8d or e, the signals returned to the radars are ultimately summed. If the signals returned from the targets TA1 and TA2 have the same amplitude V, we will obtain the summed amplitude $4V$ for the targets TA1 and TA2, but the summed amplitude $2V$ for the ghost targets. This amplitude ratio $4V/2V = 2:1$ is increased to $n/2$ if n rather than four radars are used.

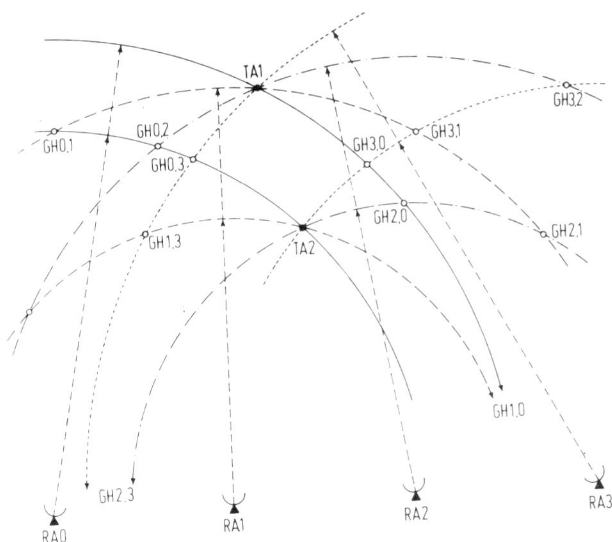

FIG. 5.2-33. Increase of the contrast between targets TA1, TA2 and ghost targets GH0,1 to GH2,3 by means of multiple measurements. The notation for the ghost targets shows which two radars cause them, e.g., GH1,3 is due to radars RA1 and RA3. All ghost targets come in pairs GH$i,k$ and GH$k,i$.

If there are more than two targets one may or may not obtain the amplitude ratio $n/2$ between targets and ghosts. Refer to Fig. 5.2-34, which shows the three targets TA1 to TA3 intersected by four circles each. There is one ghost GH0,1,3 that is intersected by three circles, and many more ghosts wherever two circles intersect.

One may readily reflect that m targets and n radars will produce ghosts by 2, 3, ..., $n - 1$ radars for $m < n$, and ghosts received by all n radars for larger values of m. However, ghosts received by more than two radars depend on certain distances being equal by coincidence, and the probability of a strong ghost—i.e., one received by almost all radars—is small. A good range resolution will decrease the probability of two or more distances being equal by coincidence.

Thus we have three methods[1] to reduce ghosts and increase the contrast range of synthetic aperture radar *images:* narrow beams, produced either by the Doppler effect or by phased antenna arrays; many measurements, done either simultaneously or as a sequence in time; and good

[1] Additional methods of ghost suppression are based on the velocity or the acceleration of targets. They are discussed in the literature in connection with multistatic radars (Caspers, 1970).

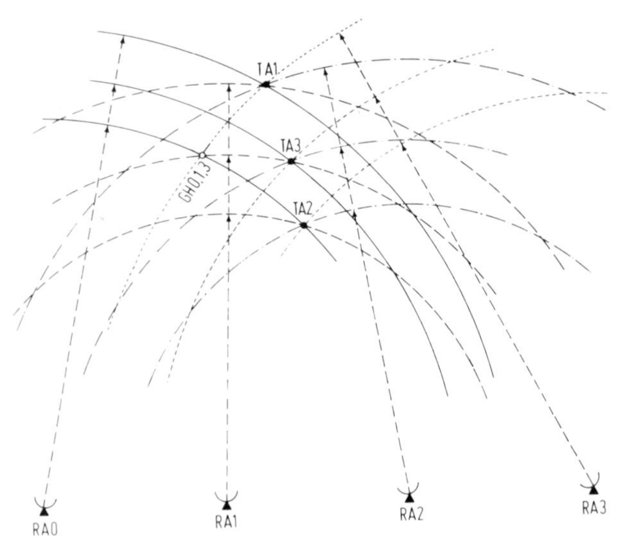

FIG. 5.2-34. Three targets TA1 to TA3 *may* cause a ghost GH0,1,3 that is received by three rather than by two of the radars RA0 to RA3.

range resolution. All three methods can be implemented by using very short pulses. Hence, we have a good incentive to reduce the pulses of synthetic aperture radars to about 0.1 ns, where the absorption of the atmosphere puts a natural rather than a technological limit.

5.2.14 Beam Forming by the MTI Technique

The Doppler effect is used in radar in two forms: (a) A relative velocity between the radar and the target changes the frequency of a sinusoidal carrier; only one pulse in the interval $-mT/2 \leqq t \leqq mT/2$ is required (Saunders, 1970). (b) A change of distance caused by a relative velocity between the radar and the target causes the round-trip travel time of a first and second pulse, radiated with a delay T_R relative to the first pulse, to differ; this is the basis for the moving target indicator (MTI) and the pulse-Doppler radar (Schrader, 1970; Mooney and Skillman, 1970; Barton, 1964, Chapter 7). We will show in this section that the MTI technique is already included in previous results on nonsinusoidal waves.

The geometric relationships for beam forming by means of the MTI technique will be derived with the help of Fig. 5.2-35. The radar RA is at the location $x = -v\tau$ at the time $t = -\tau$. A first pulse is sent to the target TA, and requires the time t_1 for the round trip; hence, it is received at the radar at the time $t = -\tau + t_1$. The time t_1 shall be so short that the move-

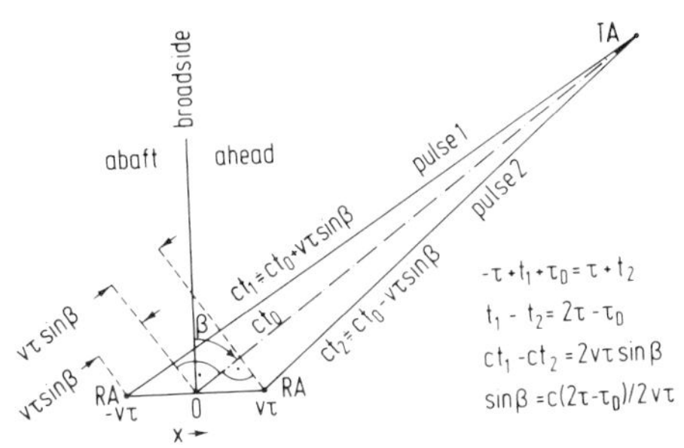

FIG. 5.2-35. Principle of beam forming by means of the time difference $2\tau - \tau_D$ due to a relative movement between the radar RA and a target TA with the azimuth β. The first pulse is transmitted at time $t = -\tau$, received at $t = -\tau + t_1$, and delayed to $t = -\tau + t_1 + \tau_D$. The second pulse is transmitted at $t = +\tau$, and received at $t = \tau + t_2$. A delay time $\tau_D = \tau_D(\beta)$ is chosen for every angle β to be resolved, so that the delayed first pulse and the received second pulse coincide.

ment of the radar during the time t_1 can be ignored. The received pulse is fed into a delay circuit with delay τ_D, and arrives thus at its output terminal at the time $t = -\tau + t_1 + \tau_D$.

At the time $t = +\tau$ the radar RA shall be at the location $x = +v\tau$. A second pulse is sent to the target. The returned pulse is received at the time $t = \tau + t_2$. Let the delay τ_D of the first pulse be chosen so that the relation

$$-\tau + t_1 + \tau_D = \tau + t_2 \tag{97}$$

holds. In analogy to Eq. (65), one may write it in the form

$$t_1 - t_2 = 2\tau - \tau_D \tag{98}$$

If the distance to the target is large compared with the distance $2v\tau$, one obtains from Fig. 5.2-35 a relation for the angle β,

$$ct_1 - ct_2 = 2v\tau \sin\beta \tag{99}$$

Insertion of Eq. (98) shows that the angle β can be determined from the time difference $2\tau - \tau_D$,

$$\sin\beta = c(2\tau - \tau_D)/2v\tau \tag{100}$$

Hence, for every value of the time difference $2\tau - \tau_D$, we obtain a different angle. For $2\tau - \tau_D > 0$, the target is ahead of the radar; for $2\tau - \tau_D = 0$, the target is broadside; and for $2\tau - \tau_D < 0$, the target is abaft.

Although we have obtained a result, it is neither new nor particularly useful. A comparison of Fig. 5.2-35 with Fig. 5.2-21 shows that the MTI technique yields the same result as making measurements from radars at different locations. The MTI technique is thus already included in the previously developed theory.

5.2.15 Beam Forming and Imaging Based on Pulse Slopes

We have discussed in Sections 5.2.6 and 5.2.7 that the time resolution of pulses can be improved by deriving the pulse amplitude A from the slope $2A\ \Delta f$ of a pulse with bandwidth Δf, and that this implies the possibility of trading aperture for signal power in beam forming. The following simple example will show how this can be done practically.

Figure 5.2-36 shows a line array of four sensors spaced with the distance d at the locations $-3d/2$, $-d/2$, $d/2$, and $3d/2$. The wavefront $F(\beta)$ produces the output voltages $v(-3d/2)$ to $v(+3d/2)$ at the sensors. Typical voltages are shown in column 1, lines a–d, of Fig. 5.2-37. The time variation of the wavefront $F(\beta)$ is that of a triangular pulse with amplitude A, followed by a second triangular pulse with amplitude $B = 2A$. Rise and fall times equal ΔT, indicating a nominal bandwidth of $\Delta f = 1/2\ \Delta T$ according to Fig. 5.2-15. The voltage $v(-d/2)$ is delayed by $\Delta T/5$ with respect to $v(-3d/2)$, while $v(+d/2)$ is delayed by $2\ \Delta T/5$ and $v(+3d/2)$ by $3\ \Delta T/5$. Since the incremental delay $\Delta T/5$ is due to the incremental distances $d \sin \beta$ to be propagated by the wavefront $F(\beta)$ in Fig. 5.2-36, we obtain the angle of incidence β that produces the voltages in Fig. 5.2-37, column 1,

$$\sin \beta = \tfrac{1}{5}(c\ \Delta T/d) \tag{101}$$

where c is the velocity of propagation of the wavefront.[1] For $\beta = 0$, the

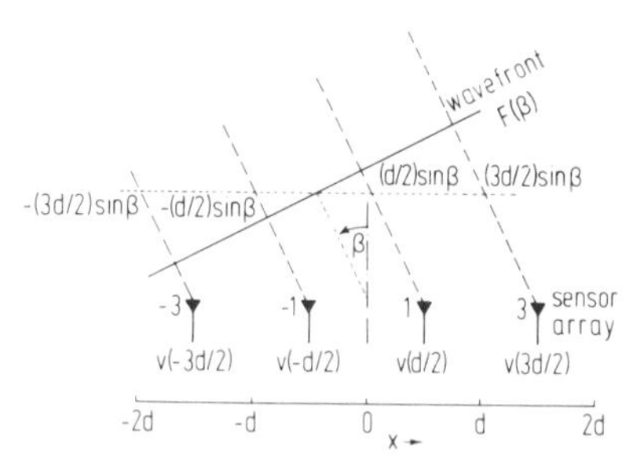

FIG. 5.2-36. Array of four sensors located at the points $x = \pm(2k - 1)(d/2)$, where $k = 1, 2, \cdots$. The aperture of the array equals $4d$. The choice of the sensor locations at odd multiples of $\pm d/2$ is convenient if the number of sensors is even.

---

[1] The propagation velocity c would be called phase velocity, if the time variation of the wavefront were sinusoidal. The term phase cannot be used properly for nonsinusoidal functions.

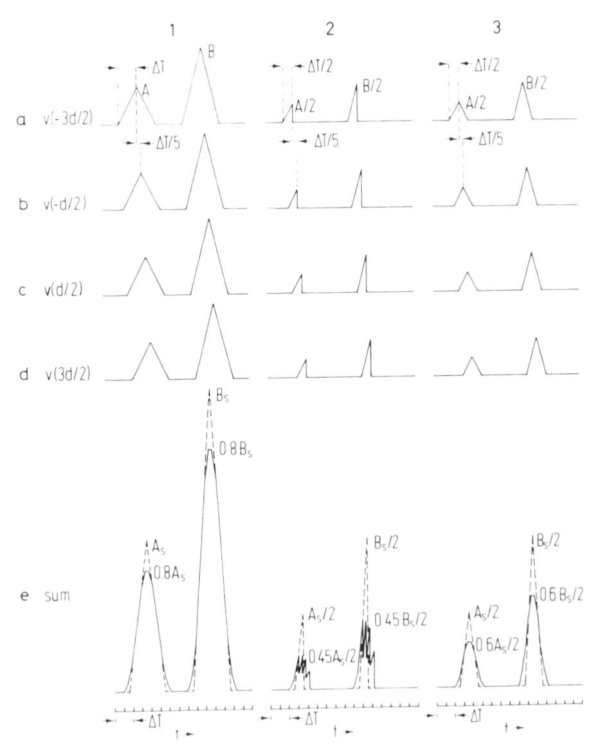

FIG. 5.2-37. Output voltages $v(-3d/2)$ to $v(3d/2)$ of the sensor array of Fig. 5.2-36 (column 1, a–d), and their sum (e). The same voltages after processing according to column c of Fig. 5.2-16 or row c of Fig. 5.2-38 (column 2, a–d), and their sum (e). Column 3 is a repetition of column 2 with the processing modified according to rows d and e of Fig. 5.2-38. The dashed lines in row e of all three columns show the sum of the voltages for the angle $\beta = 0$ in Fig. 5.2-36.

four voltages $v(-3d/2)$ to $v(+3d/2)$ would have no relative time shifts, and their sum would have produced the triangular functions with amplitude A_s and B_s shown by the dashed line in Fig. 5.2-37, column 1. The factor 0.8 by which the sums for $\beta = 0$ and β according to Eq. (101) differ gives an indication how rapidly the beam pattern varies with β.

So far, everything was done according to conventional beam forming in the broadside direction, except that a triangular rather than a sinusoidal time variation of the wavefront was assumed. Now let the processing according to Fig. 5.2-16, column c, be added. We assume that the slopes $A/\Delta T = 2A\,\Delta f$ and $B/\Delta T = 2B\,\Delta f$ of the triangular pulses are determined during a time interval of duration $\Delta T/2$ after the beginning of the pulse.

The amplitudes A and B can be computed, since Δf is known, and a predicted pulse can be computed. The predicted pulse is then subtracted from the actually arriving pulse. The result are the voltages $v(-3d/2)$ to $v(3d/2)$ in column 2 of Fig. 5.2-37. We see that pulses of duration $\Delta T/2$, amplitudes $A/2$ or $B/2$, rising slope $A/\Delta T$ or $B/\Delta T$, and falling slope $-A/0 = -B/0$ are produced. Beam forming is done as before by summing the four voltages, as shown by the solid line in column 2, row e; the dashed lines with amplitudes $A_s/2$ and $B_s/2$ hold for an angle of incidence $\beta = 0$, in which case there is no delay between the voltages $v(-3d/2)$ to $v(+3d/2)$. The factor 0.45 by which the sums for $\beta = 0$ and β according to Eq. (101) differ gives an indication of how rapidly the beam pattern varies with β. This variation is much faster for column 2 than for column 1. Evidently, it would be still faster if a value smaller than $\Delta T/2$ had been chosen for the time interval during which the slope $2A\ \Delta f$ or $2B\ \Delta f$ was determined.

We could now proceed to determine the peak values of the sum in Fig. 5.2-37, column 2, for many values of β and the resulting incremental delay times

$$\Delta t = (d/c) \sin \beta \tag{102}$$

rather than $\Delta t = \Delta T/5$ as used in Fig. 5.2-37. By squaring these peak values and plotting them—normalized by the value for $\beta = 0$—as a function of β, we would obtain a peak power reception pattern, comparable to the peak power radiation pattern in Figs. 3.2-13 or 3.2-20. An average power reception pattern would be obtained by squaring the whole sum function, rather than its peak, and integrating the squared function. We will not do so here, since the principle can be demonstrated much better without such tedious calculations.

The different shape of the pulses in columns 1 and 2 of Fig. 5.2-37 is the obstacle to a simple comparison. We avoid this obstacle by using a processing method that is somewhat worse, but preserves the pulse shape. Refer to Fig. 5.2-38 for an explanation. An arriving pulse is shown in row a. At the time $\Delta T/2$ its amplitude has been determined from the slope $2A\ \Delta f$. The rest of the pulse can be predicted; this predicted rest is shown in row b. The received pulse of row a minus the predicted pulse of row b is shown in row c. This is the pulse used in column 2 of Fig. 5.2-37. Instead of predicting the pulse of row b, one may just as well predict the pulse of row d. The received pulse of row a minus the one in row d is shown in row e. It has half the amplitude and half the duration of the pulse in row a; the shape is the same.

Column 3 in Fig. 5.2-37 uses this processing. The sum of the four voltages in row e for β defined by Eq. (101) now has the peaks $0.6A_s/2$ and

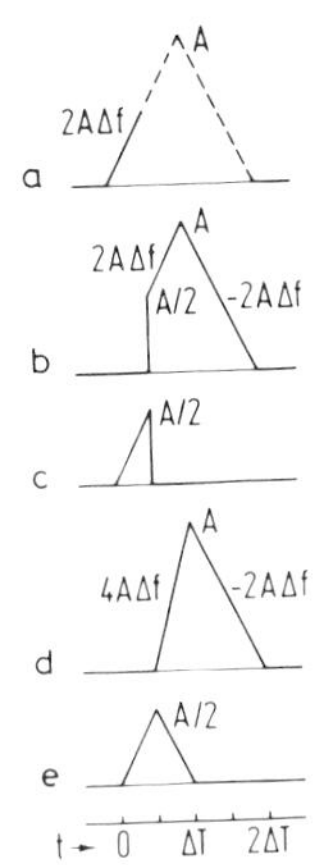

FIG. 5.2-38. Processing of voltages for beam forming. (a) Voltage produced by arriving triangular wave; (b) voltage predicted at the time $\Delta T/2$; (c) arriving voltage a minus predicted voltage b; (d) modified voltage predicted at time $\Delta T/2$; (e) arriving voltage a minus predicted modified voltage d.

$0.6B_s/2$, compared with $A_s/2$ and $B_s/2$ for $\beta = 0$. This is better than in column 1, but not as good as in column 2.

Let us calculate for which value of β the peak amplitudes in row e of column 3 in Fig. 5.2-37 drop to one half the value they have for $\beta = 0$. We will denote this angle with $\epsilon/2$, and call ϵ the resolution angle. For generality, we will assume that the amplitude A in Fig. 5.2-38 is determined at the general time ΔT_0 rather than at the time $\Delta T_0 = \Delta T/2$ shown. Furthermore, we will assume n rather than 4 sensors arranged as in Fig. 5.2-36.

Output voltages for n sensors are shown in column a of Fig. 5.2-39 for the angle of incidence $\beta = 0$. There is no time shift between the voltages. The original amplitude A of each pulse is reduced to $A\,\Delta T_0/\Delta T$; the original transient time ΔT between 0 and the peak is reduced to ΔT_0. The sum of all n pulses produces the peak amplitude $nA\,\Delta T_0/\Delta T$. A short reflection shows that the pulses should have the delay ΔT_0 between the first and the last, as shown in column b, to obtain just half this peak amplitude for the sum if n is large, $n \gg 1$. With L denoting the length of the array, we obtain

$$\sin \tfrac{1}{2}\epsilon \doteq \tfrac{1}{2}\epsilon = c\,\Delta T_0/L \tag{103}$$

In order to introduce a signal-to-noise power ratio into this formula, we investigate how the amplitude A and thus the whole time variation of the triangular pulse in Fig. 5.2-38a can be obtained. We take a sample $2A\,\Delta f\,\Delta T_1$ at the time $t = \Delta T_1$ and a second sample $2A\,\Delta f\,\Delta T_2$ at the time $t =$

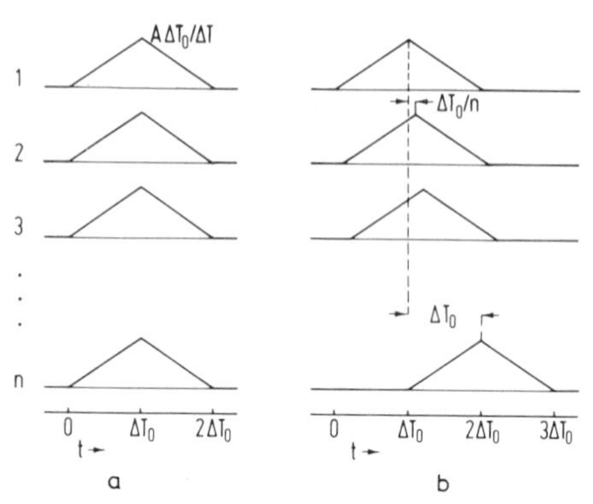

FIG. 5.2-39. Summation of n pulses to yield the peak amplitude $nA\ \Delta T_0/\Delta T$ (a), and to yield half this value (b).

ΔT_2. The difference of the samples divided by the time difference $\Delta T_2 - \Delta T_1$ yields the slope:

$$(2A\ \Delta f\ \Delta T_2 - 2A\ \Delta f\ \Delta T_1)/(\Delta T_2 - \Delta T_1) = 2A\ \Delta f \tag{104}$$

Since Δf is known, we can calculate A, and thus the time variation of the whole triangular pulse. This means we can subtract the predicted pulse from the arriving one at the time $t = \Delta T_0 \geqq \Delta T_2$. We introduce the influence of additive noise by observing that the amplitude A is reached after the time ΔT from the beginning of the triangular pulse. To give the amplitude sampled at the time ΔT_1 the same value, we need a pulse with amplitude A_0,

$$A_0 = A\ \Delta T/\Delta T_1 \tag{105}$$

Let the time ΔT_2 at which the second sample is taken equal $2\ \Delta T_1$, and let us assume that the time ΔT_0, after which the predicted pulse is subtracted from the received one, equals ΔT_2:

$$\Delta T_2 = \Delta T_0 = 2\ \Delta T_1 \tag{106}$$

We may multiply Eq. (105) on the left with ΔT_0 and on the right with $2\ \Delta T_1$, and use the approximation $\Delta T = 1/2\ \Delta f$ according to Fig. 5.2-15:

$$A_0\ \Delta T_0 = 2A\ \Delta T = A/\Delta f, \qquad \Delta T_0 = A/A_0\ \Delta f \tag{107}$$

The amplitude A_0 is proportionate to $P^{1/2}$, where P may be either the peak or the average power of the triangular pulse. Hence, we may rewrite Eq. (107),

$$\Delta T_0 = K_0/P^{1/2}\ \Delta f = K[(P/P_N)^{1/2}\ \Delta f]^{-1} \tag{108}$$

where $K_0 = K/(P_N)^{1/2}$ and K are constants, whereas P_N is the average noise power in the band $0 \leqq f \leqq \Delta f$. Hence, the resolution time ΔT_0 can be reduced either by increasing the bandwidth Δf or the signal-to-noise power ratio P/P_N. The trade between the two,

$$\Delta f(P/P_N)^{1/2} = \text{constant} \tag{109}$$

is somewhat worse than anticipated[1] by Eq. (58).

Introduction of Eq. (108) into Eq. (103) yields the resolution angle ϵ as function of the bandwidth Δf and the signal-to-noise power ratio P/P_N:

$$\epsilon = \frac{2c\,\Delta T_0}{L} = \frac{2Kc}{L(P/P_N)^{1/2}\,\Delta f} \tag{110}$$

This equation has the same form as the classical formula of Eq. (55), with T replaced by $2\,\Delta T_0$, but one can express ΔT_0 in terms of the bandwidth and the signal-to-noise power ratio, which is not possible for T. The constant K depends on a closer specification of the noise and the processor that produces the pulse in row e of Fig. 5.2-38. Its value is of little interest here, since the definition of ϵ by means of the time shift ΔT_0 in Fig. 5.2-39, and the resulting reduction of the peak of the sum of the pulses to 1/2, is quite arbitrary. One could just as well demand a reduction to some other value; a comparably arbitrary definition underlies Eq. (55). The important result is that the resolution angle ϵ can be reduced by increasing the signal power P, the bandwidth Δf, or the length L of the array, while the classical formula of Eq. (55) permits the reduction of ϵ only by an increase of the frequency f and the length L of the array.

The classical formula of Eq. (55) applies to the noise-free case, while Eq. (110) yields $\epsilon = 0$ for $P_N = 0$. Let us see how a practical circuit could be constructed for this case. Figure 5.2-40 shows in lines 1a to 3a three triangular pulses that represent the n pulses of Fig. 5.2-39. Their slope is obtained by differentiation, which can be performed practically with an operational amplifier (Graeme *et al.*, 1971, p. 218). The differentiated pulses are shown in lines 1b to 3b. A second differentiation yields the Dirac pulses of lines 1c to 3c. Amplitude reversal and suppression of the negative pulses by a rectifier finally yields the pulses of lines 1d to 3d. It is obvious that summation of these three Dirac pulses—or generally n Dirac pulses—yields a beam in the broadside direction with beamwidth $\epsilon = 0$.

---

[1] We cannot expect to obtain the theoretically best result. Processing according to column 3 of Fig. 5.2-37 is not as good as according to column 2; furthermore, the triangular pulse makes the investigation simple, but pulses with the time variation $(\sin 2\pi\,\Delta ft)/2\pi\,\Delta ft$ fill the frequency band better, and Gaussian pulses $\exp(-at^2)$ make ideal use of time–frequency intervals.

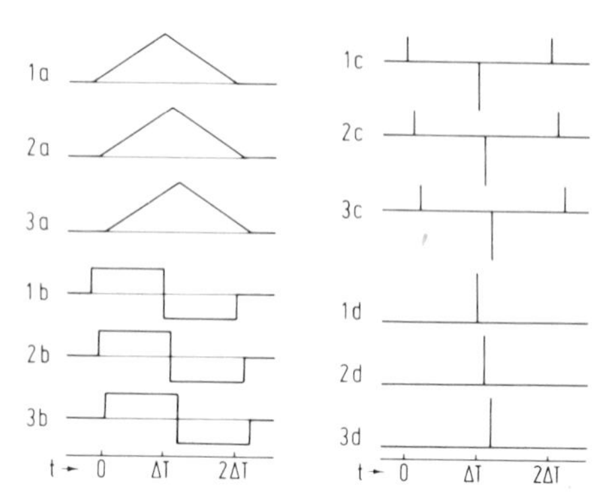

FIG. 5.2-40. Processing of triangular pulses for beam forming in the absence of noise. Received pulses (1a–3a); differentiated pulses (1b–3b); twice-differentiated pulses (1c–3c); twice-differentiated pulses after amplitude reversal and rectification (1d–3d).

Hence, we can devise a circuit that yields the angular resolution $\epsilon = 0$ if there is no noise.

There is a little flaw in this reasoning. The bandwidth Δf in Eq. (110) denotes the bandwidth "practically" occupied by the triangular pulse, which we have assumed to be $\Delta T = 1/2\,\Delta f$ in Eq. (107). However, the differentiations in Fig. 5.2-40 make it impossible to ignore the Fourier components of the triangular pulse above any finite frequency limit. In connection with Fig. 2.1-3 we had cautioned that the concept of a finite frequency bandwidth of a signal is basically wrong, and that one must be careful to use only such results that are still useful despite this wrong assumption. In our case this means that the whole frequency band $0 \leqq f < \infty$ must be noise-free, and the resolution angle $\epsilon = 0$ looks thus much less impressive. There is, however, something more fundamental to be learned. Rigorously speaking, all signals occupy an infinite bandwidth, and the relative bandwidth of all signals is $\eta = 1$. In this book we have taken the not rigorous, engineering point of view, that a "practically" finite frequency bandwidth can be assigned to signals. We have been able to find many good applications and explain them in terms understandable to engineers. However, this nonrigorous approach reaches one of its limits in beam forming. We can treat this topic in a satisfactory manner only if we abandon the concept of a "practically" finite frequency bandwidth. The triangular pulse is exactly specified by its transient time ΔT in Eq. (107), but *not* by the bandwidth Δf. Hence, ΔT should be used instead of Δf. Thermal noise can be specified exactly without reference to a frequency bandwidth (Harmuth, 1969, p. 217; 1972, p. 292). In our particular case, one multiplies noise samples with the triangular pulse used, and integrates the products. The squares of the integrals represent the energy of the received noise samples. The average of the squares taken for many noise samples yields an average noise energy or—after division by the duration of the triangular pulse—an average noise power, free of the concept of frequency bandwidth. In this way the angular resolution by means of nonsinusoidal waves can be investigated rigorously, but we cannot do so within the theory based on frequency bandwidth and relative (frequency) bandwidth upon which this book is based.[1]

The processing discussed here for the broadside beam with observation angle $\gamma = 0$ is readily extended to other values of γ and multiple beams by means by delay lines. The extension from a one-dimensional line array to a two-dimensional Cartesian or polar planar array is also possible in the conventional way (Harmuth, 1979a, Figs. 1.2-2, 2.1-4, 1.2-3, 1.2-4). The only thing new is the processor that derives the amplitude A from the slope $A/\Delta T$ of a triangular pulse, predicts the future time variation of the pulse, and modifies the actually received pulse by the predicted pulse. Since this processor is linear, it will work with superpositions of triangular pulses as well as with a single pulse. Indeed, any signal will automatically be treated as composed of a superposition of time-shifted triangular pulses, just as conventional equipment often treats signals automatically as composed of a superposition of sinusoidal functions.

5.3 Tracking Radar and Beam Rider

5.3.1 Geometric Relations

We have shown in Section 5.2-1—particularly in connection with Eqs. (5.2.1-3) and (5.2.1-5)—that beam forming based on nonsinusoidal functions does not require that the radiators or receptors of an antenna array have to be spaced at a distance of $\lambda/2$ or less. This opens a way to apply nonsinusoidal waves to tracking radars other than the low-elevation-angle radars discussed in Section 5.1. The angular resolution of the antennas of these radars is used to improve the signal-to-noise ratio and to avoid ambiguities, but the location of the target is derived strictly from the measurement of the travel time of the radar pulses. The improved resolution is primarily due to the fact that the time resolution of a radar is generally much better than the angular resolution achievable with an antenna of practical size.

---

[1] For a more detailed analysis that does not use the concept of frequency, the reader is referred to the literature (Harmuth 1980e, 1981a,b). Applications not based on a large relative bandwidth—on which essentially all of this book is based—but on an absolute bandwidth larger than zero, are discussed in a review paper (Harmuth, 1980f).

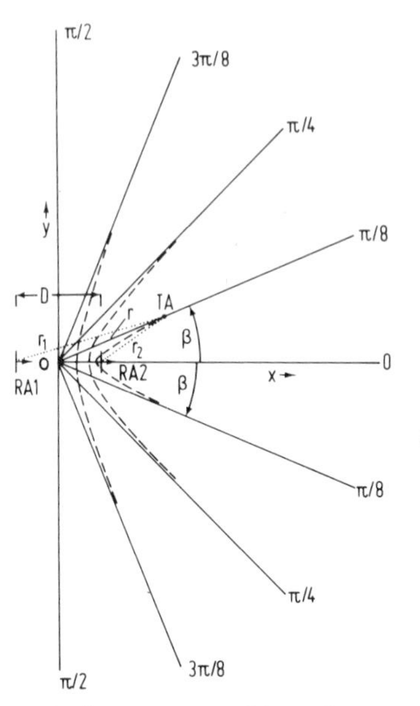

FIG. 5.3-1. Principle of a synthetic aperture tracking radar operating in a plane rather than in a three-dimensional space.

Consider Fig. 5.3-1. Two radars RA1 and RA2 are separated by a distance D. The radars shall radiate only to the right, as indicated by the arrows at RA1 and RA2. We will first discuss the principle as if the radars operated in a plane rather than in three-dimensional space.

A pulse requires the time t_1 to travel from radar RA1 to the target TA and back, and the distance

$$r_1 = t_1/2c \tag{1}$$

is obtained. Similarly, the time t_2 is required for a pulse to travel from RA2 to TA and back:

$$r_2 = t_2/2c \tag{2}$$

This method requires that both radars radiate. Instead, we can use only radar RA1 for radiation but both RA1 and RA2 for reception. Equation (1) remains unchanged, but the time t_2 is replaced by the time $t_3 = (t_1 + t_2)/2$ required by a pulse to travel from RA1 via the target TA to RA2. We obtain

$$r_1 + r_2 = t_3/c \tag{3}$$

and r_2 follows with the help of Eq. (1),

$$r_2 = (t_3 - t_1)/c \tag{4}$$

The time difference between Eqs. (2) and (1) is $t_2 - t_1$, while the time difference between Eqs. (4) and (1) equals only

$$t_3 - t_1 = \tfrac{1}{2}(t_1 + t_2) - t_1 = \tfrac{1}{2}(t_2 - t_1)$$

A fixed error of time measurement will have less effect if Eqs. (1) and (2) are used rather than Eqs. (1) and (4). The price for the reduced error will be the more complicated equipment and the doubling of the energy per distance measurement. We will assume from here on that r_1 and r_2 have been measured, and ignore whether this was done with Eq. (2) or Eq. (4).

The knowledge of r_1 and r_2 readily gives the location of the target TA, except for an ambiguity between $y > 0$ and $y < 0$. In order to form a beam, we want a quantity that shows the direction to the target. The difference $r_1 - r_2$ is such a quantity. One may infer from Fig. 5.3-1 the equation of a set of hyperbolas,

$$r_1 - r_2 = D \cos \beta \tag{5}$$

where D is the distance between the radars and β is the shown direction of the asymptotes of the hyperbolas. For distances large compared with D, we can replace the hyperbolas by their asymptotes.[1] In this case, we obtain a simple formula for the distance r of the target from the center point 0 in Fig. 5.3-1:

$$r = r_1 - \tfrac{1}{2}D \cos \beta = r_2 + \tfrac{1}{2}D \cos \beta = \tfrac{1}{2}(r_1 + r_2) \tag{6}$$

The replacement of the hyperbolas by their asymptotes permits us to advance more easily from the two-dimensional plane of Fig. 5.3-1 to a three-dimensional space in which a real tracking radar works. In three dimensions, Eq. (5) describes hyperbolic surfaces that are obtained by rotation of the hyperbolas in Fig. 5.3-1 around the x axis. The rotation of the asymptotes yields cones as shown in Fig. 5.3-2. The lower halves of the cones are cut off by the surface of the Earth, if the radars are located on the surface of the Earth.

It is evident from Fig. 5.3-2 that the knowledge of the difference $r_1 - r_2$ will tell us only on which half-cone the target is located. The additional knowledge of r_1 or $r_1 + r_2$ will yield a half-circle on this half-cone as the location of the target, according to Eq. (6). We need some additional information if we want a beam axis instead of the surface of a half-cone, and a point instead of a half-circle for the location of the target.

[1] The usual beam patterns of antenna arrays or radar dishes also hold only for distances that are large compared with their aperture.

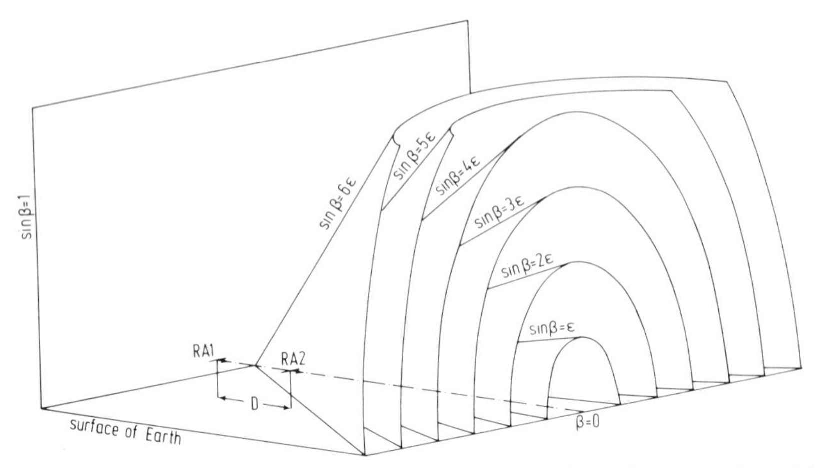

FIG. 5.3-2. Principle of a synthetic aperture tracking radar located on the surface of the Earth. The radars RA1 and RA2 radiate only into the half-space indicated by the arrows. For $\sin \beta = 1$, the cones become a plane.

5.3.2 *Beam Axis as Intersection of Two Half-Cones*

Figure 5.3-3 shows again the radars RA1 and RA2 of Fig. 5.3-1. The angle β is now denoted β_x, and the hyperbolas as well as their asymptotes are defined by the equation

$$r_1 - r_2 = D \cos \beta_x \tag{7}$$

The asymptotes for $\beta_x = 0, \pi/8, \pi/4, 3\pi/8, \pi/2$ are shown by solid lines.

A second pair of radars, RA3 and RA4, is also shown. They radiate in the direction of the positive y axis. The hyperbolas and the asymptotes associated with this pair are defined by the equation

$$r_3 - r_4 = D \cos \beta_y \tag{8}$$

The asymptotes for $\beta_y = 0, \pi/8, \pi/4, 3\pi/8, \pi/2$ are shown by dashed lines. In order to make the two sets of asymptotes more conspicuous in the first quadrant $x > 0$, $y > 0$, the dashed lines are slightly shifted parallel to the solid lines, while actually the two sets of lines coincide.

In order to transform Fig. 5.3-3 from two to three dimensions, we have to rotate the asymptotes β_x around the x axis and the asymptotes β_y around the y axis. This is shown for one value of β_x and one value of β_y in Fig. 5.3-4. (The plotting of two sets of half-cones in analogy to Fig. 5.3-2 would make the illustration too complicated.)

The two half-cones at angles β_x and β_y in Fig. 5.3-4 intersect along a straight line $\beta_x\beta_y$. This is the wanted beam—or rather the beam axis—to

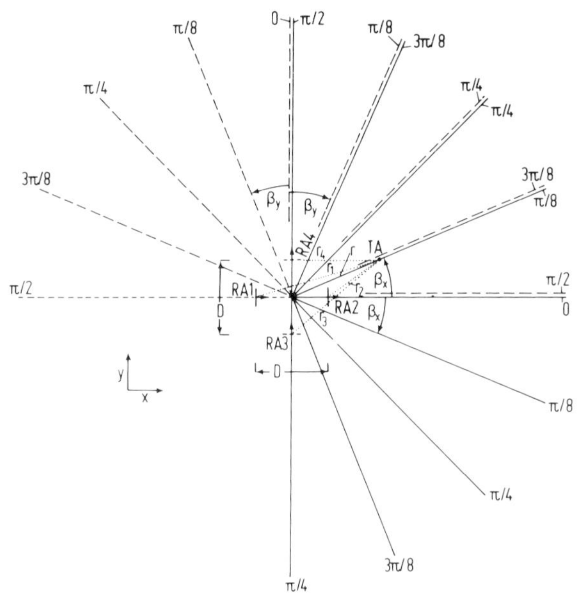

FIG. 5.3-3. The pair of radars RA1, RA2 of Fig. 5.3-1 augmented by a second pair of radars RA3, RA4 that radiate in the direction of the positive y axis, while RA1 and RA2 radiate in the direction of the positive x axis.

the target TA. The distance r from the center point 0 to TA is defined by the generalization of Eq. (6):

$$r = \tfrac{1}{4}(r_1 + r_2 + r_3 + r_4) \tag{9}$$

It is evident from Fig. 5.3-4 that we have succeeded in producing a beam axis in a way completely different from the one used by the conventional tracking radar using a parabolic dish to form a beam. In the following sections we will investigate suitable signals for such a radar and the potential resolution they provide. But let us first mention some features that can readily be recognized from Fig. 5.3-4.

Only one of the four radars in Fig. 5.3-4 has to radiate a pulse, if one wants to implement a tracking radar rather than a beam rider, but all four must receive. This leads to a generalization of the problem discussed in connection with Eqs. (1)–(4) in Section 5.3-1. A beam can only be formed in the quadrant $x > 0$, $y > 0$. To reach another quadrant, we need a method of mechanical or electronic beam steering. A mechanical movement of the beam requires the rotation of the four radars or radar receivers around an axis through the point 0. This is not impractical, since it is easier to rotate four small antennas at the ends of a large cross, than one

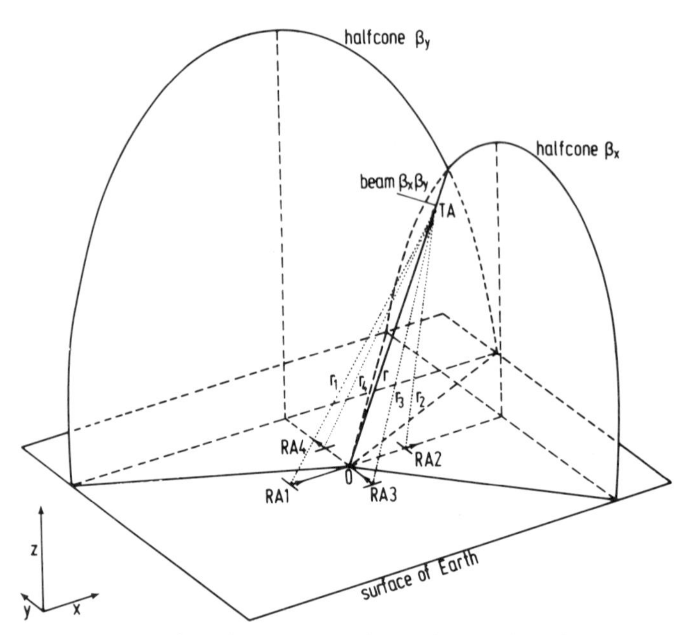

FIG. 5.3-4. Generation of a radar beam $\beta_x\beta_y$ by the intersection of two conical surfaces with angles β_x and β_y.

parabolic dish of a size comparable with that of the cross, particularly since the dish has to be rotated around two axes.

For large distances D between the radars, only electronic beam steering is practical. For an explanation refer to Fig. 5.3-5. In the upper right-hand corner the configuration of the radars in Figs. 5.3-3 and 5.3-4 is repeated. Beams will be formed in the first quadrant $x > 0$, $y > 0$.

Next let radars RA1 and RA2 radiate toward the left rather than the right, as shown in the upper left-hand corner of Fig. 5.3-5. Beams now will be formed in the second quadrant $x < 0$, $y > 0$.

Switching the direction of radiation of the radars RA3 and RA4, as shown in the lower half of Fig. 5.3-5, makes it possible to form beams in the third quadrant, $x < 0$, $y < 0$, or in the fourth quadrant $x > 0$, $y < 0$.

Our synthetic aperture radar is inherently capable of producing many simultaneous beams, which is not possible for a tracking radar with a parabolic dish. This capability implies, of course, that the signal-to-noise ratio is inherently lower, since the power is not concentrated in a narrow beam but spread fairly evenly in one quadrant.[1] We could circumvent this

[1] More precisely, the power is radiated into one octant of the three-dimensional space, e.g., $x > 0$, $y > 0$, $z > 0$ in Fig. 5.3-4. The surface of the earth eliminates the four possible octants for $z < 0$, and we may thus use the more usual term, quadrant.

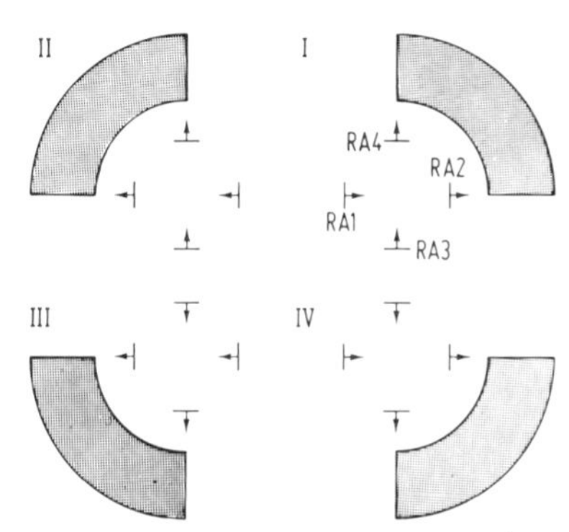

FIG. 5.3-5. Electronic switching of a radar beam, generated according to Fig. 5.3-4, into the four quadrants $x > 0$, $y > 0$ (I); $x < 0$, $y > 0$ (II); $x < 0$, $y < 0$ (III); and $x > 0$, $y < 0$ (IV).

difficulty by using parabolic dishes for the four radars in Fig. 5.3-4 to concentrate the radiated power in the direction of the target TA, and to obtain the better signal-to-noise ratio due to the antenna gain during reception. The dishes would, in this case, be used only for the improvement of the signal-to-noise ratio, and not for the determination of the direction to the target.

5.3.3 Simple Nonsinusoidal Signals

Before we can analyze the angular resolution of a beam formed by the intersection of two conical surfaces, we have to discuss the signals to be used. Figure 5.3-6 shows in line a a nonperiodic sawtooth-shaped current $i(t)$. If this current is fed to a radiator operating in the dipole mode, one obtains in the far-zone electric and magnetic field strengths varying with time like the function $f(t)$ shown in line b. Let this signal be returned by a reflector, and let it be fed into a receiver according to Fig. 5.3-7. This circuit consists of delay circuits with the delays T, $3T$, ..., $12(T + \Delta T)$, and summing circuits SUM (T), SUM($T + \Delta T$). The signal $f(t)$ at the output terminals of the delay circuits T, $3T$, $7T$, $12T$ are $f(t - T) \cdots f(t - 12T)$; these signals are shown in lines c–f in Fig. 5.3-6. The sum of the signals $f(t) \cdots f(t - 12T)$ is shown in line g; in Fig. 5.3-7 it is obtained at the output terminal g of the summer SUM(T). The signal of line a with the duration $12T$ has been compressed in line g to a pulse of duration ΔT, with an enhancement ratio 5:1 and a sidelobe-to-mainlobe ratio of 1:5. Signals with enhancement ratios of 6:1 and 11:1, and sidelobe-to-mainlobe ratios of 1:6 and 1:11 have been shown in Figs. 5.2-27, 5.2-30, and 5.2-31, and

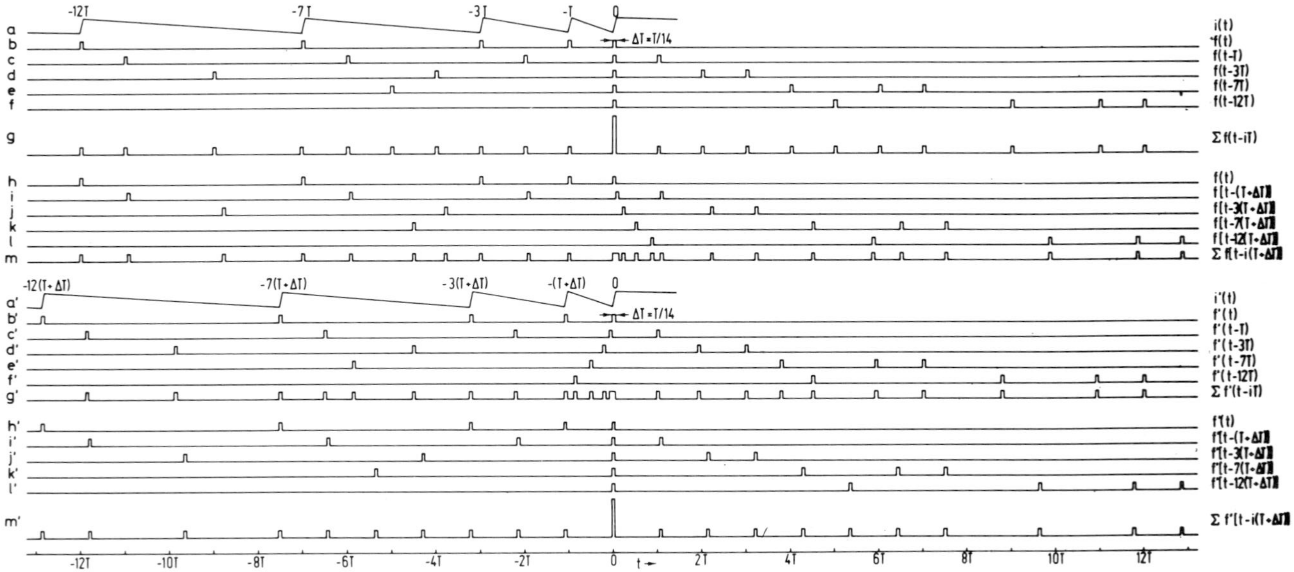

FIG. 5.3–6. Time diagram for pulse compression and selective reception of two radio signals without a sinusoidal carrier.

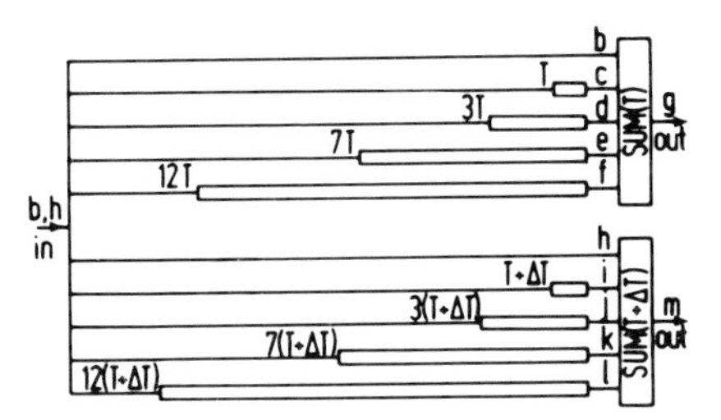

FIG. 5.3-7. Selective receiver with pulse compression for the signals of Fig. 5.3-6.

there is no known theoretical limit for the potential enhancement ratio.[1] Here we use the small enhancement ratio 5 : 1 in order to limit the width of Fig. 5.3-6 and to keep the circuit of Fig. 5.3-7 simple.

Now let the signal of Fig. 5.3-6, line b, be fed to the summer SUM($T + \Delta T$) in Fig. 5.3-7. The delay times are now $T + \Delta T$, $3(T + \Delta T)$, $7(T + \Delta T)$, and $12(T + \Delta T)$. The signals at the input terminals h ⋯ l, and their sum is shown in line m. There is no large pulse at $t = 0$ in line m.

Now let the time unit T of the antenna current $i(t)$ in Fig. 5.3-6, line a, be changed to $T + \Delta T$. The new antenna current $i'(t)$ is shown in line a′. Lines b′–f′ show the signal at the input terminals of the summer SUM(T) in Fig. 5.3-7, and line g′ the output signal of the summer. No large pulse occurs at $t = 0$. Lines h′–l′ in Fig. 5.3-6 show the input signals to the summer SUM($T + \Delta T$), and line m′ shows the output signal; there is now a large pulse at $t = 0$.

The circuit of Fig. 5.3-7 separates the signals of lines b and b′ in Fig. 5.3-6 and, in addition, compresses the signals from $12T$ to ΔT.

Let us return to Fig. 5.3-4. The four radars RA1 to RA4 can be operated so that only one radiates a signal but all four receive it, as was discussed in Section 5.3.1. In this case, we need only the antenna current of line a in Fig. 5.3-6 and the summer SUM(T) with the associated delay circuits in Fig. 5.3-7. If all four radars radiate, we need signals with time bases T, $T + \Delta T$, $T + 2\,\Delta T$, and $T + 3\,\Delta T$.

5.3.4 Beamwidth, Angular Resolution, and Accuracy of Angle

The beamwidth, i.e., the angle in which a radar according to Fig. 5.3-5 receives a return signal from the target, is[2] 2π sr, which means the azimuth

[1] There are, of course, several practical constraints. Figure 5.3-6 is used here for tutorial purposes only, since radar signals with much better enhancement ratios and sidelobe-to-mainlobe ratios will be derived in Chapter 6.

[2] sr, steradians.

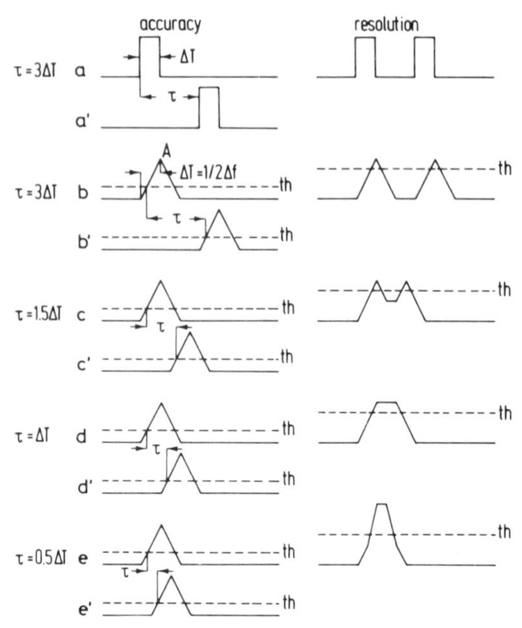

FIG. 5.3-8. Accuracy of angle measurement and resolution angle that can be achieved with signals of amplitude A and transient time $\Delta T = 1/2\,\Delta f$.

angle can have any value between 0 and 2π while the elevation angle can have any value between 0 and $\pi/2$. For this operational mode, we need four pairs of radars using different signals to avoid any ambiguity between the four quadrants. If only one pair of radars is used, as in Fig. 5.3-3, the beamwidth is $\pi/2$ sr. By means of parabolic dishes or directional antenna arrays, we can reduce this angle in order to improve the signal-to-noise ratio. The beamwidth produced by such dishes or arrays does not affect the angular resolution and the accuracy of angle measurement, except via the improvement of the signal-to-noise ratio.

Angular resolution refers to the smallest angle between two targets at "essentially" the same distance that still can be resolved. Since the synthetic aperture radar techniques discussed in Section 5.2 yield very good range resolution, it is difficult to state what "essentially" the same distance means. Thus let us turn first to the accuracy of the angle measurement, which usually refers to the determination of the azimuth and elevation angle, but will be used here for the angles β_x and β_y of the half-cones in Fig. 5.3-4; we will write β for β_x and β_y, since the same results apply to both angles.

Consider the rectangular pulses on top of the column denoted *accuracy* in Fig. 5.3-8. Line a shows the signal received by the radars RA1 or RA3 in Fig. 5.3-4, while line a′ shows the signal received by the radars RA2 or RA4. If we could produce the perfect rectangular pulses of Fig. 5.3-8, we could measure the time τ between either the leading or the trailing edges with theoretically unlimited accuracy. In reality, the pulses are not that perfect. To allow for imperfections, we assume that the measurement of τ yields actually a value between τ and $\tau + \Delta\tau$. From Eq. (5), we obtain, with the substitution

$$r_1 - r_2 = 2c\tau \tag{10}$$

the formula

$$\cos \beta = c\tau/2D \tag{11}$$

A time error $\tau - \Delta\tau$ creates an angular error $\beta + \Delta\beta$:

$$\cos(\beta + \Delta\beta) = c(\tau - \Delta\tau)/2D \tag{12}$$

The left-hand side of this equation is expanded to the quadratic term $(\Delta\beta)^2$:

$$\begin{aligned}\cos \beta \cos \Delta\beta - \sin \beta \sin \Delta\beta &= [1 - \tfrac{1}{2}(\Delta\beta)^2] \cos \beta - \Delta\beta \sin \beta \\ &= c(\tau - \Delta\tau)/2D \end{aligned}\tag{13}$$

Subtraction of Eq. (11) yields:

$$\Delta\beta = -\tan \beta + \left(\tan^2 \beta + \frac{c\ \Delta\tau}{D \cos \beta}\right)^{1/2} \tag{14}$$

For $\beta = 0$, one obtains

$$\Delta\beta = (c\ \Delta\tau/D)^{1/2} \tag{15}$$

while large values of β,

$$\beta \gg (c\ \Delta\tau/D)^{1/2} \tag{16}$$

yield the approximation

$$\Delta\beta = c\ \Delta\tau/2D \sin \beta \tag{17}$$

Figure 5.3-9 shows plots of $\Delta\beta$ as a function of β for various values of $c\ \Delta\tau/D$. In order to make the plots more readily understandable, they are also labeled D = 0.1 m ··· 100 m for a resolution time $\Delta\tau$ = 10 ps. This value of $\Delta\tau$ was chosen since it appears achievable practically in the foreseeable future. We will show later that a value $\Delta\tau$ = 1 ps can be achieved potentially by this method before severe limitations caused by nature, rather than technology, are encountered, but circuit technology has to make great advances before such a short time becomes practical.

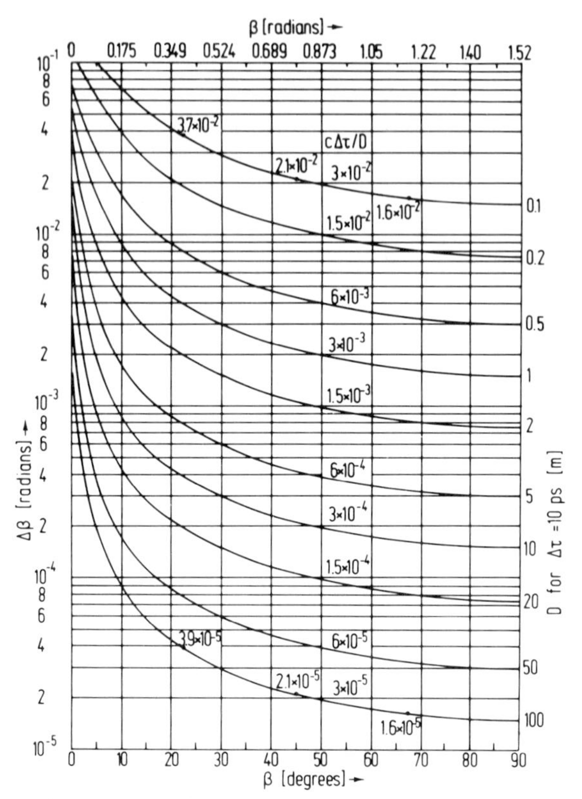

FIG. 5.3-9. Angular error $\Delta\beta$ as function of the angle β for various values of the parameter $c\,\Delta\tau/D$. The values of the distance D between the radars shown on the right hold for a time error $\Delta\tau = 10$ ps.

We can see from Fig. 5.3-9 that angular resolutions on the order of 10^{-3}–10^{-5} rad are possible, provided β is not close to zero. Let us exclude the values of β below 22.5° or $\pi/8$ rad. This implies $\beta_x \geqq \pi/8$ and $\beta_y \geqq \pi/8$ for the two pairs of radars in Fig. 5.3-3, which in turn implies that only the sector $\pi/8 \leqq \beta_x \leqq 3\pi/8$ or $\pi/8 \leqq \beta_y \leqq 3\pi/8$ is covered. Using the reversal of the direction of radiation according to Fig. 5.3-5, we obtain the four sectors shown in Fig. 5.3-10 that are covered by the radars RA1, RA2 and RA3, RA4:

$$\begin{aligned} 22.5^\circ \leq \beta_x \leq 67.5^\circ, &\qquad 112.5^\circ \leq \beta_x \leq 157.5^\circ \\ 202.5^\circ \leq \beta_x \leq 247.5^\circ, &\qquad 292.5^\circ \leq \beta_x \leq 337.5^\circ \end{aligned} \tag{18}$$

In order to cover the remaining sectors one can use a second set of four radars denoted RA5–RA8 in Fig. 5.3-10, that are rotated 45° relative to the set of radars RA1 to RA4.

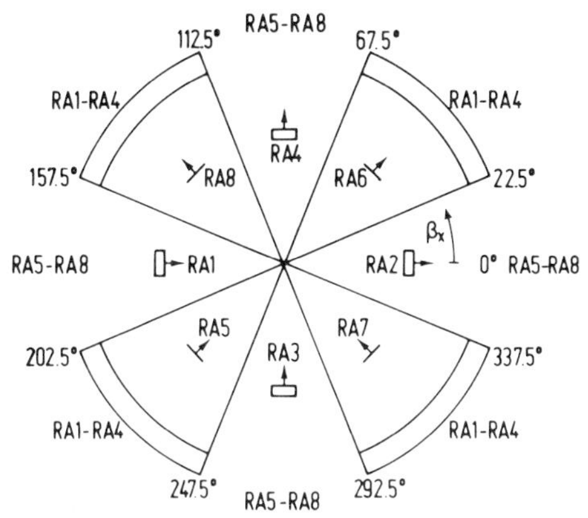

FIG. 5.3-10. Arrangement of two sets of four radars each to permit tracking in the whole half-space above the surface of the Earth without using angles β_x and β_y below 22.5° to maintain small angular errors $\Delta\beta_x$ and $\Delta\beta_y$.

The angular resolution for the arrangement of Fig. 5.3-10 may be obtained from Fig. 5.3-9. For $c\,\Delta\tau/D = 0.03$ it ranges from $\Delta\beta = 3.7 \times 10^{-2}$ for $\beta = 22.5°$ to $\Delta\beta = 1.6 \times 10^{-2}$ for $\beta = 67.5°$, etc. The ratio between $\Delta\beta$ for $\beta = 22.5°$ and $\Delta\beta$ for $\beta = 67.5°$ equals about 2.4 for any one of the curves in Fig. 5.3-9. If Fig. 5.3-10 is extended to three dimensions in analogy to Fig. 5.3-4, one obtains the following angular resolutions for β_x and β_y:

$$
\begin{aligned}
&c\,\Delta\tau/D = 0.03 \\
&\quad \beta_x = 22.5° \qquad \Delta\beta_x = 3.7 \times 10^{-2} (= 2.15°) \\
&\qquad\qquad\qquad\quad\; \Delta\beta_y = 1.6 \times 10^{-2} (= 55.6') \\
&\quad \beta_x = 45° \qquad\; \Delta\beta_x = 2.1 \times 10^{-2} (= 1.20°) \\
&\qquad\qquad\qquad\quad\; \Delta\beta_y = 2.1 \times 10^{-2} (= 1.20°) \\
&\quad \beta_x = 67.5° \qquad \Delta\beta_x = 1.6 \times 10^{-2} (= 55.6') \\
&\qquad\qquad\qquad\quad\; \Delta\beta_y = 3.7 \times 10^{-2} (= 2.15°) \\
&\qquad \vdots \\
&c\,\Delta\tau/D = 0.00003 \\
&\quad \beta_x = 22.5° \qquad \Delta\beta_x = 3.9 \times 10^{-5} (= 8.08'') \\
&\qquad\qquad\qquad\quad\; \Delta\beta_y = 1.6 \times 10^{-5} (= 3.35'')
\end{aligned}
$$

$$\begin{aligned} \beta_x &= 45^\circ & \Delta\beta_x &= 2.1 \times 10^{-5}(=4.38'') \\ & & \Delta\beta_y &= 2.1 \times 10^{-5}(=4.38'') \\ \beta_x &= 67.5^\circ & \Delta\beta_x &= 1.6 \times 10^{-5}(=3.35'') \\ & & \Delta\beta_y &= 3.9 \times 10^{-5}(=8.08'') \end{aligned} \tag{19}$$

Next let us investigate the angular error $\Delta\beta$ caused by an error in the distance D between the two radars of a pair. The actual distance shall be between D and $D + \Delta D$. In analogy to Eq. (12), we obtain:

$$\cos(\beta + \Delta\beta) = c\tau/2(D + \Delta D) \doteqdot c(\tau - \Delta D\tau/D)/2D \tag{20}$$

A comparison with Eq. (12) shows that $\Delta\tau$ has been replaced by $(\Delta D\tau/D)$. Hence, Eq. (14) assumes the following form:

$$\Delta\beta = -\tan\beta + \left(\tan^2\beta + \frac{c\tau\,\Delta D}{D^2\cos\beta}\right)^{1/2} \tag{21}$$

With the help of Eq. (11), we can eliminate τ:

$$\Delta\beta = -\tan\beta + (\tan^2\beta + 2\,\Delta D/D)^{1/2} \tag{22}$$

For $\beta = 0$, one obtains

$$\Delta\beta = (2\,\Delta D/D)^{1/2} \tag{23}$$

whereas large values of β,

$$\beta \gg (2\,\Delta D/D)^{1/2} \tag{24}$$

yield the approximation

$$\Delta\beta = \Delta D/D\tan\beta \tag{25}$$

Plots of $\Delta\beta$ as a function of β for various values of $\Delta D/D$ are shown in Fig. 5.3-11. As in Fig. 5.3-9, the error $\Delta\beta$ is largest for $\beta = 0$. The disappearance of $\Delta\beta$ for $\beta = \pi/2$ is surprising at first glance, but Fig. 5.3-1 shows that the straight line for $\beta = \pi/2$ does not depend on the distance between the radars, and Eq. (11) shows that $\beta = \pi/2$ follows from $\tau = 0$ for any value of D. Figure 5.3-11 suggests, like Fig. 5.3-9, that values of β below about 22.5° should not be used.

We return to Fig. 5.3-8 to study the influence of the pulse shape on the accuracy of angle measurement and on the resolution. Let the pulses of duration ΔT in lines a and a′ of the left column pass through an idealized low-pass filter with nominal bandwidth[1] $\Delta f = 1/2\,\Delta T$. The triangular

[1] The justification of this approximation was discussed in Section 2.1.

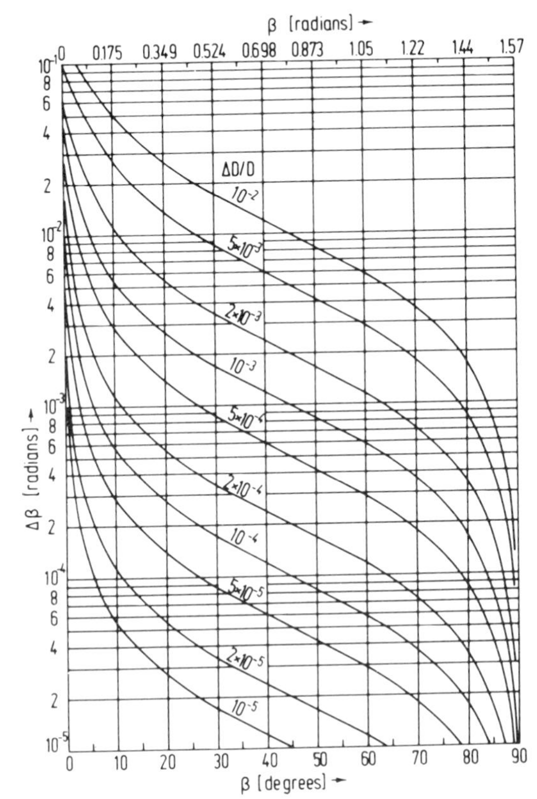

FIG. 5.3-11. Angular error $\Delta\beta$ as a function of the angle β for various values of the parameter $\Delta D/D$, where D is the distance between the radars and ΔD is the error in this distance.

pulses with transient times ΔT and peak amplitude A shown in lines b and b′ are obtained.

The time difference τ between the two pulses in lines b and b′ can be measured by means of two level detectors set to the threshold voltage th. One may see from lines c, d, and e that there is in principle no lower limit on the measurable time difference τ. Practically, the smallest resolvable time difference depends on the slope of the triangular pulses,

$$A/\Delta T = 2A\,\Delta f \tag{26}$$

which increase proportionately to the bandwidth Δf and the signal amplitude A, or proportionately to $\sqrt{P}$, where P is either the peak signal power or an average signal power. Let us assume that the smallest resolved time $\tau = \Delta\tau$ is one tenth of the transient time ΔT. For a bandwidth $\Delta f = 1$ GHz we obtain $\Delta\tau = \Delta T/10 = 1/20\,\Delta f = 50$ ps. This is about the best that current circuit technology will permit. Nature permits a bandwidth of about 10 GHz for an all-weather radar operating in rain and fog, which yields $\Delta\tau = 5$ ps. For a fair-weather radar that need operate only in weather with good visibility, we can extend Δf to about 15 GHz, before the H_2O resonance at 22.2 GHz produces signal distortions, and the corresponding value of $\Delta\tau$ drops to 3.3 ps. For a good signal-to-noise ratio that permits a resolution $\Delta\tau = \Delta T/33$ rather than $\Delta\tau = \Delta T/10$, one arrives at a potential time resolution $\Delta\tau = 1$ ps. The more conservative figure $\Delta\tau = 10$ ps is used for the values of $D = 0.1$ m $\cdots$ 100 m shown on the right margin of Fig. 5.3-9.

When we will be able to operate reliably with time intervals of 10 and 1 ps duration is rather speculative. The transition from 1 μs to 1 ns was accomplished between approximately 1940 and 1970, but it was due to the transition from electron tubes to semiconductor devices. We can currently observe the advancement to 100 ps. Josephson junction devices achieve switching times of 10 ps and less, but they operate at the temperature of liquid helium, which impedes their widespread use (Anacker, 1979).

Let us turn to the problem of resolution. The right column in Fig. 5.3-8 shows, in line a, rectangular pulses received from two targets that have almost the same distance from the radar, which may be RA1, RA2, RA3, or RA4 in Fig. 5.3-4. No plot is shown in line a′, since each radar individually must resolve two targets, while the accuracy of angle measurement was a task that required a pair of radars.

Line b shows the same pulses after passing through an idealized low-pass filter with nominal bandwidth $\Delta f = 1/2\,\Delta T$. A level detector set to the threshold th will resolve the two pulses. Line c shows that the threshold detection still resolves at a time difference $\tau = 1.5\,\Delta T$, while line d shows that the limit for the resolution by a threshold detector is reached for $\tau = \Delta T$, in which case the threshold has to be set to the peak amplitude of the two pulses.

The two signals in lines d and e can no longer be resolved by a threshold detector. However, these pulses are obviously different from a triangular pulse as shown in line b, and one will conclude that a more sophisticated processing technique than threshold detection will resolve them. Such a method was discussed in Section 5.2.6. In essence, this method yields an angular resolution that is equal to the accuracy of angle measurement shown in the left column of Fig. 5.3-8. The difference is that one needs only one threshold for accurate angle measurement, and this threshold can be set where the influence of distortions and noise is minimal, e.g., at half the peak amplitude of the pulses in Fig. 5.3-8, lines b, b′ to e, e′. For the resolution of two or more targets, on the other hand, one needs two thresholds and they have to be set at a fraction of the peak amplitude. Hence, a higher signal-to-noise ratio is required and deviations of the actual pulse shape from the theoretical one will be more important.

5.3.5 *Modes of Operation*

A configuration for a radar that covers the half-space above the surface of the Earth or a spherical angle of 2π sr has been shown in Fig. 5.3-10. From Fig. 5.3-4, which shows only four rather than eight radars, one may infer that this is inherently a surveillance radar which observes continuously in all directions without scanning. There is, however, little incentive to use a radar with very good angular resolution for surveillance.

In order to transform the configuration of Fig. 5.3-10 into that of a tracking radar, we must introduce directional antennas in order to use the antenna gain both for radiation and reception. The directional antennas may either be mechanically steerable parabolic dishes or electronically steerable radiator arrays. The use of directional antennas reduces immediately the problem of ghosts produced by multiple targets.

The problem of ghosts does not occur if the configuration of radars in Fig. 5.3-10 is used to guide a beam rider to a target. In this case, we can radiate into the spherical angle of 2π sr, since the signal power required for the beam rider is much smaller than that required for a tracking radar, and produce

$$4(\tfrac{1}{2}\pi/\Delta\beta_x)(\tfrac{1}{2}\pi/\Delta\beta_y) = \pi^2/\Delta\beta_x\,\Delta\beta_y$$

beams. The beam rider follows a course on which the time difference τ_x between the arrival of the signal from two transmitters RA1 and RA2 in Fig. 5.3-3, as well as the difference τ_y between the arrival of the signals from RA3 and RA4, is constant. This course is a straight line—except very close to the transmitters, as shown in Fig. 5.3-1—which is desirable for a beam rider due to Newton's law of inertia. Hence, an arrangement of pulse radiators without sinusoidal carriers according to Fig. 5.3-10 is an all-weather control guidance system with great angular accuracy. For instance, a resolution of 2 m at a distance of 40 km requires a spacing of the transmitters of 40 m for $\Delta\tau = 10$ ps. In conjunction with the previously discussed tracking radar, which has the same resolution, we arrive at an all-weather command/control system that may need no seeker guidance in applications like ICBM defense.[1]

[1] It has been pointed out previously that pulses with a duration between 1 and 0.1 ns can discriminate between a scatterer, such as the nose of an ICBM, and a reflector, such as a radar decoy (Harmuth, 1977a, Section 3.3.6).

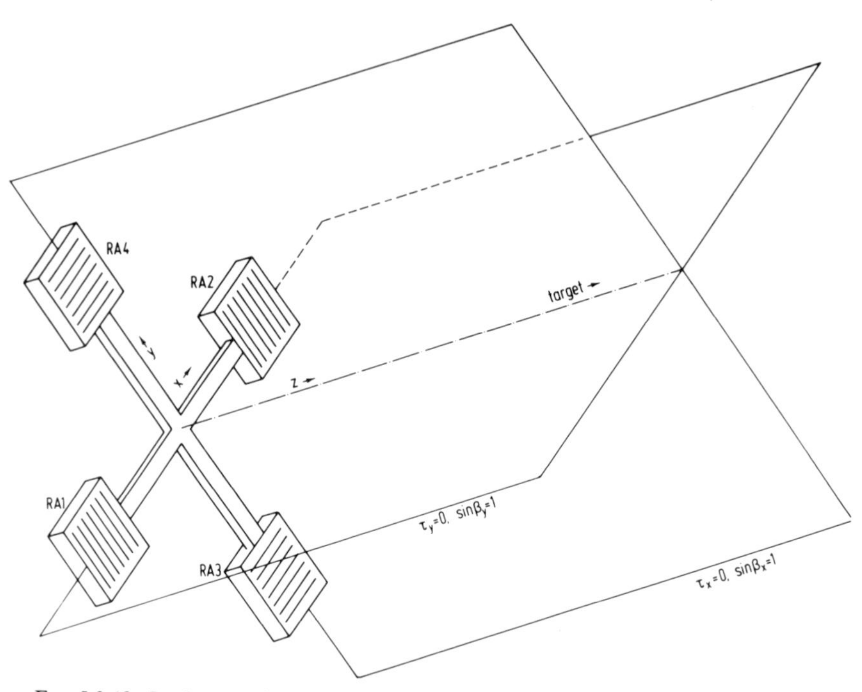

FIG. 5.3-12. Implementation of the principle of Fig. 5.3-4 by means of the four radar transmitters and receivers mounted at the ends of a cross with arms of length D. The beam axis z to the target is formed by the intersection of the plane $\sin\beta = \sin\beta_x = 1$ shown in Fig. 5.3-2 for RA1 and RA2, with a similar plane for RA3 and RA4 denoted $\sin\beta_y = 1$. The radiators/receptors RA1, RA2 are shown polarized normal to radiators/receptors RA3, RA4.

The radars and transmitters discussed so far were assumed to be mounted on the ground as shown in Fig. 5.3-4. However, the principle can also be applied to a movable radar as shown in Fig. 5.3-12. The four radars or radiators/receptors RA1 to RA4 are mounted at the ends of the arms of a cross. The "beam axis" to the tracked target is the intersection of the planes on which the time differences for the directions x and y vanish, which means $\tau_x = \tau_y = 0$. In terms of Fig. 5.3-2, one may also write with the help of Eq. (11), $\sin\beta_x = \sin\beta_y = 1$. It follows from Fig. 5.3-11 that the tolerance ΔD of the length of the arms of the cross does not affect the accuracy of angle measurements, while Fig. 5.3-9 shows that the angular error $\Delta\beta$ as a function of the time error $\Delta\tau$ is at its minimum.

Let us compare a radar according to Fig. 5.3-12 radiating pulses without a sinusoidal carrier with one that uses a sinusoidal carrier. The radiators/receptors shall be in the same location in both cases. If the highest significant frequency used, as well as the signal power, are the same for both radars, we must get the same angular resolution, since the angle is measured in both cases by the time differences of the signals radiated and received by RA1 to RA4. In the case of the sinusoidal carrier, we can use the term phase difference rather than time difference, but phase difference and time difference are the same thing, except for the ambiguity of the phase difference for multiples of the period of the sinusoid. The practical way to make the time differences τ_x and τ_y zero for the sinusoidal signal is to use the monopulse technique. This means the difference of the signals received by RA1 and RA2 is produced, and the structure of Fig. 5.3-12 is rotated so that the difference vanishes. We had discussed a different method of the measurement of time differences in connection with Fig. 5.3-8 for a radar without sinusoidal carrier. However, it is obvious from Fig. 5.3-8, left column, that we could subtract the signal in line e′ from that in line e to obtain no output for $\tau = 0$, and also an indication whether τ is positive or negative for $\tau \neq 0$. This technique is thus a way to transfer the monopulse principle to a radar without a sinusoidal carrier.

Since the angular resolution is the same, we can ask, "What is the significant difference between using a sinusoidal carrier and not using one with the structure of Fig. 5.3-12?" The answer is, for an all-weather radar, that the transmitted information is about two orders of magnitude greater if no sinusoidal carrier is used. If we choose the highest usable significant frequency to be about 10–15 GHz, we are restricted to a signal bandwidth of about 100–150 MHz with a sinusoidal carrier, but we can use a bandwidth of about 9–14 GHz without a sinusoidal carrier. The difference

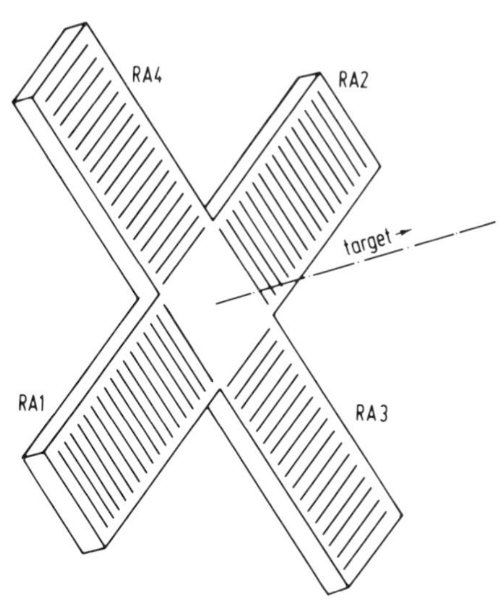

FIG. 5.3-13. Variation of the structure of Fig. 5.3-12 to resemble two crossed-line arrays.

shows up primarily in an improvement of the range resolution by two orders of magnitude, but this implies secondary effects like multiple-target discrimination or a target signature with more information.

There are many possible variations of the structure of Fig. 5.3-12. The radiators/receptors do not have to be concentrated at the ends of the arms of the cross but can fill up the arms as shown in Fig. 5.3-13. This structure looks like two crossed line arrays for sinusoidal waves. The difference between the two arrays may be inferred from the discussion of dipole arrays in Section 3.2.2. The structure of Fig. 5.3-12 will give more power in the mainlobe, but the structure of Fig. 5.3-13 will yield a smaller rest-lobe.

5.4 Look-Down Radar

5.4.1 Problem of Resolution

One of the most challenging tasks in radar is the detection of low-flying aircraft and cruise missiles. With the help of a terrain follower, a cruise missile can fly at an altitude of 20 m above water and 40 m above land. We have seen in Section 5.1, that the performance of ground-based radar against such low-flying targets can be improved by means of nonsinusoidal electromagnetic waves. Two principles were involved in this application: (a) Pulses as short as 0.1 ns can be transmitted without exceeding a highest significant frequency of about 10 GHz. (b) The first-order multipath effect can be combated due to the difference between an amplitude-reversed and a delayed electromagnetic wave.

The range of a ground-based radar is severely limited by topographic features and the curvature of the Earth, which makes such a radar suitable for fire control but not for surveillance. Furthermore, the cross section of a cruise missile seen head-on is very small, and the returned radar signal is thus very weak. The solution to the problem of a weak signal as well as a limited surveillance range is to mount the radar on a high-flying aircraft, and to look down.

Two major problems have to be overcome to operate a radar successfully in this mode. The highest significant frequency used should not be higher than about 10 GHz to provide all-weather capability. For an aircraft flying at an altitude of 30 km, the horizon is about 600 km away, and the area that can be surveyed is more than 10^6 km^2 large. One cannot expect fair weather to prevail over such a large area very often, and clouds or rain would be used routinely as protection against a radar that has no all-weather capability.

The second problem is to resolve the tracked target from the huge clutter signal produced by the surface of the Earth. The preferred means to do so is the Doppler effect.[1] Nonsinusoidal electromagnetic waves yield about two orders of magnitude better time resolution than waves with a sinusoidal carrier, if the highest significant frequency used is limited. Since the Doppler effect is caused by the compression or expansion of time intervals due to a relative motion, one will expect that the good time resolution can be turned into a good Doppler resolution.

Let us derive some formulas and numerical values that characterize the use of the Doppler effect to overcome the clutter problem. Figure 5.4-1 shows a radar RA at an altitude H_R above the surface of the earth. The radar shall be stationary relative to the earth. This assumption is inherently unrealistic, but it simplifies the problem greatly, and it can be approximated if the airplane carrying the radar is much slower than the target. The method itself does not depend on this assumption, only its explanation becomes more difficult for a moving radar. The target TA flies at the altitude H_T with the velocity v, but the component of this

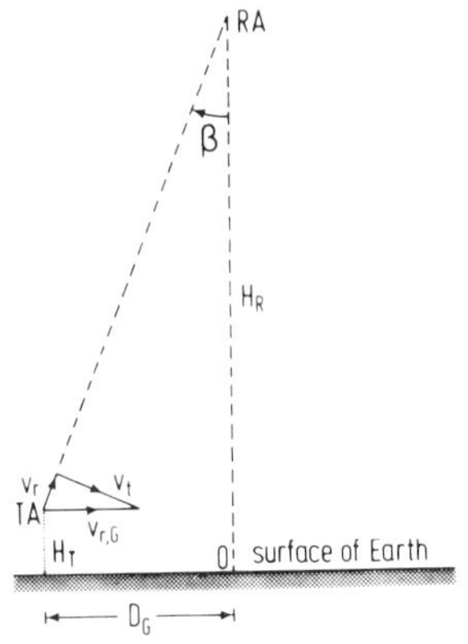

FIG. 5.4-1. Radar RA and target TA in the vertical plane shown by the trace D_G in Fig. 5.4-2 in the ground plane. H_R and H_T, altitudes of the radar and the target; $v_{r,G}$, velocity relative to the ground in the direction of the radar (see Fig. 5.4-2); v_r, velocity of target relative to the radar in the direction of the radar; v_t, velocity component of $v_{r,G}$ transversal to the direction of the target.

[1] More precisely, the longitudinal Doppler effect. The transversal Doppler effect (Born, 1962, p. 300; 1964, p. 259) is a theoretical possibility when the longitudinal Doppler effect vanishes due to the geometric relations, but it is too small for practical use. Another effect is the modulation of the radar signal by the vibrations of the target; this effect was already used in World War II to distinguish aircrafts from chaff. A new and not yet experimentally investigated effect is the difference between the Brewster angle of properly polarized waves for a metallic target and the surface of the sea or land as discussed in Section 5.1.2.

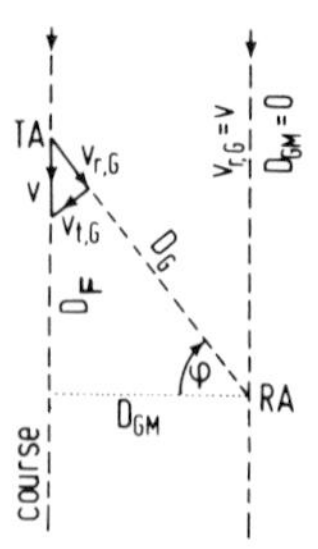

FIG. 5.4-2. Top view of the radar RA and the target TA of Fig. 5.4-1. v, velocity of the target with the surface of the Earth as reference; $v_{r,G}$, relative velocity in the ground plane in the direction of the radar; $v_{t,G}$, component of v in the ground plane transversal to the direction of the target. Note that the course of the target is defined by the minimum ground distance D_{GM} from the radar RA, regardless of the compass heading of the target. This simplification is the reason for the assumption of a negligible ground speed of the radar.

velocity in the direction of the vertical line from the radar RA to the point 0 on the surface of the earth is the relative ground velocity $v_{r,G}$. The relation between v and $v_{r,G}$ may be seen from Fig. 5.4-2, which is a top view of the radar RA and the target TA. The course of the target passes the radar RA at the minimum ground distance D_{GM}. For a course passing directly under the radar, $D_{GM} = 0$, one has $v_{r,G} = v$.

The connection between v_r and $v_{r,G}$ follows from Fig. 5.4-1:

$$v_r = v_{r,G} \sin \beta = v_{r,G} D_G[(H_R - H_T)^2 + D_G^2]^{-1/2} \tag{1}$$

From Fig. 5.4-2 we obtain the relation between v and $v_{r,G}$:

$$v_{r,G} = v \sin \varphi = v(1 - \cos^2 \varphi)^{1/2} = v[1 - (D_{GM}/D_G)^2]^{1/2} \tag{2}$$

From Eqs. (1) and (2) follows v_r as function of v:

$$v_r = vD_G[1 - (D_{GM}/D_G)^2]^{1/2}[(H_R - H_T)^2 + D_G^2]^{-1/2} \tag{3}$$

The Doppler shift of the frequency f of a sinusoidal wave transmitted from the radar RA to the target TA and returned to the radar in first order of v_r/c is denoted f':

$$f' = f(1 - 2v_r/c) \tag{4}$$

The choice of the sign $-2v_r/c$ indicates that a velocity toward the radar is negative while a velocity away from the radar is positive.

We will use the periods or time intervals $T_D = k/f$ and $T_D' = k/f'$ instead of the frequencies, where k is the number of oscillations in the time intervals T_D or T_D'. With the same approximation as Eq. (4) one obtains

$$T_D' = T_D(1 + 2v_r/c) \tag{5}$$

This equation does not only hold for time intervals T_D and T'_D that happen to be multiples of the periods of sinusoidal waves, but for any time interval.[1]

Since the change of the period T_D to T'_D is a function of the relative velocity v_r only, we express the ground distance D_G in Eq. (3) as function of v_r, or rather of v_r/v:

$$D_G = [D_{GM}^2 + (H_R - H_T)^2 v_r^2/v^2]^{1/2}(1 - v_r^2/v^2)^{-1/2} \tag{6}$$

For the purpose of plotting it is convenient to use the distance D_F instead of D_G in Fig. 5.4-2:

$$D_F = (D_G^2 - D_{GM}^2)^{1/2} \tag{7}$$

$$D_F = [D_{GM}^2 - (H_R - H_T)^2]^{1/2}(v^2/v_r^2 - 1)^{-1/2} \tag{8}$$

Figure 5.4-3 shows for which points, with the Cartesian coordinates D_{GM} and D_F according to Fig. 5.4-2, the ratio v_r/v has the same value,

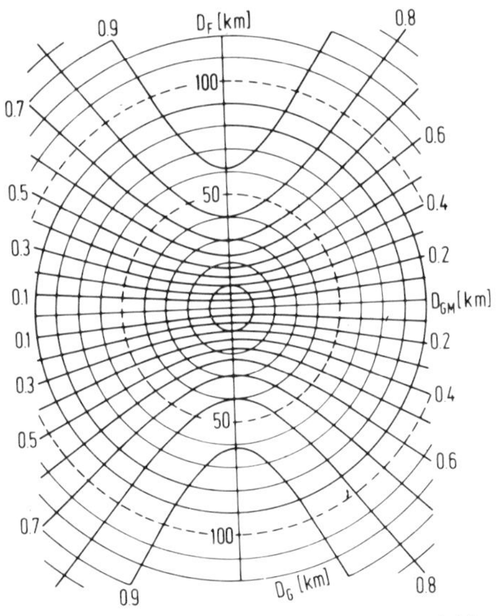

FIG. 5.4-3. Lines of constant relative velocity $v_r/v = 0.1 \cdots 0.9$ between a target and the radar located in the center ($D_{GM} = D_F = 0$). The altitude of the radar is 30 km, that of the target may be 20 or 40 m, or any other value that is small compared with 30 km. The distances D_G, D_F, and D_{GM} are in kilometers. See Fig. 5.4-2 for the definition of these distances.

[1] For a more formal, explicit derivation of this result see Harmuth (1977a, p. 272). It follows readily from the transform equations of relativistic mechanics for the time and space variables.

$v_r/v = 0.1 \cdots 0.9$. The circles show the ground distance D_G. The radar is located in the center, $D_{GM} = D_F = 0$, at an altitude $H_R = 30$ km. The altitude H_T of the target is assumed to be either 20 or 40 m, but it is of no significance due to the much larger value of H_R.

The interpretation of Fig. 5.4-3 is as follows. If a target with velocity[1] v is in the area above the curve $v_r/v = 0.1$ in the positive half-plane ($D_F > 0$) or below the curve $v_r/v = 0.1$ in the negative half-plane ($D_F < 0$), it can be discriminated from the ground clutter if the radar can resolve a relative velocity $v_r > 0.1v$. Only in the area between the two lines $v_r/v = 0.1$ will the target disappear in the ground clutter. Corresponding statements hold for the other values $v_r/v = 0.2 \cdots 0.9$ shown in Fig. 5.4-3.

It is evident from Fig. 5.4-3 that values $v_r/v = 0.9, 0.8, \ldots$ permit detection and tracking of a target only in a small fraction of the observed area, but a value $v_r/v = 0.1$ appears to give a satisfactory coverage. This is not quite so, since the horizon of an airplane at an altitude of 30 km is about 600 km away, while Fig. 5.4-3 extends only to $D_{GM} = 90$ km. For large values of D_{GM} we may ignore $H_R - H_T$ in Eq. (8), and we may also ignore 1 compared with v^2/v_r^2 for $v_r/v \leqq 0.1$. One obtains

$$D_F \doteqdot D_{GM} v_r/v \tag{9}$$

At a distance $D_{GM} = 600$ km we obtain with $v_r/v = 0.1$ the value $D_F = 60$ km. The distance between the two lines for $v_r/v = 0.1$ in Fig. 5.4-3 is thus 120 km at the horizon $D_{GM} = 600$ km. A target flying with a velocity $v = 300$ m/s would require 400 s to travel these 120 km. This is a serious gap for a radar. For $v_r/v = 0.01$ the gap is reduced to 12 km and 40 s. This implies the Doppler resolution of a relative velocity $v_r = 3$ m/s for the typical velocity $v = 300$ m/s. For a reason immediately to be seen, we will investigate a Doppler resolution of $v_r = 1.5$ m/s, which implies a surveillance gap of 6 km or 20 s at the horizon. It follows from Eq. (5) that we must resolve a relative time difference

$$(T_D' - T_D)/T_D = 2v_r/c = 2 \times 1.5/3 \times 10^8 = 10^{-8} \tag{10}$$

to discriminate the relative velocity $v_r = 1.5$ m/s.

We will investigate how this time resolution can be achieved with an all-weather radar. The long range, up to 600 km in our example, also requires the consideration of adequate signal power as well as angular resolution and accuracy of angle measurement. The problem of angular resolution and accuracy was investigated in Section 5.3, and the extension of

---

[1] Low-flying aircraft or cruise missiles have typically a velocity just below the velocity of sound for a variety of reasons. We will generally assume $v = 300$ m/s.

the synthetic aperture techniques discussed there to an airborne radar will be tacitly assumed.[1]

5.4.2 *Digitally Controlled Waves*

In order to resolve relative time differences of the order 10^{-8} one must radiate the electromagnetic wave with relative short-time errors of about 10^{-9}. Digital circuits controlled by a quartz clock will yield such small relative short-time errors. The jitter of pulses produced in this way is currently about 100 ps, while the very best equipment is approaching 10 ps. For a relative short-time error of 10^{-9} we must thus use time intervals T_D of the order of 100–10 ms. For a velocity of $v = 300$ m/s the target will move during this time 30–3 m, which means it will advance about its own length.

A typical signal to be radiated will be explained with reference to Fig. 5.4-4. A periodic sequence of pulses with period T_D is shown on top. If the duration of these pulses is in the order of 0.1–1 ns and the period T_D is in the order of 10–100 ms, we can in principle observe the Doppler shift with an error of 10^{-8} s or less. However, the peak pulse power would have to be extremely high to achieve a practical signal-to-noise ratio. In order to reduce the peak power—or the peak-to-average power ratio—we introduce two levels of coding for the pulses of line a.

The first level is a coding of pulse positions. Each pulse in line a is replaced by the 15 pulses of line b. The pulse on the extreme left, denoted

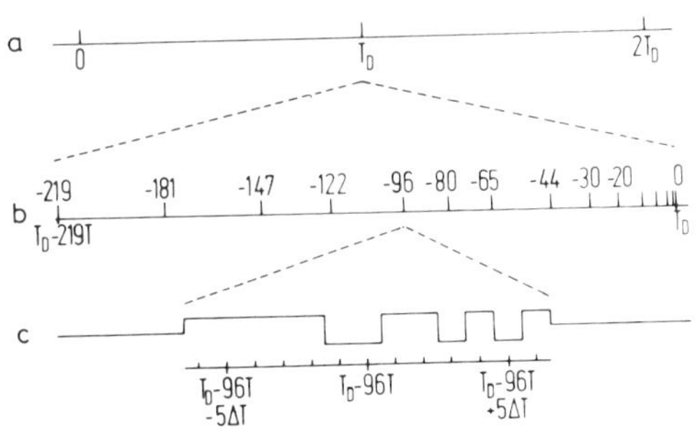

FIG. 5.4-4. Periodic, nonsinusoidal electromagnetic pulses with period T_D for good Doppler resolution (a). Fine structure of these pulses with time unit T for enhancement of the peak-to-average power ratio (b), and hyperfine structure with time unit ΔT (c). $T_D \gg T \gg \Delta T$.

[1] Figure 5.3-12 shows four radiators/receptors mounted at the ends of the arms of a cross. The fuselage and the wings of an airplane provide such a cross.

-219, occurs at the time $T_D - 219T$, the next pulse at the time $T_D - 181T$, etc., with the last pulse—denoted 0—occurring at the time T_D.

The second level of coding is a coding of pulse shapes. Each pulse in line b is replaced by the sequence of pulses in line c, which is a Barker code with 13 pulses (Barker, 1953; Lange, 1971, Vol. 3, p. 315). The duration of these pulses is denoted ΔT, and they are centered at the pulse location specified in line b, e.g., at $T_D - 96T$ for the pulse -96 in line b.

Let ΔT be about 1 ns, T about 100 ns and T_D about 100 ms. The pulses in lines a and b would then appear as shown. Only on a much larger scale would the pulses of line a show the structure due to the coding of pulse position and pulse shape, while the pulses of line b would show the structure due to the coding of the pulse shape.

What has been gained? Each pulse in line a has been replaced by 15 pulses in line b, and by $15 \times 13 = 195$ pulses in line c. Hence, the peak power is reduced to 1/195 for the same radiated energy. Of course, we must be able to compress these 195 pulses into a time interval of duration ΔT if we want to retain the Doppler resolution.

Before we investigate this pulse compression, we discuss a more complicated case of pulse position and pulse shape coding with reference to Fig. 5.4-5. Line a in Fig. 5.4-5 is equal to line a in Fig. 5.4-4. However, line b now contains twice as many pulses. This line is broken up into the two lines b_1 and b_2. Line b_2 is equal to line b in Fig. 5.4-4 for $T = T_1$, while line b_2 differs by a shift $-35T_1$ and a different time unit $T_2 = 1.1T_1$ from

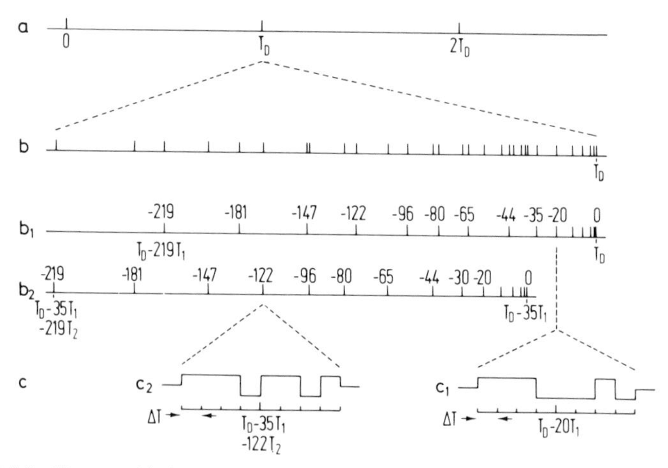

FIG. 5.4-5. More sophisticated pulse position and pulse shape coding than in Fig. 5.4-4. Pulses with period T_D for good Doppler resolution (a). Fine structure of these pulses (b), consisting of two pulse sequences with time units T_1 (b_1) and T_2 (b_2), and different hyperfine structures for the two pulse sequences (c_1, c_2).

line b_1. Hence, the pulse position coding in Fig. 5.4-5 is more sophisticated than in Fig. 5.4-4. There is still another difference. Each pulse in line b_1 has the pulse shape shown in line c_1, and each pulse in line b_2 the shape shown in line c_2. Lines c_1 and c_2 are not equal. The sequence of pulses in these two lines is usually referred to as a *complementary code* (Erickson, 1961; Golay, 1961). The important features of complementary codes are that the sum of their autocorrelation functions has no sidelobes, that their maximum known length is *not* 13 as in the case of the Barker codes, and that there are many more complementary codes than Barker codes.[1]

Figures 5.4-4 and 5.4-5 are examples of digitally controlled electromagnetic waves that are useful for the detection of small Doppler shifts. It is quite obvious that digital circuits permit us to produce such sequences with the required length as well as the required jitter. Before we go into a more detailed discussion of pulse position and pulse shape coding, it appears appropriate to make some comments about the influence of the Doppler shift on Barker and complementary codes.

Barker codes have been known since 1953, and complementary codes since 1961. Both are reputed to give poor performance in the presence of a Doppler shift. However, this statement refers to codes that are long—of the order of T_D—and use a sinusoidal carrier with a period of order ΔT for fine structure. We are here using Barker codes and complementary codes for the fine structure, their length is of the order ΔT rather than T_D in Figs. 5.4-4 and 5.4-5. The Doppler shift is insignificant for such short intervals since we have chosen T_D as short as possible for an observable Doppler effect, and ΔT is eight orders of magnitude smaller. Hence, we need not be concerned here about the effect of a Doppler shift on Barker codes or complementary codes.

5.4.3 Pulse Generation and Compression

The generation and compression of signals according to Fig. 5.4-4 has already been discussed in Section 4.6. Thus we turn to the more complicated principle of the receiver for an electromagnetic wave according to Fig. 5.4-5 with the two interlaced fine-structure pulse sequences of line b and the complementary codes of line c for hyperfine structure. The currents $i_1(t)$ and $i_2(t)$ required to flow in a dipole mode radiator to produce

---

[1] A large number of codes implies better protection against electronic countermeasures. Those familiar with electronic countermeasures will have no trouble recognizing advantages of an electromagnetic wave according to Fig. 5.4-5. The signals are inherently spread-spectrum signals, and their digital generation is ideally suited for frequent changes.

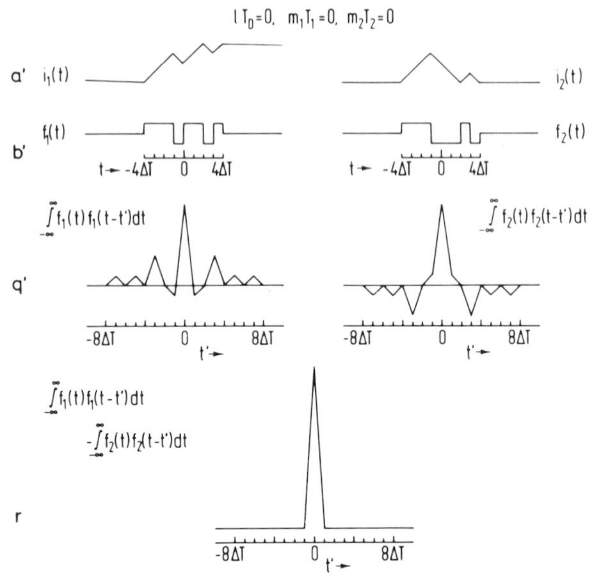

FIG. 5.4-6. Hyperfine structure of the pulses in line a of Fig. 5.4-5. Sawtooth currents fed into the radiator (a'), the electric and magnetic field strengths in the far zone produced by it (b'), autocorrelation functions of the field strengths (q'), and the sum of the autocorrelation functions (r).

the complementary codes $f_1(t)$ and $f_2(t)$ are shown in Fig. 5.4-6, line a', while $f_1(t)$ and $f_2(t)$ are shown in line b'. The autocorrelation functions of $f_1(t)$ and $f_2(t)$ are shown in line q', and their sum in line r. One advantage of complementary codes compared with Barker codes, the absence of sidelobes, is apparent from line r.

Figure 5.4-7 shows the principle of the receiver. The pulses of line b_1, Fig. 5.4-5, are compressed by delay circuits with delays $219T_1$, $181T_1$, $147T_1$, ..., $3T_1$, T_1, 0, and the summer SUM1. The pulses of line b_2 are compressed by the circuits with delays $219T_2$, $181T_2$, $147T_2$, ..., $3T_2$, T_2, 0, and the summer SUM2. The circuit with delay $35T_1$ assures that the compressed pulses at the output terminals q and q' of the summers coincide.

The output signals of the summers SUM1 and SUM2 are fed to sliding correlators that are set for the signals $f_1(t)$ and $f_2(t)$ of Fig. 5.4-6, line b'. One obtains at their output terminals q_1' and q_2' the autocorrelation functions of Fig. 5.4-6, line q', if the signals $f_1(t)$ and $f_2(t)$ were received. The sidelobe-free signal of line r, Fig. 5.4-6, is produced at the output terminal r of the summer SUM5 in Fig. 5.4-7.

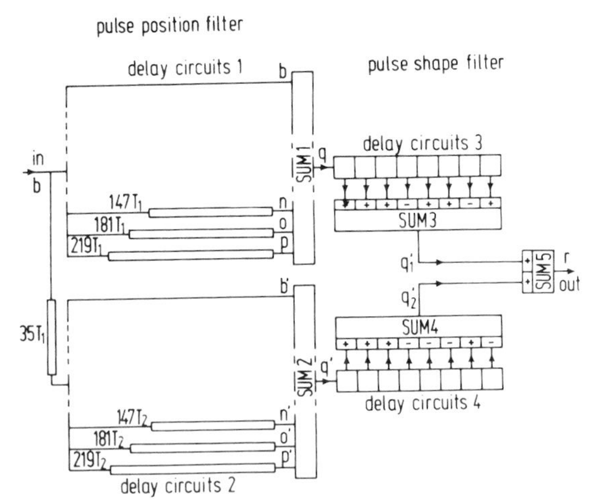

FIG. 5.4-7. Principle of pulse compression for the reception of signals according to Fig. 5.4-5. The pulse position filter compresses the 15 pulses of line b_1, Fig. 5.4-5, into one at the terminals q, and those of line b_2 into one at the terminal q′. The pulse shape filter compresses the 2 × 8 pulses of line c, Fig. 5.4-5, by means of cross correlation and summation.

5.4.4 *Clutter Cancellation*

The signal of Fig. 5.4-6, line r, is obtained at the times lT_D at the output terminal r of the circuit of Fig. 5.4-7; the times lT_D are shown in Fig. 5.4-5, line a, where $l = 0, 1, 2, \ldots$. In reality, one obtains a superposition of many such signals. Almost all of this superposition is due to the signals returned from the surface of the earth, and only a minute fraction comes from the moving target. The task is to eliminate the ground return completely but leave the return from the moving target essentially unchanged. The signal returned from the target does not only provide information that an object is moving with a certain relative velocity in a certain direction at a certain distance; it also contains the target signature, which contains information about its size, the relative contribution of reflected, scattered or creep waves to the signal, etc. The conservation of this information permits one to distinguish between an airplane and a cruise missile; it also helps against electronic countermeasures.

Let the output terminal r of the circuit in Fig. 5.4-7 be connected to the input terminal of the circuit in Fig. 5.4-8. The hybrid coupler HYC1 splits the input signal $f(t)$ into two components, each with half the power. The one component goes through a delay circuit DEL with the exact delay time T_D, the other runs through an amplitude-reversing amplifier IA. The resulting signals $f(t - T_D)$ and $-f(t)$ are summed by the hybrid coupler HYC2 to yield the output signal $f(t - T_D) - f(t)$.

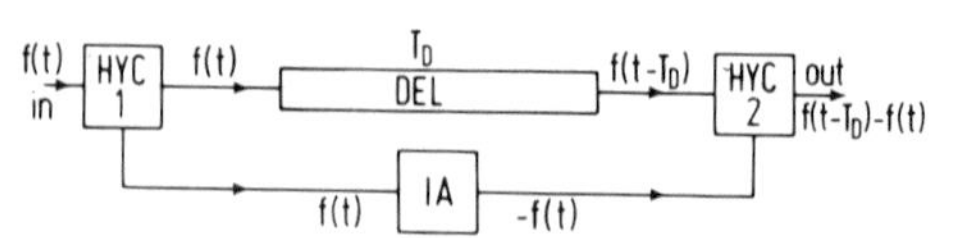

FIG. 5.4-8. Clutter cancellation circuit. HYC, hybrid coupler; IA, amplitude-reversing amplifier; DEL, delay circuit with delay time T_D.

If $f(t - T_D)$ and $f(t)$ are equal, the output signal $f(t - T_D) - f(t)$ will be zero. This is the principle of clutter cancellation (Schrader, 1970). Consider now the signal returned by a target that moves away from the radar (opening target). A reflector or a point-like scatterer will produce the pulse $f(t - T_D)$ in Fig. 5.4-9a, while the pulse $-f(t)$ will be delayed by the time τ and amplitude-reversed, but otherwise equal to $f(t - T_D)$. The sum $f(t - T_D) - f(t)$ does not vanish. Figure 5.4-9 shows the output signals $f(t - T_D) - f(t)$ for various values of τ. For $\tau \geqq \Delta T$ one obtains the full amplitude of the pulse; for $\tau < \Delta T$ one obtains a signal with reduced signal-to-noise ratio.

An opening target is characterized by a positive value of τ, which means that $-f(t)$ is later than $f(t - T_D)$. According to Fig. 5.4-9 a positive value of τ is indicated by a positive pulse followed by a negative pulse. A negative value of τ implies an approaching or closing target, and the signal $f(t - T_D)$ will be later than the signal $-f(t)$. The sums $f(t - T_D) - f(t)$ in Fig. 5.4-9 will in this case be amplitude-reversed. Hence, a negative value of τ is indicated by a negative pulse followed by a positive pulse.

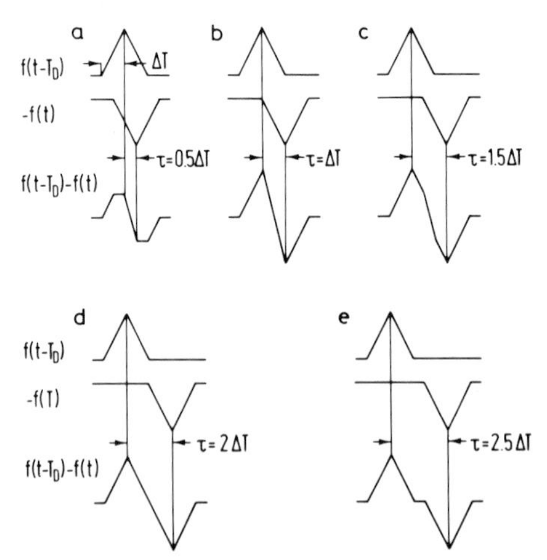

FIG. 5.4-9. Effect of the clutter cancellation circuit of Fig. 5.4-8 on a signal returned by a moving reflector or point-like scatterer.

The relative velocity v_r follows from τ with the help of Eq. (5):

$$T'_D = T_D + \tau = T_D(1 + 2v_r/c), \qquad v_r/c = \tau/2T_D \tag{11}$$

5.4.5 *Technical Problems*

The timing for the transmitter can be derived by means of digital circuits from a precision clock. Although it is not easy to build such circuits with a jitter of 100 ps, it is a task that is within current technology. The receiver is much more of a problem, primarily because of the delay circuits. If we had CCD delay circuits[1] operating at a clock rate of about 10^9 pulses per second and using voltages of about 1 μV, we could build digitally controlled delay circuits having the same stability as the transmitter. For the near future, we will have to be satisfied with analog delay circuits such as coaxial cables, strip lines on ferrite substrates, or surface acoustic wave devices.

Consider the delay circuits 1 and 2 in Fig. 5.4-7. If T is in the order of 10 ns, the largest delay will be in the order of 1 μs. For a jitter of 100 ps one thus needs a stability of (100 ps)/(1 μs) = 10^{-4}. This is manageable.

The delay circuits 3 and 4 in Fig. 5.4-7 require delays on the order of 1 ns. Hence, the whole *pulse-shape filter* can be built as a strip-line circuit.

Figure 4.6-5 contains the same type of circuits as Fig. 5.4-7, and the same comments apply. The real problem is the delay circuit DEL in Fig. 5.4-8. The delay time T_D is on the order of 10 ms, and its stability must be on the order of (100 ps)/(10 ms) = 10^{-8} or better. Furthermore, the clutter cancellation requires that $f(t - T_D)$ and $-f(t)$ at the input terminal of the hybrid coupler HYC2 have the same magnitude with great accuracy. The practical way to maintain accuracy of delay and gain seems to be to frequently feed a test pulse $f(t)$, derived from the same stable clock as the transmitter pulses, instead of the radar return signal to the input terminal. An automatic control circuit can then adjust the delay time T_D and the gain of the reversing amplifier IA so that the output signal $f(t - T_D) - f(t)$ produced by the test signal vanishes.

Even with such an automatic adjustment one still faces the problem that the delay line DEL is very long, since it must not only give a delay in the order of 10 ms but it must resolve 10^8 time intervals. The problem can be resolved by letting a signal circulate s times in a feedback delay circuit with delay T_D/s to obtain a total delay T_D.

Refer to Fig. 5.4-10 for an explanation of the principle. This circuit is

[1] CCD denotes charge-coupled devices

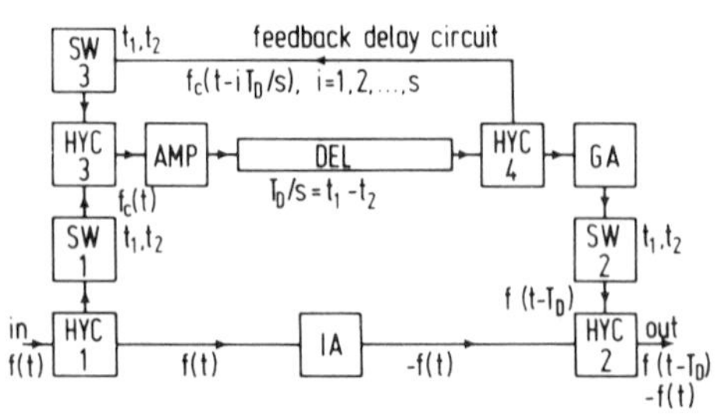

FIG. 5.4-10. Modification of the clutter cancellation circuit of Fig. 5.4-8 that permits the use of a delay circuit with shorter delay time T_D/s. HYC, hybrid coupler; IA, amplitude-reversing amplifier; GA, nonreversing gain control amplifier; AMP, nonreversing amplifier; SW, analog switch.

equal to the one in Fig. 5.4-8, except that the delay circuit is replaced by a delay feedback circuit. The radar return signal $f(t)$ is fed via the hybrid coupler HYC1, the inverting amplifier IA, and the hybrid coupler HYC2 to a display or detection circuit. The switches SW1 and SW2 disconnect the delay feedback loop, except during a short time interval $t_1 < t < t_2$. Hence, the return signal is displayed most of the time without clutter cancellation. The clutter cancellation is only needed during the short time when returns from the surface of the Earth are received. From the altitude of the radar and the direction of observation one can predict a time interval $t_1 < t < t_2$ during which a clutter signal can be received. This time interval will be on the order of 1–10 μs, rather than on the order of 10 ms. The switches SW1 and SW2 connect the feedback delay circuit during this time to the hybrid couplers HYC1 and HYC2. The gated signal $f_c(t)$ is fed to the delay feedback loop and is stored in the delay circuit. One has to choose

$$T_D/s = t_2 - t_1 \tag{12}$$

When the signal $f_c(t)$ has run s times through the feedback loop, it will have become the delayed signal $f_c(t - T_D)$. Switches SW1 and SW2 now connect during the time interval $t_1 + T_D < t < t_2 + T_D$ the feedback loop to the hybrid couplers HYC1 and HYC2, while the switch SW3 interrupts the feedback loop. The clutter-cancelling signal $f_c(t - T_D)$ is thus fed to the hybrid coupler HYC2, while a new gated signal $f_c(t)$ is fed from the hybrid coupler HYC1 to the feedback delay loop.

The feedback loop requires exactly unit gain, since even a gain of 0.999 would reduce a signal amplitude A to $0.999^{1000}A = 0.37A$ after 1000 circulations. The way to accomplish this is to add a test pulse in front of each new signal $f_c(t)$ which adjusts the gain of either the amplifier AMP or of the gain control amplifier GA so that the amplitude of the test signal is neither increased nor decreased.

6 Advanced Signal Design and Processing

6.1 Carrier Coding and Baseband Signal Coding

6.1.1 Pulse Position and Pulse Shape Coding

We have repeatedly used sequences of short pulses that could be compressed into a single large pulse with many small pulses as sidelobes. The design of the pulse sequences was referred to as pulse position coding. Furthermore, the individual pulses of the sequences had sometimes received a finer structure, with the time variation of Barker codes or complementary codes, that was referred to as pulse shape coding.

We will now investigate pulse position and pulse shape coding in more detail, and clarify the resolution that can be obtained in the range-Doppler domain. Woodward (1953) introduced the ambiguity function for the comparison of the resolution achievable by various radar signals in the range-Doppler domain. Numerous computer plots of ambiguity functions have been published (Rihaczek, 1969; Deley, 1970; Cook and Bernfeld 1967; Barton, 1975). All the radar signals investigated in these publications are amplitude- or frequency-modulated sinusoidal functions, and they all have a small relative bandwidth. The coding is applied to the baseband signals, while the carrier is always a sinusoidal function. This type of coding is referred to as baseband signal coding.

A typical ambiguity function is shown in Fig. 6.1-1. The radar signal that yields this function is a sinusoidal carrier amplitude-modulated by a Barker code. This ambiguity function is acceptable, if one receives a return signal only from one target, or if the ambiguity functions due to several targets do not significantly overlap. If the ambiguity functions overlap, it is difficult to tell whether a particular peak is the mainlobe of a weak signal or a sidelobe of a strong signal.

From the standpoint of resolution in the range-Doppler domain, an ambiguity function according to Fig. 6.1-2 would be ideal, if ΔT is as small as permitted by the bandwidth used. Ambiguity functions of this type are called thumbtack functions. A great deal of effort has gone into finding radar signals that yield thumbtack ambiguity functions. The many computer plots in the literature show best to what extent these attempts have succeeded (Rihaczek, 1969; Deley, 1970; Cook and Bernfeld, 1967; Barton, 1975).

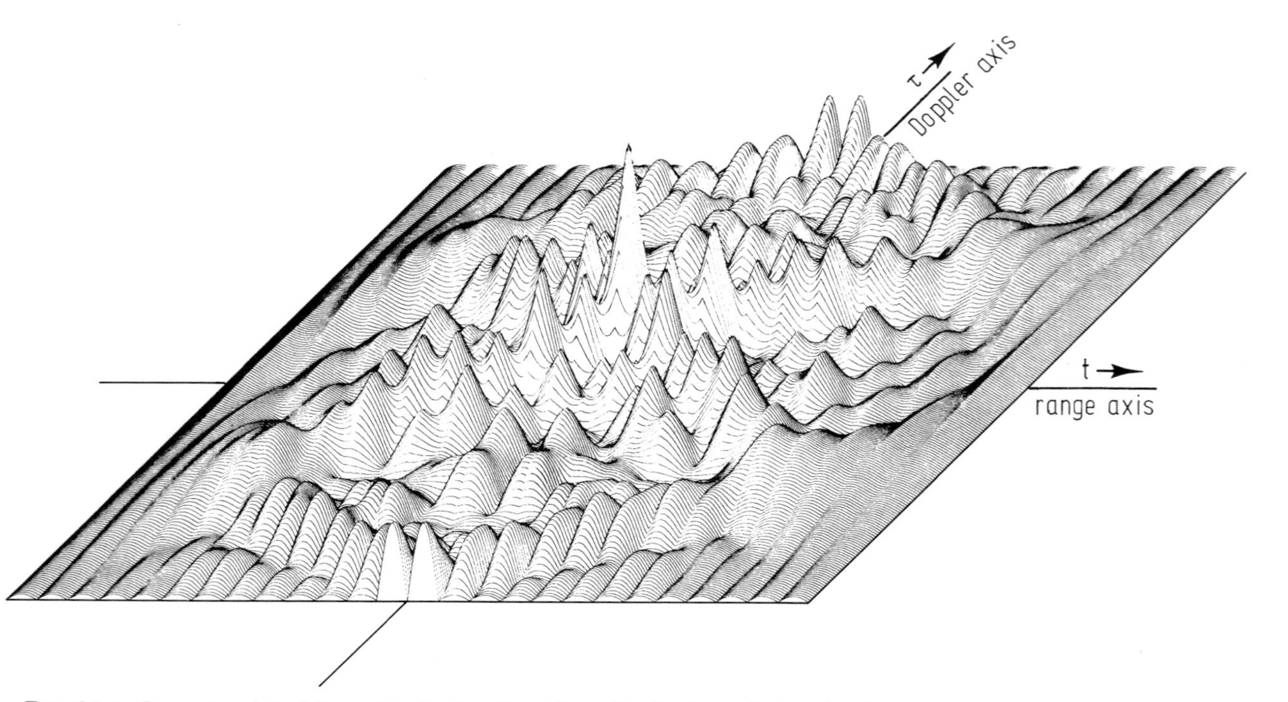

FIG. 6.1-1. Computer plot of the ambiguity function of a radio signal consisting of a sinusoidal function amplitude-modulated by the Barker code with 13 pulses shown in line b of Fig. 4.6-1. (Courtesy of A. W. Rihaczek, Mark Resources Inc., Marina del Ray, California.)

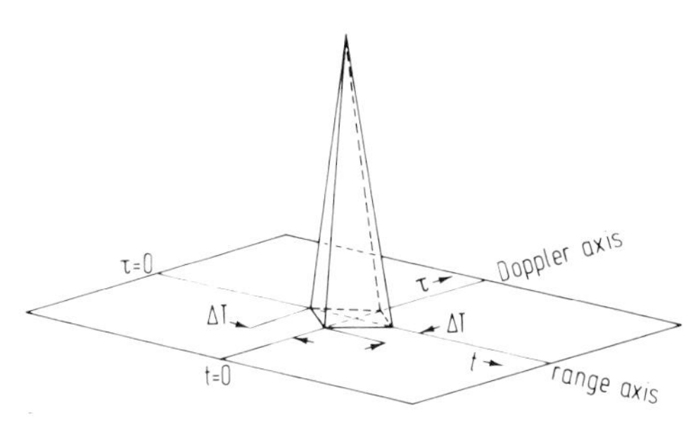

FIG. 6.1-2. Thumbtack ambiguity function with rise and fall times ΔT along the range and Doppler axes.

The plotting of the time variable t along the range axis in Figs. 6.1-1 and 6.1-2 is conventional, but the plotting of a time variable τ along the Doppler axis is not. The Doppler effect in radar is usually expressed as a change of the frequency f of the radiated wave to the frequency $f' = f + \Delta f$ of the returned and received wave. With v as the velocity of the target and c as the velocity of light, one obtains in first order in v/c:

$$f' = f(1 - 2v/c), \qquad \Delta f = -2fv/c \tag{1}$$

A negative velocity implies a decreasing distance, i.e., an approaching or closing target, while a positive velocity implies an increasing distance or an opening target. The frequency shift Δf is usually plotted instead of τ in illustrations like Figs. 6.1-1 and 6.1-2.

Instead of the change of frequency one may use the change of a time interval T_D defined by the radiated wave[1] to a time interval $T'_D = T_D + \tau$ defined by the received wave. In first order in v/c one obtains

$$T'_D = T_D(1 + 2v/c), \qquad \tau = 2T_D v/c \tag{2}$$

We will use the change τ of a time interval rather than the change Δf of a frequency, since the change of a time interval applies to any function while a change of frequency refers to sinusoidal functions only. We will have to investigate very complicated pulse sequences, and anyone trying to do this investigation via Fourier series or Fourier transforms will quickly see why we do not follow this path.

[1] Such a time interval may be defined as the time between two zero crossings of a sinusoidal function, or between $n = 1, 2, \ldots$ zero crossings of a sinusoidal or a more general function, such as a Walsh or a Haar function. There are many other points of a function that can define a time interval, but the convenience and accuracy of measurement usually depends on the choice of points.

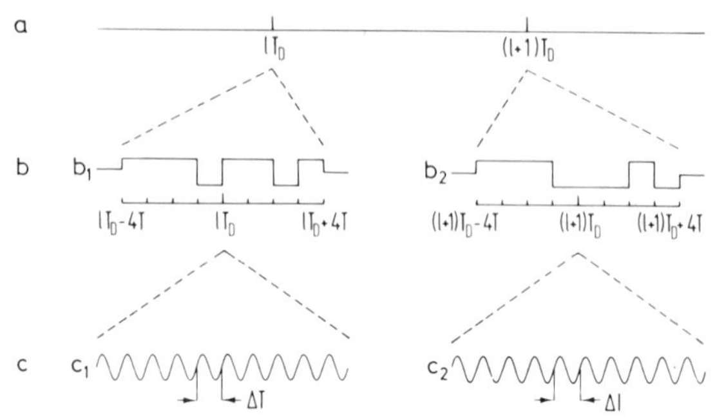

FIG. 6.1-3. Signals of a pulsed radar (a), with alternating fine structures of a pair of complementary codes (b) on an expanded time scale, and the hyperfine structure (c) shown on a still more expanded time scale. $T_D \gg T \gg \Delta T$.

Consider a conventional radar that sends out pulses at the times 0, T_D, $2T_D$, ... as shown in Fig. 4.6-1, line a. Each pulse has a fine structure, e.g., that of a Barker code shown in line b with an expanded time scale. The hyperfine structure, which is the sinusoidal carrier, is shown in line c on a still more expanded time scale. Note that this separation into fine structure and hyperfine structure also is applicable to radar signals produced by frequency modulation; instead of frequency modulation of a sinusoidal carrier by a simple baseband signal one can amplitude-modulate a carrier with a complicated baseband signal to obtain the same radio signal.

As a further example consider the radar signal of Fig. 6.1-3. Again, pulses are radiated at the times lT_D, $l = 0, 1, 2, \ldots$. The fine structure at the time lT_D is that of one character of a complementary code[1] (Erickson, 1961; Golay, 1961). At the time $(l + 1)T$ the fine structure is that of the other character of the complementary code. The hyperfine structure is at both times lT_D and $(l + 1)T_D$ a sinusoidal function with the same frequency. Alternately, one could transmit both characters at the time lT_D, but use two sinusoidal functions with different frequencies for the hyperfine structure.

We will in the following section use the distinction between fine and hyperfine structure to derive a class of nonsinusoidal signals with good features for radar. However, before doing so, let us assure ourselves that such an investigation can have a useful result, and judge what improvement can be expected.

It is generally accepted that a radar signal with high resolution capability must have a large time–bandwidth product; resolution may refer to

[1] The sum of the autocorrelation functions of a pair of complementary codes has no sidelobes.

range, Doppler shift, angle, or to the variety of features summarized under the term target signature. The reason for the need for a large time–bandwidth product is that the information that a signal can potentially transmit increases proportionately to its duration and bandwidth. The duration of the signal is limited by the fact that a radar cannot receive as long as it radiates; this restriction is independent of the type of signal used. The bandwidth of the signal is limited by the propagation and absorption features of the atmosphere, at least for the vast majority of radars.[1] An over-the-horizon radar can use a bandwidth of about 3 MHz located between a lower limit of a few megahertz and an upper limit of some 30 MHz. An all-weather line-of-sight radar can use a bandwidth of about 10 GHz, located between a few hundred megahertz and about 10 GHz. Attenuation by rain and fog prevents the use of frequencies significantly higher than 10 GHz. Even a fair-weather radar is band-limited, since the increasing noise temperature and the attenuation by molecular absorption make it difficult to use bandwidths larger than some 15 GHz.

If a sinusoidal carrier is used, the bandwidth of the radio signal must be on the order of 1% or less of the highest usable frequency. The bandwidth of the baseband signal can have only half this value for double sideband modulation. If the bandwidth of the radio signal is larger, the resonating circuits and structures will not resonate, amplitude and frequency modulation will not work if done in the usual way, etc. The result is that the time–bandwidth product for a line-of-sight all-weather radar using a sinusoidal carrier can only be about 1% of what it could be without a sinusoidal carrier. This statement, in a less definite form, also applies to a fair-weather line-of-sight radar. For an over-the-horizon radar, the possible gain in the time–bandwidth product is not as great, more like one order of magnitude rather than two.

The increase of the possible time–bandwidth product provides the incentive for the investigation of ambiguity functions of nonsinusoidal radio signals; a significant improvement must be possible, and we only have to find out how to exploit this fact.

6.1.2 Structure of Nonsinusoidal Radio Signals

There are many ways of constructing nonsinusoidal radio signals. We will investigate a class of signals for which the role of fine and hyperfine structures in Figs. 4.6-1 and 6.1-3 is essentially reversed. The reason for

[1] There are radars that look into the ground, and others that operate in the essentially empty space above the atmosphere. They have bandwidth limitations too, but for other reasons.

studying these particular signals is that the chance of finding something unconventional is increased by studying something that is as different as possible from the conventional.

Figure 4.6-2 shows a pulsed radar signal on top that is radiated at the times 0, T_D, $2T_D$, ... Line b shows a fine structure that is clearly very different from the one in Fig. 4.6-1. A sequence of 15 pulses at the locations $T_D - 219T$, $T_D - 181T$, $T_D - 147T$, ..., T_D is used. Later on we will discuss why these particular pulse locations were chosen and how they were arrived at. Line c in Fig. 4.6-2 shows the hyperfine structure, which is the structure of each one of the 15 pulses of line b. The hyperfine structure in Fig. 4.6-2 is equal to the fine structure in Fig. 4.6-1 except for the changed time scale.

A more complicated signal is shown in Fig. 5.4-5. Again there is a pulsed radar signal on top that is radiated at the times 0, T_D, $2T_D$, ... Line b shows a fine structure with 30 pulses. A closer analysis shows that it consists of the sum of the two sequences of 15 pulses each shown in lines b_1 and b_2. The sequence in line b_1 is the same as the one in line b of Fig. 4.6-2 for $T = T_1$, while the sequence in line b_2 is shifted by $35T_1$ and has the time unit $T_2 \neq T_1$. The hyperfine structure in line c consists of the two characters of the complementary code used as fine structure in Fig. 6.1-3.

Complementary codes have three important advantages compared with Barker codes. The sum of their autocorrelation functions has no sidelobes, their length is not limited to 13 pulses, and there are very many of them. A problem for their practical use is that both characters have to be transmitted in essentially the same time–frequency interval, or the variations of the transmission properties of two different time–frequency intervals may negate the sidelobe elimination. The two characters in lines c_1 and c_2 of Fig. 5.4-5 very much occupy the same time–frequency interval. Hence, we will concentrate our investigation on complementary codes, and try to make the time–frequency intervals used for their transmission even more equal than in Fig. 5.4-5.

6.1.3 Fine Structure or Pulse Position Coding

The pulse sequence of lines b_1 and b_2 in Fig. 5.4-5 are also shown in line b of Fig. 4.6-3. The current that must be fed into a radiator operating in the electric dipole mode, to produce electric and magnetic field strengths in the far zone with this time variation, is shown in line a. Note that the pulse sequence in line b is the first derivative of the sawtooth function in line a. Hence, it consists not only of the short positive pulses with large amplitude shown, but also of long negative pulses with small amplitude, which are not visible because the amplitude is too small.

The function $f(t)$ of line b delayed by T, $3T$, ..., $219T$ is shown in lines c–p, and the sum of lines b–p is shown in line q. A large pulse is obtained at the time $t = 0$, and many small pulses—or sidelobes—at other times. Line q is continued symmetrically for $t > 0$.

Let us see how the times $t/T = -1, -3, -7, \ldots, -219$, at which the current $i(t)$ in line a of Fig. 4.6-3 has jumps, were found. Table 6.1-1a shows in the first line the numbers -1 and 0. This is the location of the pulses $f(t)$ in Fig. 4.6-3, line b, at the times $t/T = -1$ and 0. The location of the shifted pulses $f(t - T)$ in line c is shown by the numbers 0 and 1 in the second line of Table 6.1-1a. Hence, this table states that there is one pulse at $t/T = -1$, two pulses at $t/T = 0$, and one pulse at $t/T = +1$. If we want a large mainlobe and small sidelobes, we must chose the pulse positions so that 0 occurs very often while every other number should occur only once. With this in mind, we chose the three pulse positions $t/T = -3, -1, 0$ in Table 6.1-1b. The shift $f(t - T)$ produces pulses at the positions $t/T = -2, 0, 1$, while the shift $f(t - 3T)$ produces pulses at the positions $t/T = 0, 2, 3$. The number 0 occurs three times, the numbers $-3, -2, -1, 1, 2, 3$ occur once each. If we had chosen instead $t/T = -2$, -1, 0 we would have obtained for $f(t - T)$ the positions $t/T = -1, 0, 1$ and for $f(t - 2T)$ the positions $t/T = 0, 1, 2$. The number 0 occurs three times, the numbers -2, 2 once each, but the numbers -1, 1 twice each; hence, the sidelobes at $t/T = -1$ and $+1$ are twice as large as for the choices $t/T = -3, -1, 0$ for the pulse locations.

The next pulse position that we can choose without increasing one of the sidelobes is $t/T = -7$; for $t/T = -4$, -5 or -6 one obtains at least one sidelobe that is enhanced. The locations $t/T = -7, -3, -1, 0$ of the pulses $f(t)$ are shown in Table 6.1-1c; also shown are the pulse locations $-6, -2, 0, 1$ for the shifted pulses $f(t - T)$, $-4, 0, 2, 3$ for the pulses $f(t - 3T)$, and 0, 4, 6, 7 for the pulses $f(t - 7T)$.

TABLE 6.1-1

DERIVATION OF THE LOCATION OF THE FIRST FOUR PULSES AT THE TIMES $t = 0$, $-T$, $-3T$, $-7T$ IN LINE b OF FIG, 4.6-3. THE NUMBERS $i = -7, -6, -4, \ldots, +7$ INDICATE THAT THE PULSE IS LOCATED AT $t = iT$

a		b			c			
−1	0	−3	−1	0	−7	−3	−1	0
0	1	−2	0	1	−6	−2	0	1
		0	2	3	−4	0	2	3
					0	4	6	7

TABLE 6.1-2
DERIVATION OF THE LOCATION OF ALL 15 PULSES IN LINE b OF FIG. 4.6-3. THE NUMBERS $i = -219, -218, \ldots, +219$ INDICATE THAT THE PULSE IS LOCATED AT $t = iT$. THE 15 ROWS OF THE TABLE SHOW THE LOCATION OF THE PULSES IN THE 15 LINES b-p IN FIG. 4.6-3

−219	−181	−147	−122	−96	−80	−65	−44	−30	−20	−12	−7	−3	−1	0
−218	−180	−146	−121	−95	−79	−64	−43	−29	−19	−11	−6	−2	0	1
−216	−178	−144	−119	−93	−77	−62	−41	−27	−17	−9	−4	0	2	3
−212	−174	−140	−115	−89	−73	−58	−37	−23	−13	−5	0	4	6	7
−207	−169	−135	−110	−84	−68	−53	−32	−18	−8	0	5	9	11	12
−199	−161	−127	−102	−76	−60	−45	−24	−10	0	8	13	17	19	20
−189	−151	−117	−92	−66	−50	−35	−14	0	10	18	23	27	29	30
−175	−137	−103	−78	−52	−36	−21	0	14	24	32	37	41	43	44
−154	−116	−82	−57	−31	−15	0	21	35	45	53	58	62	64	65
−139	−101	−67	−42	−16	0	15	36	50	60	68	73	77	79	80
−123	−85	−51	−26	0	16	31	52	66	76	84	89	93	95	96
−97	−59	−25	0	26	42	57	78	92	102	110	115	119	121	122
−72	−34	0	25	51	67	82	103	117	127	135	140	144	146	147
−38	0	34	59	85	101	116	137	151	161	169	174	178	180	181
0	38	72	97	123	139	154	175	189	199	207	212	216	218	219

The positions $t/T = -8, -9, -10, -11$ for the fifth pulse turn out to increase one or several sidelobes. The first value of t/T that does not increase a sidelobe is $t/T = -12$. By choosing always the smallest absolute value of t/T for the next pulse position that does not increase a sidelobe, one arrives at the pulse sequence of line b in Fig. 4.6-3. The generalization of Table 6.1-1 for this sequence of 15 pulses is Table 6.1-2. The pulse position 0 occurs 15 times, while every other pulse position $t/T = -219 \cdots +219$ occurs only once.

The choice of the first two pulse locations, $t/T = -1$ and 0, determined the pulse sequence $f(t)$ in Fig. 4.6-3 and all the numbers in Table 6.1-2. A different pulse sequence is obtained by choosing the first two pulse positions differently, e.g., $t/T = -2$ and 0. Figure 6.1-4 shows the resulting sequence $f(t)$ in line b, while Table 6.1-3 shows its derivation in complete analogy to Tables 6.1-1 and 6.1-2.

The pulse sequences $f(t)$ in both Figs. 4.6-3 and 6.1-4 contain 15 pulses; the sums in lines q show the enhancement ratio 15:1 for the pulse at the time $t/T = 0$, and a sidelobe-to-mainlobe ratio of 1/15. However, the sequence in Fig. 6.1-4 only requires the time $204T$ while the one in Fig. 4.6-3 requires the longer time $219T$, even though the shortest time interval between the pulses is T in both illustrations. This implies that the average-to-peak power ratio of the radio signal can be influenced by coding. Before investigating this possibility, let us clean up another undesirable feature of the pulse position codes in Figs. 4.6-3 and 6.1-4.

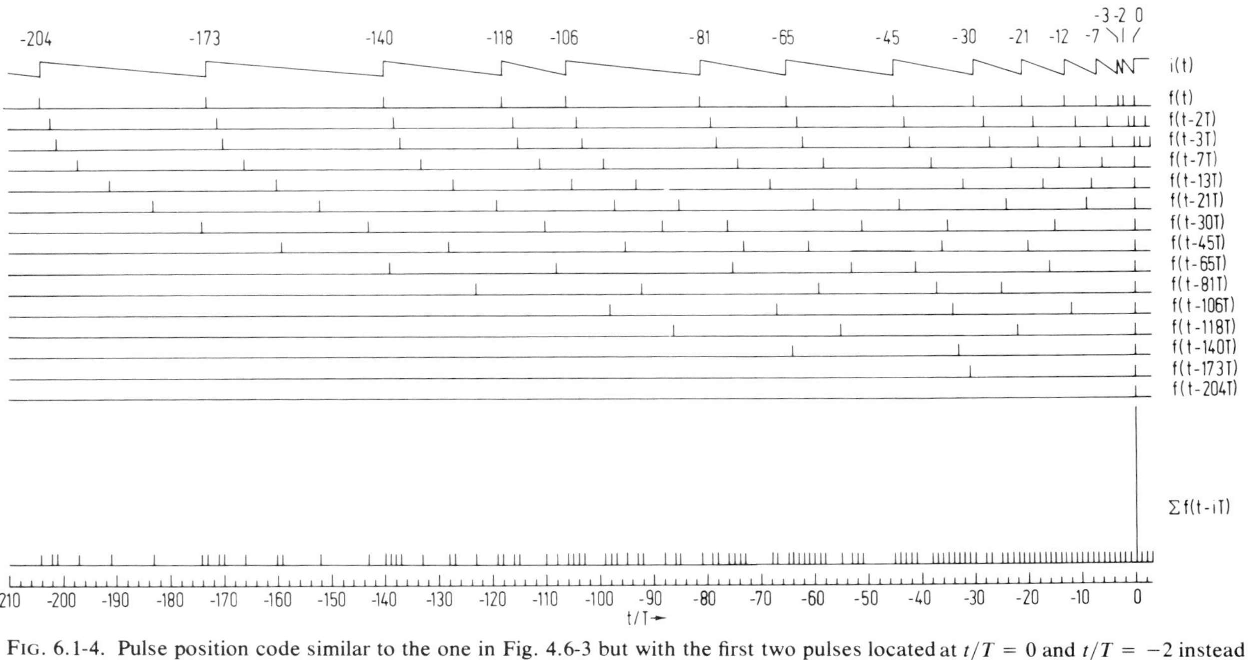

FIG. 6.1-4. Pulse position code similar to the one in Fig. 4.6-3 but with the first two pulses located at $t/T = 0$ and $t/T = -2$ instead of $t/T = 0$ and $t/T = -1$.

TABLE 6.1-3

DERIVATION OF THE LOCATION OF THE 15 PULSES IN LINE b OF FIG. 6.1-4. THE NUMBERS $i = -204, -202, \ldots, +204$ INDICATE THAT THE PULSE IS LOCATED AT $t = iT$. THE 15 ROWS OF THE TABLE SHOW THE LOCATION OF THE PULSES IN THE 15 LINES b–p IN FIG. 6.1-4

−204	−173	−140	−118	−106	−81	−65	−45	−30	−21	−13	−7	−3	−2	0
−202	−171	−138	−116	−104	−79	−63	−43	−28	−19	−11	−5	−1	0	2
−201	−170	−137	−115	−103	−78	−62	−42	−27	−18	−10	−4	0	1	3
−197	−166	−133	−111	−99	−74	−58	−38	−23	−14	−6	0	4	5	7
−191	−160	−127	−105	−93	−68	−52	−32	−17	−8	0	6	10	11	13
−183	−152	−119	−97	−85	−60	−44	−24	−9	0	8	14	18	19	21
−174	−143	−110	−88	−76	−51	−35	−15	0	9	17	23	27	28	30
−159	−128	−95	−73	−61	−36	−20	0	15	24	32	38	42	43	45
−139	−108	−75	−53	−41	−16	0	20	35	44	52	58	62	63	65
−123	−92	−59	−37	−25	0	16	36	51	60	68	74	78	79	81
−98	−67	−34	−12	0	25	41	61	76	85	93	99	103	104	106
−86	−55	−22	0	12	37	53	73	88	97	105	111	115	116	118
−64	−33	0	22	34	59	75	95	110	119	127	133	137	138	140
−31	0	33	55	67	92	108	128	143	152	160	166	170	171	173
0	31	64	86	98	123	139	159	174	183	191	197	201	202	204

The sidelobes in Figs. 4.6-3 and 6.1-4 are small but they are all positive. If returns from several targets are received simultaneously, the sidelobes caused by different targets will add algebraically and rather large peaks can be produced under adverse conditions. A major improvement would be to have about equally many sidelobes with positive and negative signs in lines q of Figs. 4.6-3 and 6.1-4. Sidelobes from several, statistically independent targets will then sum statistically only, and yield the average value zero. To accomplish this, we have only to replace the sawtooth currents $i(t)$ in Figs. 4.6-3 and 6.1-4 by the two-valued current $i(t)$ in line a of Fig. 6.1-5. The jumps of this current are at the same locations as those of the current $i(t)$ in Fig. 4.6-3, but the one at $t/T = -219$ was left out in order to have the same value of the current $i(t)$ at the beginning and the end of the signal in Fig. 6.1-5.

The electric and magnetic field strength in the far zone has the time variation of the pulse sequence $f(t)$ in line b of Fig. 6.1-5. In order to produce an enhanced pulse at $t/T = 0$ we must produce the delayed sequences with alternating signs $-f(t - T)$, $+f(t - 3T)$, $-f(t - 7T)$, ..., $+f(t - 147T)$, $-f(t - 181T)$ shown in lines c–o. The sum of all pulses in lines b–o is shown in line p. The 14 pulses of $f(t)$ yield a pulse enhanced by 14:1 at $t/T = 0$, and a sidelobe-to-mainlobe ratio of $\pm 1/14$. The sum of all positive and negative pulses in line p—including the enhanced pulse at $t/T = 0$—equals zero.

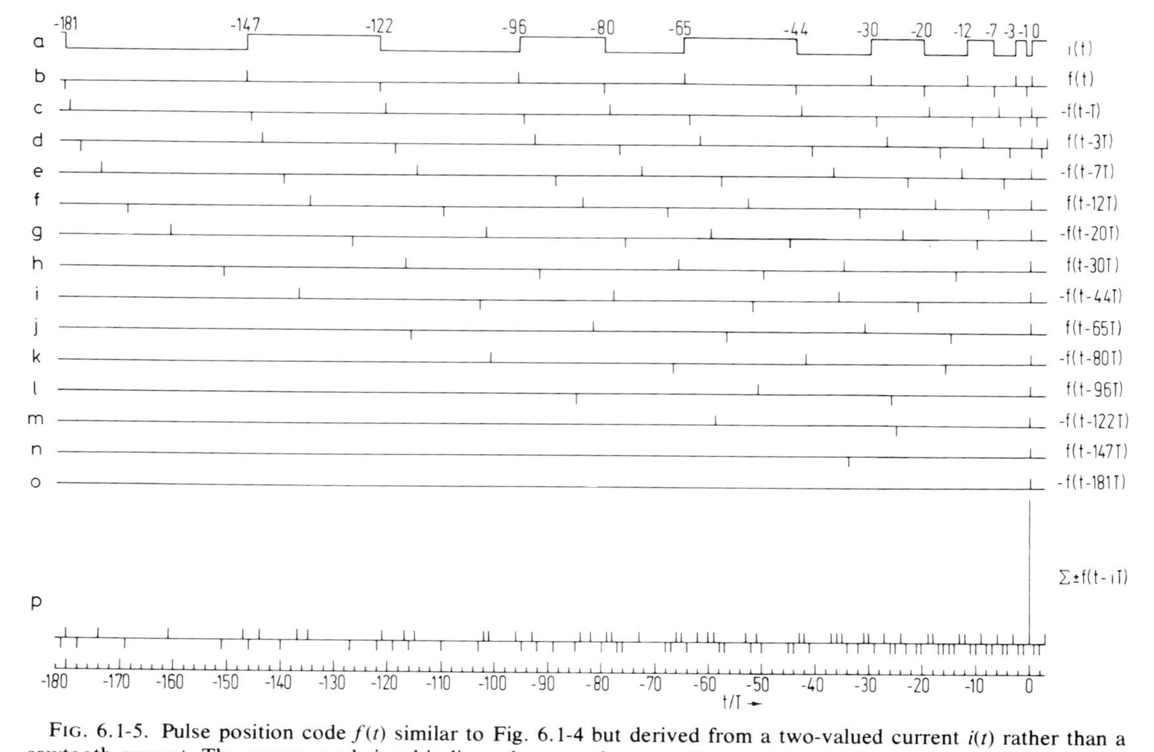

FIG. 6.1-5. Pulse position code $f(t)$ similar to Fig. 6.1-4 but derived from a two-valued current $i(t)$ rather than a sawtooth current. The compressed signal in line p has negative as well as positive sidelobes.

The solution of the problem of algebraically summed sidelobes by means of the two-valued current $i(t)$ in Fig. 6.1-5 is very desirable since such a current is often easier to produce than the sawtooth currents of Figs. 4.6-3 and 6.1-4. The advantage of the sawtooth current is that the pulse sequences $f(t)$ in lines b of Figs. 4.6-3 and 6.1-4 are not polarity symmetric. This means that the amplitude-reversed sequence looks different from the original sequence with a certain time shift. A sinusoidal function shifted by half a period looks like the amplitude-reversed original function. It has been shown that the distinction between an amplitude-reversed and a time-shifted signal can be used for low-flying aircraft tracking, and there are other applications as well. We may readily recognize that the compressed signal in line p of Fig. 6.1-5 is also not polarity symmetric. If $f(t)$—and thus also all the delayed pulse sequences in lines c–o of Fig. 6.1-5—is time-shifted, the compressed signal in line p will also be time-shifted; but if $f(t)$ is amplitude-reversed, the compressed signal will also be amplitude-reversed and look totally different from the nonreversed, time-shifted signal. We have thus learned to produce *not* polarity-symmetric signals generated by two-valued currents.

6.1.4 Maximizing the Average-to-Peak Power Ratio

According to line c of Fig. 4.6-2, each pulse of the sequence $f(t)$ in Fig. 6.1-5, line b, represents a character. The maximal duration of this character equals T; longer characters centered at the pulses of $f(t)$ in Fig. 6.1-5 located at $t/T = 0$ and -1 would overlap. Since $f(t)$ has 14 pulses, energy can thus be radiated at most during a total time $14T$, but radiation during this time is at peak power if a two-valued character is used for the hyperfine structure of Fig. 4.6-2, line c.

The distance between the first and the last pulse of $f(t)$ in Fig. 6.1-5 is $181T$. To obtain the length of the signal we must add $T/2$ on the right of the pulse located at $t/T = 0$ and on the left of the pulse located at $t/T = -181$, to allow for the duration T of the character of the hyperfine structure. Hence, the signal length is $181T + T = 182T$. Power is radiated during the fraction $14T/182T$ of the signal duration. Hence, the average power is $(14/182)P$, where P is the peak power, and $14/182 = 0.077$ is the average-to-peak power ratio. A frequency-modulated sinusoidal wave has always an average-to-peak power ratio of 0.5, which is almost an order of magnitude larger, and this is the reference one has to use in radar. We must find a pulse position code that increases our average-to-peak power ratio at least to 0.5.

The reason for the small ratio 0.077 is the variation of the distance between two adjacent pulses of $f(t)$ in Fig. 6.1-5. The smallest distance is

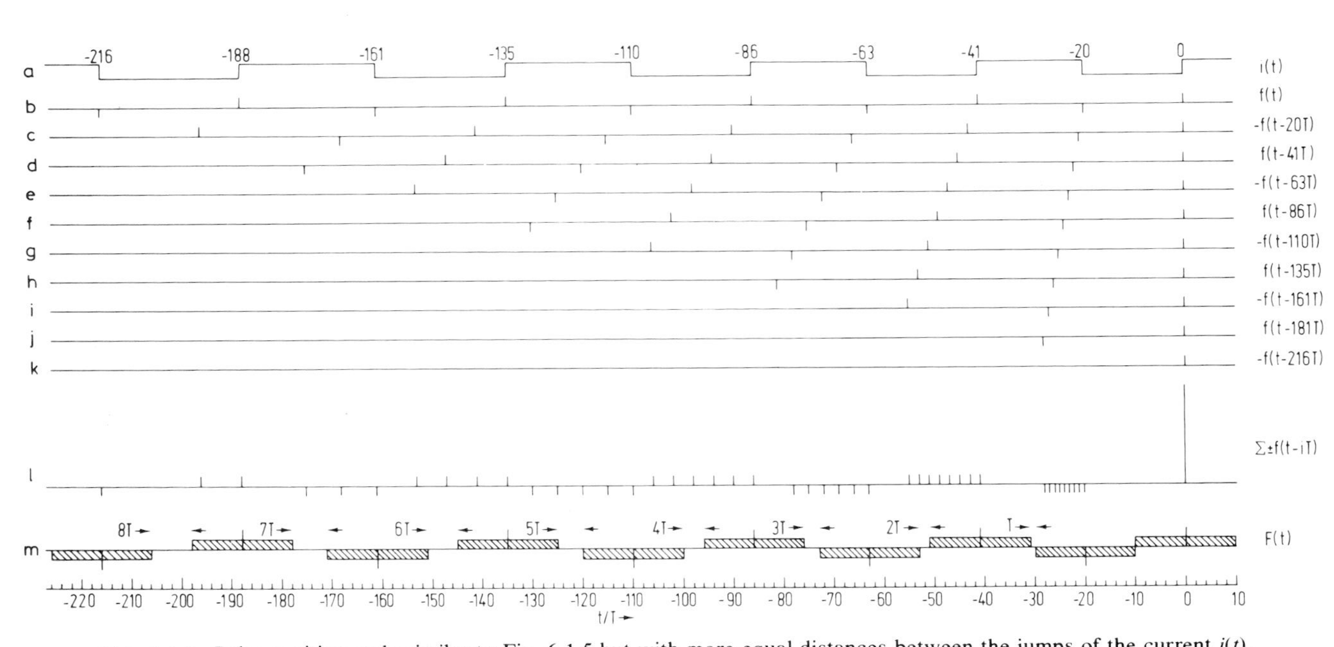

FIG. 6.1-6. Pulse position code similar to Fig. 6.1-5 but with more equal distances between the jumps of the current $i(t)$.

TABLE 6.1-4

DERIVATION OF THE LOCATION OF THE 10 PULSES IN LINE b OF FIG. 6.1-6. THE NUMBERS $i = -216, -196, \ldots, +216$ INDICATE THAT THE PULSE IS LOCATED AT $t = iT$. THE 10 ROWS OF THE TABLE SHOW THE LOCATION OF THE PULSES IN THE 10 LINES b–k IN FIG. 6.1-6

−245	−216	−188	−161	−135	**−110**	−86	−63	−41	−20	0
	−196	−168	−141	−115	−90	−66	−43	−21	0	20
	−175	−147	−120	−94	−69	−45	−22	0	21	41
	−153	−125	−98	−72	−47	−23	0	22	43	63
	−130	−102	−75	−49	−24	0	23	45	66	86
	−106	−78	−51	−25	0	24	47	69	90	110
−110	−81	−53	−26	0	25	49	72	94	115	135
−84	−55	−27	0	26	51	75	98	120	141	161
−57	−28	0	27	53	78	102	125	147	168	188
−29	0	28	55	81	106	130	153	175	196	216
0	29	57	84	110						

T, the largest is $38T$. Consider now Fig. 6.1-6. The distances between adjacent jumps of $i(t)$ in line a or between the pulses of $f(t)$ in line b—starting with the pulse at $t/T = 0$—are $20T$, $21T$, $22T$, ..., $28T$. They are much more equalized than in Fig. 6.1-5.

For the derivation of this pulse position code refer to Table 6.1-4. We start by writing zeros in the main diagonal, just as in Tables 6.1-2 and 6.1-3. Then we chose the distances $20T$, $21T$, $22T$, ... and write the corresponding pulse locations -20, -21, -22, ... just above the main diagonal. The position of the pulses of $f(t)$ in Fig. 6.1-6 follows then from the additions $-20 + (-21) = -41$, $-41 + (-22) = -63$, $-63 + (-23) = -86$, ... The top row and the column on the extreme right of Table 6.1-4 are obtained in this way. The complete second row is obtained by adding 20 to the top row, the third row follows by adding 41 to the top row, etc.

It all works well for the ten columns headed by $0, -20, -41, \ldots, -216$. In the column headed -245 we obtain the number -110, which occurs already in the top row. This means that the pulse originally located at $t/T = -245$ would be at the location $t/T = -110$ after the shift $f(t - 135T)$, and coincide with the pulse at that location of the sequence $f(t)$. Hence, the pulse position code terminates with the position $t/T = -216$.

Figure 6.1-6 shows the pulse sequences $f(t)$, $-f(t - 20T)$, $f(t - 41T)$, ..., $-f(t - 216T)$, according to Table 6.1-4 with alternating signs. The sum of these sequences gives the signal of line 1 with an enhancement ratio 10:1, sidelobe-to-mainlobe ratio 1/10, and sidelobes with positive or negative sign.

Let us now add the hyperfine structure to the pulse sequence $f(t)$. We assume tentatively that the Barker code of Fig. 4.6-2 is used. It may have the length $20T$, since $20T$ is the shortest distance between any adjacent pulses of $f(t)$. The pulse sequence $f(t)$ with the Barker code hyperfine structure added is shown as $F(t)$ in line m of Fig. 6.1-6.

The duration of $F(t)$ is $216T + 20T = 236T$. Peak power is radiated during this time, except during the gaps of duration $T, 2T, 3T, \ldots, 8T$. The gaps sum up to $36T$. The average-to-peak power ratio becomes $(236 - 36)/236 = 0.847$. This is an improvement of $100(0.847 - 0.5)/0.5 = 69\%$ over the average-to-peak power ratio of a radar using frequency modulation.

Let us now refine our tentative assumption that the hyperfine structure of the pulse sequence $f(t)$ in Fig. 6.1-6 is a Barker code. Instead we want to use one character of a complementary code—as shown in Fig. 5.4-5—as hyperfine structure for the positive pulses of $f(t)$, and the other character as hyperfine structure for the negative pulses of $f(t)$. This assures an extremely good interlacing of the two characters in the time–frequency interval.[1]

Complementary codes can be constructed with 2, 4, 8, 16, ..., 2^n digits. A good matching between a pulse position code and the complementary code for its hyperfine structure is achieved, if we choose the distances between the pulses of $f(t)$ in Table 6.1-4 *not* $20T, 21T, 22T, \ldots$ but 2^nT, $(2^n + 1)T$, $(2^n + 2)T, \ldots$. The reason for this choice will become evident when the receiving equipment is being discussed. The duration ΔT of the pulses of the hyperfine structure and the duration T of the time unit for the fine structure become equal in this case, $\Delta T = T$.

Let us derive a pulse position code with distances $2^n = 32$, $2^n + 1 = 33$, $2^n + 2 = 34, \ldots$ between the pulses. In analogy to Table 6.1-4 we write in Table 6.1-5 zeros along the main diagonal. The numbers -32, -33, $-34, \ldots$ are then written above the main diagonal. The sums $-32 - 33 = -65$, $-65 - 34 = -99$, $-99 - 35 = -134, \ldots$ yield the row on top and the column on the right. We could now fill in all the other numbers by adding 32, 65, 99, ... to the first row to obtain the rows 2, 3, 4, ..., but this is not really needed. Instead we make the guess that the code will terminate in the row headed by -494. We calculate the numbers $-494 + 450 = -44$, $-494 + 407 = -87$, $-494 + 365 = -129$, and $-494 + 324 =$

[1] This will work particularly well if both characters of the complementary code have the same number of positive pulses, and thus necessarily also the same number of negative pulses. The characters in Fig. 5.4-5 are not of this type, since one has six positive pulses and the other four. Erickson (1961) does not list any characters with an equal number of positive pulses for $2^n = 8$, but he lists them for $2^n = 4$ $(+ + + -, + + - +)$ and for $2^n = 16$ (e.g., $+ + + + + - - + + + - - + - + -$, $+ + + + + - - + - - + + - + - +$).

TABLE 6.1-5
DERIVATION OF THE LOCATION OF THE 13 PULSES THAT ARE DEFINED BY THE DISTANCES $32T$, $33T$, $34T$, ... BETWEEN ADJACENT PULSES

−494	−450	−407	−365	−324	−284	−245	−207	**−170**	−134	−99	−65	−32	0
	−418							−138			−33	0	32
	−385						−142			−34	0		65
	−351					−146			−35	0			99
	−316				−150			−36	0				134
	−279			−154			−37	0					170
	−243		−158			−38	0						207
	−205	−162			−39	0							245
	−166			−40	0								284
−170	−126		−41	0									324
−129	−85	−42	0										365
−87	−43	0											407
−44	0												450
0													494

−170. The number −170 appears already in the first row. To check whether our guess was right, we calculate the second column, headed by −450. The magnitude of the number −166 is less than that of −170, the magnitude of −205 is less than the magnitude of −207, etc. The written minor diagonal −134, −138, −142, ..., −170 shows why −166, −205, −243, ... have to be compared with −170, −207, −245, ... to assure that no two equal numbers can occur in this table, even though we are not calculating all the numbers.[1]

The 13 numbers 0, −32, −65, ..., −407, −450 are the pulse position code we wanted. For a complementary code we need an even number of pulse positions, and we chose 12 positions 0, −32, −65, ..., −407. Using a complementary code of length $2^n = 32$ we arrive at a signal length of $407T + 32T = 439T$. The number of gaps, during which no power is radiated, follows in analogy to Fig. 6.1-6, line m, as $1 + 2 + \dots + (12 - 2) = 55$. The average-to-peak power ratio becomes $(439 - 55)/439 = 0.875$. This is $100(0.875 - 0.5)/0.5 = 75\%$ better than for a frequency-modulated sinusoidal wave. The pulse enhancement is 12:1 due to the pulse position code, but the complementary code with 32 pulses gives another enhancement of 32:1 for a total of $(12 \times 32):1 = 384:1$. The sidelobe-to-mainlobe ratio is 1/12, the pulse compression ratio 439:1.

One may readily extend Table 6.1-5 from $2^n = 32$ to 64, 128, etc. The results for $2^n = 64$ and 128 are summarized in Tables 6.1-6 and 6.1-7. In

[1] The reader interested in this subject should calculate a table like Table 6.1-4, but with the distances 20, 21, 22, ... replaced by 16, 17, 18, ..., and plot an illustration according to Fig. 6.1-6 lines b–l. In this way he will learn the reason for the simplification of Table 6.1-5 and many other details much faster than one could from a lengthy explanation.

TABLE 6.1-6
PULSE POSITIONS OF FINE STRUCTURE CODES ACCORDING TO TABLE 6.1-5 FOR 2^n = 8, 16, 32, 64, 128

Pulse No.	Pulse position $[t/T]$				
	$2^n = 8$	$2^n = 16$	$2^n = 32$	$2^n = 64$	$2^n = 128$
1	0	0	0	0	0
2	−8	−16	−32	−64	−128
3	−17	−33	−65	−129	−257
4	−27	−51	−99	−195	−387
5	−38	−70	−134	−262	−519
6	−50	−90	−170	−330	−651
7	−63	−111	−207	−399	−784
8		−133	−245	−469	−918
9		−156	−284	−540	−1053
10			−324	−612	−1189
11			−365	−685	−1326
12			−407	−759	−1464
13			−450	−834	−1603
14				−910	−1743
15				−987	−1885
16				−1065	−2027
17				−1144	−2170
18					−2314
19					−2459
20					−2605
21					−2752
22					−2900
23					−3049
24					−3199

order to make these tables easier to understand, they also list the values derived above for $2^n = 32$.

6.1.5 *Principles of Receiver Design*

The generation and reception of radio signals according to Figs. 4.6-2 and 5.4-5 was discussed in Sections 5.4-3 to 5.4-5. Hence, we need to discuss only the changes brought about by the replacement of the saw-tooth currents $i(t)$ in Figs. 4.6-3 and 6.1-4 by the two-valued currents in Figs. 6.1-5 and 6.1-6. A positive and a negative step of the currents $i(t)$ in Figs. 6.1-5 and 6.1-6 are shown in line a of Fig. 6.1-7. Line b shows the two pulses of the pulse sequence $f(t)$ produced by it. Lines a′ and b′ are a repetition of lines a and b with a larger time scale. The radiator current $i(t)$ in line a′ is designed to generate the Barker codes with alternating ampli-

TABLE 6.1-7

CHARACTERISTIC NUMBERS FOR PULSE POSITION CODING FOR FINE STRUCTURE CODES COMBINED WITH COMPLEMENTARY CODES FOR HYPERFINE STRUCTURE HAVING $2^n = 32$, 64, AND 128 DIGITS

	$2^n = 32$	$2^n = 64$	$2^n = 128$
Available pulse positions	13	17	24
Usable pulse positions	12	16	24
Signal duration	$(407 + 32)T = 439T$	$(1065 + 64)T = 1129T$	$(3199 + 128)T = 3327T$
Duration of gaps	$10 \times 11T/2 = 55T$	$14 \times 15T/2 = 105T$	$22 \times 23T/2 = 253T$
Pulse amplitude enhancement	$12 \times 32 = 384$	$16 \times 64 = 1024$	$24 \times 128 = 3072$
Sidelobe-to-mainlobe amplitude ratio	1/12	1/16	1/24
Pulse compression ratio	439:1	1129:1	3327:1
Average-to-peak power ratio	$(439 - 55)/439 = 0.875$	$(1129 - 105)/1129 = 0.907$	$(3327 - 253)/3327 = 0.924$
Improvement of average-to-peak power ratio over frequency-modulated sinusoidal carrier	75%	81%	85%

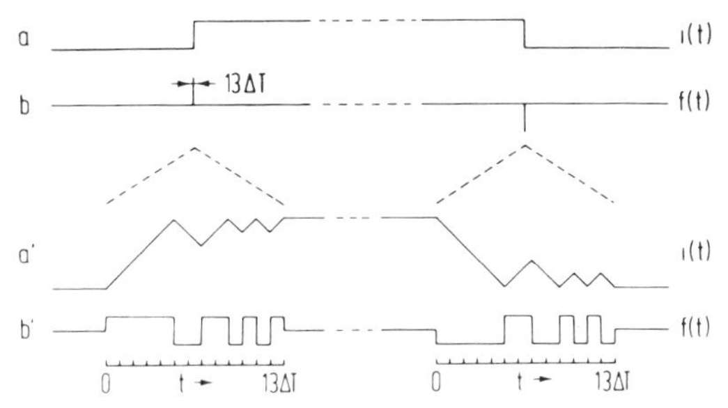

FIG. 6.1-7. Two-valued current (a) and its derivative (b) shown on expanded time scales (a′, b′) for a Barker code with 13 digits as hyperfine structure.

tudes in line b′. A practical way to produce currents according to line a′ has been mentioned in Section 4.6.

Figure 6.1-8 is a repetition of Fig. 6.1-7, except that the Barker code is replaced by the 2 characters of a complementary code with 16 digits. The characters are the same as shown in the footnote on p. 271, and credited to Erickson there.

A basic receiver for the pulse sequence $f(t)$ in Fig. 6.1-5, line b, is shown in Fig. 6.1-9. The functions $f(t - T), f(t - 3T), \ldots, f(t - 181T)$ are produced by means of delay lines with delays T, $3T$, …, $181T$ from the input signal $f(t)$. The functions $f(t)$, $f(t - 3T)$, …, $f(t - 147T)$—having positive signs in Fig. 6.1-5—are summed in the summer SUM1; the functions $f(t - T), f(t - 7T), \ldots, f(t - 181T)$—having negative signs in Fig. 6.1-5—are summed in the summer SUM2. The summer SUM3 with an adding input terminal (+) and a subtracting input terminal (−) produces the sum $\Sigma \pm f(t - iT)$ of line p in Fig. 6.1-5. This sum may be fed to a sliding correlator to recover a Barker code signal according to Fig. 6.1-7, as was discussed with Fig. 4.6-5.

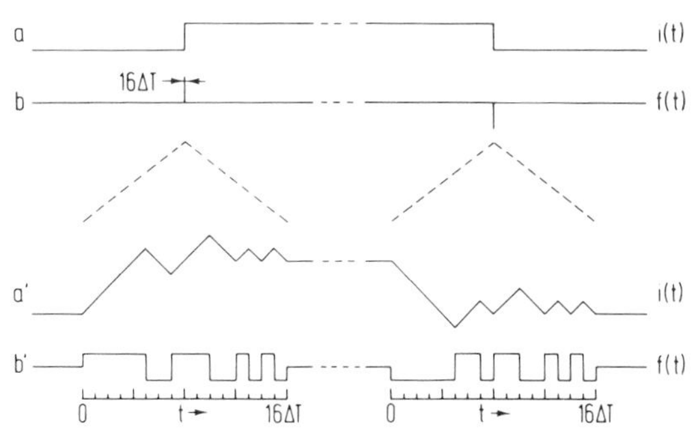

FIG. 6.1-8. Two-valued current (a) and its derivative (b) shown on expanded time scales (a′, b′) for a pair of complementary codes with 16 digits each.

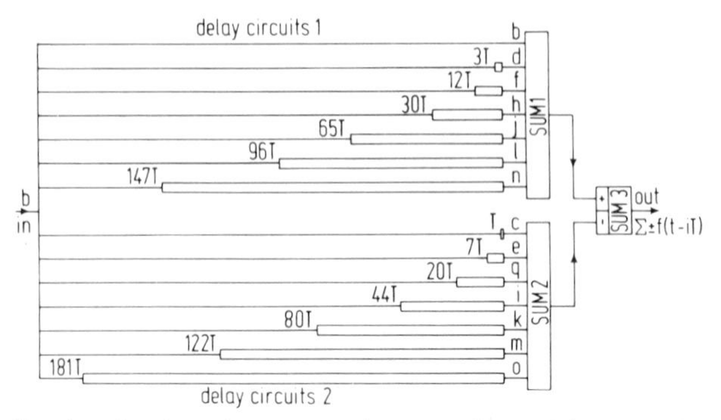

FIG. 6.1-9. Receiver for the pulse compression according to Fig. 6.1-5. SUM, summer.

A receiver for the signal $f(t)$, with an eight-digit complementary code as hyperfine structure, according to Fig. 6.1-6 is shown in Fig. 6.1-10. The delay circuits 1 and 2 as well as the summing circuits SUM1 and SUM2 work like the respective circuits in Fig. 6.1-9. Sliding correlators set for the two characters of the complementary code follow these two summers as in Fig. 5.4-7; they are denoted pulse shape filters. The outputs of the sliding correlators are summed as in Fig. 6.1-9 by a summer with adding (+) and subtracting (−) input terminal.

Typical output signals produced by circuits according to Fig. 6.1-10 are shown in Fig. 6.1-11. The one on top, denoted $2^n = 32$, is due to the use of the fine structure defined by Tables 6.1-5 to 6.1-7, and a complementary code with 32 pulses as hyperfine structure.[1] According to Table 6.1-7, the sidelobe-to-mainlobe amplitude ratio is 1/12, which explains the amplitudes 12 and 1 in Fig. 6.1-11. The location of the first set of sidelobes follows from Table 6.1-5 as $\pm 32T$, $\pm 33T$, ..., $\pm 42T$; these sidelobes are shown in Fig. 6.1-11. A second set of sidelobes is located at $\pm 65T$, $\pm 67T$, ..., a third one at $\pm 99T$, $\pm 102T$, ..., and the last one at $\pm 407T$. The location of all these sidelobes either follows from Table 6.1-5 or may be ob-

[1] We are making here an assumption that simplifies the sidelobes in Fig. 6.1-11 as well as in the plots for the range axis $\tau = 0$ in Figs. 6.1-13 and 6.1-14, but yields a somewhat worse result than obtainable without this assumption. A study of Fig. 6.1-6 shows that the first, third, etc. set of sidelobes—shown by pulses with negative amplitudes—is due to a sum of cross-correlation functions of the complementary code with various time shifts, while the second, fourth, etc. set of sidelobes is due to a sum of autocorrelation functions with various time shifts. Only the mainlobe is due to a sum of autocorrelation functions without a time shift. We proceed here as if the sidelobes were like the mainlobe, except for the compression factor due to the fine structure. This makes the sidelobes on the range axis $\tau = 0$ larger than in reality, but the sidelobes for $\tau \neq 0$ are not affected in the approximation used here. The matter will be discussed in detail in Section 6.2.

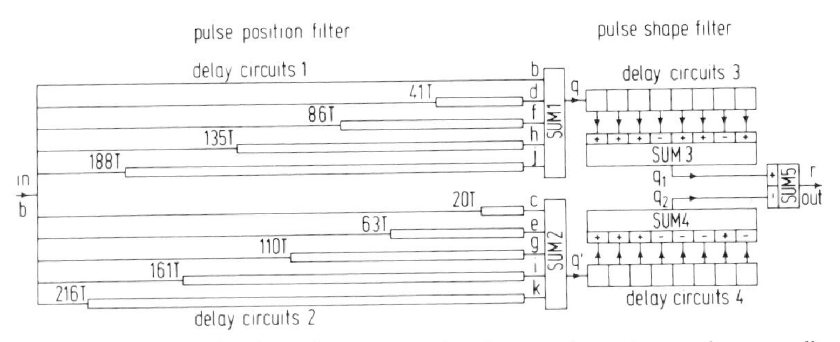

FIG. 6.1-10. Receiver for the pulse compression for complementary codes according to Fig. 6.1-6. SUM, summer.

tained by completing that table. The sets of sidelobes beyond the first one are too far out to be plotted conveniently.

The output signals for $2^n = 16$ and 8 in Fig. 6.1-11 may be inferred from Table 6.1-6. For $2^n = 16$ one can use eight pulses—since their number must be even—and one obtains the sidelobe-to-mainlobe ratio 1/8 used in Fig. 6.1-11. For $2^n = 8$ one obtains from Table 6.1-6 six usable pulses and the sidelobe-to-mainlobe ratio 1/6 used in Fig. 6.1-11. The location of the sidelobes is obtained by writing the complete tables corresponding to Table 6.1-5 for $2^n = 8$ and 16. All the sidelobes are shown for the signal $2^n = 8$ in Fig. 6.1-11, but only the first two sets of sidelobes are shown for the signal $2^n = 16$.

6.1.6 Doppler Shift Resolvers

The Doppler resolution is based on the pulse period T_D in Figs. 4.6-2 and 5.4-5. We will discuss two methods to achieve the Doppler resolution. The first uses many pulses and thus many pulse periods T_D, while the second uses only two pulses and thus only one pulse period T_D.

For an explanation of the first method, let the output signal from terminal r in Fig. 6.1-10 be fed to the input terminal r in Fig. 6.1-12. We assume that a complementary code with $2^n = 32$ digits is used, which means the signal shown on top of Fig. 6.1-11 is fed to the input terminal in Fig. 6.1-12.

The circuit of Fig. 6.1-12 consists of a number of feedback delay circuits that differ only by the delay provided. Consider the circuit in the center characterized by $s(0)$ and $v/c = 0$. The input signal passes through a hybrid coupler HYC5, a nonreversing amplifier AMP3, another hybrid coupler HYC6, and is fed back via the delay circuit DEL3, having the delay T_D, to the hybrid coupler HYC5. Let the signal of the top line of Fig.

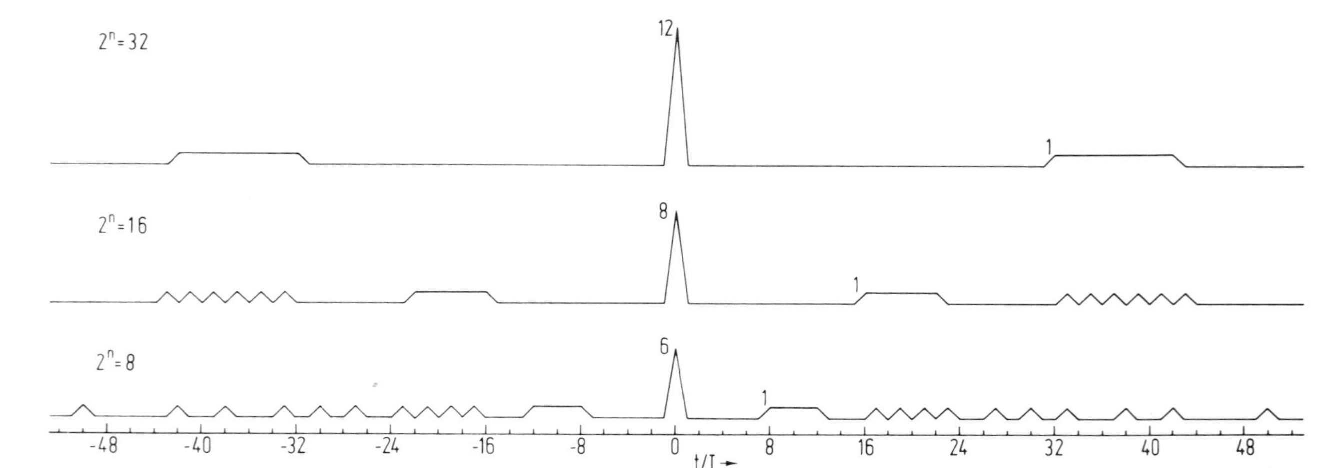

FIG. 6.1-11. Typical output signals of circuits according to Fig. 6.1-10 for complementary codes with $2^n = 8$, 16 or 32 digits.

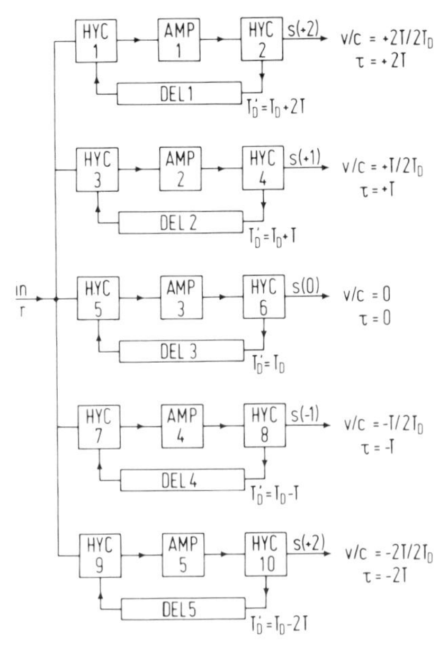

FIG. 6.1-12. Doppler resolver based on feedback delay lines. HYC, hybrid coupler; AMP, nonreversing amplifier; DEL, delay circuit.

6.1-1 be fed to this circuit, and let the velocity of the target by $v/c = 0$. If the amplification of the amplifier AMP3 is set to yield exactly unit gain in the feedback loop, and n signals with period T_D according to the top line of Fig. 6.1-11 are received, we will obtain the input signal with the amplitude multiplied by n at the output terminal $s(0)$ in Fig. 6.1-12 after the nth signal has been received. Let us choose $n = 12$, because the mainlobe-to-sidelobe ratio of the signal in the top line of Fig. 6.1-11 equals 12. The output signal will then be the one shown for $s(0)$ in Fig. 6.1-13, with the mainlobe having the peak amplitude $12^2 = 144$ and the two sidelobes having the peak amplitude $144/12 = 12$.

Let now the same signal be fed to the circuit with the delay time $T'_D = T_D + T$ and the output terminal $s(+1)$ in Fig. 6.1-12. The mainlobe in Fig. 6.1-11, top line, will now produce 12 triangular pulses at the times $t/T = 0, T, 2T, \ldots, 11T$. These pulses add up to the trapezoidal pulse with peak amplitude 12 in the interval $-1 \leqq t/T \leqq 12$ in line $s(+1)$ of Fig. 6.1-13. The sidelobes in Fig. 6.1-11, top line, add to the trapezoidal pulses with peak amplitude 12 in the two intervals $-43 \leqq t/T \leqq -20$ and $33 \leqq t/T \leqq 56$.

If 12 repetitions of the signal of Fig. 6.1-11, top line, are fed to the circuit with delay time $T'_D = T_D + 2T$ in Fig. 6.1-12, one obtains at the output

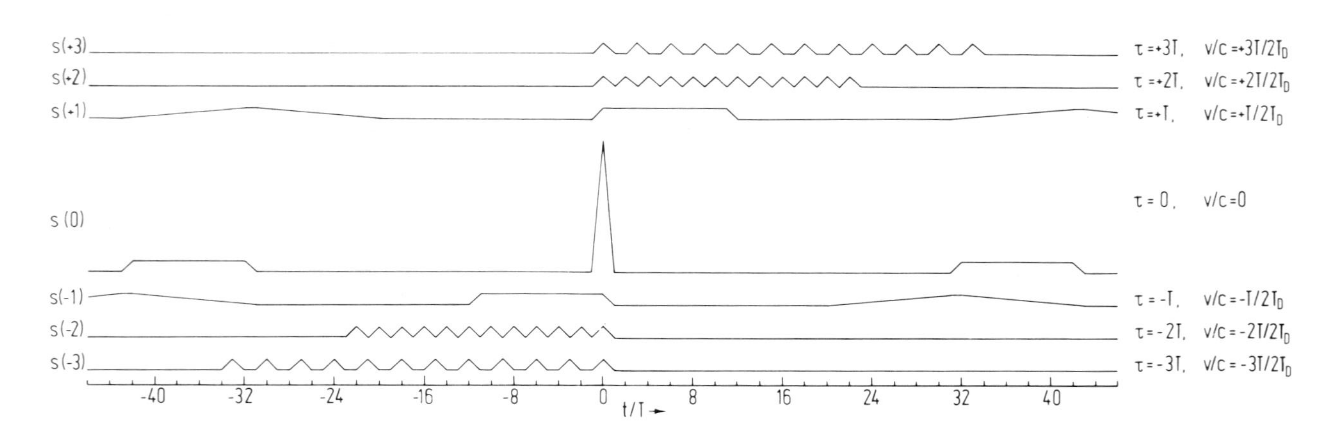

FIG. 6.1-13. Output signals of the circuit of Fig. 6.1-12 at the time $t = 11T_D$, if the signal $2^n = 32$ of Fig. 6.1-11 is fed to the input terminal at the times $t = 0, T_D, 2T_D, \ldots, 11T_D$.

terminal $s(+2)$ the signal shown in line $s(+2)$ of Fig. 6.1-13. The mainlobe produces the 12 triangular pulses with peak amplitude 12, while the sidelobes produce triangular pulses with peak amplitude 1, which are too small to show.

The output signals $s(+3)$, $s(-1)$, $s(-2)$, and $s(-3)$ in Fig. 6.1-13 may be derived in the same way. It is obvious from Fig. 6.1-13 why we choose 12 pulses—or 11 periods T_D according to Fig. 5.4-5. This choice produces sidelobes with the same peak amplitude 12 along the range axis $s(0)$, as well as anywhere else in the range-Doppler domain. To emphasize this point, we show Fig. 6.1-13 again in Fig. 6.1-14, but in the axonometric representation usual for ambiguity functions. It is apparent that this ambiguity function comes very close to the ideal thumbtack function shown in Fig. 6.1-2.

The ambiguity function of Fig. 6.1-14 can be made to approach the ideal thumbtack function as closely as one wants by means of the following two steps. In order to reduce the relative peak amplitude of the sidelobes *not* on the range axis $\tau = 0$, one must increase the number of pulses and pulse periods T_D used. The use of 100 pulses—and thus of 99 pulse periods T_D—will make the largest peak amplitudes off the range axis $\tau = 0$ equal to 1% of the mainlobe.

In order to reduce the relative amplitude of the sidelobes *on* the range axis $\tau = 0$ one must advance from the code $2^n = 32$ to codes $2^n = 64, 128, \ldots$. According to Table 6.1-6, the sidelobe-to-mainlobe ratio decreases from 1/12 to 1/16, 1/24, etc. In addition, the sidelobes are pushed further away from the mainlobe, as one may see for $2^n = 8$, 16, and 32 from Fig. 6.1-11.

The main drawback of this Doppler processor is that many pulses and pulse periods T_D are required. We have seen in Section 5.4.4 that a clutter canceller requires only two pulses and one period T_D, and we will try to generalize this clutter canceller to a Doppler processor. The difference

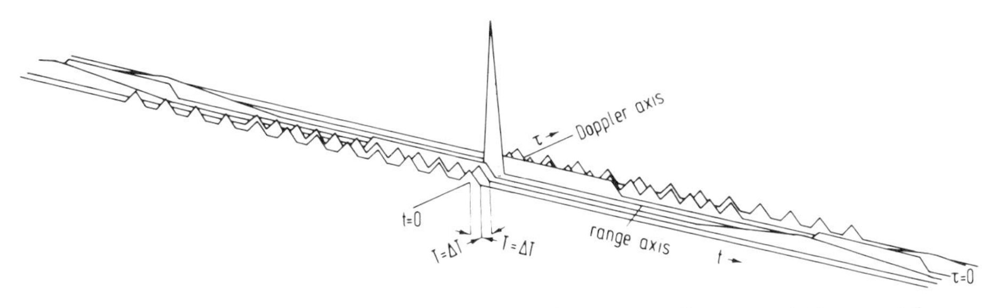

FIG. 6.1-14. The seven functions of Fig. 6.1-13 in axonometric representation usual for ambiguity functions.

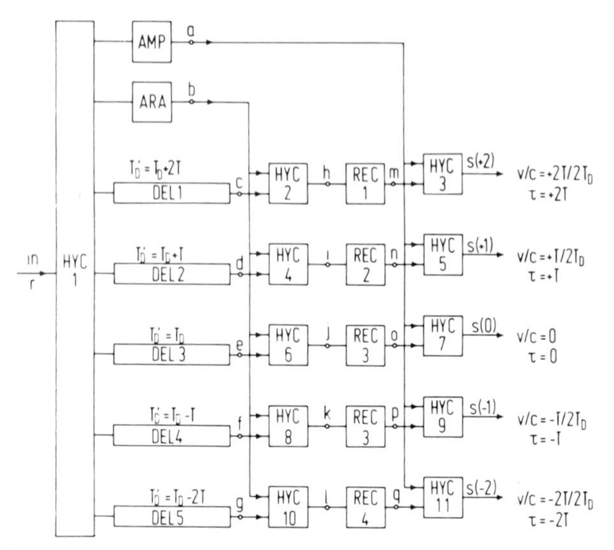

FIG. 6.1-15. Doppler resolver based on the clutter cancellation principle. HYC, hybrid coupler; AMP, nonreversing amplifier; ARA, amplitude-reversing amplifier; DEL, delay circuit; REC, rectifier.

between the two is that a clutter canceller eliminates all signals of targets which have a certain velocity v/c, while the Doppler processor eliminates all signals of targets which *do not* have a certain velocity v/c.

Refer to the circuit diagram of Fig. 6.1-15 and the timing diagram of Fig. 6.1-16. The output terminal r in Fig. 6.1-10 is connected to the input terminal r in Fig. 6.1-15. The hybrid coupler HYC1 feeds this signal to the amplifier AMP, the amplitude-reversing amplifier ARA, and to five delay circuits DEL1–DEL5. The signals[1] at the points r, a, and b are shown in the first column of Fig. 6.1-16. The signals at the output terminals c–g of the delay circuits DEL1–DEL5 are shown in the second column of Fig. 6.1-16.

The hybrid couplers HYC2, HYC4, ..., HYC10 sum the amplitude-reversed signal from the amplifier ARA and the delayed signals from the delay circuits DEL1–DEL5. The signals produced at the terminals h–l by a stationary target are shown in the third column of Fig. 6.1-16. The signal

[1] We tacitly assume that the amplitude of the input signal to the hybrid coupler HYC1 is the same as that of the seven output signals. Furthermore, we use only the mainlobe of any one of the signals in Fig. 6.1-11, since we have already explained how the relative amplitude of the sidelobes can be made arbitrarily small and how these sidelobes can be pushed away from the mainlobe along the range axis.

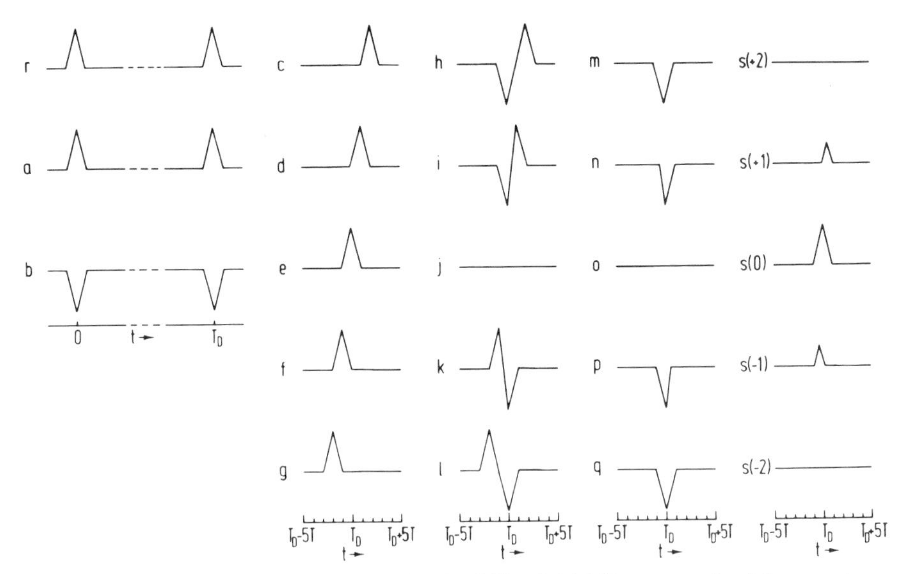

FIG. 6.1-16. Timing diagram for the circuit of Fig. 6.1-15 if a return signal from a stationary target, $v/c = 0$, is received.

at terminal j, corresponding to the delay line with delay $T_D' = T_D$, vanishes. What we have so far is a generalized clutter canceller. For $v/c = 0$ the signal at the terminal j is cancelled, for $v/c = +T/2T_D$ the signal at terminal i is cancelled, for $v/c = -T/2T_D$ the signal at terminal k is cancelled, etc.

In order to obtain a Doppler processor we must invert the action of the generalized clutter canceller. First, rectifiers are used to suppress the positive pulses at the terminals h–l. The output voltages of the rectifiers at the terminals m–q are shown in the fourth column of Fig. 6.1-16. The hybrid couplers HYC3, HYC5, ..., HYC11 produce the sums of the rectified signals with the output signal of the amplifier AMP. The resulting signals are shown in the fifth column of Fig. 6.1-16. We now get a large signal at the terminal for $v/c = 0$, smaller ones at the terminals for $v/c = \pm T/2T_D$, and no signals at the terminals for $v/c = \pm 2T/2T_D$.

The Doppler resolution of the circuit of Fig. 6.1-15 is much better than the one of the circuit of Fig. 6.1-12, if both use the same length of time to resolve two targets with different velocity. On the other hand, the circuit of Fig. 6.1-15 uses only two pulses, and thus much less information than the circuit of Fig. 6.1-12, which uses many more than two pulses. This difference in used information makes it easier to construct situations in which the circuit of Fig. 6.1-15 fails, while the circuit of Fig. 6.1-12 works correctly. One can, of course, combine the good features of both circuits in a more complicated circuit.

6.1.7 Fourier Series Representation of Complicated Radio Signals

So far we have carried out our investigation without the usual Fourier representation of radio signals. We will now show that a Fourier series expansion of the simple signal in Fig. 4.6-2 is already too complicated for any practical use, and this holds true even more for the more complicated signals of Fig. 6.1-6 and Table 6.1-6.

The Fourier series of the signal defined by Fig. 4.6-2 has the form

$$F(t) = \sqrt{2} \sum_{k=1}^{\infty} [a_c(k) \cos(2\pi kt/T_D) + a_s(k) \sin(2\pi kt/T_D)] \tag{3}$$

The terms $a_c(k)$ and $a_s(k)$ consist of integrals $\sqrt{2}\int \cos(2\pi kt/T_D)\, d(t/T_D)$ and $\sqrt{2}\int \sin(2\pi kt/T_D)\, d(t/T_D)$, which offer no problem except that there are so many of them. For instance, the fine structure of line c, Fig. 4.6-2, yields

$$\int_a^b y\, dx - \int_b^c y\, dx + \int_c^d y\, dx - \int_d^e y\, dx + \int_f^g y\, dx - \int_g^h y\, dx + \int_h^i y\, dx \tag{4}$$

where y stands for either $\sqrt{2} \cos(2\pi kt/T_D)$ or $\sqrt{2} \sin(2\pi kt/T_D)$, and dx stands for $d(t/T_D)$. The limits $a \cdots i$ follow from Fig. 4.6-2, line c:

$$a = T_D - 96T - 6.5\, \Delta T, \qquad b = T_D - 96T - 1.5\, \Delta T, \ldots, \qquad i = T_D - 96T + 6.5\, \Delta T \tag{5}$$

Since there are 9 different limits $a \cdots i$ in Eq. (4) we would obtain 9 components if we actually performed the integrations. Line b in Fig. 4.6-2 shows 15 pulses at the locations $t/T = -219, -181, \ldots, 0$. Since we get 9 components for each one of them, we arrive at $15 \times 9 = 135$ components for $a_c(k)$ and 135 components for $a_s(k)$ in Eq. (3). Clearly, one cannot work with such a series. The more complicated signals of Table 6.1-6 have hundreds to thousands of components in each term of their Fourier series. The concepts of pulse position and pulse shape coding permit us to investigate and design complicated radio signals with manageable effort.

For an estimate of the number of terms in a Fourier series required to yield an acceptable approximation of the signals, we observe that the period in Fig. 4.6-2 and Eq. (3) is T_D; the Fourier series thus has terms with frequency $f = k/T_D$, $k = 1, 2, \ldots$. The shortest time interval in Fig. 4.6-2 is ΔT. Hence, the largest required value of k is about one order of magnitude larger than $T_D/\Delta T$. For the signals discussed here, the number of significant terms in a Fourier series expansion is typically between a million and a billion. This not only shows that a Fourier series is not a practical means for the representation of the signals, but it also shows that the signals are ideal for spread-spectrum transmission and frequency sharing.

Let us observe that one could write the signal of Fig. 4.6-2 line b, with the hyperfine structure of line c as a ternary signal with digits +1, 0, and −1. However, this signal would consist of a string of $15 \times 9 = 135$ digits +1, $15 \times 4 = 60$ digits −1, and at least $13(219 - 15) = 2652$ digits 0; such a ternary number would be almost as unmanageable as the Fourier series.

6.2 Thumbtack Ambiguity Function

6.2.1 Cross-Correlation Function for 16 Pulses

We investigated in Section 6.1 how the concepts of pulse position and pulse shape coding of radio signals could be used to approximate ideal thumbtack ambiguity functions. The exact structure of the sidelobes along the range axis turned out to be rather complicated, and an approximation was used in Figs. 6.1-11, 6.1-13, and 6.1-14; a more detailed investigation had been postponed.

We derived a pulse position code for $2^n = 32$ in Table 6.1-5, but only

TABLE 6.2-1
DERIVATION OF THE LOCATION OF THE SEVEN PULSES THAT ARE DEFINED BY THE DISTANCES $8T$, $9T$, $10T$, ... BETWEEN ADJACENT PULSES

−77	−63	**−50**	−38	−27	−17	−8	0
	−55	−42	−30	−19	−9	0	8
	−46	−33	−21	−10	0	9	17
−50	−36	−23	−11	0	10	19	27
−39	−25	−12	0	11	21	30	38
−27	−13	0	12	23	33	42	50
−14	0	13	25	36	46	55	63
0							77

TABLE 6.2-2

DERIVATION OF THE LOCATION OF THE NINE PULSES THAT ARE DEFINED BY THE DISTANCES $16T$, $17T$, $18T$, ... BETWEEN ADJACENT PULSES

−180	−156	−133	−111	**−90**	−70	−51	−33	−16	0
	−140	−117	−95	−74	−54	−35	−17	0	16
	−123	−100	−78	−57	−37	−18	0	17	33
	−105	−82	−60	−39	−19	0	18	35	51
	−86	−63	−41	−20	0	19	37	54	70
−90	−66	−43	−21	0	20	39	57	74	90
−69	−45	−22	0	21	41	60	78	95	111
−47	−23	0	22	43	63	82	100	117	133
−24	0	23	45	66	86	105	123	140	156
0									180

listed the pulse positions for $2^n = 8$ and 16 in Table 6.1-6 without deriving them. The derivations for these two cases are shown in Tables 6.2-1 and 6.2-2. We need these rather small values of 2^n to facilitate plotting, as will soon become apparent.

The nominal antenna current $i(t)$ with jumps at the locations defined by Table 6.2-1 is shown in line a of Fig. 6.2-1. Only the six jumps at the locations $t = 0, -8T, \ldots, -50T$ are used, since we want to use complementary codes and thus need an even number of pulse positions. The derivative $f(t)$ of the nominal antenna current $i(t)$ is shown in line c of Fig. 6.2-1; this function $f(t)$ is the nominal electric or magnetic field strength in the far zone.

Let us now replace the nominal field strengths by the actual ones. We use the complementary code 1 with eight digits shown in Table 6.2-3 (Erickson, 1961). Each positive pulse in Fig. 6.2-1, line c, is replaced by the character $+++---+-$, while each negative pulse is replaced by the character $+++-++-+$; in order to allow for the negative sign of the pulse we use the reversed signs $---+--+-$ for the time being. In this way one obtains line d from line c in Fig. 6.2-1; it shows the actual time variation of the electric and magnetic field strengths defined by the pulse position code of the top row of Table 6.2-1 and the pulse shape code of the top row of Table 6.2-3.

The actual antenna current $I(t)$ is obtained by integration of line d in Fig. 6.2-1. The integral is plotted in line b. This current varies considerably, which implies large peak currents, but we will see later on how they can be avoided.

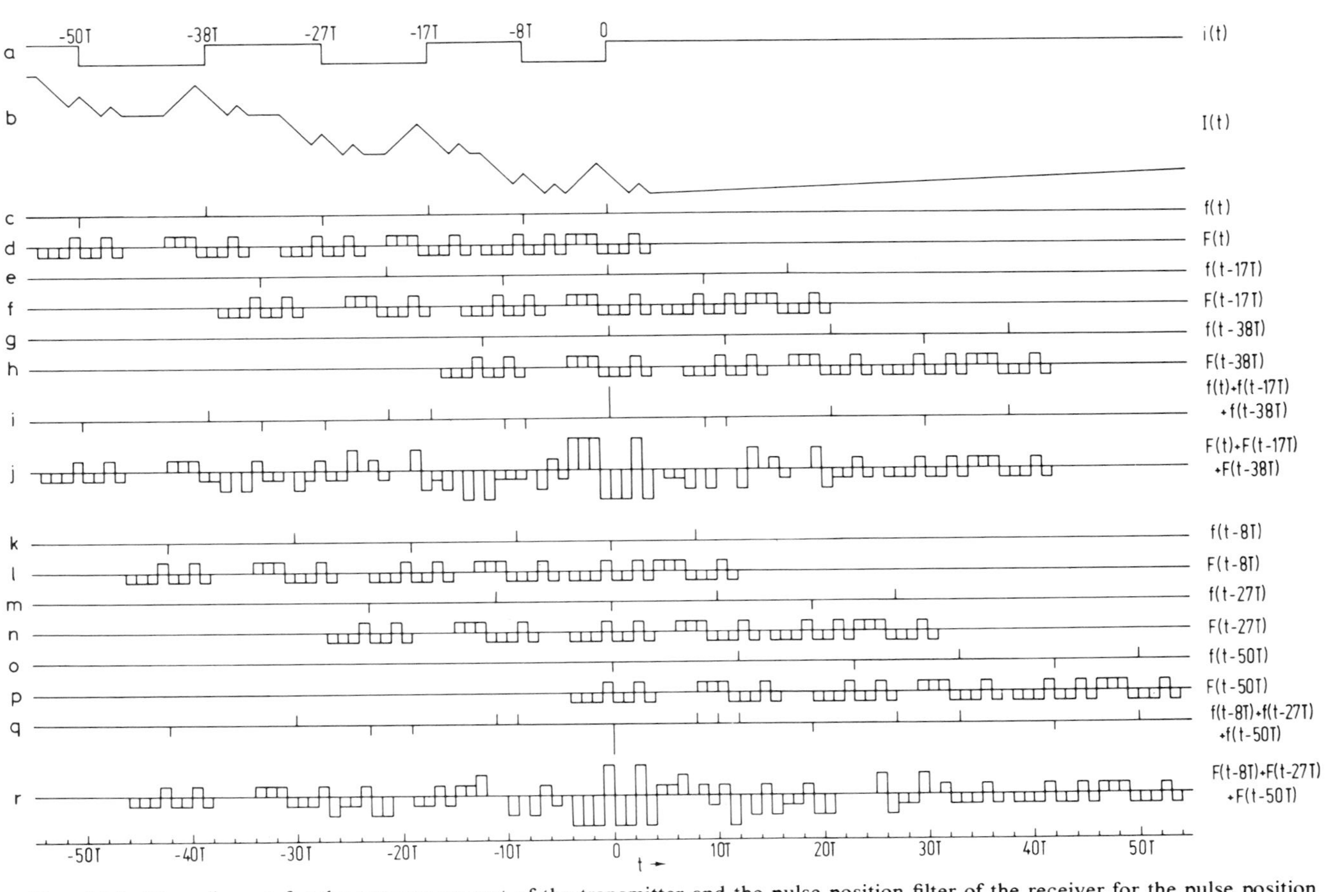

FIG. 6.2-1. Time diagram for the antenna current of the transmitter and the pulse position filter of the receiver for the pulse position code of Table 6.2-1 and the complementary code 1 of Table 6.2-3 for the pulse shape code. The duration of the individual pulses is T.

TABLE 6.2-3

EXAMPLES OF COMPLEMENTARY CODES WITH EIGHT DIGITS. THE DIGITS +1 AND −1 ARE REPRESENTED BY THE SIGNS + AND −. THE LEFT HALF OF THE AUTOCORRELATION FUNCTION IS SHOWN ON THE RIGHT[a]

1	+	+	+	−	−	−	+	−	−1	0	−1	0	−3	0	+1	+8
	+	+	+	−	+	+	−	+	+1	0	+1	0	+3	0	−1	+8
2	+	+	+	+	+	−	−	+	+1	0	−1	0	+1	0	+3	+8
	+	+	−	−	+	−	+	−	−1	0	+1	0	−1	0	−3	+8
3	+	+	+	+	−	+	+	−	−1	0	+1	0	+3	0	+1	+8
	+	+	−	−	−	+	−	+	+1	0	−1	0	−3	0	−1	+8
4	+	+	−	+	+	+	+	−	−1	0	+3	0	+1	0	+1	+8
	+	+	−	+	−	−	−	+	+1	0	−3	0	−1	0	−1	+8

[a] From Erickson, 1961.

Let us turn to the reception of the radio signal of Fig. 6.2-1, line d. We delay this signal by $17T$ and $38T$, which is shown for the nominal field strength $f(t)$ in lines e and g, while the delayed actual field strengths $F(t)$ are shown in lines f and h. The positive pulses at $t = 0, -17T$, and $-38T$ are lined up at $t = 0$ in lines c, e, and g, and so is the center of the three characters + + + − − − + − in lines d, f, and h. The sum of the nominal signals $f(t)$, $f(t - 17T)$, and $f(t - 38T)$ is shown in line i, while the sum of the actual signals $F(t)$, $F(t - 17T)$, and $F(t - 38T)$ is shown in line j. A circuit performing these delays and the summation is shown in Fig. 6.2-2 by the delay circuits 1 and the summer SUM1.

The negative pulses in line c of Fig. 6.2-1 and the characters − − − + − − + − represented by them are lined up at $t = 0$ by the delays $8T$, $27T$, and $50T$, as shown in lines k–p of Fig. 6.2-1. The sums of the de-

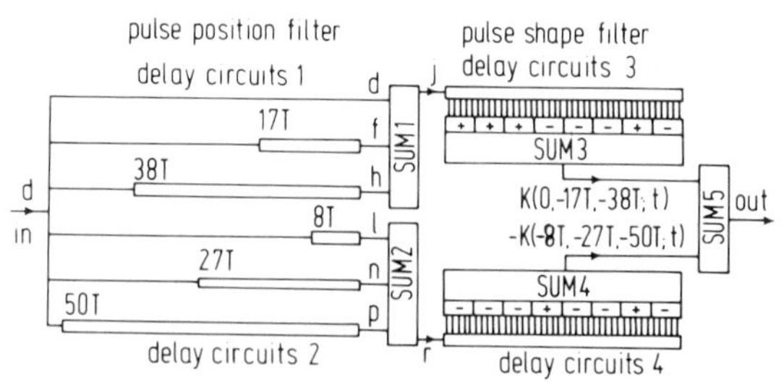

FIG. 6.2-2. Receiver for the signal of line d in Fig. 6.2-1. The pulse position filter is implemented by delay circuits and summers, while the pulse shape filter is implemented by two sliding correlators and a summer (SUM).

layed signals are shown in lines q and r. The delay circuits 2 and the summer SUM2 perform these operations in the circuit of Fig. 6.2-2.

The delay circuits 1 and 2 with the summers SUM1 and SUM2 use the position of the signals in order to enhance them, as is readily apparent from the enhancement around $t = 0$ in lines j and r of Fig. 6.2-1. This part of the circuit is thus called pulse position filter. A further enhancement is possible by exploiting the known shape of the pulses.

The accepted method for the detection of a pulse or a character of known shape is by means of cross correlation of the received signal with the pulse shape one wants to detect. The cross correlation can be performed by means of a sliding correlator. Such a sliding correlator consists of the delay circuits 3 with many taps, shown in Fig. 6.2-2, and a summing circuit SUM3 with adding (+) and subtracting (−) input terminals. Let the adding and subtracting terminals be set for the detection of the character $+++----+-$ that was already enhanced by the summer SUM1 according to line j of Fig. 6.2-1. The output signal $K(0, -17T, -38T; t)$ of the summer SUM3 then has the time variation shown[1] in Fig. 6.2-3, line a.

The adding and subtracting input terminals of the summer SUM4 in Fig. 6.2-2 are set according to the signal[2] $---+--+-$ already enhanced by the pulse position filter, as shown in line p of Fig. 6.2-1. The output signal $-K(-8T, -27T, -50T; t)$ of the summer SUM4 has the time variation shown in Fig. 6.2-3, line b.

The sum of the two cross-correlation functions $K(0, -17T, -38T; t)$ and $-K(-8T, -27T, -50T; t)$ is produced in the summer SUM5 of Fig. 6.2-2 and its time variation is shown in Fig. 6.2-3, line c. One may readily see that the sidelobe-to-mainlobe ratio is almost exactly 1/6, as we have assumed previously for the code $2^n = 8$ in Fig. 6.1-11. The structure of the sidelobes is not yet sufficiently distinct to permit comparison with the previous plot, and we have to advance to the code $2^n = 16$ to recognize the structure better.

Figure 6.2-4 shows in line a the nominal antenna current $i(t)$ with six jumps at $t = 0, -16T, -33T, \ldots, -133T$ according to Table 6.2-2. The

---

[1] When computing this plot one must shift the function of Fig. 6.2-1, line j, from right to left past the function $+++----+-$ shown in the upper left corner of Fig. 6.2-3, multiply and sum for all values of t. The shifting from right to left is confusing since it is the opposite of the shift from left to right in the delay circuits 3 in Fig. 6.2-2. However, it is standard practice to draw circuits with a signal flow from left to right, so that an earlier signal is further to the *right*, and it is an equally standard practice to plot time functions with the time increasing toward the right, so that an earlier signal is further to the *left*.

[2] One could also use the sequence $+++-++-+$ shown in row 1 of Table 6.2-3, but the summer SUM5 in Fig. 6.2-2 would then need one adding and one subtracting input terminal, while two adding input terminals are assumed in Fig. 6.2-2.

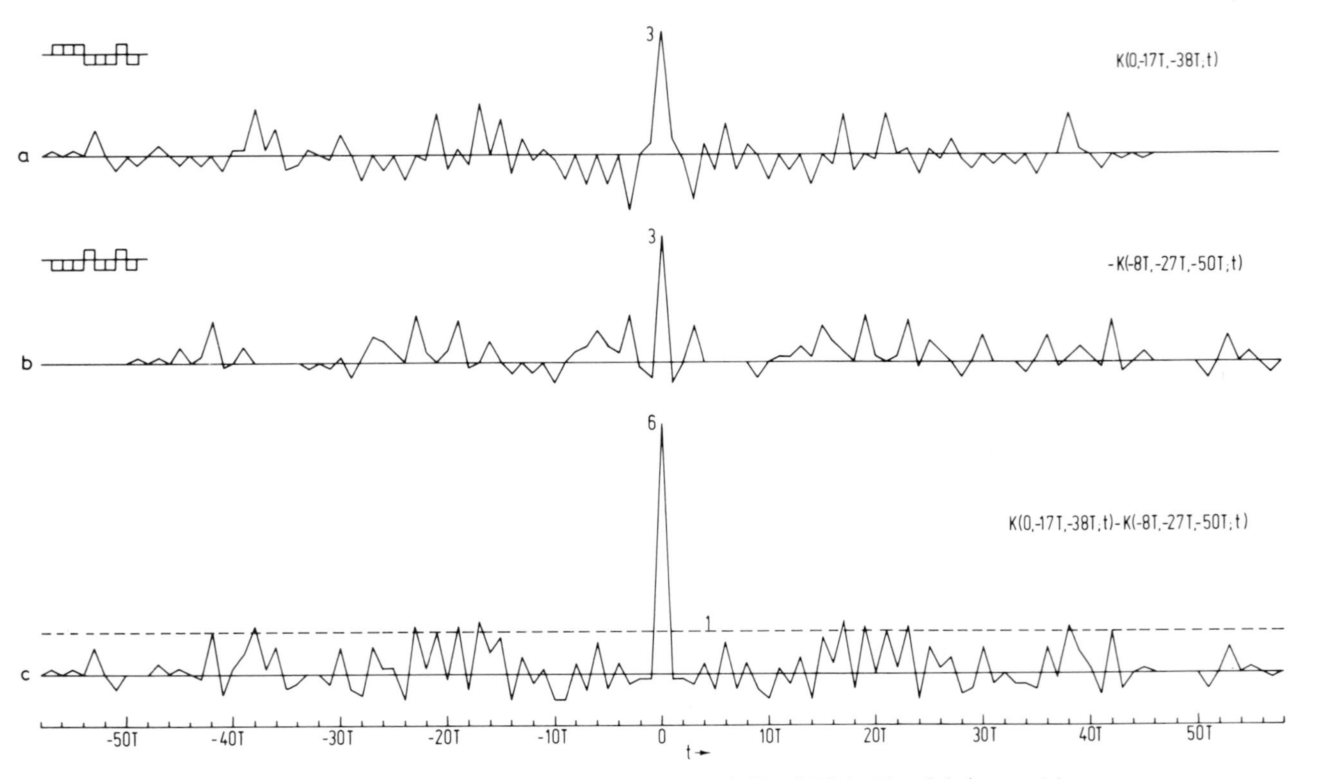

FIG. 6.2-3. Output voltages of the sliding correlators in Fig. 6.2-2 (a, b) and their sums (c).

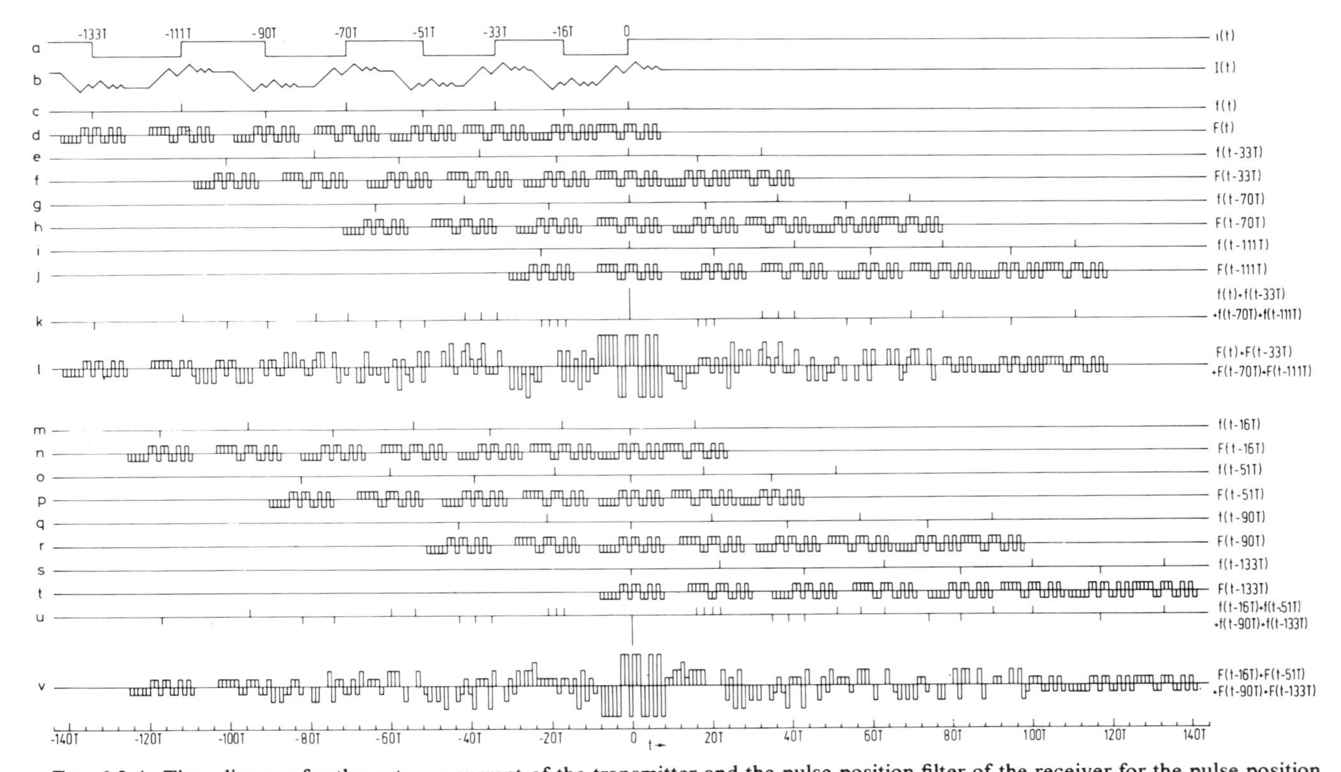

FIG. 6.2-4. Time diagram for the antenna current of the transmitter and the pulse position filter of the receiver for the pulse position code of Table 6.2-2 and the complementary code 1 of Table 6.2-4 for the pulse shape code.

TABLE 6.2-4

EXAMPLES OF COMPLEMENTARY CODES WITH 16 DIGITS. THE DIGITS +1 AND −1 ARE REPRESENTED BY THE SIGNS + AND −. THE LEFT HALF OF THE AUTOCORRELATION FUNCTION IS SHOWN BELOW EACH CHARACTER[a]

1	+	+	+	+	+	−	−	+	+	+	−	−	+	−	+	−
	−1	0	−1	0	−1	0	−1	0	+1	0	+5	0	−3	0	+1	+16
	+	+	+	+	+	−	−	+	−	−	+	+	−	+	−	+
	+1	0	+1	0	+1	0	+1	0	−1	0	−5	0	+3	0	−1	+16
2	+	+	+	+	−	+	+	−	+	+	−	−	−	+	−	+
	+1	0	+1	0	−3	0	−3	0	+3	0	−1	0	+3	0	−1	+16
	+	+	+	+	−	+	+	−	−	−	+	+	+	−	+	−
	−1	0	−1	0	+3	0	+3	0	−3	0	+1	0	−3	0	+1	+16
3	+	+	−	−	+	−	+	−	+	+	+	+	+	−	−	+
	+1	0	−3	0	+5	0	+1	0	−1	0	−1	0	−1	0	−1	+16
	+	+	−	−	+	−	+	−	−	−	−	−	−	+	+	−
	−1	0	+3	0	−5	0	−1	0	+1	0	+1	0	+1	0	+1	+16

[a] From Erickson, 1961.

seventh jump at $t = -156T$ is not used since we need an even number of jumps. The derivative $f(t)$ of the nominal antenna current is shown in line c. We substitute the first character of the 16-digit complementary code 1 of Table 6.2-4 for each positive pulse of $f(t)$, and the second character with reversed sign for each negative pulse of $f(t)$ to obtain the actual time variation $F(t)$ of the electric and magnetic field strengths in the far zone, as shown in line d. The actual antenna current $I(t)$—without regard to the propagation time of the electromagnetic wave—is obtained by integration of $F(t)$; it is shown in line b.

The actual antenna current $I(t)$ in Fig. 6.2-4 has a much lower peak value than the current $I(t)$ in Fig. 6.2-1, line b. The reason is that the two characters of the complementary code 1 in Table 6.2-4 have the same number of digits +1, while this is not so for any one of the four complementary codes in Table 6.2-3. One may see that the characters of the complementary code 2 in Table 6.2-4 also have the same number of digits +1. This is not so for the characters of the code 3, but the number of digits −1 of the second character now equals the number of digits +1 of the first character. When using this code 3, one may substitute the second character without amplitude reversal for the negative pulses of $f(t)$ to avoid large peaks of the antenna current.

Without proof, we mention that complementary codes with $4^n = 4, 16, 64, \ldots$ digits permit the avoidance of large peak antenna currents in the

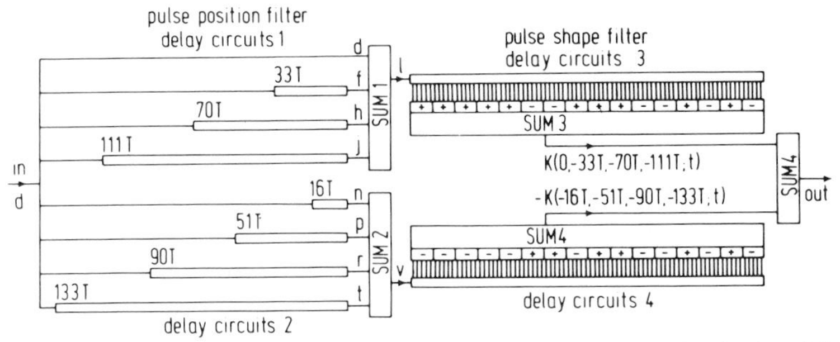

FIG. 6.2-5. Receiver for the signal of line d in Fig. 6.2-4. The delay circuits 3 and 4 with the summers SUM3 and SUM4 form two sliding correlators.

way just described, while a different method must be found for complementary codes with $4^n/2 = 2, 8, 32, \ldots$ digits.

Lines e, g, and i of Fig. 6.2-4 show the nominal signal $f(t)$ shifted by $-33T$, $-70T$, and $-111T$, while lines f, h, and j show the shifted actual signal $F(t)$. The sums of the nominal and the actual signals are shown in lines k and l. The shifting and summing is implemented in Fig. 6.2-5 by the summer SUM1 and the delay circuits 1 connected to its input terminals.

Lines m–v show the same processes of shifting and summing applied so that the second character of the complementary code 1 in Table 6.2-4 is enhanced. The practical implementation is done by the summer SUM2 and the delay circuits 2 in Fig. 6.2-5.

The output signals of the summers SUM1 and SUM2 in Fig. 6.2-5 are again fed to sliding correlators consisting of the delay circuits 3 and 4, as well as the summers SUM3 and SUM4. The adding (+) and subtracting (−) input terminals of the summers are set for the first character and the amplitude-reversed second character of the complementary code 1 in Table 6.2-4. The cross-correlation functions $K(0, -33T, -70T, -111T; t)$ and $-K(-16T, -51T, -90T, -133T; t)$ produced are plotted in lines a and b of Fig. 6.2-6. Their sum is produced by the summer SUM4 in Fig. 6.2-5 and it is plotted in line c of Fig. 6.2-6.

Line d in Fig. 6.2-6 shows the sum of the function in line k of Fig. 6.2-4 and the amplitude-reversed function of line u. This is the nominal cross-correlation function, since this function would have been produced instead of the actual cross-correlation function of line c, Fig. 6.2-6, if the nominal signal $f(t)$ of Fig. 6.2-4, line c, rather than the actual signal $F(t)$ of line d had been transmitted. One may immediately see that the positive peaks of the actual cross-correlation function in line c of Fig. 6.2-6 coincide with the positive pulses of the nominal cross-correlation function in line d. This is so because these peaks are due to the autocorrelation func-

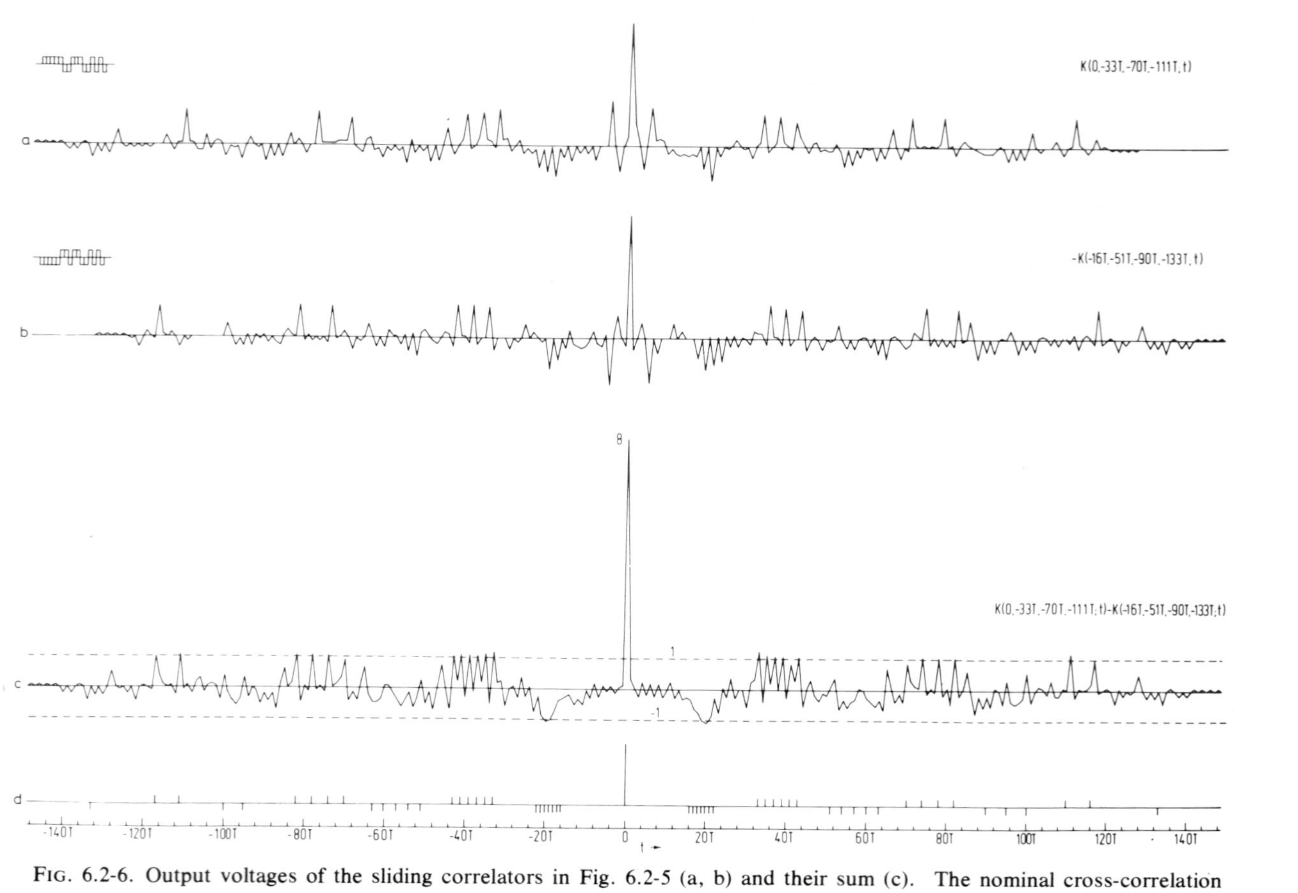

FIG. 6.2-6. Output voltages of the sliding correlators in Fig. 6.2-5 (a, b) and their sum (c). The nominal cross-correlation function (d) is the sum of the function in line k of Fig. 6.2-4 and the amplitude-reversed function of line u.

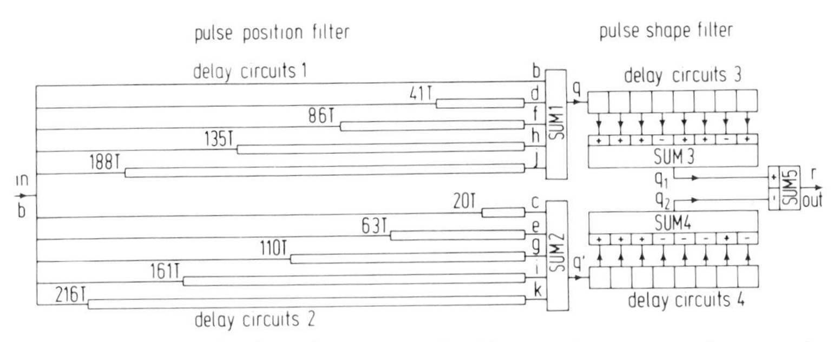

FIG. 6.1-10. Receiver for the pulse compression for complementary codes according to Fig. 6.1-6. SUM, summer.

tained by completing that table. The sets of sidelobes beyond the first one are too far out to be plotted conveniently.

The output signals for $2^n = 16$ and 8 in Fig. 6.1-11 may be inferred from Table 6.1-6. For $2^n = 16$ one can use eight pulses—since their number must be even—and one obtains the sidelobe-to-mainlobe ratio 1/8 used in Fig. 6.1-11. For $2^n = 8$ one obtains from Table 6.1-6 six usable pulses and the sidelobe-to-mainlobe ratio 1/6 used in Fig. 6.1-11. The location of the sidelobes is obtained by writing the complete tables corresponding to Table 6.1-5 for $2^n = 8$ and 16. All the sidelobes are shown for the signal $2^n = 8$ in Fig. 6.1-11, but only the first two sets of sidelobes are shown for the signal $2^n = 16$.

6.1.6 Doppler Shift Resolvers

The Doppler resolution is based on the pulse period T_D in Figs. 4.6-2 and 5.4-5. We will discuss two methods to achieve the Doppler resolution. The first uses many pulses and thus many pulse periods T_D, while the second uses only two pulses and thus only one pulse period T_D.

For an explanation of the first method, let the output signal from terminal r in Fig. 6.1-10 be fed to the input terminal r in Fig. 6.1-12. We assume that a complementary code with $2^n = 32$ digits is used, which means the signal shown on top of Fig. 6.1-11 is fed to the input terminal in Fig. 6.1-12.

The circuit of Fig. 6.1-12 consists of a number of feedback delay circuits that differ only by the delay provided. Consider the circuit in the center characterized by $s(0)$ and $v/c = 0$. The input signal passes through a hybrid coupler HYC5, a nonreversing amplifier AMP3, another hybrid coupler HYC6, and is fed back via the delay circuit DEL3, having the delay T_D, to the hybrid coupler HYC5. Let the signal of the top line of Fig.

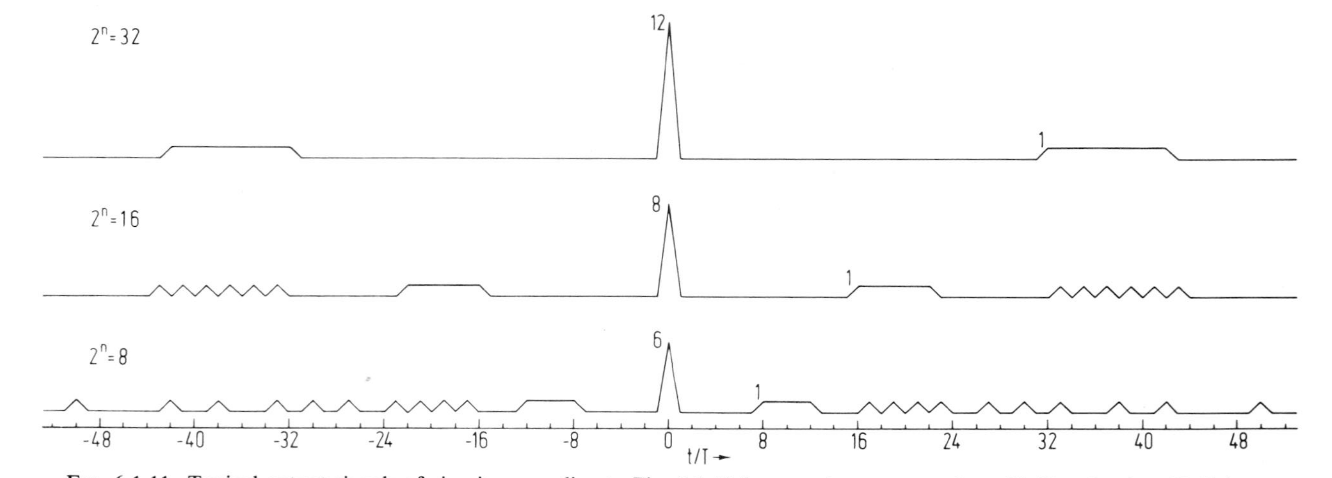

FIG. 6.1-11. Typical output signals of circuits according to Fig. 6.1-10 for complementary codes with $2^n = 8$, 16 or 32 digits.

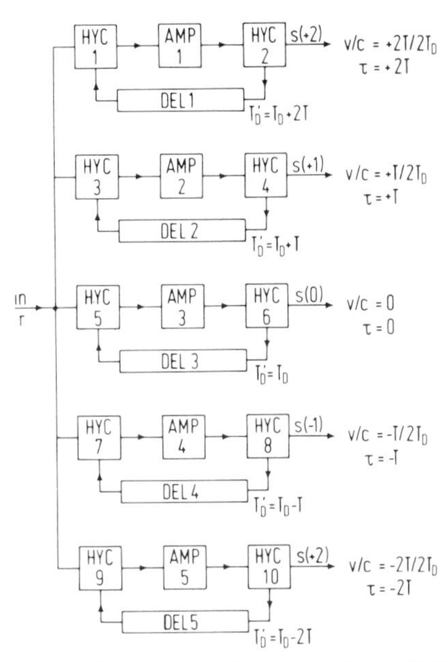

FIG. 6.1-12. Doppler resolver based on feedback delay lines. HYC, hybrid coupler; AMP, nonreversing amplifier; DEL, delay circuit.

6.1-1 be fed to this circuit, and let the velocity of the target by $v/c = 0$. If the amplification of the amplifier AMP3 is set to yield exactly unit gain in the feedback loop, and n signals with period T_D according to the top line of Fig. 6.1-11 are received, we will obtain the input signal with the amplitude multiplied by n at the output terminal $s(0)$ in Fig. 6.1-12 after the nth signal has been received. Let us choose $n = 12$, because the mainlobe-to-sidelobe ratio of the signal in the top line of Fig. 6.1-11 equals 12. The output signal will then be the one shown for $s(0)$ in Fig. 6.1-13, with the mainlobe having the peak amplitude $12^2 = 144$ and the two sidelobes having the peak amplitude $144/12 = 12$.

Let now the same signal be fed to the circuit with the delay time $T_D' = T_D + T$ and the output terminal $s(+1)$ in Fig. 6.1-12. The mainlobe in Fig. 6.1-11, top line, will now produce 12 triangular pulses at the times $t/T = 0, T, 2T, \ldots, 11T$. These pulses add up to the trapezoidal pulse with peak amplitude 12 in the interval $-1 \leqq t/T \leqq 12$ in line $s(+1)$ of Fig. 6.1-13. The sidelobes in Fig. 6.1-11, top line, add to the trapezoidal pulses with peak amplitude 12 in the two intervals $-43 \leqq t/T \leqq -20$ and $33 \leqq t/T \leqq 56$.

If 12 repetitions of the signal of Fig. 6.1-11, top line, are fed to the circuit with delay time $T_D' = T_D + 2T$ in Fig. 6.1-12, one obtains at the output

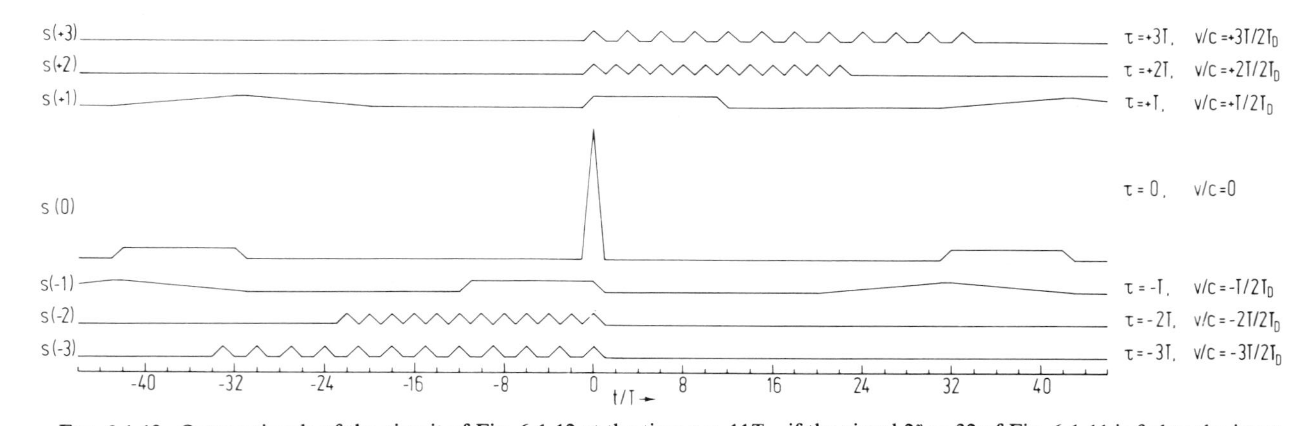

FIG. 6.1-13. Output signals of the circuit of Fig. 6.1-12 at the time $t = 11T_D$, if the signal $2^n = 32$ of Fig. 6.1-11 is fed to the input terminal at the times $t = 0, T_D, 2T_D, \ldots, 11T_D$.

tions of the two characters of the complementary code used, as one may infer from a closer study of Fig. 6.2-4. The sections of the actual cross-correlation function coinciding with the negative pulses of the nominal cross-correlation function show no such close correspondence, except for the first set of sidelobes at $\pm 20T$. The reason is that the negative pulses indicate that the cross-correlation functions rather than the autocorrelation functions of the two characters of the complementary code used cause the respective section of the function of line c, Fig. 6.2-6.[1] The peaks of these cross-correlation functions are smaller than the peaks of the autocorrelation functions, and they do not coincide with the pulses of line d, Fig. 6.2-6, as do the peaks of the autocorrelation functions.

We have previously constructed an approximation to the cross-correlation function of line c, Fig. 6.2-6, by substituting triangles with base length T and amplitude 1 for the pulses in line d, Fig. 6.2-6, and plotting their sum in Fig. 6.1-11, line $2^n = 16$. We can see now that this approximation is excellent where the autocorrelation functions of the two characters of the complementary code are formed, which means at the location of the second, fourth, sixth, ... sets of sidelobes in line d, Fig. 6.2-6; but the approximation gives a worse than actual result where the cross-correlation functions are formed, which means at the location of the first, third, fifth, ... sets of sidelobes.

6.2.2 *Avoidance of Large Peak Currents*

The complementary code for $2^n = 8$ produced very large peak antenna currents $I(t)$ in line b of Fig. 6.2-1. The code for $2^n = 16 = 4^2$ produced a much smaller peak current $I(t)$ in line b of Fig. 6.2-4, and we had stated that this was generally so for codes with 4^n digits. Let us investigate now how large peak currents can be avoided for codes with 2, 8, 32, ... $= 4^n/2$ digits.

Figure 6.2-7 shows on top the nominal antenna current according to Table 6.2-2 rather than according to Table 6.2-1 as used in Fig. 6.2-1. This pulse position code gives 8 pulse positions while the one used in Fig. 6.2-1 gave only 6 positions. We need a number of pulse positions that is a multiple of 4.

---

[1] We had also plotted all sidelobes with positive sign, while the first sidelobe around $\pm 20T$ in Fig. 6.2-6, line c, is definitely negative. However, negative sidelobes are usually as undesirable as positive ones, and only their magnitude is thus of interest.

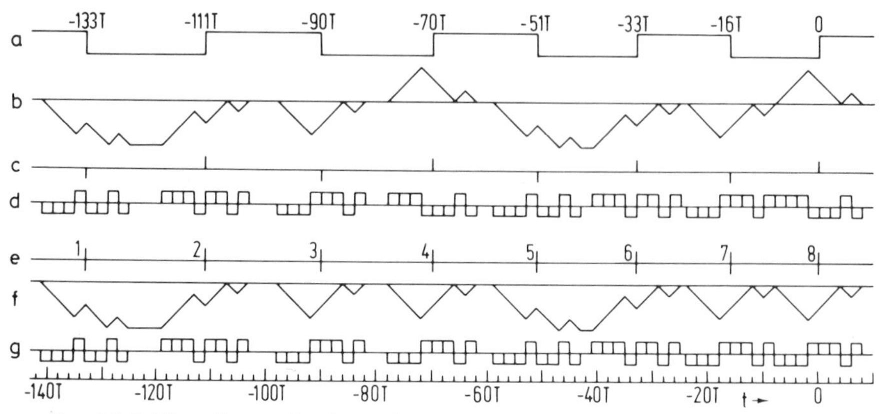

FIG. 6.2-7. Time diagram for the avoidance of large peak antenna currents for complementary codes with $4^n/2$ digits: (a) nominal antenna current; (b, f) actual antenna currents; (c) nominal signal in the far zone; (d, g) actual signal in the far zone; (e) nominal signal in the far zone reduced to a pure pulse position code without positive and negative signs as in line c.

The nominal signal in the far zone is shown in line c of Fig. 6.2-7. We transmit the same character of the complementary code 1 of Table 6.2-3 at two adjacent pulse positions of the nominal signal, but with alternating signs. For instance, the second character $(+++-++-+)$ is transmitted at the time $t = -133T$ with negative sign $(---+--+-)$ as shown in line d, and at the time $t = -111T$ with positive sign $(+++-++-+)$. The integral of these signals is shown in line b; it necessarily returns to its initial value at the end of the second character ($t = -111T + 8T = -103T$) because the number of positive and negative pulses is the same, regardless of how many positive and negative pulses the individual character has.

The first character of the complementary code 1 of Table 6.2-3 is transmitted at the time $t = -90T$ with negative sign $(---+++-+)$ and at the time $t = -70T$ with positive sign $(+++----+-)$. Since this character has equally many positive and negative signs, the current returns to its initial value at the end of each character.

A circuit for the reception of the signal of line d in Fig. 6.2-7 is shown in Fig. 6.2-8a. The delay circuits 1 with the summers SUM1–SUM4 work like in Fig. 6.2-5, but the adding (+) and subtracting (−) input terminals of the summers SUM5 and SUM6 correct the alternating signs of the transmitted signals. The sliding correlators operate as in Fig. 6.2-2, but the signs at the input terminals of the summer SUM8 in Fig. 6.2-8 are the reversed signs of the summer SUM4 in Fig. 6.2-2.

The antenna current in line b of Fig. 6.2-7 has positive as well as nega-

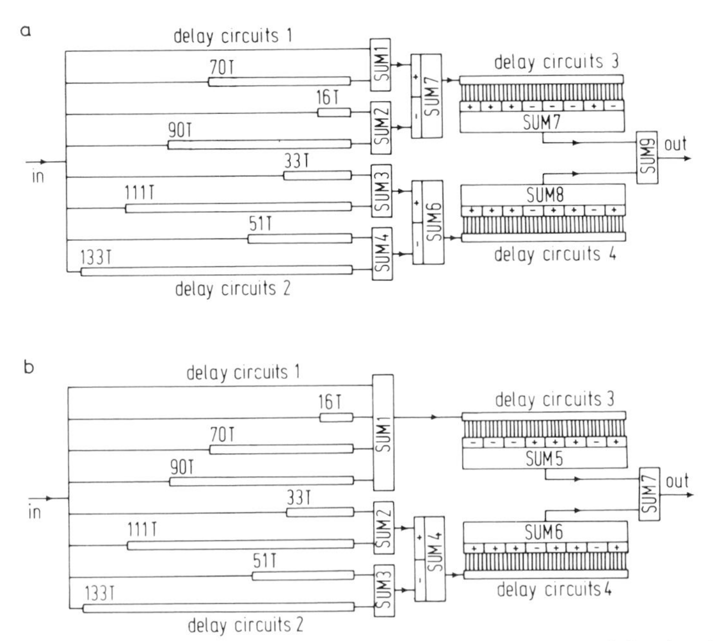

FIG. 6.2-8. Receivers for the signals in line d of Fig. 6.2-7 (a) and for the signal in line g (b).

tive values. One may eliminate either the positive or the negative currents. This is shown for the elimination of the positive currents in lines e–g. The signals at the times $-133T$, $-111T$, $-90T$, $-51T$, $-33T$, and $-16T$ in line g have the same signs as in line d, but the signals at the times $-70T$ and 0 are amplitude-reversed. The positive currents at the times $-70T$ (pulse position 4) and 0 (pulse position 8) are thus avoided. Figure 6.2-8b shows the receiving circuit for the signal of line f. It is somewhat simpler than the circuit in Fig. 6.2-8a; note the sign reversal at the input terminals of the summers SUM5 in Fig. 6.2-8b and SUM7 in Fig. 6.2-8a.

6.2.3 Improvement of Cross-Correlation Functions

The cross-correlation functions in Figs. 6.2-3 and 6.2-6 are remarkably good, since the mainlobe-to-sidelobe ratio equals essentially the number of pulse positions used (6 and 8) rather than the square root of this number. Nevertheless, one can find better cross-correlation functions, usually at the expense of a reduced average-to-peak power ratio.

Figure 6.2-9 shows a modification of Fig. 6.2-1. The pulse positions $-50T$, $-38T$, ..., 0 in line a of either illustration are the same, but the

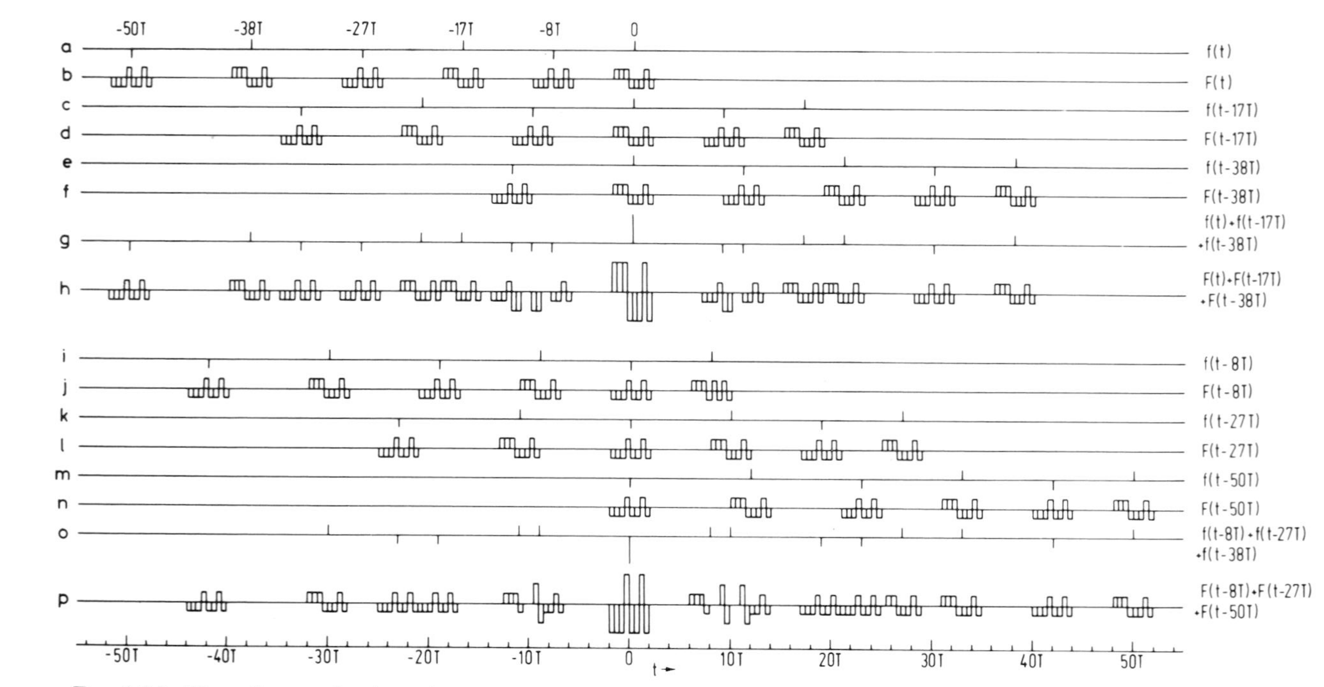

FIG. 6.2-9. Time diagram for the pulse position filter of the receiver for the pulse position code of Table 6.2-1 and the complementary code 1 of Table 6.2-3 for the pulse shape code. The duration of the individual pulses is $T/2$.

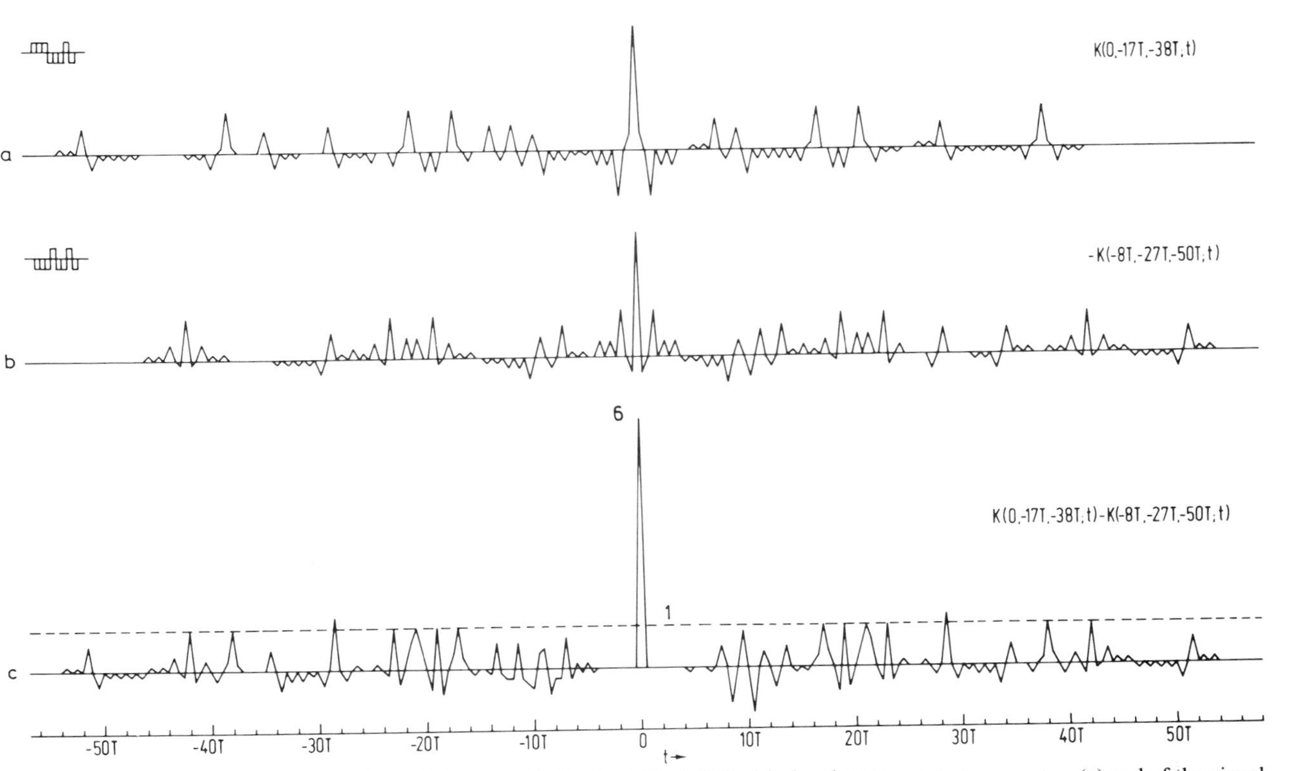

FIG. 6.2-10. Cross-correlation functions of the signal in line h of Fig. 6.2-9 with the character + + + − − − + − (a) and of the signal in line p with the character − − − + − − + − (b). The sum of these two cross-correlation functions (c) has no sidelobes in the immediate neighborhood of the mainlobe.

duration of the pulses in line b of Fig. 6.2-9 equals $T/2$ while it equals T in line d of Fig. 6.2-1. Except for this change in pulse duration, and the elimination of the nominal as well as the actual antenna current, the pulse diagrams in Figs. 6.2-1 and 6.2-9 are the same.

The receiver shown in Fig. 6.2-2 will receive the signal of Fig. 6.2-9, line b, if all delay times in the delay circuits 3 and 4 are reduced to one half. The output voltages $K(0, -17T, -38T; t)$ and $-K(-8T, -27T, -50T; t)$ of the sliding correlators as well as their sum are shown in Fig. 6.2-10. One may see that a sidelobe-free region is created in the neighborhood of the mainlobe, but the largest peaks of the sidelobes are as high as in Fig. 6.2-3. The price for the improvement is the reduction of the average-to-peak power ratio of the signal in line d of Fig. 6.2-1 to one half for the signal in line b of Fig. 6.2-9.

The type of coding discussed so far was based on autocorrelation functions, both for the mainlobe and for the largest peaks of the sidelobes. By using n signals we could achieve essentially a sidelobe-to-mainlobe ratio $1/n$; by spacing the signals sufficiently far apart one could produce this ratio exactly, but one would have to pay for this result with a reduction of the average-to-peak power ratio. An improvement would be possible if the largest peaks of the sidelobes were produced by *cross*-correlation functions while the mainlobe was produced by the superposition of n *autocorrelation* functions. Since the peaks of all the cross-correlation functions of the eight characters in Table 6.2-3 are necessarily smaller than the peak $+8$ of the autocorrelation functions, one must be able to reduce the sidelobe-to-mainlobe ratio by the substitution of cross-correlation for autocorrelation functions in the sidelobes. The process is best explained with the help of an example. Let us observe ahead of time that our example will not provide an improvement, but it will show the principle—and the pitfall—one has to avoid.

Once more Fig. 6.2-11 shows in line a the pulse positions $-50T$, $-38T$, ..., 0 of Table 6.2-1. In Fig. 6.2-1, line d, we had used the two characters of the complementary code 1 of Table 6.2-3 at these positions. Since we used two characters for six pulse positions, we had to use each character three times. This repeated use of the same character at different positions caused the largest peaks of the sidelobes of the signal in Fig. 6.2-3, line c, to be due to autocorrelation functions.

We avoid sidelobes due to autocorrelation functions by using each character only once. An autocorrelation function can then appear only at the location of the mainlobe. We use the six characters of the complementary codes 2–4 of Table 6.2-3. The first character $(+++++--+)$ is used at the pulse position $-50T$ in Fig. 6.2-11, line b. The second character $(++--+-+-)$ is used at the position $-38T$, the third character

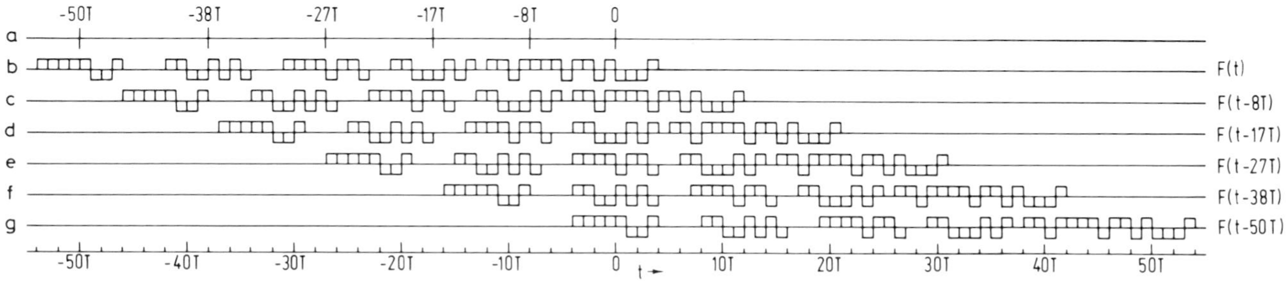

FIG. 6.2-11. Time diagram for the reception of a signal consisting of three different complementary codes in order to avoid autocorrelation functions in the sidelobes.

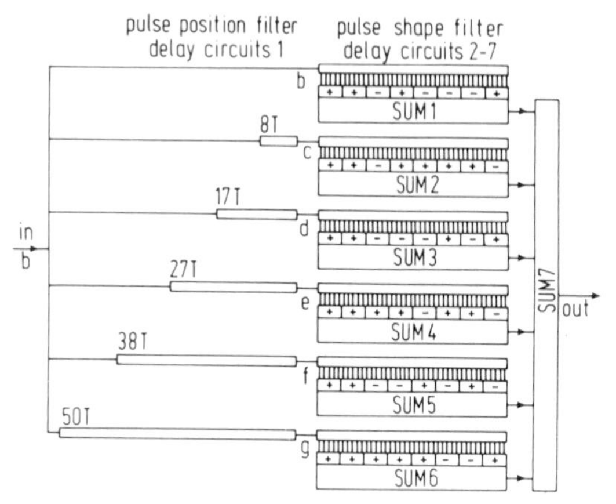

FIG. 6.2-12. Receiver for the signal of Fig. 6.2-11 requiring six sliding correlators. SUM, summer.

$(+ + + + - + + -)$ at the position $-27T$, and so on until the sixth character $(+ + - + - - - +)$ is used at the position 0.

For the reception, the signal $F(t)$ of line b in Fig. 6.2-11 is shifted to become $F(t - 8T)$ in line c, $F(t - 17T)$ in line d, etc. The delay circuits 1 in Fig. 6.2-12 provide these delays. The delayed signals must be fed to six different sliding correlators, each one set to receive one of the six characters of the complementary codes 2–4 of Table 6.2-3. One may readily recognize that the signs at the input terminals of the summers SUM1–SUM6 are equal to the signs of the characters of the complementary codes 2–4 in Table 6.2-3.

The output voltages of the sliding correlators are shown in lines b–g of Fig. 6.2-13, and their sum is shown in line h. A comparison with line c of Fig. 6.2-3 shows that the large sidelobes around $t = -38T$ and $t = -20T$ have been reduced significantly, but the largest sidelobes are larger in Fig. 6.2-13 than in Fig. 6.2-3. The reason is that we have used the pulse position code of Table 6.2-1, which was designed to avoid the coincidence of autocorrelation functions in Fig. 6.2-3, and thus the large sums produced by coinciding peaks of autocorrelation functions. The peaks of the cross-correlation functions are not located where the peaks of the autocorrelation functions are. Hence, the pulse position code used in Fig. 6.2-11 is wrong, and it causes the coincidence of some large peaks in lines b–g in Fig. 6.2-13 that produce large sidelobes in line h.

A closer study of Fig. 6.2-11 shows that the coincidence of large peaks of cross-correlation functions in lines b–g in Fig. 6.2-13 can be avoided, if one spaces the pulse positions in line a of Fig. 6.2-11 sufficiently far apart. The price for doing so is again a reduction of the average-to-peak power

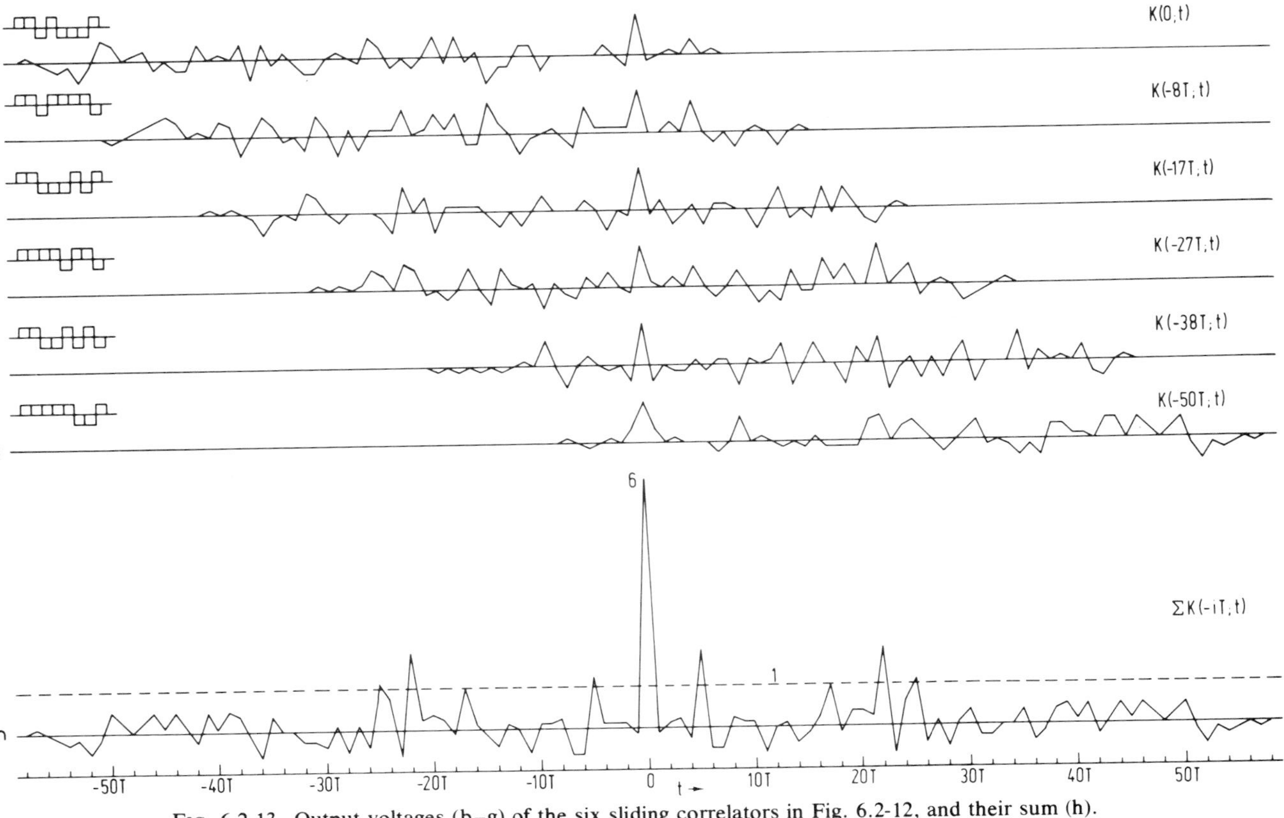

FIG. 6.2-13. Output voltages (b–g) of the six sliding correlators in Fig. 6.2-12, and their sum (h).

ratio. Hence, pulse positions have to be traded for average-to-peak power ratios. Furthermore, one can use characters—which do not even have to belong to complementary codes—other than those listed[1] in Table 6.2-3, and one can arrange them in different permutations.[2] This abundance of possibilities to find signals with a smaller sidelobe-to-mainlobe ratio makes it impossible to pursue the topic any further here.

6.2.4 *Perfect Cross-Correlation and Ambiguity Functions*

One can produce cross-correlation functions and ambiguity functions that have no apparent sidelobes whatsoever. The structure of signals that accomplish this is shown in Fig. 6.2-14. A radar transmits periodically pulses with period T_D. This period shall be sufficiently long to permit a Doppler resolution with two pulses, as discussed in Section 6.1.6. The fine structure—or pulse position code—of the radar signal consists of only two pulses with a distance T, as shown in line b of Fig. 6.2-14. The time T must be so short that the behavior of the target and the transmission path remain predictable during this time. This condition is met if neither the location of the target nor the features of the transmission path change significantly, but also if the change of position of a fast-moving target can be predicted from prior measurements. The shortest permissible value for T is determined by the length of the ambiguity-free range one wants to observe.

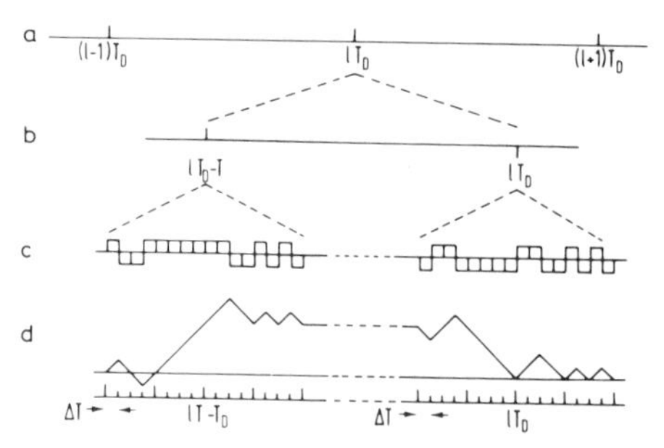

FIG. 6.2-14. Structure of a radar signal (a) that yields a sidelobe-free ambiguity function. (b) The fine structure or pulse position code, (c) the hyperfine structure or pulse shape code, and (d) the antenna current.

[1] Erickson (1961) lists 1 complementary code with 4 digits, 6 with 8 digits, and 48 with 16 digits. The number of codes increases rapidly for larger numbers of digits, and there are actually more codes than listed by Erickson, since the list only includes characters that yield different autocorrelation functions.

[2] There are $6! = 720$ permutations of the six characters in line b of Fig. 6.2-11. It is evident that we have stumbled across a method to produce cryptosecure spread-spectrum radar signals.

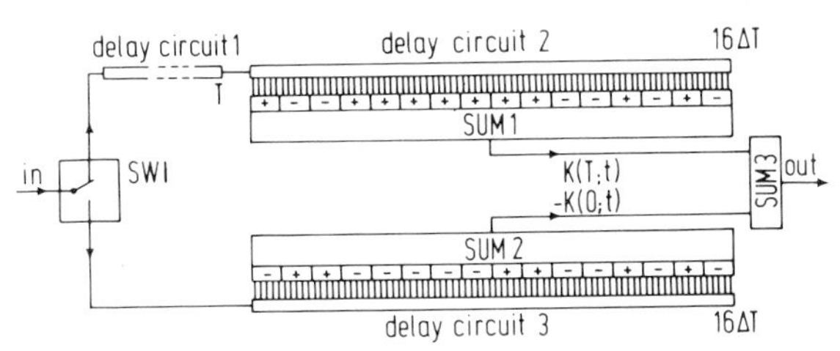

FIG. 6.2-15. Receiver for the signal of Fig. 6.2-14. SUM, summer; SWI, switch.

The hyperfine structure or the pulse shape code of the radar signal consists of the pair of characters of a complementary code, as shown in line c of Fig. 6.2-14 for signals with 16 digits. The duration of one digit is ΔT, and the duration of each character is 16 ΔT. The duration of a character cannot exceed T. We obtain from this condition the maximum value of ΔT for a character with 2^n rather than 16 digits:

$$\Delta T \leqq T/2^n \tag{1}$$

The antenna current required to produce the signals of line c, Fig. 6.2-14, is shown in line d. It returns to its initial value at the end of the second character.

Figure 6.2-15 shows a receiver for the signal of Fig. 6.2-14. The first character is fed through the switch SWI and the delay circuit 1 with delay time T to a sliding correlator, which produces the cross-correlation function $K(T;\, t)$. After the first character has been received and before the second character arrives, the switch SWI connects the receiver input terminal to the second sliding correlator, which produces the cross-correlation function $-K(0;\, t)$; the discussion of the operation of the switch SWI is postponed until later. The functions $K(T;\, t)$, $-K(0;\, t)$ and their sum $K(T;\, t) - K(0;\, t)$ are shown in Fig. 6.2-16. The sum yields a perfect cross-correlation function with no apparent sidelobes, although this statement will need some clarification.

The sidelobe-free signal of Fig. 6.2-16, line c, is centered at the time $t = lT_{\mathrm{D}}$. According to Fig. 6.2-14, an equal signal will be received at the time $t = (l + 1)T_{\mathrm{D}}$, if there is no Doppler shift. In the presence of a Doppler shift, the distance between the two signals will be somewhat smaller or larger than T_{D}. The two signals are fed to a Doppler processor based on the clutter canceller of Fig. 6.1-15, and an ambiguity function is obtained from the two autocorrelation functions as in Fig. 6.1-16. The result is the perfect, sidelobe-free ambiguity function of Fig. 6.2-17. This ambiguity function is a pyramid over a rhomboid with side lengths $2\,\Delta T$ and $[(2\,\Delta T)^2 + (2\,\Delta T)^2]^{1/2} = 2\sqrt{2}\,\Delta T$. The sides of length $2\,\Delta T$ are parallel to the Doppler axis at the distance $\pm\Delta T$ from the center of the rhomboid. Two

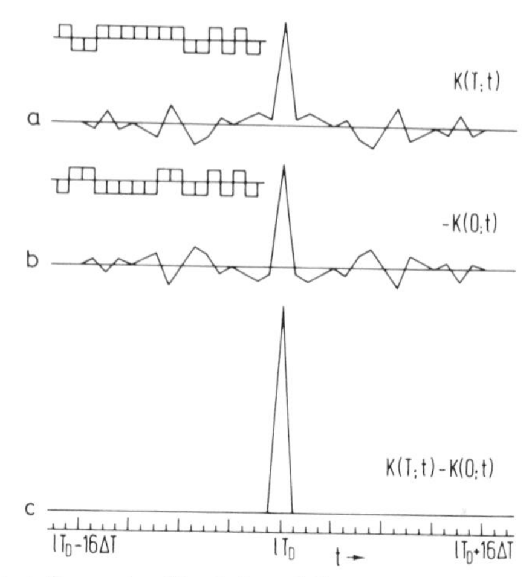

FIG. 6.2-16. Output voltages (a, b) of the sliding correlators in Fig. 6.2-15 and their sum (c).

of the corner points of the rhomboid are located on the range axis, the other two have the distance $\pm 2\,\Delta T$ from the range axis.

The Doppler processor based on the clutter canceller principle uses subtraction as well as rectification, and this causes problems if returns from many closely spaced targets are received; the many targets may, of course, be many resolvable points on the same physical structure. These problems are not encountered if the Doppler resolver of Fig. 6.1-12, based on feedback delay lines, is used. Many more than two radar signals at the times lT_D and $(l+1)T_D$ are needed in this case, and the ambiguity function of Fig. 6.2-17 is only approached closer and closer as the number of radar signals used increases. This Doppler processor makes poor use of the received information, but is relatively easy to implement[1]. A Doppler processor using only two radar signals is so difficult to implement because one cannot afford to use the information inefficiently.

Let us now turn to the operation of the switch SWI in Fig. 6.2-15, the statement about "no apparent sidelobes," and typical applications of perfect thumbtack ambiguity functions. All three topics are closely related.

Let d_{max} be the largest distance from which we expect a radar signal to return, and d_{min} the shortest distance. The time interval between the two

[1] If one receives two signals a and b, both carrying the same amount of information, and one forms the sum a + b but not the difference a − b, one reduces the information to one half. For k rather than two signals, the reduction of the information is $1/2^{k-1}$.

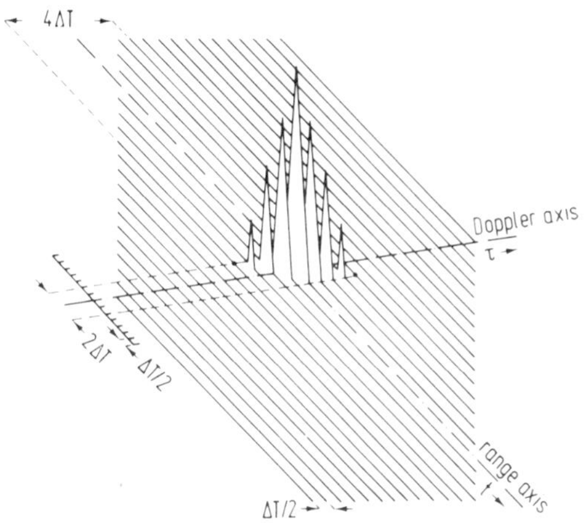

FIG. 6.2-17. Ambiguity function based on the sidelobe-free function $K(T; t) - K(0; t)$ of Fig. 6.2-16 and a Doppler processor discussed in Section 6.1.6.

characters in line c of Fig. 6.2-14 must then be at least $2(d_{max} - d_{min})/c$, and the time T follows with the help of Eq. (1):

$$T \geqq 2(d_{max} - d_{min})/c + 2^n \Delta T \tag{2}$$

The switch SWI in Fig. 6.2-15 must be switched to the delay circuit 3 at the time $(lT_D - T) + 2d_{max}/c + 2^n \Delta T/2$, if the center of the first character in line c of Fig. 6.2-14 is radiated at the time $lT_D - T$. The second character in line c will thus not be fed to the delay circuits 2 in Fig. 6.2-15 and this is the reason why there are no apparent sidelobes in Fig. 6.2-16, line c, or along the range axis, $\tau = 0$, in Fig. 6.2-17. If the switch SWI is not used in Fig. 6.2-15, one obtains the cross-correlation functions of the two characters in line c of Fig. 6.2-14 centered at $lT_D + T$ or $lT_D - T$ in line c of Fig. 6.2-16.

In order to get a large average-to-peak power ratio one must make T small. This implies, according to Eq. (2), a small difference $d_{max} - d_{min}$. Such a small distance is encountered, e.g., in the detection of low-flying aircraft and cruise missiles by a look-down radar. This is obviously a case where one wants a perfect thumbtack ambiguity function, whose actual sidelobes are only due to imperfections and not due to theoretical shortcomings. Another typical application is the distinction of an ICBM from the cloud of decoys that masks it. The distance and velocity of such a cloud of targets can be measured accurately, and the velocity is very constant over a time of duration T. The decoys—being reflectors—yield thumbtack functions as shown in Fig. 6.2-17, while the ICBM yields a function that is spread in the direction of the range axis, provided the geometric dimensions of the ICBM are large compared with $c\,\Delta T$. For an all-weather radar that uses no sinusoidal carrier one can make ΔT theoretically on the order of 0.1 ns—implying a range resolution of 1.5 cm—which is more than sufficient.

The potential range resolution of 1.5 cm for an all-weather radar shows that the requirement for a small value of $d_{max} - d_{min}$ in Eq. (2) is a restriction imposed already by a more fundamental requirement. For a range interval $d_{max} - d_{min} = 1.5$ km and a resolution of 1.5 cm one obtains 100,000 resolved points in this range interval, which still have to be multiplied by the number of resolved Doppler intervals. Hence, the potential information flow becomes overwhelming for any equipment or user, and one must restrict the high resolution to the smallest possible section of the range-Doppler domain.

6.3 Velocity and Acceleration Processing

6.3.1 *Distance and Its First and Second Derivatives*

We have repeatedly discussed the Doppler effect of radar signals that do not use a sinusoidal carrier. These discussions were very much influenced by the concepts of Doppler processing for conventional radar with a sinusoidal carrier. We will now try to cut loose from the traditional Doppler processing and to develop concepts that apply specifically to carrier-free radar.

Since it is usually easier to develop a concept with the help of a practical example, we will start with the look-down radar discussed in Section 5.4 and see how its performance can be improved by better signal processing. This example is chosen because the look-down radar presents an

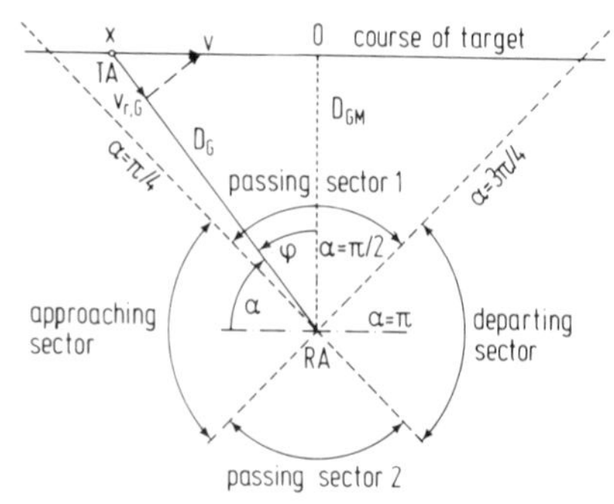

FIG. 6.3-1. Geometric relations in the ground plane for a look-down radar. RA, radar; TA, target.

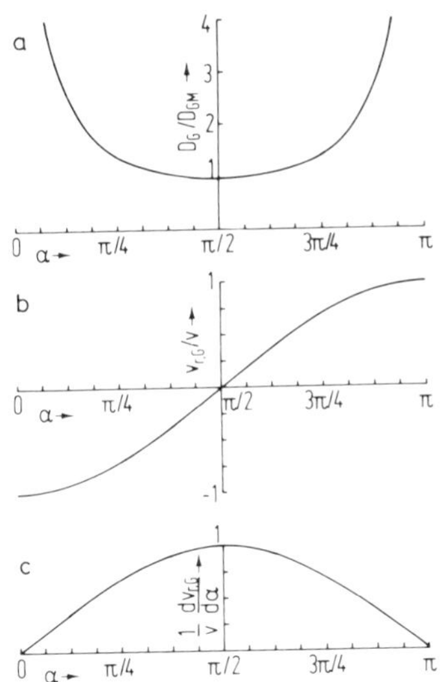

FIG. 6.3-2. Range (a), relative ground velocity (b), and relative ground acceleration (c) as function of the angle α in Fig. 6.3-1. The velocity (b) varies less than the acceleration (c) in the approaching and departing sectors of Fig. 6.3-1, while the acceleration varies less than the velocity in the passing sectors. Less variation is good for the summation of multiple return signals.

extreme case of a weak signal from a moving target being masked by a huge signal from the stationary surface of the Earth.

Figure 6.3-1 shows the geometric relations in the ground plane. As in Section 5.4, the look-down radar RA is assumed to be stationary with respect to the ground to simplify the investigation. The target TA with velocity v has the relative velocity $v_{r,G}$ in the ground plane with reference to the radar RA. The relative velocity v_r between target and radar differs from $v_{r,G}$, as was discussed in connection with Figs. 5.4-1 and 5.4-2, but the relative velocity $v_{r,G}$ in the ground plane is sufficient for our current purpose.

The distance D_G in the ground plane between the radar and the target is plotted as a function of the angle α in Fig. 6.3-2a; the normalization D_G/D_{GM} is used, where D_{GM} is the minimum distance in the ground plane according to Fig. 6.3-1:

$$D_G = D_{GM}/\sin\alpha = D_{GM}/\cos\varphi \tag{1}$$

The velocity $v_{r,G}$ normalized by v is plotted[1] in Fig. 6.3-2b:

$$v_{r,G} = -v\cos\alpha = v\sin\varphi \tag{2}$$

---

[1] Note that $v_{r,G}$ is the derivative dD_G/dt with respect to time and not the derivative $dD_G/d\alpha$ with respect to the angle α. Hence, the infinite slope in Fig. 6.3-2a at $\alpha = 0, \pi$ is not in contradiction to the finite values of $v_{r,G}/v$ in Fig. 6.3-2b.

The velocity $v_{r,G}$ is zero at $\alpha = \pi/2$, which makes the target indistinguishable from the stationary surface of the earth.

The angular acceleration

$$dv_{r,G}/d\alpha = v \sin \alpha = v \cos \varphi \tag{3}$$

which is plotted in Fig. 6.3-2c, has its maximum where $v_{r,G}$ equals zero. Hence, it appears to be reasonable to measure the acceleration rather than the velocity to distinguish the target from the surface of the earth. This will turn out to be correct, but for a different reason: The acceleration in Fig. 6.3-2c is not only large around $\alpha = \pi/2$, but also almost constant. Hence, one may make several measurements and take their average to get a more accurate and reliable result in the presence of statistical disturbances. The velocity $v_{r,G}$ in Fig. 6.3-2b, on the other hand, changes its sign at $\alpha = \pi/2$. If one takes the average velocity of several measurements around $\alpha = \pi/2$, one will get a velocity $v_{r,G}$ that is essentially zero, and the target will be less distinguishable from the stationary surface of the Earth than if no averaging is used.

For the conventional radar with a sinusoidal carrier it is usually not considered worthwhile to measure acceleration. The reason is that the range resolution is determined by the resolution of the radar pulse, and the velocity resolution is determined by the resolution of the carrier period, which is essentially independent of the pulse resolution. Hence, one has two independent measurements, one yielding range and the other yielding velocity. Acceleration and higher derivatives of the velocity are based on the measurement of the velocity and can thus bring no fundamental improvement over the velocity resolution. This point is discussed in detail by Rihaczek (1969, Section 4.3, particularly pp. 107, 110). A carrier-free radar measures only range with one pulse or one *return*. Two returns are needed to derive velocity based on the change of range[1] $(d_2 - d_1)/(t_2 - t_1)$, where t_2 and t_1 are the times at which the target has the distances[2] d_2 and d_1:

$$v = v_1 = (d_2 - d_1)/(t_2 - t_1) \tag{4}$$

Let the first signal be radiated at the time t_1' and the returned signal be received at the time $t_1' + 2\,\Delta T_1$. The time at which the signal hit the target

---

[1] The distance d_2 is measured later than the distance d_1. A closing target, $d_2 < d_1$, will thus yield a negative velocity $(d_2 - d_1)/(t_2 - t_1)$ while an opening target, $d_2 > d_1$, will yield a positive velocity $(d_2 - d_1)/(t_2 - t_1)$. This corresponds to the sign of $v_{r,G}$ in Fig. 6.3-2b. The choice of $(v_2 - v_1)/T_D$ for the acceleration in Eq. (7) corresponds to the sign of $dv_{r,G}/d\alpha$ in Fig. 6.3-2c.

[2] Relativistic effects are ignored since we are interested only in terms of order v/c.

is thus $t_1 = t'_1 + \Delta T_1$. The second signal is radiated at the time $t'_1 + T_D$; it hits the target at $t_2 = t'_1 + T_D + \Delta T_2$, and it is received at the time $t'_1 + T_D + 2\,\Delta T_2$. The time difference $t_2 - t_1$ becomes:

$$t_2 - t_1 = T_D + \Delta T_2 - \Delta T_1 \tag{5}$$

The time difference $\Delta T_2 - \Delta T_1$ is due to the target having advanced the distance $v(t_2 - t_1)$ during the time between the hits by the first and the second signal. The radar signal has to travel a round-trip distance shortened by $2v(t_2 - t_1)$ for the second hit. The time difference $\Delta T_2 - \Delta T_1$ is thus $2(v/c)(t_2 - t_1)$. Hence, $\Delta T_2 - \Delta T_1$ is of the order $(v/c)T_D$, and we may ignore $\Delta T_2 - \Delta T_1$ in Eq. (5) compared with T_D. Equation (4) becomes

$$v = v_1 = (d_2 - d_1)/T_D \tag{6}$$

where T_D is the time between the radiation of the radar pulses.

Three radar pulses are needed to derive an acceleration a based on a change of velocity, or on the second difference quotient of the range:

$$a = a_1 = \frac{v_2 - v_1}{T_D} = \frac{1}{T_D}\left(\frac{d_3 - d_2}{T_D} - \frac{d_2 - d_1}{T_D}\right) = \frac{d_3 - 2d_2 + d_1}{T_D^2} \tag{7}$$

In this manner one may advance to four pulses, which yield a change of acceleration $(a_2 - a_1)/T_D$, or the third difference quotient of the range, $(d_4 - 3d_3 + 3d_2 - d_1)/T_D^3$, etc.

It is immediately apparent that there is no inherent reason why one should use the velocity $(d_2 - d_1)/T_D$, but not the acceleration $(d_3 - 2d_2 + d_1)/T_D^2$ with a carrier-free radar. Both are derived from range measurements, but the number of measurements and thus the received information increases as we proceed from range to velocity to acceleration to change of acceleration, etc.

The same considerations would apply, of course, to a radar with sinusoidal carrier if acceleration, change of acceleration, etc., were derived from several returned radar pulses rather than from the carrier period of—essentially—one radar pulse. However, this is not of great practical interest since the time resolution of the radar pulse is so much poorer than the time resolution of the period of the radar carrier.

6.3.2 Simple Velocity and Acceleration Processor

Let a first radar pulse be radiated at t'_1 and a second one at $t'_1 + T_D$. The target advances the distance

$$d_2 - d_1 = vT_D \tag{8}$$

between the hit by the first and the second pulses. The difference in the round-trip distance the two pulses must travel is $2(d_2 - d_1)$, and the difference τ of the travel times is thus

$$\tau = 2(d_2 - d_1)/c = 2T_D(v/c) \tag{9}$$

If we delay the first pulse after reception by $T_D + \tau$, it will coincide with the second received pulse which is not delayed. Note that τ and v are negative for a closing target, $d_2 < d_1$, and positive for an opening target, $d_2 > d_1$.

A circuit using this principle to resolve the velocities $v/c = \tau/2T_D$ for $\tau = 0, \pm T, \pm 2T$ is shown in Fig. 6.3-3. There are five delay feedback circuits with delays $T_D' = T_D + \tau$. Pulses returned from a closing target with velocity $v/c = \tau/2T_D = -2T/2T_D$ will coincide at the output terminal of the feedback delay circuit with delay $T_D' = T_D - 2T$, those returned from a closing target with velocity $v/c = -T/2T_D$ will coincide at the output terminal of the feedback delay circuit with delay $T_D' = T_D - T$, etc. After n pulses have been received, the coinciding and summed pulses will have n times the amplitude of the sum of pulses that do not coincide.

To determine acceleration, we need three pulses radiated at the times t_1', $t_1' + T_D$, and $t_1' + 2T_D$. The relations of Eqs. (8) and (9) apply again to

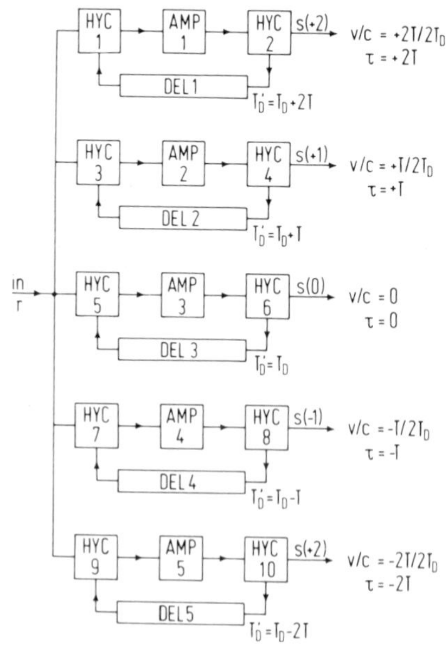

FIG. 6.3-3. Velocity or Doppler resolver based on feedback delay circuits. HYC, hybrid coupler; AMP, nonreversing amplifier; DEL, delay circuit.

the first two pulses, but they have to be modified for the third pulse. The target advances the distance

$$d_3 - d_2 = (v + \Delta v)T_D \tag{10}$$

between the hit by the second and the third pulses. The difference in the round-trip distance the two pulses must travel is $2(d_3 - d_2)$, and the difference $\tau + \Delta\tau$ is thus

$$\tau + \Delta\tau = 2(d_3 - d_2)/c = 2T_D(v/c + \Delta v/c), \qquad \Delta\tau = 2T_D\,\Delta v/c \tag{11}$$

A change Δv of velocity during the time T_D implies an acceleration $\Delta v/T_D$. Hence, one may rewrite Eq. (11) either in the form

$$\Delta v/c = \Delta\tau/2T_D \tag{12}$$

or in the form

$$(\Delta v/T_D)/(c/T_D) = \Delta\tau/2T_D \tag{13}$$

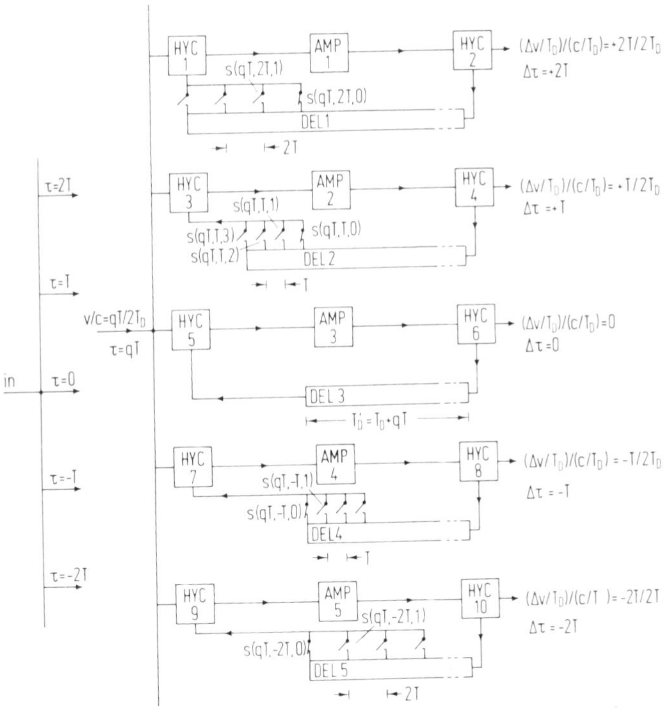

FIG. 6.3-4. Acceleration resolver based on feedback delay lines. HYC, hybrid coupler; AMP, nonreversing amplifier; DEL, delay circuit.

where $\Delta v/T_{\mathrm{D}}$ is the acceleration and c/T_{D} is the proper normalization factor.

If Δv in Eq. (11) is positive, which holds for all values of $v_{\mathrm{r,G}}$ in Fig. 6.3-2b, one obtains a positive value for $\Delta\tau$. The relative acceleration is always positive for a target flying with constant (ground) velocity on a straight course. An example of negative relative acceleration $\Delta v/T_{\mathrm{D}}$ is a target approaching the radar with increasing ground velocity.

Let us turn to the circuit design. The first received pulse must be delayed by $T_{\mathrm{D}} + \tau$ to make it coincide with the second pulse returned from the target with velocity $v/c = \tau/2T_{\mathrm{D}}$. The sum of these two pulses must then be delayed by $T_{\mathrm{D}} + \tau + \Delta\tau$ to make the sum coincide with the third pulse returned from a target with velocity $v/c = \tau/2T_{\mathrm{D}}$ and acceleration $(\Delta v/T_{\mathrm{D}})/(c/T_{\mathrm{D}}) = \Delta\tau/2T_{\mathrm{D}}$.

Figure 6.3-4 shows a circuit for the five velocities $v/c = \tau/2T$ with $\tau = 0, \pm T, \pm 2T$, and five accelerations $(\Delta v/T_{\mathrm{D}})/(c/T_{\mathrm{D}}) = \Delta\tau/2T_{\mathrm{D}}$ with $\Delta\tau = 0, \pm T, \pm 2T$. Actually, one needs five of the circuits shown, but this drafting job was avoided by denoting the velocity $v/c = qT/2T_{\mathrm{D}}$, where $q = 0, \pm 1, \pm 2$. The important difference between Figs. 6.3-3 and 6.3-4 is that the delay circuits have variable delays that can be changed in increments $\Delta\tau$.

For an explanation of the operation of the circuit consider a closing target with velocity $v/c = -2T/2T_{\mathrm{D}}$ and acceleration $(\Delta v/T_{\mathrm{D}})/(c/T_{\mathrm{D}}) = +T/2T_{\mathrm{D}}$. The value of q in Fig. 6.3-4 is thus -2. The basic length of the delay circuits is $T_{\mathrm{D}}' = T_{\mathrm{D}} - 2T$. A first received pulse is delayed in all five delay circuits by $T_{\mathrm{D}} - 2T$. If the second pulse is returned from a target with velocity $v/c = -2T/2T_{\mathrm{D}}$ it arrives at the time $T_{\mathrm{D}} - 2T$ later than the first pulse, and it will coincide with the delayed first pulse at the output terminals of all five feedback delay circuits in Fig. 6.3-4. The third pulse, returned at the time $T_{\mathrm{D}} - 2T + T$ later than the second pulse, will coincide with the sum of the first two pulses if they have been delayed by $T_{\mathrm{D}} - 2T + T$. This requires a change of the delay times, which is accomplished by means of the switches. When the third pulse is radiated, the switches $s(qT, 2T, 0)$, $s(qT, T, 0)$, ..., $s(qT, -2T, 0)$ are opened and the adjacent switches $s(qT, 2T, 1)$, $s(qT, T, 1)$, ..., $s(qT, -2T, 1)$ are closed. The delay times are thus changed from $T_{\mathrm{D}} - 2T$ to $T_{\mathrm{D}} - 2T + 2T$, $T_{\mathrm{D}} - 2T + T$, $T_{\mathrm{D}} - 2T$, $T_{\mathrm{D}} - 2T - T$, and $T_{\mathrm{D}} - 2T - 2T$. The second delay circuit has the correct delay time to make the third received pulse coincide with the sum of the first two pulses. Hence, the output signal at the terminal $\Delta\tau = +T$ will be 3/2 times as large as the sum of the first two pulses obtained at the other output terminals; generally, the amplitude ratio will be $n/2$ after n pulses have been received, if the acceleration remains constant.

The summation of coinciding pulses is shown more clearly in Fig. 6.3-5.

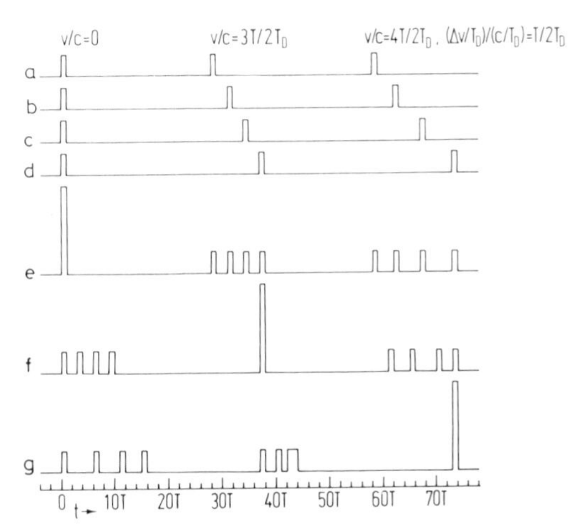

FIG. 6.3-5. Resolution of a return from a stationary target $v/c = 0$, a return from a target with constant velocity $v/c = 3T/2T_{\mathrm{D}}$, and a return from a target with initial velocity $v/c = 4T/2T_{\mathrm{D}}$ and a constant acceleration $(\Delta v/T_{\mathrm{D}})/(c/T_{\mathrm{D}}) = T/2T_{\mathrm{D}}$ by means of the circuit of Fig. 6.3-4.

Line a shows three returns due to the first radiated radar pulse. Line b shows the returns due to the second pulse. A comparison of the two lines yields the velocities $v/c = 0$, $3T/2T_{\mathrm{D}}$, and $4T/2T_{\mathrm{D}}$ for the three targets. The returns due to the third pulse in line c confirm that two targets have the velocities $v/c = 0$ and $3T/2T_{\mathrm{D}}$, and in addition yield the acceleration $(\Delta v/T_{\mathrm{D}})/(c/T_{\mathrm{D}}) = T/2T_{\mathrm{D}}$ for the third target. The returns of the fourth pulse in line d confirm the previously measured velocities and acceleration. For the practical exploitation of the pulses in lines a–d one sums them to produce the large signal for $v/c = 0$ in line e. Three times repeated delay by $T_{\mathrm{D}} + 3T$ and summation yields the large signal for $v/c = 3T/2T_{\mathrm{D}}$ in line f. Delays of $T_{\mathrm{D}} + 4T$, $T_{\mathrm{D}} + 4T + T$, $T_{\mathrm{D}} + 4T + 2T$, and summation yield the large signal for $v/c = 4T/2T_{\mathrm{D}}$ and $(\Delta v/T_{\mathrm{D}})/(c/T_{\mathrm{D}}) = T/2T_{\mathrm{D}}$ in line g.

The circuits of Figs. 6.3-3 and 6.3-4 have the advantage of being readily understandable, but they require the summation of many coinciding pulses to provide a large amplitude ratio between the output signal of the "correct" terminal and the others. The velocity or the acceleration must remain fairly constant during the reception of these many pulses in order to make the pulses coincide. From the theoretical development in Section 6.3.1, one would expect that two pulses should suffice for velocity resolution, and three pulses for acceleration resolution, but our results do not come close to this expectation. One obvious reason is that the circuits of Figs. 6.3-3 and 6.3-4 use only a small part of the received information. Let

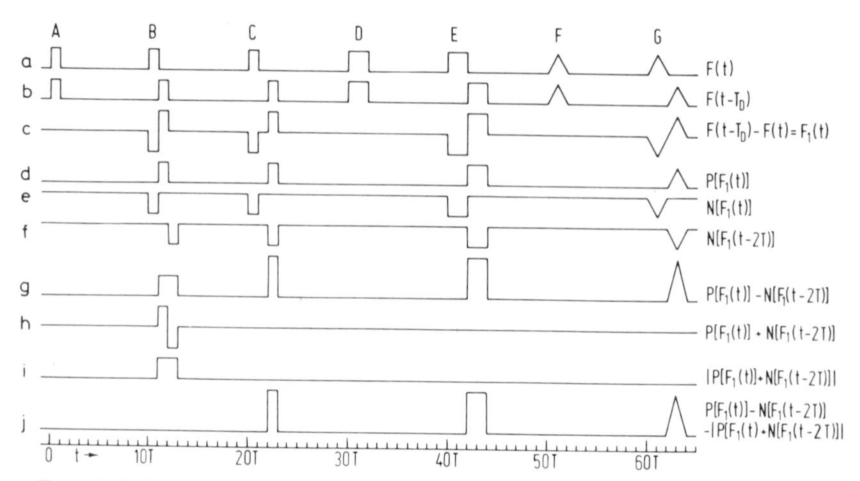

FIG. 6.3-6. Principle of the elimination of returns due to stationary targets or targets with wrong velocity by means of two radar pulses.

us denote a first pulse by p_1 and a second pulse by p_2. We formed the sum $p_1 + p_2$, but not the difference $p_1 - p_2$, which means that half the information is discarded. If the sum $p_1 + p_2 + p_3$ of three pulses is formed, but not the linearly independent combinations $p_1 - p_2 + p_3$, $p_1 + p_2 - p_3$, and $p_1 - p_2 - p_3$, one discards three quarters of the information. One will expect that a processing technique that discards information not quite as freely will yield better results.

6.3.3 Principle of Efficient Processing

Figure 6.3-6 shows in line a the returns A ··· G of a first radar pulse radiated at the time t_1'. The returns A ··· E are rectangular pulses, but F and G are triangular in order to show that the processing method to be discussed does not depend on the rectangular shape of the pulses.[1] Line b shows the returns due to a second radar pulse radiated at the time $t_1' + T_D$. It is evident that the returns A, D, and F are due to stationary targets, while the others are due to moving targets with various velocities. Since we can see that the pulses C, E, and G are returned from targets moving with the same velocity, there must be a way to separate these pulses com-

[1] The triangular shape was chosen because the compressed signals used in Sections 5.4, 6.1, and 6.2 had a triangular time variation. The radar returns in Fig. 6.3-6a may, of course, be signals that have been compressed in time. Rectangular pulses are generally used in this section to simplify the drafting of the illustrations.

pletely from all the others, but we do not yet know how complicated this separation might be.

As a first step we eliminate the returns from stationary targets by forming the difference $F(t - T_D) - F(t) = F_1(t)$ shown in line c of Fig. 6.3-6. This is in accord with the concept of the conventional moving target indicator. The next step is unconventional. The positive and the negative parts of $F_1(t)$ are separated, e.g., by means of diodes. The resulting signals are denoted $P[F_1(t)]$ and $N[F_1(t)]$; they are shown in lines d and e. Note that $P[F_1(t)]$ is equal to $F(t - T_D)$ in line b, except that the returns from stationary targets are missing. Furthermore, $-N[F_1(t)]$ is equal to $F(t)$ in line a, again without the returns from stationary targets.

The next step is the elimination of the return B, which is due to a target with velocity $+T/2T_D$, while the wanted returns C, E, and G are due to targets with velocity $+2T/2T_D$. To do so, we shift $N[F_1(t)]$ by $-2T$ as shown in line f. The returns C, E, and G in lines d and f are thus made to coincide. The difference $P[F_1(t)] - N[F_1(t - 2T)]$ of line g is formed. The returns C, E, and G are emphasized over the return B, but B is not yet eliminated. To eliminate B, we form the sum $P[F_1(t)] + N[F_1(t - 2T)]$ of line h, and its absolute value $|P[F_1(t)] + N[F_1(t - 2T)]|$ of line i. The returns C, E, and G are eliminated, since they do not change position from line d to f and are thus "stationary." Return B is equal in lines g and i. Hence, the difference of lines g and i eliminates B and leaves C, E, and G. This difference $P[F_1(t)] - N[F_1(t - 2T)] - |P[F_1(t)] + N[F_1(t - 2T)]|$ is shown in line j. Hence, we have achieved complete elimination of returns due to stationary targets or returns from targets with wrong velocity, using two radar signals only. This is a conspicuous improvement over the reduction of the "wrong" signals in Fig. 6.3-5.

Figure 6.3-7 shows a variation of this principle for the elimination of either all returns from nonstationary targets, or of all returns from targets

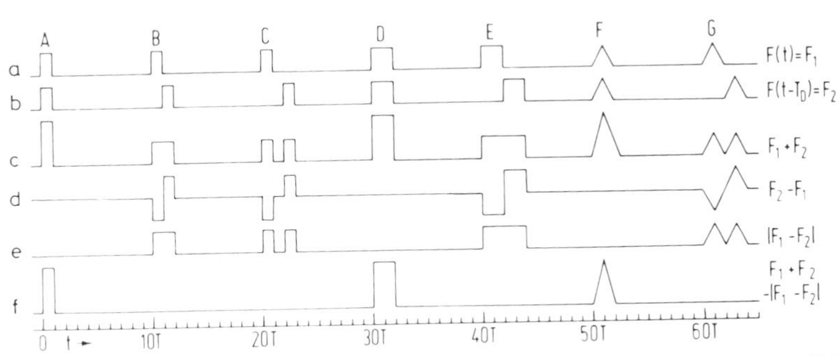

FIG. 6.3-7. Principle of the elimination of all returns in the signals $F(t)$ and $F(t - T_D)$ that have changed their position.

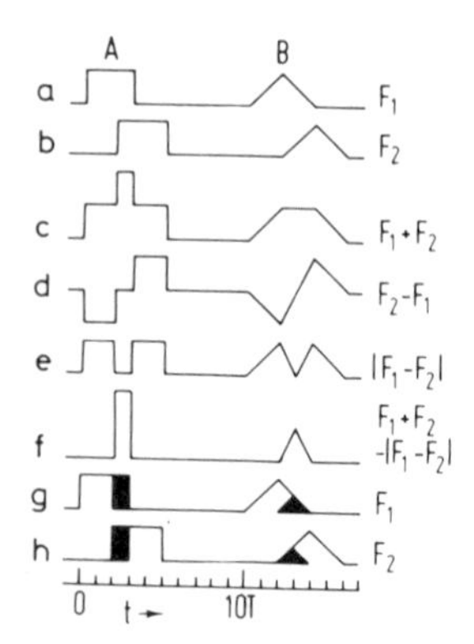

FIG. 6.3-8. Uncertainty of the velocity of targets. Signals A and B may be returns from two slow-moving targets or superpositions of returns from stationary and moving targets; the stationary parts are shown black in lines g and h.

that do not have a certain velocity. Line a shows again the returns due to a radar signal radiated at the time t_1', and line b the returns due to a radar signal radiated at the time $t_1' + T_D$. The sum $F_1 + F_2$ and the difference $F_1 - F_2$ are shown in lines c and d. The absolute value $|F_1 - F_2|$, shown in line e, subtracted from $F_1 + F_2$ yields only the returns from stationary targets A, D, and F of line a, as shown in line f. This illustration demonstrates the improvement over the sum $F_1 + F_2$ of line c, achieved by using the information contained in the difference $F_1 - F_2$.

The processes used in Figs. 6.3-6 and 6.3-7 to recognize motion sometimes operate differently from what one intuitively might expect. Refer to Fig. 6.3-8 for an example. Returns from two moving targets are shown in lines a and b. The expression $F_1 + F_2 - |F_1 - F_2|$ is formed in lines c–f. Since this expression should eliminate all returns from nonstationary targets according to Fig. 6.3-7, one might expect to see no pulses in line f. However, the two pulses in line f of Fig. 6.3-8 are quite proper. To show this, we plot F_1 and F_2 again in lines g and h, but blacken the sections of the pulses that are equal in F_1 and F_2. These blackened parts show up in line f. The mathematical process $F_1 + F_2 - |F_1 - F_2|$ interprets the signals F_1 and F_2 to consist of a stationary part, shown black in lines g and h, and a moving part shown white. This is a correct interpretation. Our intuitive interpretation is that of a moving rectangle and triangle without any stationary part, but there is not enough information in F_1 and F_2 to make this interpretation necessary.

Figure 6.3-9 shows a similar discrepancy between intuitive and mathematical interpretation for the extraction of moving signals according to Fig. 6.3-6. The three returns A, B, and C in lines a and b are processed in

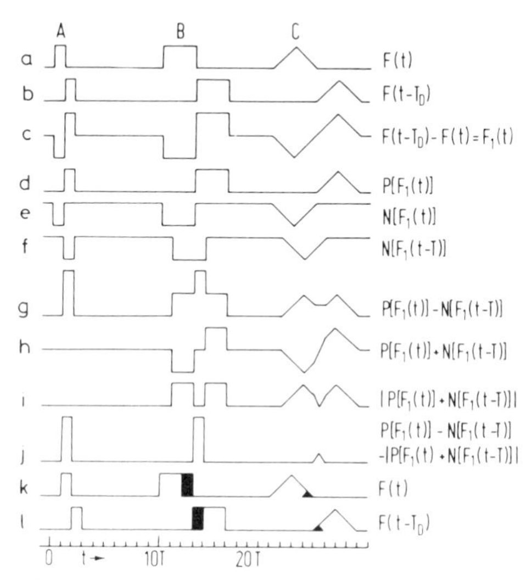

FIG. 6.3-9. Uncertainty of the velocity of targets. The signals B and C may be returns from two slow-moving targets or superpositions of returns from slow and fast-moving targets; the slow-moving part is shown black in lines k and l.

lines c–e as in the same lines of Fig. 6.3-6. In line f the shift $t - T$ rather than $t - 2T$ is introduced. From there on, the processing in lines g–j is again like in Fig. 6.3-6. The time shift $t - T$ should show the return A in line j, but parts of the returns B and C are also shown. The reason becomes clear when we plot $F(t)$ and $F(t - T_D)$ of lines a and b again in lines k and l. The blackened sections of the returns B and D move with the same velocity $v/c = +T/T_D$ as the return A. The processing method shows this by decomposing the returns B and C into a part that moves with velocity $v/c = +T/T_D$ and into a part that moves with another velocity. There is not enough information in $F(t)$ and $F(t - T_D)$ to require the conclusion that the returns B and C do *not* consist of two parts moving with different velocities.

6.3.4 Velocity Processors

The two conspicuous features of the processing principles just discussed are the separation of signals into a positive and a negative part, and the generation of the absolute value of the signal.[1] In the presence of noise, one will want to perform these operations at a high signal-to-noise

[1] Absolute-value circuits are precision full-wave rectifiers. For a detailed discussion see Graeme (1973).

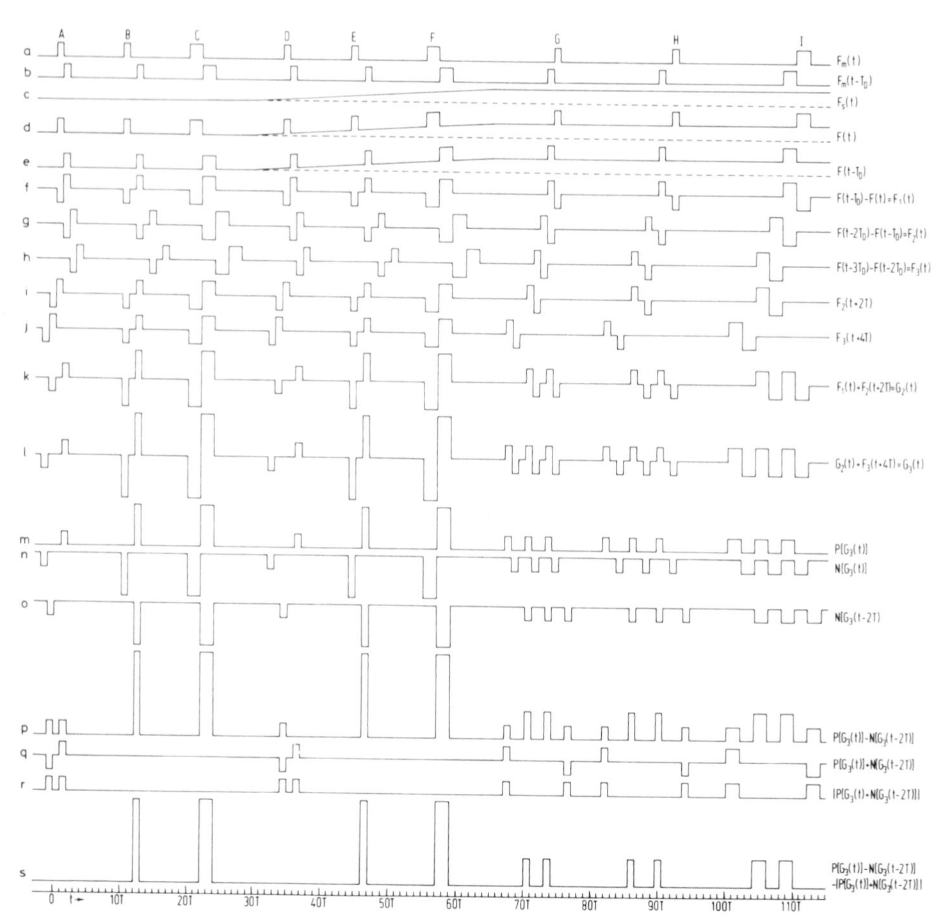

FIG. 6.3-10. Time diagram for the separation of returns $F_m(t)$ from moving targets and $F_s(t)$ from stationary targets, and a following separation of the returns from moving targets according to their velocity.

ratio. Hence, we will discuss two variations of the processing principles. The one will produce the sum of several returns before splitting the signal into a positive and a negative part; this method gives the best performance from the standpoint of noise, yields moderate velocity resolution, and requires relatively much time. The second variation will be designed to yield the best velocity resolution in the shortest time, but it will be poor for low signal-to-noise ratios.

Figure 6.3-10 shows in line a the signal $F_m(t)$ returned from moving targets A ... I, and in line b the signal $F_m(t - T_D)$ due to a second radar pulse radiated with the delay T_D relative to the first one. A signal $F_s(t)$ due

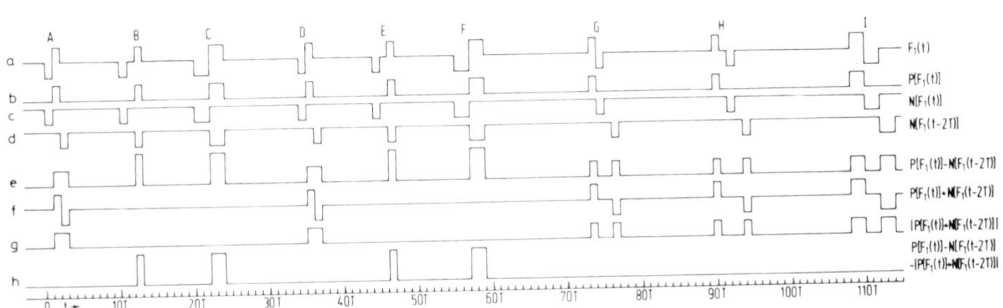

FIG. 6.3-11. Variation of the time diagram of Fig. 6.3-10 that yields a better velocity discrimination in less time, but requires a higher signal-to-noise ratio.

to stationary targets is shown in line c. The sums $F_m(t) + F_s(t) = F(t)$ and $F_m(t - T_D) + F_s(t) = F(t - T_D)$ in lines d and e are the actually received signals. The difference $F(t - T_D) - F(t) = F_1(t)$ is shown in line f. The signals $F_2(t)$ and $F_3(t)$ in lines g and h follow from line f if all the targets move with constant velocity.

In order to extract approaching targets moving with velocity $v/c = -2T/2T_D$ we advance $F_2(t)$ and $F_3(t)$ by $2T$ and $4T$ to obtain the signals $F_2(t + 2T)$ and $F_3(t + 4T)$ of lines i and j. The sums $F_1(t) + F_2(t + 2T) = G_2(t)$ and $F_1(t) + F_2(t + 2T) + F_3(t + 4T) = G_3(t)$ are shown in lines k and l. The signal-to-noise ratio of these sums is better than that of the signals $F_1(t)$, $F_2(t)$, or $F_3(t)$.

The sum $G_3(t)$ is split into a positive part $P[G_3(t)]$ and a negative part $N[G_3(t)]$ in lines m and n. The negative part is delayed by $-2T$ in line o. Line p shows the difference $P[G_3(t)] - N[G_3(t - 2T)]$. The returns B, C, E, and F from targets with velocity $v/c = +2T/2T_D$ are clearly emphasized over the returns A and D with velocity $v/c = +T/2T_D$, return G with velocity $v/c = -T/2T_D$, and returns H and I with velocity $v/c = -2T/2T_D$. The stationary return $F_s(t)$ is completely eliminated.

An improvement of the velocity resolution is still possible. Following the principle of Fig. 6.3-6, we produce $P[G_3(t)] + N[G_3(t - 2T)]$ in line q of Fig. 6.3-10, the absolute value of this expression in line r, and the difference $P[G_3(t)] - N[G_3(t - 2)] - |P[G_3(t)] + N[G_3(t - 2T)]|$ in line s. The signals from some of the targets with wrong velocity in line p have been suppressed in line s.

Let us turn to the second variation of our velocity-processing method. The signal $F_1(t)$ in line f of Fig. 6.3-10 is shown again in line a of Fig. 6.3-11. The signal is split into its positive and negative parts in lines b and c. The negative part $N[F_1(t)]$ is shifted by $-2T$ in line d. The difference $P[F_1(t)] - N[F_1(t - 2T)]$ is formed in line e, the sum $P[F_1(t)] + N[F_1(t - 2T)]$ in line f, and its magnitude in line g. Finally, the expression

$P[F_1(t)] - N[F_1(t-2T)] - |P[F_1(t)] + N[F_1(t-2T)]|$ is formed in line h.

Line h of Fig. 6.3-11 shows evidently a better velocity resolution than line s of Fig. 6.3-10, and only the two signals $F(t)$ and $F(t - T_D)$ rather than four signals $F(t)$, $F(t - T_D)$, $F(t - 2T_D)$, and $F(t - 3T_D)$ are required to produce it. However, the splitting into positive and negative parts has to be done at a lower signal-to-noise ratio. The signal-to-noise ratio of the signal in line h of Fig. 6.3-11 can, of course, be improved by repeating the processing for $F(t - 2T_D) - F(t - T_D) = F_2(t)$, $F(t - 3T_D) - F(t - 2T_D) = F_3(t)$, and summing the resulting signals.[1]

A circuit implementing the time diagram of Fig. 6.3-10 is shown in Fig. 6.3-12. The moving-target extractor on the left implements lines a–h in Fig. 6.3-10. The multiple-return exploiters implement lines i–l, and the correct-velocity extractors lines m–s. The quantities produced by the various blocks in Fig. 6.3-12 are marked for the top channel $v/c = +2T/2T_D$ in correspondence with Fig. 6.3-10 in order to obviate a long discussion of the operation of the circuit.

Figure 6.1-13 shows a circuit that implements the processing according to the time diagram of Fig. 6.3-11. The moving-target extractor produces the signal $F_1(t) = F_i(t)$ of line a, the correct-velocity extractors produce the signal of line h. The multiple-return exploiters produce a sum of signals according to line h from multiple returns, in order to obtain an improved signal-to-noise ratio.

6.3.5 Acceleration Processor

Figure 6.3-14 shows a time diagram for a processor that separates targets according to their velocity determined by two radar pulses, and according to their acceleration determined by a third radar pulse. Returns from moving targets due to three successive radar pulses are shown in lines a–c. Return A is due to an opening target with constant relative velocity; B is due to a closing target with constant relative velocity; C is due to a larger opening target with greater constant velocity; A and E are due to opening targets with constant relative acceleration; F is due to a larger closing target with constant relative velocity; G and H are due to

[1] A still better, and more complicated way, is based on the production of the linear combinations $F_1(t) - F_2(t) + F_3(t)$, $F_1(t) + F_2(t) - F_3(t)$, and $F_1(t) - F_2(t) - F_3(t)$ in addition to the sum $F_1(t) + F_2(t) + F_3(t)$ in order to conserve information in Fig. 6.3-10. Hence, the nonlinear process of splitting the signal into positive and negative parts, and producing the magnitude, can be done after the signal-to-noise ratio has been improved by means of multiple returns, but the processing becomes very complicated.

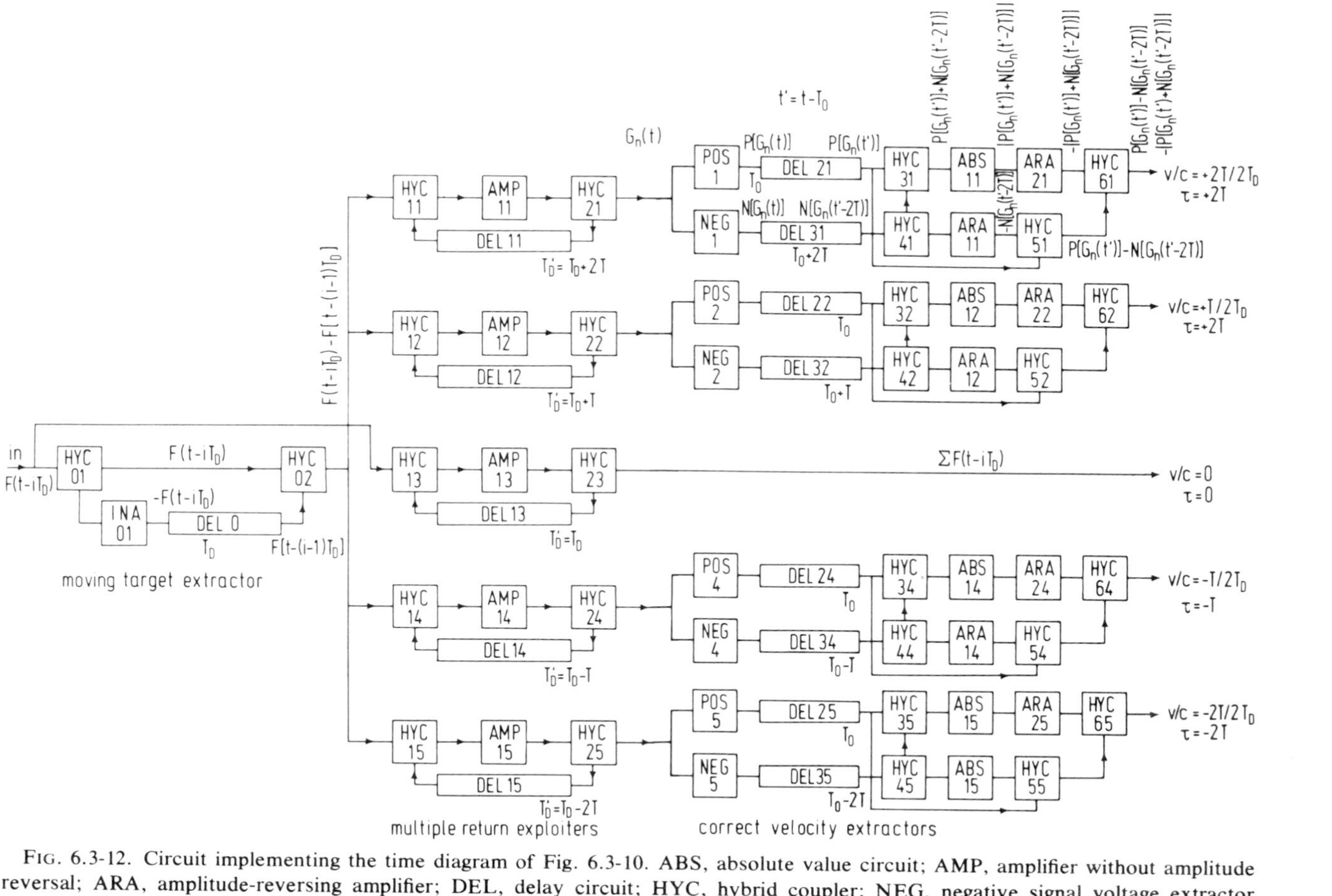

FIG. 6.3-12. Circuit implementing the time diagram of Fig. 6.3-10. ABS, absolute value circuit; AMP, amplifier without amplitude reversal; ARA, amplitude-reversing amplifier; DEL, delay circuit; HYC, hybrid coupler; NEG, negative signal voltage extractor (diode); POS, positive signal voltage extractor (diode).

targets that first close and then open with the relative acceleration being constant, as a target with constant ground speed would do in the neighborhood of $x = 0$ in Fig. 6.3-1.

Line d in Fig. 6.3-14 shows a stationary signal $F_s(t)$ that is superimposed on the returns E–H in lines e–g. The signals $F(t)$, $F(t - T_D)$, and $F(t - 2T_D)$ are received by the radar. The change from $F(t)$ to $F(t - T_D)$ is used to eliminate the stationary part $F_s(t)$ in line h, while the change from $F(t - T_D)$ to $F(t - 2T_D)$ is used to do the same in line i; the signals without stationary component are denoted $F_1(t)$ and $F_2(t)$.

Signal $F_1(t)$ is separated into its positive and negative parts in lines j and k. The negative part is advanced by $+T$ in line l, which means that all targets with relative velocity $v_1/c = -T/2T_D$ are to be extracted, while targets with different velocity are to be eliminated. The processing of $P[F_1(t)]$ and $N[F_1(t + T)]$ in lines m–p is as discussed in connection with Fig. 6.3-11. Only the returns from closing targets with relative velocity $v_1/c = -T/2T_D$ remain in line p. The processed signal in this line is denoted $G_1(t)$.

The signal $F_2(t)$ of line i is processed in lines q–w in the same way as $F_1(t)$ was processed in lines j–p, but the returns from opening targets with relative velocity $v_2/c = +T/2T_D$ are now extracted, and all others are suppressed. That the chosen velocity is now $v_2/c = +T/2T_D$ is shown by the delay of $N[F_2(t)]$ by $-T$ to $N[F_2(t - T)]$ in lines r and s. The processed signal in line w is denoted $G_2(t)$.

A close comparison of the signals $G_1(t)$ in line p and $G_2(t)$ in line w shows that no returns coincide. A change of velocity from $v_1/c = -T/2T_D$ to $v_2/c = +T/2T_D$ requires an acceleration a according to Eq. (7):

$$a = \frac{\Delta v}{T_D} = \frac{v_2 - v_1}{T_D} = \frac{c}{T_D}\left(\frac{T}{2T_D} + \frac{T}{2T_D}\right) = \frac{cT}{T_D^2} \tag{14}$$

In analogy to Eq. (12) we may also write

$$\Delta v/c = +2T/2T_D, \qquad \Delta\tau = +2T \tag{15}$$

In order to make the returns G in the processed signals $G_1(t)$ and $G_2(t)$ in lines p and w coincide one must either shift $G_2(t)$ by $+T$ or $G_1(t)$ by $-T$. We chose the shift $G_1(t - T)$, which is shown in line x. The returns G in lines w and x are now "stationary" in the sense discussed in connection with Fig. 6.3-7, and we may use the processing method of Fig. 6.3-7 to suppress all the other returns. The sum $G_1(t - T) + G_2(t)$ is shown in line y, the difference $G_1(t - T) - G_2(t)$ in line z, and its absolute value in line aa. The final expression $G_1(t - T) + G_2(t) - |G_1(t - T) - G_2(t)|$ is shown in line ab. All returns are suppressed, except the return G from a target

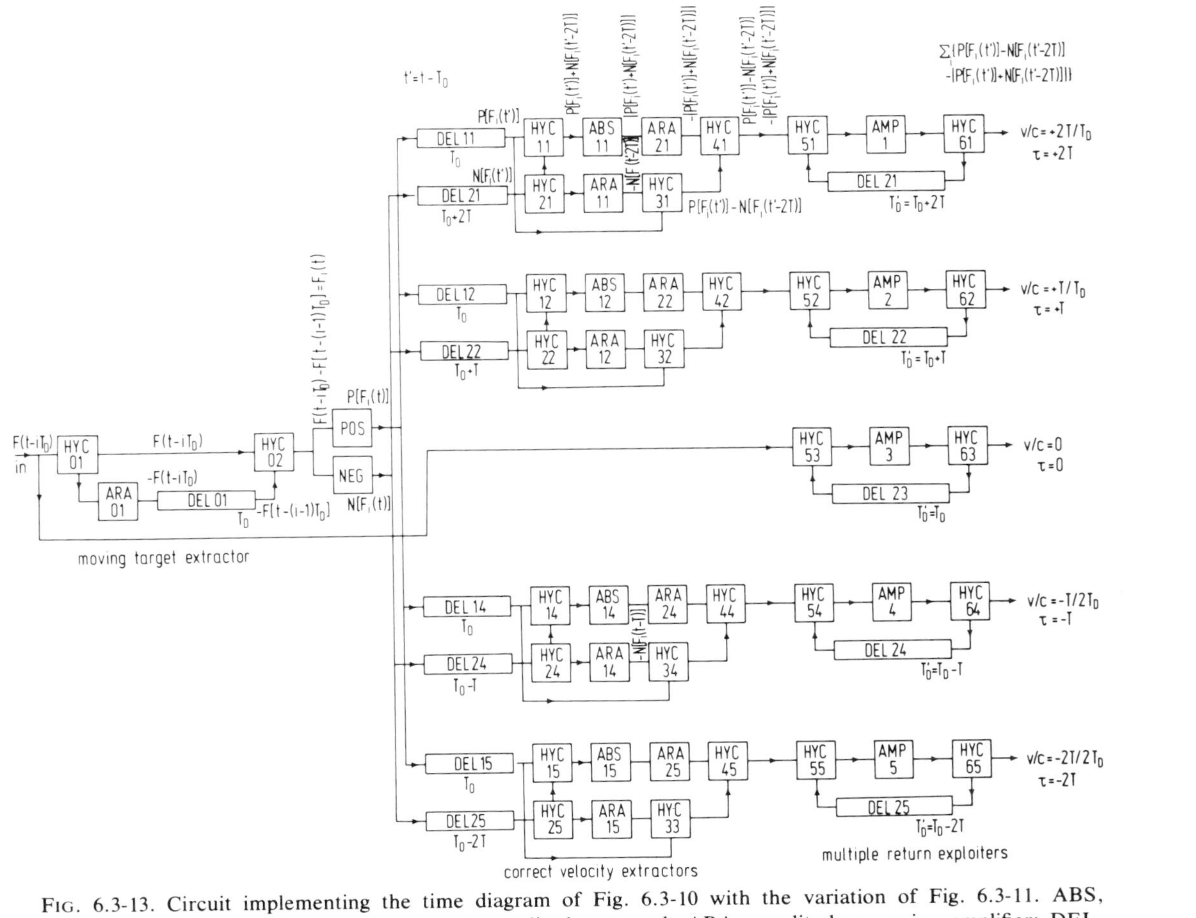

FIG. 6.3-13. Circuit implementing the time diagram of Fig. 6.3-10 with the variation of Fig. 6.3-11. ABS, absolute value circuit; AMP, amplifier without amplitude reversal; ARA, amplitude-reversing amplifier; DEL, delay circuit; HYC, hybrid coupler; NEG, negative signal voltage extractor (diode); POS, positive signal voltage extractor (diode).

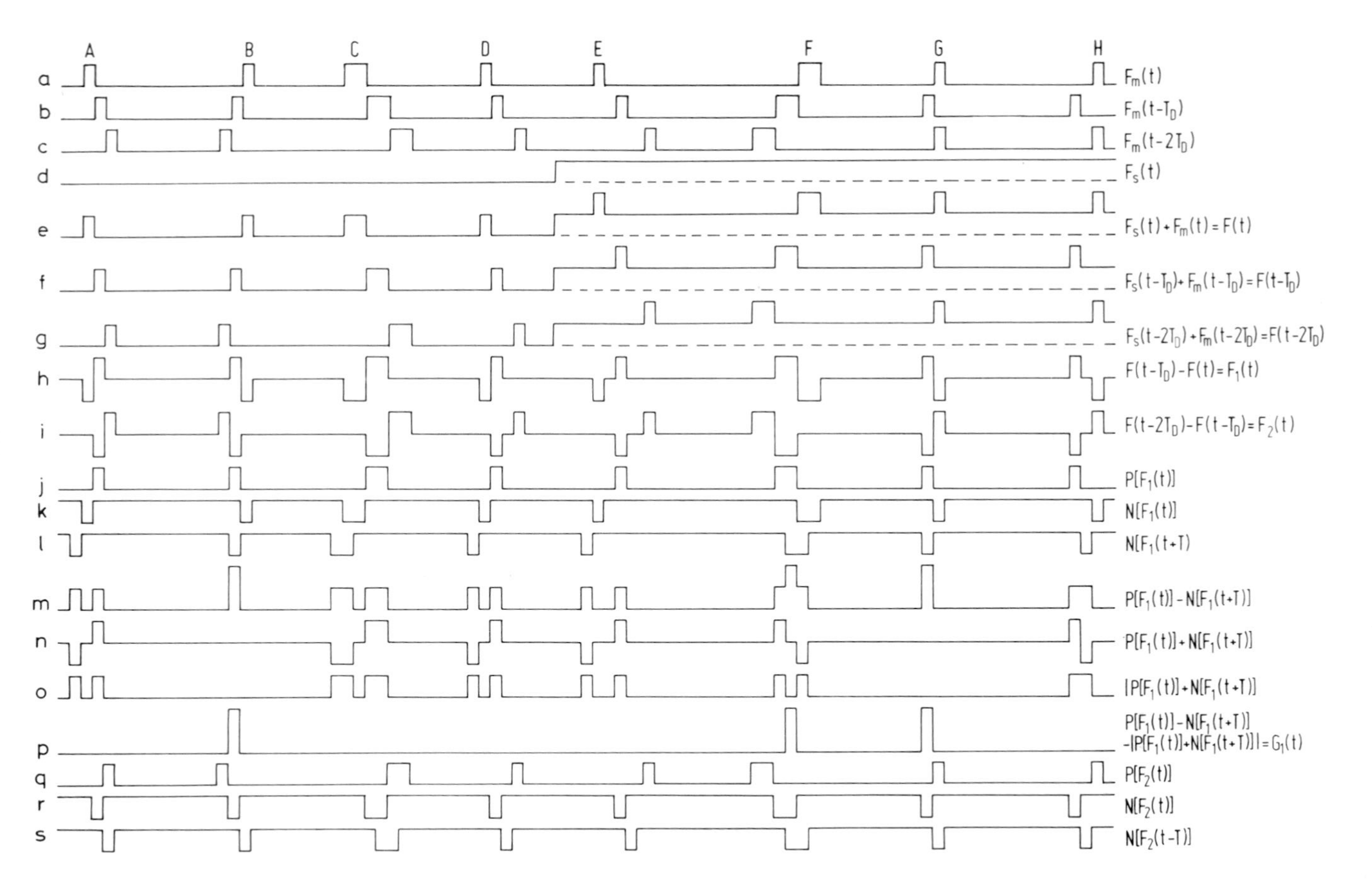

A
B
C
D
E
F
G
H
a
b
c
d
e
f
g
h
i
j
k
l
m
n
o
p
q
r
s
F_m(t)
F_m(t-T_D)
F_m(t-2T_D)
F_s(t)
F_s(t)+F_m(t)=F(t)
F_s(t-T_D)+F_m(t-T_D)=F(t-T_D)
F_s(t-2T_D)+F_m(t-2T_D)=F(t-2T_D)
F(t-T_D)-F(t)=F_1(t)
F(t-2T_D)-F(t-T_D)=F_2(t)
P[F_1(t)]
N[F_1(t)]
N[F_1(t+T)
P[F_1(t)]-N[F_1(t+T)]
P[F_1(t)]+N[F_1(t+T)]
|P[F_1(t)]+N[F_1(t+T)]
P[F_1(t)]-N[F_1(t+T)]
-|P[F_1(t)]+N[F_1(t+T)]|=G_1(t)
P[F_2(t)]
N[F_2(t)]
N[F_2(t-T)]

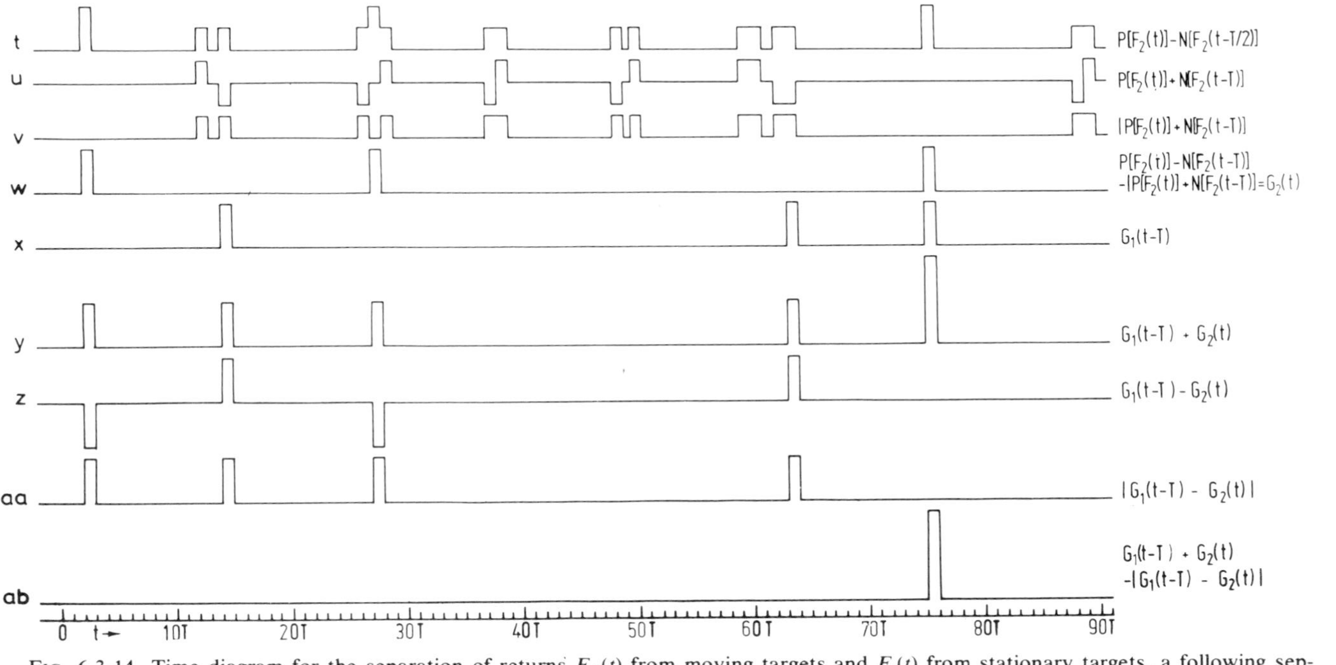

FIG. 6.3-14. Time diagram for the separation of returns $F_m(t)$ from moving targets and $F_s(t)$ from stationary targets, a following separation of the moving targets according to their velocity (lines p and w), and according to their acceleration (line ab).

with initial velocity $v_1/c = -T/2T_D$ and change of velocity $\Delta v/c = +2T/2T_D$.

Let us turn the time diagram of Fig. 6.3-14 into the circuit diagram of Fig. 6.3-15. The moving target extractor and the correct velocity extractors implement lines a–w in Fig. 6.3-14. These circuits are equal to the circuits with the same names in Fig. 6.3-13. Hence, we do not show again the processed signals at the terminals of each block as was done on top of Fig. 6.1-13, but we show the output signals $P[F_i(t')] - N[F_i(t' - 2T)] - |P[F_1(t')] + N[F_1(t' - 2T)]|$, etc., of the correct-velocity extractors only.

The return $G_1(t)$ of line p in Fig. 6.1-14 is due to the velocity $v/c = -T/2T_D$, and will thus be obtained at the terminal $v/c = -T/2T_D, \tau = -T$ in Fig. 6.3-15. The return $G_2(t)$ in line w is due to the velocity $v/c = +T/2T_D$, and it will be obtained at the terminal $v/c = +T/2T_D, \tau = +T$ in Fig. 6.3-15.

The output voltages of the correct-velocity extractors in Fig. 6.3-15 are fed to several sets of correct-acceleration extractors. Only one, denoted $v_0/c = T/2T_D$, is shown. It contains five extractors for the accelerations—or normalized velocity differences—$\Delta v/c = -2T/T_D$ to $\Delta v/c = +2T/T_D$. The return[1] $G_1(t + T_D)$ from the terminal $v/c = -T/2T_D$ is fed to all extractors of the set $v_0/c = T/2T_D$. At the output terminals of the delay circuits DEL31–DEL35 one obtains the delayed returns $G_1(t - T)$ to $G_1(t + 3T)$. The return $G_1(t - T)$ fed to the extractor denoted $\Delta v/c = +2T/2T_D$ equals the signal shown in line x of Fig. 6.3-14.

The signal $G_2(t)$ of line w in Fig. 6.1-14 is obtained at the terminal $v/c = +T/2T_D$. In the set $v_0/c = T/2T_D$ of acceleration extractors shown, it is fed to the extractor on top, denoted $\Delta v/c = +2T/2T_D$. This extractor thus forms the sum $G_1(t - T) + G_2(t)$, the difference $G_1(t - T) - G_2(t)$, its absolute value, and finally the expression $G_1(t - T) + G_2(t) - |G_1(t - T) - G_2(t)|$ according to lines y to ab in Fig. 6.3-14. Hence, the return due to the target with (initial) velocity $v/c = v_0/c = -T/2T_D$ and acceleration $(\Delta v/T_D)/(c/T_D) = +2T/2T_D$ is obtained at the terminal $\Delta v/c = +2T/2T_D$ of the set $v_0/c = -T/2T_D$ of correct acceleration extractors.

---

[1] The time diagram of Fig. 6.3-14 contains shifts $-T_D$ and $-2T_D$ to cause the returns from the first, second, and third radar pulses to be in the same time interval. As a result, the processed signal $G_2(t)$ occurs at the time T_D later than the signal $G_1(t)$. Hence we write $G_1(t + T_D)$ and produce $G_1(t)$ with the help of a delay circuit with delay T_D. The derivation of the circuit diagram would be easier if we had not used the delayed signals $F(t - T_D)$ and $F(t - 2T_D)$ as shown in lines f and g of Fig. 6.3-14 but had shown three signals at different times; however, the time diagram would have become at least three times as wide.

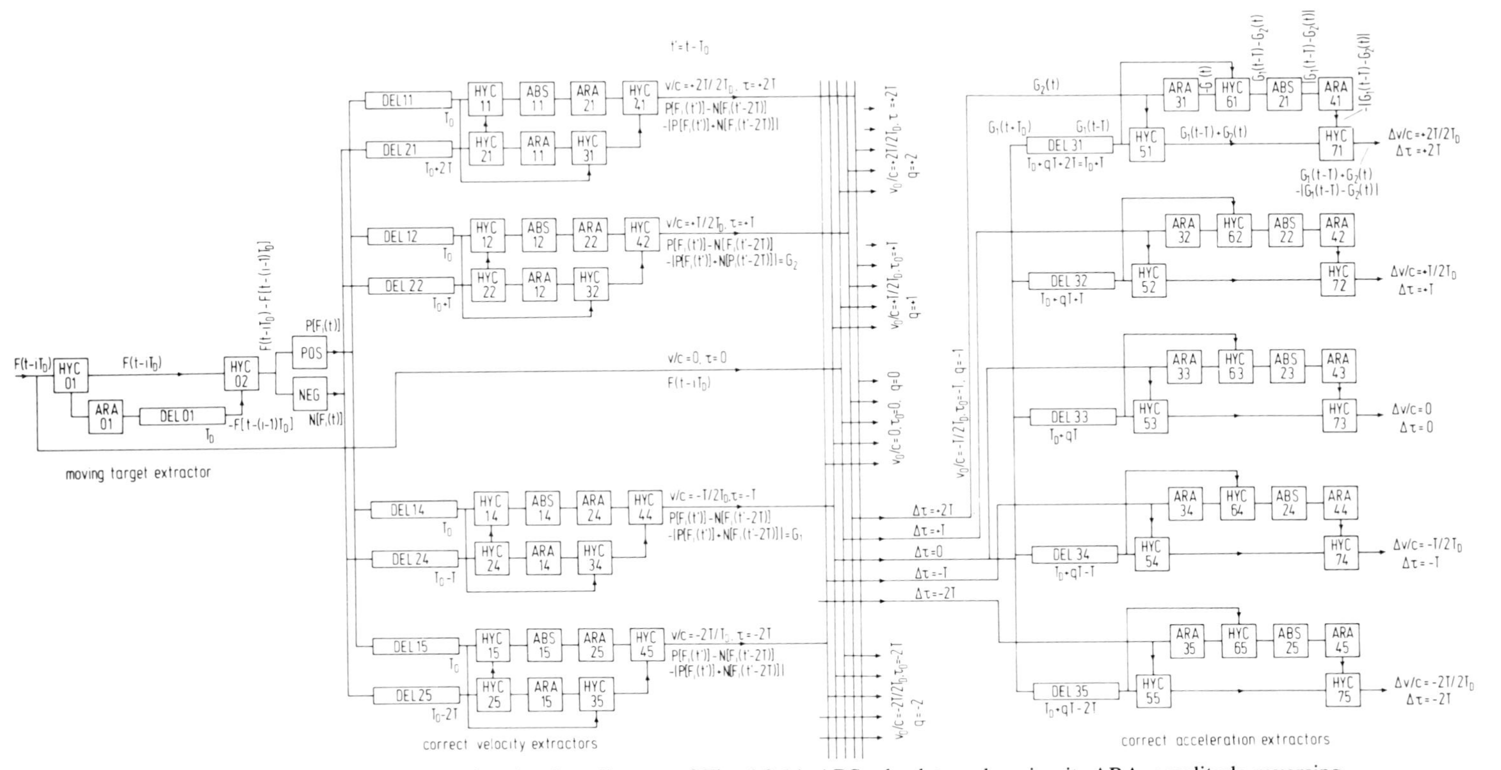

FIG. 6.3-15. Circuit implementing the time diagram of Fig. 6.3-14. ABS, absolute value circuit; ARA, amplitude-reversing amplifier; DEL, delay circuit; HYC, hybrid coupler; NEG, negative signal voltage extractor (diode); POS, positive signal voltage extractor (diode).

There are many improvements of the circuit of Fig. 6.3-15 that come to mind after a short study. For instance, one could connect feedback delay loops with variable delay times according to Fig. 6.3-4 to the output terminals $\Delta v/c = -2T/2T_{\mathrm{D}} \cdots +2T/2T_{\mathrm{D}}$ to obtain multiple-return exploiters in analogy to Fig. 6.3-13. However, we will not go into any more details.

7 Radio Communication with Submarines

7.1 Bandwidth Required for Teletype and Data Links

Radio communications with deeply submerged submarines provides a textbook case for the need to use a large relative bandwidth. The attenuation in the water increases exponentially with the square root of the frequency. Hence, the attenuation increases enormously if a carrier and a small relative bandwidth are used. Consider a baseband signal in the band $0 \leqq f \leqq \Delta f$ and a sinusoidal carrier with frequency $f_c = 10\,\Delta f$. Using single-sideband modulation one obtains for the lower sideband the relative bandwidth $\eta = \Delta F/(9\,\Delta f + 10\,\Delta f) \doteqdot 0.05$. According to Fig. 1.4-1, this is about the largest value of η for which one still obtains a resonance effect; this value of η is the same as used in Section 1.5 for a radar operating in the band 89 GHz $\leqq f \leqq$ 99 GHz. The increase of the highest frequency from Δf to $10\,\Delta f$ implies an increase of the attenuation in seawater of about 60 dB at a depth of 300 m for teletype transmission. One cannot realize this whole gain, since the efficiency of an antenna of fixed length increases typically with the square of the frequency at low frequencies. Hence, an increase of the frequency by a factor 10 implies an increase of the radiated power by 20 dB, and this reduces the realizable gain from 60 to 40 dB.

Another way of looking at the same numbers is to say that the attenuation of the actually radiated power without a carrier is of the order of 60 dB less than with a carrier. This is the significant number if one is concerned about radiation hazards. Transmitters for radio communication with submarines are very powerful, and a reduction of the radiated power by 60 dB—or by the ratio of one million to one—is the best safeguard against proven or suspected radiation hazards.

The first step in the design of a communications system for submarines, based on the limitations imposed by nature rather than those of a theory based on the need for sinusoidal carriers, is to determine the minimum frequency bandwidth required for certain data links. We will investigate two of them. The first is a teletype link based on the international CCIT standard of characters with a duration of 150 ms and a data rate of 33.3 bits/s. The second is a data link transmitting 2500 bits/s; such a link would be capable of transmitting voice signals by pulse-code modulation (PCM) if a vocoder is used.

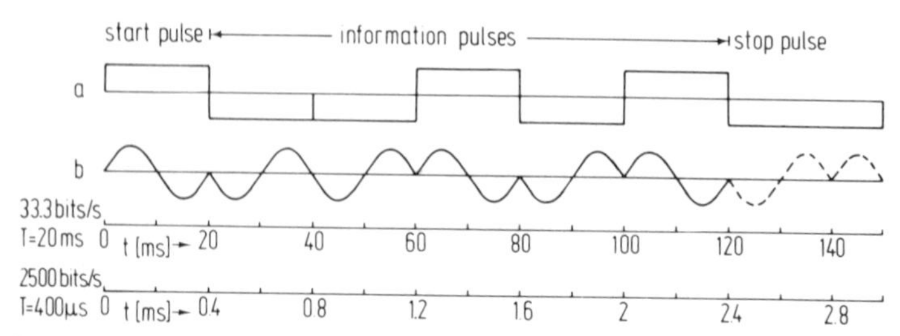

FIG. 7.1-1. Teletype character consisting of rectangular pulses (a) and sinusoidal pulses with one cycle (b). A data link transmitting 2500 bits/s can use the same pulses with a duration $T = 400$ μs rather than $T = 20$ ms; no stop pulse is used in this case.

Figure 7.1-1 shows on top a typical teletype character consisting of two-valued pulses. The character begins with a start pulse of 20 ms duration, then follow five information pulses of the same duration and finally a stop pulse of 30 ms duration. This signal can be transmitted via a wire line or any other link that can transmit a dc component. For a radio link we must remove any dc component. There are many ways of doing so.

A readily understandable way to produce pure ac signals is shown in line b of Fig. 7.1-1. The rectangular pulses of line a are replaced by sinusoidal pulses having one cycle with a duration of 20 ms. The pulses have a positive or negative amplitude according to the positive or negative amplitude of the corresponding rectangular pulse in line a. The stop pulse is represented by no pulse at all, since one only needs the stop pulse as a waiting period to assure that signals are not transmitted faster than they can be received.

The frequency energy spectrum of the sinusoidal pulses in Fig. 7.1-1 is

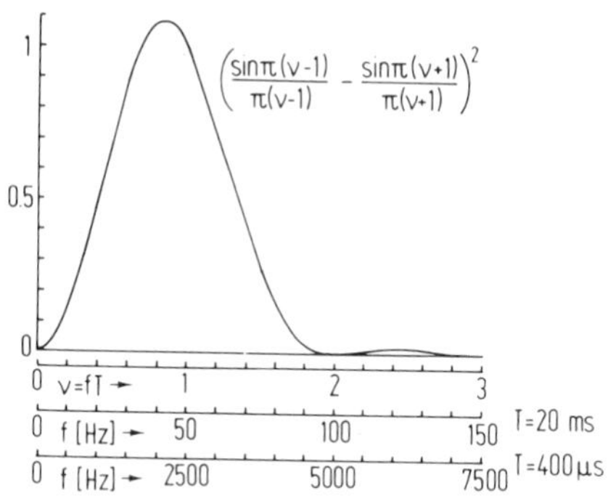

FIG. 7.1-2. Energy or power spectrum of a sinusoidal pulse with one cycle of duration T according to Fig. 7.6-1. The scales denoted $T = 20$ ms and 400 μs show the frequency in a nonnormalized fashion for these particular pulse durations T.

shown in Fig. 7.1-2. This spectrum holds for an individual pulse having one cycle. If several pulses with the same amplitude follow in succession—such as the second and the third pulses in Fig. 7.1-1—one obtains a narrower frequency energy spectrum; for an infinite succession of such pulses one obtains a discrete spectral line at $fT = 1$ or $f = 50$ Hz. The scale for $\nu = fT$ in Fig. 7.1-2 holds for an arbitrary duration T of the sinusoidal pulses while the scales for f denoted $T = 20$ ms and $T = 400$ μs hold for these respective pulse durations.

One may readily see from Fig. 7.1-2 that a teletype character consisting of sinusoidal pulses according to line b of Fig. 7.1-1 has practically all its energy concentrated at frequencies of less than 100 Hz. A data link transmitting 2500 bits/s with binary pulses requires pulses of 400 μs duration according to Fig. 7.1-1. Practically all energy of sinusoidal pulses with a duration of 400 μs is concentrated below 5000 Hz according to Fig. 7.1-2.

We can radiate and receive electromagnetic waves with the time variation shown in line b of Fig. 7.1-1. This means we can transmit teletype signals at the usual rate to submarines without using frequencies above 100 Hz or we can operate a data link transmitting 2500 bits/s without using frequencies above 5000 Hz. This appears rather good compared with current practice that calls for a radio carrier having about 10 times as high a frequency. However, we can do substantially better by using more sophisticated signals than the sinusoidal pulses in Fig. 7.1-1. A practical method to reduce the required highest frequency below the value of Fig. 7.1-2 will be explained with the help of Fig. 7.1-3.

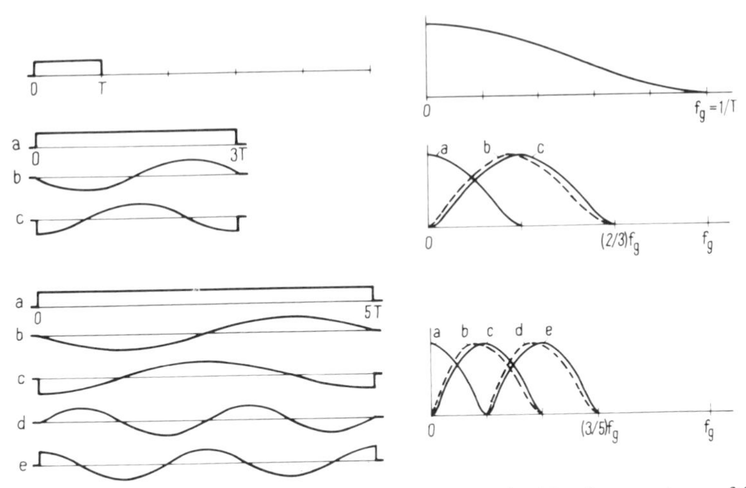

FIG. 7.1-3. Comparison of the frequency bandwidth required by three systems of functions.

Three systems of functions are shown on the left in Fig. 7.1-3; their energy spectra are shown on the right. The first system consists of a rectangular pulse of duration T only. Its frequency energy or power spectrum drops from $f = 0$ to $f_g = 1/T$ like the function $(\sin x/x)^2$. The energy beyond f_g is very small and will be ignored here.

We can use three rectangular pulses of duration T during a time interval of duration $3T$. Alternately, we may use the second system of functions in Fig. 7.1-3 consisting of a rectangular pulse (a), a sine pulse (b), and a cosine pulse (c), all having the duration $3T$. The energy spectrum of the rectangular pulse (a) has the same shape as before but it is compressed to one third the bandwidth. The energy spectrum of the sine pulse (b) is the same as in Fig. 7.1-2, but the first zero at $\nu = fT = 2$ or $f = 2/T$ is compressed to the value $f = 2f_g/3 = 2/3T$. The energy spectrum of the cosine pulse (c) is very similar to that of the sine pulse but shifted to slightly higher frequencies due to the jumps at the beginning and end of the cosine pulse. The important feature of the energy spectra of the three pulses is that practically all energy is concentrated below $f = 2f_g/3 = 2/3T$.

Let us now turn to the third system of functions. It consists of one rectangular pulse (a), a sine pulse with one cycle (b), a cosine pulse with one cycle (c), a sine pulse with two cycles (d), and a cosine pulse with two cycles (e). The duration of these five pulses is $5T$. Hence, they can transmit as much information as five rectangular pulses of duration T each. The energy spectra of the rectangular pulse and the sine and cosine pulses with one cycle are the same as in the second system of functions except for a compression of the frequency scale by a factor 3/5. The sine and cosine pulses with two cycles have the energy spectra d and e; we will show more accurate plots of them later on. The important point is that almost all energy is now concentrated at frequencies below $3f_g/5 = 3/5T$.

We may generalize and consider a system of functions consisting of one rectangular pulse, i sine and i cosine pulses of duration $(2i + 1)T$. The sine and cosine pulses have 1, 2, ..., i cycles in this interval. The energy of these $2i + 1$ pulses is concentrated in the frequency band $0 \leqq f \leqq (i + 1)/(2i + 1)T$. The time–bandwidth product equals $fT = (i + 1)/(2i + 1)$. This means that multiplication with $(2i + 1)/(i + 1)$ yields $fT = 1$, and this implies that $(2i + 1)/(i + 1)$ pulses are transmitted per second and hertz. The fraction $(2i + 1)/(i + 1)$ approaches 2 for large values of i, which is the Nyquist limit. The use of sine and cosine pulses as discussed here is a practical way to operate *essentially* at the Nyquist limit, while the use of pulses of the form $(\sin 2\pi t/T)/2\pi T$ is a theoretical but not a practical way to operate *exactly* at the Nyquist limit.

For radio communications we cannot use the rectangular pulse in Fig. 7.1-3 due to its dc component. As a result, only $2i/(i + 1)$ rather than

TABLE 7.1-1

CHARACTERISTIC NUMBERS FOR A TELETYPE SYSTEM TRANSMITTING 33.3 bits/s AND A DATA LINK TRANSMITTING 2500 bits/s USING i SINE AND i COSINE PULSES AS SHOWN FOR $i = 4$ IN FIG. 7.1-4

			Highest frequency required $f_g = (i+1)/2iT$		
Number of sine *or* cosine pulses i	Number of sine *and* cosine pulses $2i$	Transmission rate per second and hertz $2i/(i+1)$	Time–frequency product f_gT $(i+1)/2i$	Teletype bandwidth $T = 30$ ms (Hz)	2500 bits/s data link bandwidth $T = 400$ μs (Hz)
1	2	1	1	33.33	2500
2	4	1.33	0.75	25	1875
3	6	1.5	0.667	22.22	1666.7
4	8	1.6	0.625	20.83	1562.5
5	10	1.67	0.6	20	1500
6	12	1.71	0.583	19.44	1458.3
7	14	1.75	0.571	19.05	1428.6
∞	∞	2	0.5	16.67	1250

$(2i + 1)/(i + 1)$ pulses are transmitted per second and hertz. We then have two choices. Either we are satisfied with a reduced transmission rate and want to maintain the used frequency band $0 \leqq f \leqq (i + 1)/(2i + 1)T$ unchanged; or we maintain the transmission rate by multiplying the original pulse duration T with $2i$ rather than with $2i + 1$. The used bandwidth is then increased to $0 \leqq f \leqq (i + 1)/2iT$.

To obtain some idea about the achievable numerical values consider Table 7.1-1. The number of sine *and* cosine pulses used equals $2i$. Figure 7.1-4 shows these pulses for $2i = 8$. The pulse transmission rate per second and hertz is given by $2i/(i + 1)$. One may see from Table 7.1-1 that for $2i = 14$ one transmits at a rate only 12.5% below the Nyquist limit 2, while for $2i = 8$ one is 20% below this limit. The time–frequency product is given by $(i + 1)/2i$. Its limit for $2i = \infty$ equals 0.5.

In order to explain the numbers in the column "teletype bandwidth T = 30 ms" we have to return to Fig. 7.1-1. The teletype character shown there has a start and a stop pulse, which is the common standard, but these pulses do not have to be transmitted in a sophisticated system. They can be eliminated at the transmitter and reinserted at the receiver. As a result, we have in the theoretical limit the full time of 150 ms available to transmit the five information pulses, or the time T = 30 ms/pulse. This is the time T used in Table 7.1-1. One can see that the highest frequency that must be transmitted drops from 33.33 Hz for $2i = 2$ to 19.05 Hz for $2i =$

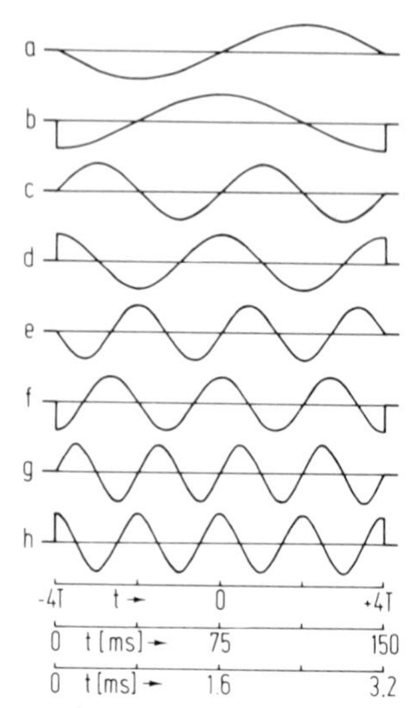

FIG. 7.1-4. Sine and cosine pulses without dc components with $i = 1, 2, 3, 4$ cycles.

14. The theoretical limit for $2i = \infty$ is 16.67 Hz, which is only 12.5% below the value for $2i = 14$.

The data link transmitting 2500 bits/s (T = 400 μs) requires a highest frequency or a bandwidth of 2500 Hz for $2i = 2$ that drops to 1250 Hz for $2i = \infty$ according to Table 7.1.1.

The primitive teletype system using one sine pulse and operating with start-stop pulses according to Fig. 7.1-1 requires the highest frequency of 100 Hz for transmission as shown by Fig. 7.1-2 for $T = 20$ ms. A sophisticated system, not using start–stop pulses and operating close to the Nyquist limit, requires a highest frequency of less than 20 Hz according to Table 7.1-1, column 5. The reduction for the data link transmitting 2500 bits/s is not quite so dramatic since no start–stop pulses were introduced. Nevertheless, the highest frequency of 5000 Hz in Fig. 7.1-2 is reduced to close to a quarter in Table 7.1-1 for $T = 400$ μs.

The values of Table 7.1-1 are lower limits for the highest frequency required since no bandwidth whatsoever was allocated to synchronization. To see how close this assumption is to reality consider the case of $2i = 10$ sine and cosine pulses for the teletype system with $T = 30$ ms. The highest frequency is 20 Hz, which implies a period of 50 ms. A synchronization error of 1% of this shortest period requires that the synchronization must not deviate by more than 500 μs from its nominal value. A quartz oscillator with a frequency stability $\Delta f/f = 10^{-6}$ producing the period of 50 ms requires $10^{-2}/10^{-6} = 10^4$ periods of 50 ms duration, which is equal to 500 s to deviate by 500 μs from the nominal synchro-

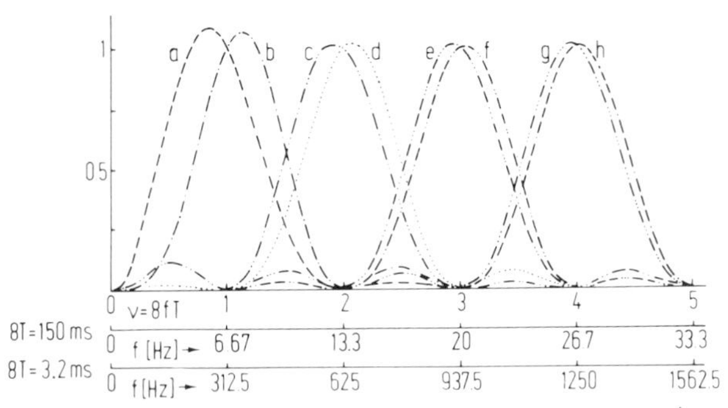

FIG. 7.1-5. Frequency power spectra of the sinusoidal and cosinusoidal pulses in Fig. 7.1-4. The scale for $8T = 150$ ms holds for pulses of 150-ms duration while the scale $8T = 3.2$ ms holds for pulses with 3.2-ms duration.

nization. The teletype link may thus transmit $500 \times 33.3 = 16{,}667$ bits or 3333 characters before a correction of the synchronization is necessary. Putting it differently, an increase of the highest frequency by a fraction of 1% or a decrease of the transmission rate for a fixed highest frequency by a fraction of 1% will allow for synchronization. A fringe benefit for covert communications is that the synchronization signal can be transmitted regularly every 10 min or so, completely independent from the information signals. If the transmission of the information signals can be hidden from detection by conventional receivers one will thus not compromise covertness by the synchronization signal.

Since the frequencies for a teletype link according to Table 7.1-1 are extremely small compared with the usual frequencies of radio communications, we can afford to be rather wasteful with bandwidth in order to arrive at simple equipment. Let us consider a teletype system that uses the sinusoidal function a in Fig. 7.1-4 as synchronization signal that is continuously transmitted. The power spectrum a in Fig. 7.1-5 then becomes a discrete spectral line with a frequency of 6.67 Hz. The cosine pulse b in Fig. 7.1-4 is not used at all to avoid interference with the synchronization signal. The five sine and cosine pulses c–g in Fig. 7.1-4 are used for the five information pulses of the teletype signal. The cosine pulse h is not used. The highest frequency of the resulting signals becomes 33.33 Hz according to Fig. 7.1-5, which is twice the theoretical limit of 16.67 Hz shown in Table 7.1-1. Such equipment has been built and tested successfully.[1] Hence we will not discuss the construction of the

[1] This equipment was built and tested by Allen-Bradley Co., but never marketed due to problems of marketing rather than technical problems. For a discussion of circuits and test results of this and similar equipment known under the names Kineplex, Rectiplex, and Digiplex see Harmuth (1972, pp. 173–176).

equipment but accept that a teletype system for the international CCIT standard, which is slightly faster than 60 words/min, can operate in the band $0 \leqq f \leqq 33.3$ Hz, and a 2500 bits/s data link can operate in the band $0 \leqq f \leqq 1562.5$ Hz, as shown by Fig. 7.1-5 using the functions of Fig. 7.1-4.

7.2 Attenuation in Seawater and Antenna Gain

The attenuation of an electromagnetic wave in seawater is given by the following formula in decibels per unit length as a function of the frequency f (Horvat, 1969):

$$\begin{aligned} \alpha(f) &= 54.58f\{\tfrac{1}{2}\mu\epsilon([1 + (\sigma/2\pi f\epsilon)^2]^{1/2} - 1)\}^{1/2} \\ \mu &= 4\pi \times 10^{-7} \quad [\mathrm{Vs/Am}] \\ \epsilon &\doteqdot 7.08 \times 10^{-10} \quad [\mathrm{As/Vm}] \\ \sigma &\doteqdot 4 \quad [\mathrm{A/Vm}] \end{aligned} \tag{1}$$

Insertion of the numerical values of μ, ϵ, and σ yields the following numerical equation in which the frequency f must be inserted in hertz. The equation holds in the range from $f = 0$ to about 1 GHz:

$$\alpha(f) = 0.0345\sqrt{f} \quad [\mathrm{dB/m}] \tag{2}$$

The group velocity v, which represents with good approximation the

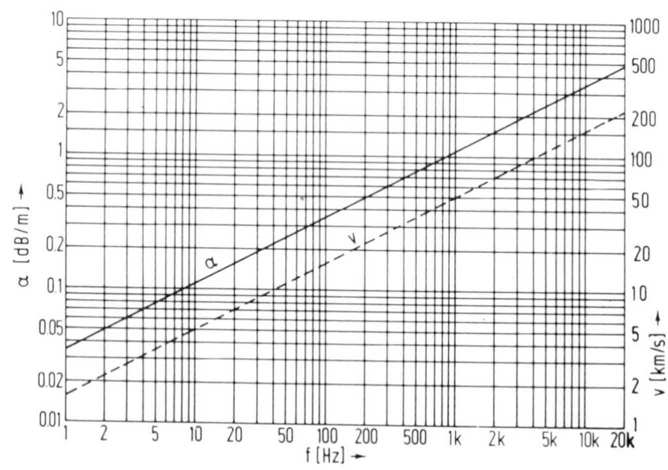

FIG. 7.2-1. Attenuation α in dB/m and group velocity v in km/s in seawater as function of frequency f.

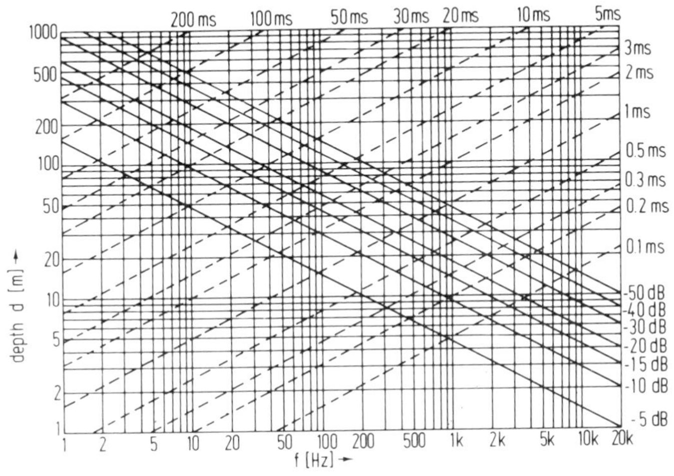

FIG. 7.2-2. Penetration depth d as function of frequency f for an attenuation in the range from -5 to -50 dB (solid lines), or for a propagation time t_p in the range from 0.1 to 20 ms (dashed lines).

velocity of propagation of a signal, has the following value in seawater:

$$v(f) = (4\pi f/\sigma\epsilon)^{1/2} \tag{3}$$

The following numerical equation holding for the frequency f expressed in hertz is obtained:

$$v(f) = 1581\sqrt{f} \quad [\text{m/s}] \tag{4}$$

Figure 7.2-1 shows α and v for the frequency range from 1 Hz to 20 kHz. The penetration depth d for a certain attenuation A is defined by the formula

$$d = A/\alpha \tag{5}$$

The depth d reached by a signal during the propagation time t_p in the water is defined by the formula

$$d = t_p v \tag{6}$$

Figure 7.2-2 shows d as function of the frequency f for values of the attenuation A ranging from -5 to -50 dB and for propagation times t_p ranging from 0.1 to 200 ms. The penetration depth d for a certain attenuation A decreases rapidly with increasing frequency; on the other hand, the penetration depth for a certain propagation time t_p increases rapidly with increasing frequency. This implies that a signal occupying a wide relative frequency bandwidth will suffer both attenuation and delay distortions.

We have shown in Table 7.1-1 that a data link transmitting 2500 bits/s requires a frequency bandwidth of about 1500 Hz. If we locate this bandwidth in the band from $f = 0$ to 1500 Hz we can reach a depth d of not quite 40 m with an attenuation in the water of $A = -50$ dB or less according to Fig. 7.2-2. On the other hand, if we locate the very same bandwidth in the band from $f = 18.5$ to 20 kHz we can reach a depth d of about 10 m only. It is standard practice in communications to shift a signal produced in a frequency band $0 \leqq f \leqq 1500$ Hz by means of a radio carrier to a frequency band of the same width but with absolutely higher frequencies for radiation. Why is this done? We can list the following reasons:

(1) The relative bandwidth of the signal is made small, and this *may* reduce the attenuation—as well as the delay—distortions.

(2) The efficiency of a transmitter antenna that is short compared with the wavelength increases with the square of the frequency.

(3) Our usual radio receivers only work for signals with a small relative bandwidth; amplitude or frequency modulation of a carrier is the typical means to achieve this small relative bandwidth.

We will see in Section 7.3 how we can use a large relative bandwidth and still get around the distortion problem. In Section 7.4 we will investigate which frequency band is best suited for operation, considering the reduction of the efficiency of the transmitter antenna and of the attenuation in seawater with decreasing frequency. In Section 7.5 we will show that it is quite easy to build receivers for radio signals that do not use a carrier.

Let us return for a moment to the teletype system that requires a bandwidth of about 20 Hz according to Table 7.1-1. If we locate this bandwidth from $f = 0$ to 20 Hz and permit an attenuation in the water of -50 dB or less, we can reach a depth d of slightly more than 300 m according to Fig. 7.2-2. We will see later on that it is theoretically possible to do even better.

7.3 Distortion of Signals in Seawater

Consider the two signals a and b in Fig. 7.1-4. Their power spectra in Fig. 7.1-5 are practically concentrated for $8T = 150$ ms in the band $0 \leqq f \leqq 13.3$ Hz. Hence, the bandwidth Δf equals 13.3 Hz and the highest frequency f_H also equals 13.3 Hz. The relative bandwidth assumes its largest possible value $\eta = 1$. The signals g and h in Fig. 7.1-4 have almost all of their energy in the band $3 \leqq \nu \leqq 5$, $20\text{ Hz} \leqq f \leqq 33.3\text{ Hz}$ or $937.5\text{ Hz} \leqq f \leqq 1562.5\text{ Hz}$, and the relative bandwidth in all three cases equals $(5 - 3)/(5 + 3) = 0.25$. The relative bandwidths encountered here are large compared with the relative bandwidths typically used in radio communications. This suggests a problem of signal distortion. We investigate first attenuation distortions and then delay or propagation time distortions.

The analysis of attenuation distortions starts with the transmitter antenna. The frequencies we are interested in are so low that the length L of the transmitter antenna is short compared with the wavelength $\lambda = c/f$ in air, where we assume the transmitter antenna to be located. For instance, a representative frequency $f = 30$ Hz for teletype transmission yields a wavelength $\lambda = 10{,}000$ km, while a representative frequency $f = 1500$ Hz for a 2500 bits/s data link yields $\lambda = 200$ km. A sinusoidal current $I \sin 2\pi ft$ flowing in such an antenna of length $L \ll \lambda$ radiates[1] the average power S:

$$S(f) = (2\pi Z_0/3)(L/\lambda)^2 I_{\text{rms}}^2$$
$$I_{\text{rms}}^2 = I^2/2, \qquad Z_0 = (\mu_0/\epsilon_0)^{1/2} \doteqdot 377\ \Omega, \qquad \lambda = c/f \gg L \tag{1}$$

The power radiated by this antenna at the frequency f relative to the power $S(f')$ radiated at the frequency f' yields the relative antenna gain:

$$10 \log[S(f)/S(f')] = 20 \log(f/f') \quad [\text{dB}] \tag{2}$$

The geometric transmission losses and the losses due to absorption in the atmosphere are essentially equal for the frequencies of interest here. The losses in the water for a penetration depth d are $d\alpha(f)$ for the frequency f and $d\alpha(f')$ for the frequency f', where $\alpha(f)$ is defined by Eq. (7.1-1) or (7.1-2). The difference of attenuation $\Delta A_{f'}$ of a wave with frequency f relative to one with frequency f' at a depth d equals the difference of attenuation in the water plus the relative antenna gain:

$$\begin{aligned} \Delta A_{f'} &= d\alpha(f') - d\alpha(f) + 20 \log(f/f') \quad [\text{dB}] \\ &\doteqdot 0.0345d(\sqrt{f'} - \sqrt{f}) + 20 \log(f/f') \end{aligned} \tag{3}$$

The term $\sqrt{f'} - \sqrt{f}$ decreases with increasing f but the term $\log(f/f')$

[1] A number of simplifying assumptions are tacitly made in Sections 7.3–7.5 in order to avoid obscuring the principle by too many details. Those familiar with the transmission of electromagnetic waves to submarines are begged to be patient until Section 7.6, where the tacit assumptions are shown to be quite justified.

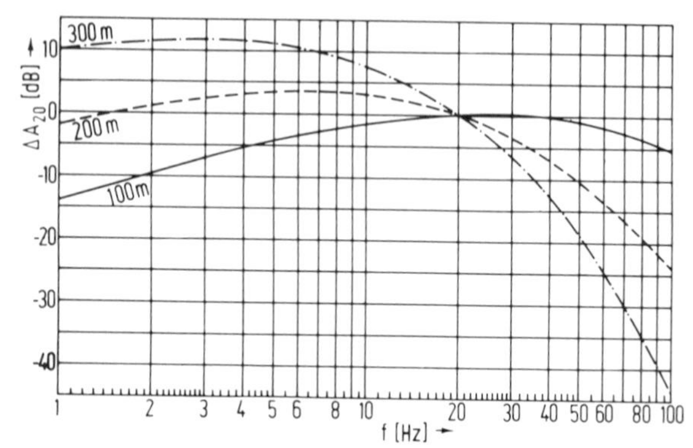

FIG. 7.3-1. Relative attenuation ΔA_{20} referred to the frequency $f' = 20$ Hz for penetration depths of 100, 200, and 300 m. The absolute attenuation in the water for 20 Hz equals -15.4 dB for 100 m, -30.8 dB for 200 m, and -46.2 dB for 300 m.

increases with increasing f. Hence, the distortions due to the variation of the attenuation in the water and the variable antenna gain partly compensate each other. Figure 7.3-1 shows ΔA_{20} for $f' = 20$ Hz in the range 1 Hz $\leqq f \leqq$ 100 Hz for $d = 100$, 200, and 300 m. The variation of ΔA_{20} for depths d between 100 m and 200 m is quite small considering the fact that the frequency changes by a factor 100 and the relative bandwidth equals $(100 - 1)/(100 + 1) = 0.98$. Figure 7.3-2 shows ΔA_{120} for the same absolute bandwidth of 100 Hz but a relative bandwidth of $(200 - 100)/(200 + 100) = 0.33$. The variation of ΔA_{120} is smaller than that of ΔA_{20}. Figure 7.3-3 shows ΔA_{1000} for the range 200 Hz $\leqq f \leqq$ 2000 Hz. The relative bandwidth equals $1800/2200 = 0.82$, which is very large, but the smaller values of d permit a good balance of the terms $\sqrt{f'} - \sqrt{f}$ and $\log(f/f')$ over this large relative bandwidth.

Let us turn to delay or propagation time distortions. The propagation time of a wave from the transmitter until it enters the seawater is practi-

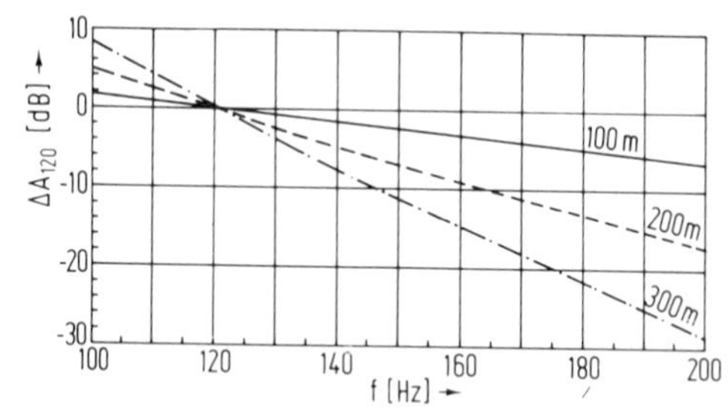

FIG. 7.3-2. Relative attenuation ΔA_{120} referred to the frequency $f' = 120$ Hz for penetration depths of 100, 200, and 300 m. The absolute attenuation in the water for 120 Hz equals -37.8 dB for 100 m, -75.6 dB for 200 m, and -113 dB for 300 m.

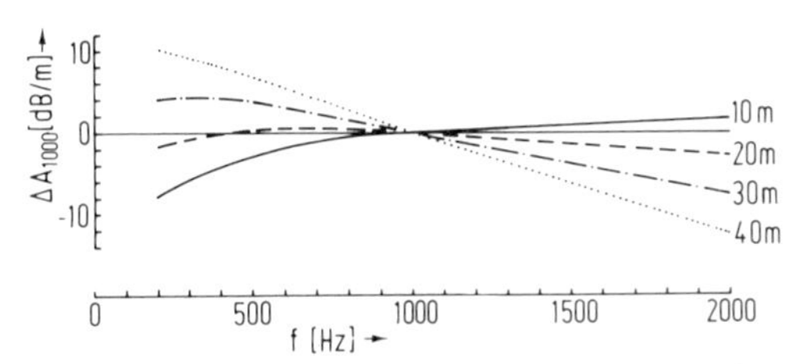

FIG. 7.3-3. Relative attenuation ΔA_{1000} referred to the frequency $f' = 1000$ Hz for penetration depths of 10, 20, 30, and 40 m. The absolute attenuation in the water for 1000 Hz equals -10.9 dB for 10 m, -21.8 dB for 20 m, -32.7 dB for 30 m, and -43.64 dB for 40 m.

cally independent of the frequency. The propagation time t_p to a depth d follows from Eqs. (7.2-4) and (7.2-6):

$$t_p = d/v \doteqdot d/1851\sqrt{f} \tag{4}$$

The relative propagation time referred to a frequency f' is defined as $\Delta T_{f'}$:

$$\Delta T_{f'} = \frac{d}{1851}\left(\frac{1}{\sqrt{f'}} - \frac{1}{\sqrt{f}}\right) \tag{5}$$

The delay distortions are entirely due to the water; there is no compensating term due to the transmitter antenna as in the case of the attenuation distortions. Figure 7.3-4 shows ΔT_{20} for $f' = 20$ Hz in the range 1 Hz $\leqq f \leqq$ 100 Hz for d = 100, 200, and 300 m. The variation of ΔT_{20} is substantial if one considers that the typical length of a teletype character is 150 ms. The delay distortions are reduced by an order of magnitude in

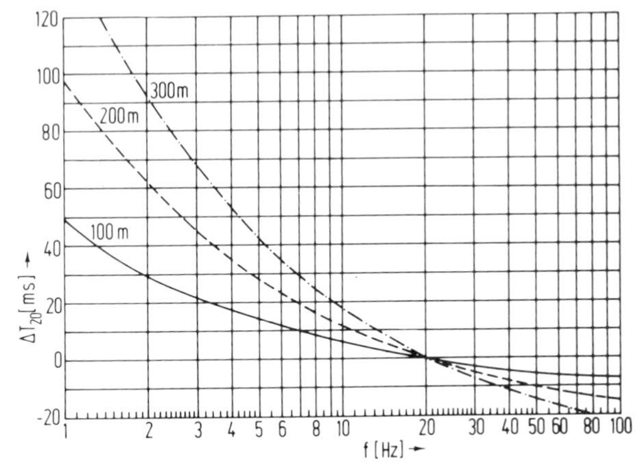

FIG. 7.3-4. Relative delay ΔT_{20} referred to the frequency $f' = 20$ Hz for penetration depths of 100, 200, and 300 m. The absolute delay in the water for 20 Hz equals 14.14 ms for 100 m, 28.28 ms for 200 m, and 42.43 ms for 300 m.

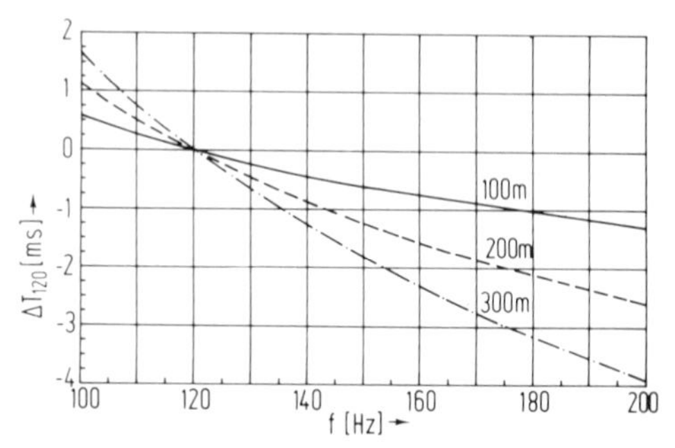

FIG. 7.3-5. Relative delay ΔT_{120} referred to the frequency $f' = 120$ Hz for penetration depths of 100, 200, and 300 m. The absolute delay in the water for 120 Hz equals 5.77 ms for 100 m, 11.55 ms for 200 m, and 17.32 ms for 300 m.

Fig. 7.3-5, which shows ΔT_{120} in the range 100 Hz $\leqq f \leqq$ 200 Hz. However, the attenuations of -75.6 to -113 dB listed in the legend of Fig. 7.3-2 show that the reduction of delay distortions costs an incredible price in signal power. Hence, we will have to find a way to live with the delay distortions of Fig. 7.3-4 if we want to penetrate several hundred meters of water with a teletype link. Figure 7.3-6 shows the relative propagation times ΔT_{1000} referred to a frequency $f' = 1000$ Hz. The values of ΔT_{1000} are again reduced by an order of magnitude compared with those of ΔT_{120} in Fig. 7.3-5. However, ΔT_{1000} is of interest for a data link transmitting 2500 bits/s. The typical signal length for this link is 3.2 ms according to Fig. 7.1-4, and the values of ΔT_{1000} are by no means small compared with 3.2 ms.

To show how the distortion problem can be solved, we specify our signals more precisely. We use the sinusoidal and cosinusoidal pulses of Fig. 7.1-4 with a duration T_s. The first function a is transmitted periodi-

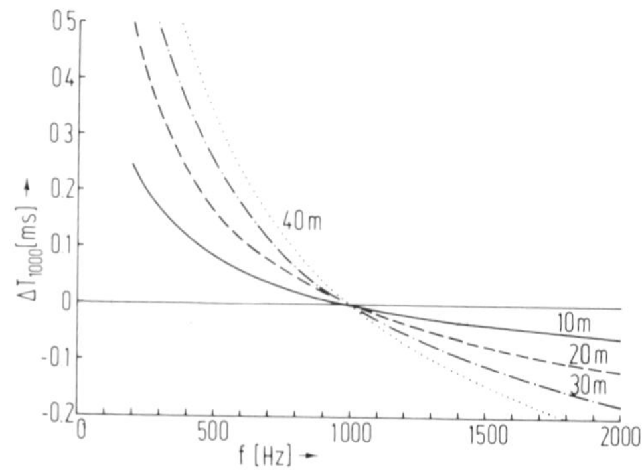

FIG. 7.3-6. Relative delay ΔT_{1000} referred to the frequency $f' = 1000$ Hz for penetration depths of 10, 20, 30, and 40 m. The absolute delay in the water for 1000 Hz equals 0.2 ms for 10 m, 0.4 ms for 20 m, 0.6 ms for 30 m, and 0.8 ms for 40 m.

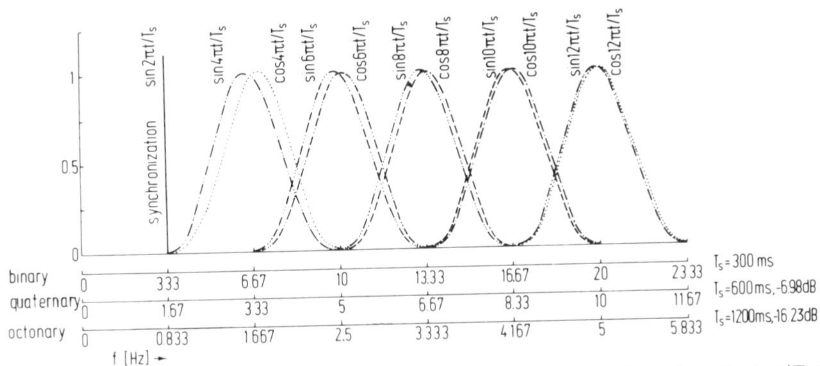

FIG. 7.3-7. Frequency power spectra for teletype systems using the pulses sin $4\pi t/T_s$ to cos $12\pi t/T_s$ for information transmission and the periodic function sin $2\pi t/T_s$ for synchronization.

cally as synchronization signal. This synchronization signal yields a discrete line in the frequency power spectrum with the frequency $1/T_s$. Figure 7.3-7 shows this line at $f = 3.33$ Hz for $T_s = 300$ ms on the scale denoted "binary." The cosinusoidal pulse (b) in Fig. 7.1-4 cannot be used without risk of interference with synchronization. We then use the 10 pulses with 2, 3, 4, 5 or 6 cycles during the time T_s, and either sinusoidal or cosinusoidal variation. Only the six pulses with 2, 3, and 4 cycles are shown in Fig. 7.1-4, but the power spectra for all ten pulses are shown in Fig. 7.3-7. The relative bandwidth of the whole signal equals (23.33 − 3.33)/(23.33 + 3.33) = 0.75. However, this relative bandwidth is unimportant from the standpoint of distortions since we can synchronize the propagation time and compensate the attenuation of the individual pulses. For instance, the relative bandwidth of the pulses sin $4\pi t/T_s$ and cos $4\pi t/T_s$ equals (10 − 3.33)/(10 + 3.33) = 0.5, while that of the pulses sin $12\pi t/T_s$ and cos $12\pi t/T_s$ is only (23.33 − 16.67)/(23.33 + 16.67) = 0.17. The relative attenuation $\Delta A_{f'}$ and the relative delay $\Delta T_{f'}$ are only of interest for bands of the width 6.67 Hz centered at $f' = 6.67$, 10, 13.33, 16.67, and 20 Hz. A look at the curves of Fig. 7.3-1 shows that the attenuation varies only about 5 dB over these bands. The delay distortions are more important according to Fig. 7.3-4. Hence, they are plotted in more detail in Fig. 7.3-8. This illustration shows the delay distortion $\Delta T_{f'}$ as fraction of the signal duration T_s for the reference frequencies $f' = 6.67$ and 20 Hz over bands from $f' - 3.33$ Hz to $f' + 3.33$ Hz. One may see that $\Delta T_{f'}/T_s$ varies from +12% to −4.5% for $f' = 6.67$ Hz and $d = 300$ m, and over a lesser range for $d = 200$ and 100 m. The variation for the reference frequency $f' = 20$ Hz is from +1.67% to −1.23% for $d = 300$ m. The reference frequencies $f' = 10$, 13.3, and 16.67 Hz yield curves intermediate to those shown for $f' = 6.67$ and 20 Hz.

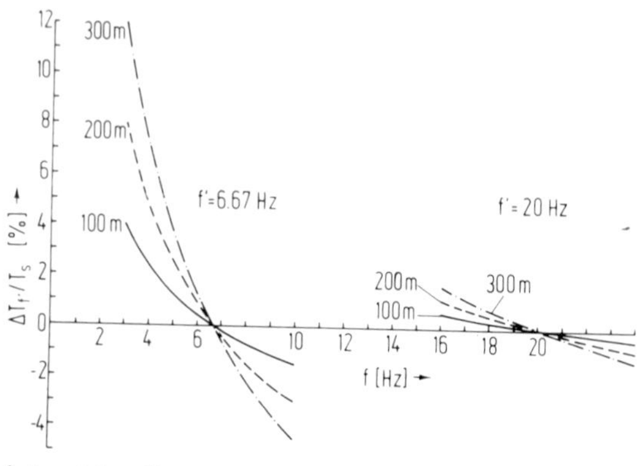

FIG. 7.3-8. Relative delay distortions $\Delta T_{f'}/T_s$ as function of frequency for the pulses sin $4\pi t/T_s$, cos $4\pi t/T_s$ and sin $12\pi t/T_s$, cos $12\pi t/T_s$ for $T_s = 300$ ms according to Fig. 7.3-7.

Let us return to Fig. 7.3-7. We show there the power spectra of the 10 pulses sin $4\pi t/T_s$ to cos $12\pi t/T_s$ for a signal duration $T_s = 300$ ms. A teletype character has five information-carrying pulses and a duration of 150 ms. Hence, we must transmit two teletype characters at a time to permit a signal duration of 300 ms and to make use of the 10 pulses available for information transmission. This is readily done by inserting a buffer storage between teletype writer and transmitter. We can, of course, combine four teletype characters to obtain a signal duration of $T_s = 600$ ms and use 20 pulses sin $4\pi t/T_s$ to cos $22\pi t/T_s$ to transmit the $4 \times 5 = 20$ information pulses. The frequency scale marked "quaternary" in Fig. 7.3-7 applies in this case, but only the power spectra of the first 10 of the required 20 pulses are shown. The bandwidth of the pulses is reduced to 3.33 Hz. The relative bandwidth of the pulses sin $4\pi t/T_s$ and cos $4\pi t/T_s$ is unchanged $(5 - 1.67)/(5 + 1.67) = 0.5$ but the pulses sin $22\pi t/T_s$ and cos $22\pi t/T_s$ have a relative bandwidth of $(20 - 16.67)/(20 + 16.67) = 0.09$.

The delay distortions $\Delta T_{f'}$ as fraction of the signal duration $T_s = 600$ ms for the reference frequencies $f' = 3.33$ and 10 Hz are shown over bands from $f' - 1.67$ Hz to $f' + 1.67$ Hz in Fig. 7.3-9. The distortions are reduced compared with Fig. 7.3-8, even though we have done nothing more than increased T_s by a factor 2 and reduced all frequencies by a factor 1/2. The reason for the reduction of the distortions is that $\sqrt{f}$ rather than f occurs in Eq. (5). We could reduce the distortions still more by not using the pulses sin $4\pi t/T_s$ and cos $4\pi t/T_s$ but the pulses from sin $6\pi t/T_s$ and cos $6\pi t/T_s$ to sin $24\pi t/T_s$ and cos $24\pi t/T_s$. The signal would then occupy the band from 3.33 to 21.67 Hz, which means that the upper band limit is

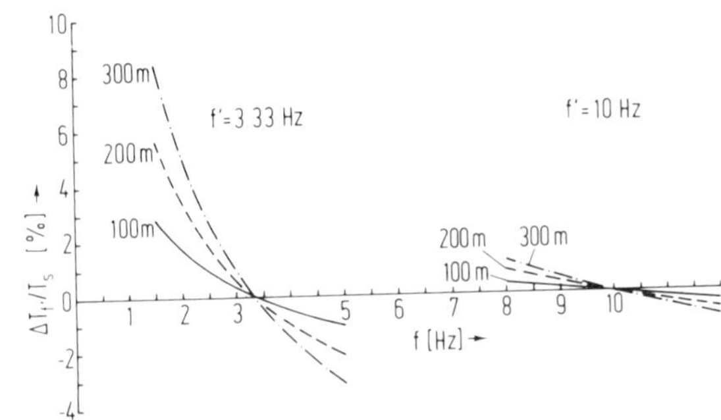

FIG. 7.3-9. Relative delay distortions $\Delta T_{f'}/T_s$ as function of frequency for the pulses sin $4\pi t/T_s$, cos $4\pi t/T_s$ and sin $12\pi t/T_s$, cos $12\pi t/T_s$ for T_s = 600 ms according to Fig. 7.3-7.

still below the 23.33 Hz shown on the scale marked "binary" for T_s = 300 ms in Fig. 7.3-7.

One can go further and combine eight teletype characters into one signal. The signal duration then becomes $8 \times 150 = 1200$ ms and the frequency scale marked "octonary" in Fig. 7.3-7 applies. One now needs 40 pulses sin $4\pi t/T_s$ to cos $42\pi t/T_s$, and one can further reduce the distortions by not using sin $4\pi t/T_s$, cos $4\pi t/T_s$ and perhaps sin $6\pi t/T_s$, cos $6\pi t/T_s$, and adding the pulses sin $44\pi t/T_s$, cos $44\pi t/T_s$ and sin $46\pi t/T_s$, cos $46\pi t/T_s$. The problem of distortion control is thus reduced to a problem of equipment complexity and cost which will be addressed later on when circuits are discussed.

We have so far tacitly assumed that binary signals are to be transmitted. This is in line with the common experience that binary signals are best when the signal-to-noise ratio is of primary interest. However, a closer analysis shows that the rapid increase of the attenuation with frequency in seawater makes quaternary and octonary signals yield a better signal-to-noise ratio at depths of several hundred meters.

Consider a binary signal producing the voltages $+E$ or $-E$ at the output of a receiver. A quaternary signal must produce the voltages $-3E$, $-E$, $+E$, and $+3E$ to yield the same error rate due to thermal and many other types of noise, while an octonary signal must produce the voltages $\pm 7E$, $\pm 5E$, $\pm 3E$, and $\pm E$. The ratio of the average power of the quaternary and the binary signals is as follows, if all output voltages occur equally often:

$$\tfrac{1}{2}[(3E)^2 + E^2]/E^2 = 5$$

Hence, the average power must be five times the average power of the binary signal for equal error rate. In other words, one needs 6.98 dB more power for quaternary signals, but one needs only half the bandwidth.

Let us return to Fig. 7.3-7. If we use binary signals we need the 10 pulses sin $4\pi t/T_s$ to cos $12\pi t/T_s$ to transmit 10 bits of information in T_s = 300 ms. Using quaternary signals we can transmit 20 bits of information with the same pulses, or four rather than two teletype characters. Since teletype characters have a duration of 150 ms, we now have $T_s = 600$ ms for the transmission of four of them. The frequency scale marked "quaternary" in Fig. 7.3-7 applies in this case. Practically all the energy of the quaternary signals is concentrated below 11.67 Hz, while the respective number for binary signals is 23.33 Hz. The attenuation of the binary signal after penetrating water of depth d is less than the attenuation at the frequency of 23.33 Hz:

$$A = d\alpha(23.33) \doteqdot 0.0345 \times 23.33^{1/2}d \tag{6}$$

where the units of d are meters and those of A are decibels. Reading Eq. (6) from right to left yields the penetration depth as function of the attenuation A:

$$d \doteqdot 6.00A \tag{7}$$

Figure 7.3-10 shows d by the solid line for the range A = 20–50 dB.

Consider quaternary signals. The frequency of 23.33 Hz in Eq. (6) has to be replaced by the frequency 11.67 Hz. Since we need 6.98 dB more power, we can tolerate only the attenuation $A - 6.98$ dB rather than A in Eq. (7). Hence, we obtain the penetration depth of quaternary signals:

$$d \doteqdot 8.49(A - 6.98) \tag{8}$$

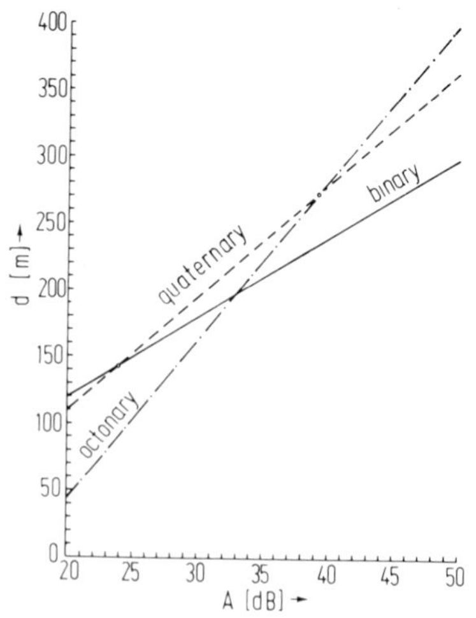

FIG. 7.3-10. Penetration depth d of water as function of attenuation A for binary, quaternary, and octonary signals according to Fig. 7.3-7. The crossover of binary and quaternary is at 24 dB and 140 m, the crossover of quaternary and octonary at 39 dB and 275 m.

Figure 7.3-10 shows d by the dashed line. This line crosses the solid line at a depth of about 140 m. Hence, binary signals have a better signal-to-noise ratio than quaternary signals for the same average power entering the water for depths from 0 to 140 m; beyond 140 m the quaternary signals have a higher signal-to-noise ratio. A closer inspection of Fig. 7.3-10 shows that quaternary signals yield a signal-to-noise ratio 8 dB better than binary signals at a depth of $d = 300$ m.

Consider octonary signals. The ratio of the average power of the octonary and the binary signals is as follows, if all output voltages occur equally often:

$$\tfrac{1}{4}[(7E)^2 + (5E)^2 + (3E)^2 + E^2]/E^2 = 42$$

The average power must be 42 times the average power of the binary signals, or 16.23 dB higher, for the same error rate. The 10 pulses sin $4\pi t/T_s$ to cos $4\pi t/T_s$ can now transmit 40 bits of information or 8 teletype characters with a total duration of $8 \times 150 = 1200$ ms. The frequency scale marked "octonary" in Fig. 7.3-7 applies. Practically all energy is concentrated below 5.833 Hz. We have to replace 23.33 Hz in Eq. (6) by 5.833 Hz and substitute $A - 16.23$ for A. The penetration depth of octonary signals corresponding to Eq. (8) becomes:

$$d \doteqdot 12.00(A - 16.23) \tag{9}$$

Figure 7.3-10 shows d by the dash–dot line. The crossover with the dashed line is at about 275 m. Hence, at depths greater than 275 m octonary signals yield a better signal-to-noise ratio than either quaternary or binary signals for the same average power entering the water.

Let us briefly consider delay distortions for quaternary and octonary signals. The plots of Fig. 7.3-9 apply directly to quaternary pulses with the time variation of sin $4\pi t/T_s$ to cos $12\pi t/T_s$ according to Fig. 7.3-7. We have discussed before that the distortions in Fig. 7.3-9 are lower than in Fig. 7.3-8, which applies to binary signals, and have stated why this is so. It follows that the delay distortions for octonary signals will be still less. The same applies to attenuation distortions, but they are less significant due to the compensating effect of the transmitter antenna as previously discussed.

7.4 The Best Frequency Band as Function of Depth

Figure 7.2-2 shows that the penetration depth d increases for a certain attenuation as the frequency decreases. The efficiency with which an antenna current I_{rms} is converted into radiated power decreases with decreasing frequency according to Eq. (7.3-1):

$$S(f) = (2\pi Z_0/3)(fL/c)^2 I_{\mathrm{rms}}^2 \tag{1}$$

TABLE 7.4-1
FREQUENCIES AND CORRESPONDING WAVELENGTHS IN VACUUM OF AIR

Frequency (Hz)	Wavelength (km)
10000	30
1000	300
100	3000
10	30000
1	300000

Hence, there must be a best frequency to work with. For the derivation of this best frequency we will ignore the frequency dependence of the propagation through the air and of the reflection at the air–water boundary.

Equation (1) assumes that the antenna is short compared with the wavelength of the radiated wave. According to Table 7.4-1 this condition will always be satisfied below 100 Hz but not necessarily above 1000 Hz. Hence, the following calculation is sure to apply to the low frequencies shown for teletype systems in Figs. 7.3-7 and 7.1-5, but we have to be somewhat careful in applying it to the 2500 bits/sec data link according to Fig. 7.1-5.

The power radiated by an antenna of fixed length L and a sinusoidal current of fixed rms value I_{rms} at a frequency f relative to the power radiated at a reference frequency f' follows from Eq. (1):

$$20 \log[S(f)/S(f')] = 20 \log(f/f') \tag{2}$$

The attenuation of a layer of seawater of depth d equals $d\alpha(f)$, where $\alpha(f)$ is defined by Eqs. (7.2-1) and (7.2-2). Let us assume that we can tolerate a certain attenuation A in the water. The penetration depth then follows by solving the equation

$$d\alpha(f) = A + 20 \log(f/f')$$

for the depth d:

$$d = \frac{A + 20 \log(f/f')}{\alpha(f)} \doteq \frac{A - 20 \log f' + 20 \log f}{0.0345 f^{1/2}} \tag{3}$$

where the units of d are meters and those of f are hertz. The reference frequency f' may be eliminated by denoting $A - 20 \log f'$ by A'. This implies that the family of curves one obtains for $d = d(f)$ is independent of the reference frequency f'; only the value of A that we use as parameter for these curves depends on f'. Figure 7.4-1 shows such a family of curves denoted a, b, ..., h. The value of A for various reference frequencies f' is

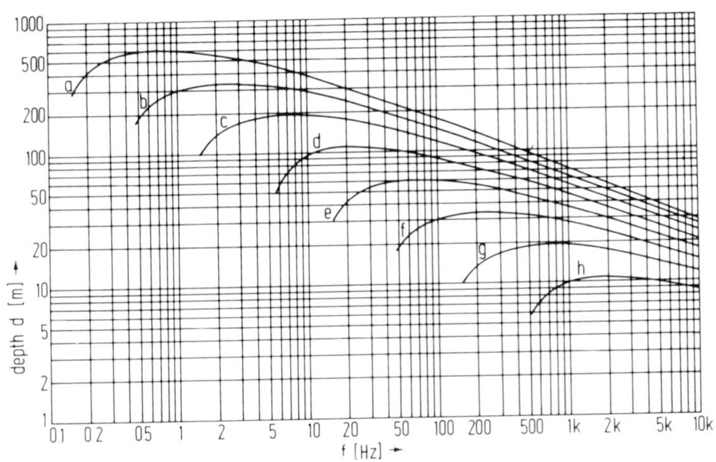

FIG. 7.4-1. The best frequency bands for operation to a certain depth *d*. The curves are spaced 10 dB apart, curve a requiring the most power and curve h the least power.

shown in Table 7.4-2, but it is of no consequence here. The negative values of *A* are of no significance since they only mean that the power is below rather than above an arbitrarily chosen reference level. Let us turn to the discussion of the curves in Fig. 7.4-1.

The attenuation for curve a is 10 dB more than for curve b, which in turn holds for 10 dB more attenuation than curve c, etc. Let us assume we want to reach a depth *d* of 300 m. The curves c–h do not reach this depth, but curves b and a do. Since curve a holds for 10 dB more attenuation than curve b we can save 10 dB by choosing curve b. It exceeds 300 m in

TABLE 7.4-2

THE PERMISSIBLE ATTENUATION *A* [dB] IN THE WATER FOR CURVES a, b, ..., h IN FIG. 7.4-1 FOR VARIOUS REFERENCE FREQUENCIES f'

	Reference frequency f' (Hz)			
Curve	10,000	1000	100	10
a	90	70	50	30
b	80	60	40	20
c	70	50	30	10
d	60	40	20	0
e	50	30	10	−10
f	40	20	0	−20
g	30	10	−10	−30
h	20	0	−20	−40

the frequency range from 1 to 7 Hz. If we want a wider frequency band, we can add 10 dB of power and use curve a. It exceeds 300 m in the frequency range from 0.15 to 20 Hz, and it exceeds 500 m in the range from 0.3 to 3 Hz. These optimal frequency ranges correspond very well to the frequency bands occupied by teletype signals according to Fig. 7.3-7. For instance, the octonary system occupies the band from 0.833 to 5.833 Hz. Curve a in Fig. 7.4-1 exceeds 420 m in the range from 0.2 to 6 Hz. Hence, the octonary teletype system is almost ideal for a depth of 420 m. Actually, a curve with about 1 to 2 dB less attenuation than curve a would be ideal. The quaternary system in Fig. 7.3-7 requires the band from 1.67 to 11.67 Hz. Curve b in Fig. 7.4-1 exceeds 275 m in the range from 0.8 to 12 Hz; this or a somewhat reduced depth is ideal for the quaternary system. Finally, the binary system in Fig. 7.3-7 occupies the band from 3.33 to 23.33 Hz. Curve c in Fig. 7.4-1 exceeds 170 m in the range from 3.2 to 24 Hz; this is the best depth for the binary teletype system. For reduced depths one should use higher frequencies. Curve e shows that for a depth of about 60 m the preferred frequency band is somewhat below 100 Hz etc.

It seems like the teletype standard accidently happens to be ideal for great penetration depths. This is not quite so. We have picked the international standard of 150 ms/character. The fast U.S. standard of 100 ms/character would not be so well matched. The slow U.S. standard of 166.7 ms/character would be slightly better than the international standard for depths of 300 m and beyond. The reason why everything comes out so ideally is that we could choose from several standards and do our calculation for a depth that fitted this standard. The best transmission rate for a depth of 500 m would be lower than any teletype standard.

Let us try to learn from Fig. 7.4-1 something for the data link operating at 2500 bits/s. According to Fig. 7.1-5 we have to multiply the frequencies in Fig. 7.3-7 by a factor 150/3.2. The band limits of the binary scale in Fig. 7.3-7 become $3.33 \times 150/3.2 = 156.25$ Hz and $23.33 \times 150/3.2 = 1093.75$ Hz. According to Fig. 7.4-1, a curve between f and g would be ideal to reach a depth of about 25 m. To reach a depth of 70 m would require curve a, which implies 50 dB more power than curve f and operation that is far from the ideal of this curve. A change to quaternary signals would be of no use since they become worthwhile only beyond 140 m according to Fig. 7.3-10.

One may readily recognize from Fig. 7.4-1 how the useful bandwidth decreases with depth. Curve h shows that a 10-kHz wide band can be used for a penetration depth of about 10 m efficiently, while the usable bandwidth for efficient penetration to about 500 m is only a few hertz wide according to curve a.

The relative bandwidth for all "good" cases according to Fig. 7.4-1 is very large. For instance, the band 1 Hz $< f <$ 7 Hz implies a relative bandwidth $\eta = (7 - 1)/(7 + 1) = 0.75$, the band 0.15 Hz $< f <$ 20 Hz implies $\eta = 0.99$, and the band 156.25 Hz $< f <$ 1093.75 Hz implies $\eta = 0.75$. Since the submarine communication system Seafarer uses a relative bandwidth of only a few percent, one can expect to achieve significant improvements by using the concepts developed here.

7.5 Typical Circuits

Figure 7.5-1 shows the block diagram of the transmitter of a teletype link that operates to a depth of 300 m and more. A teletype transmitter operated either manually or by a papertape reader feeds binary characters with five information digits each into a series-parallel register SPR. Ten information digits are read in for binary transmission, twenty for quaternary transmission, and forty for octonary transmission. These digits are transferred via the buffer BUF to the converter BQO, which converts the binary input digits into quaternary or octonary output digits if these modes of operation are chosen.

A function generator FUG produces *periodic* sinusoidal and cosinusoidal functions sin $4\pi t/T_s$, cos $4\pi t/T_s$ to sin $12\pi t/T_s$ and cos $12\pi t/T_s$. The time base T_s equals 300 ms for binary transmission, 600 ms for quaternary transmission, and 1200 ms for octonary transmission as shown in Fig. 7.3-7. The additional function sin $2\pi t/T_s$ is produced for synchronization. The only good way to produce these sinusoidal functions with the required stability of frequency, phase, and amplitude is by digital circuits.

The functions sin $4\pi t/T_s$ to cos $12\pi t/T_s$ are multiplied in the multiplier

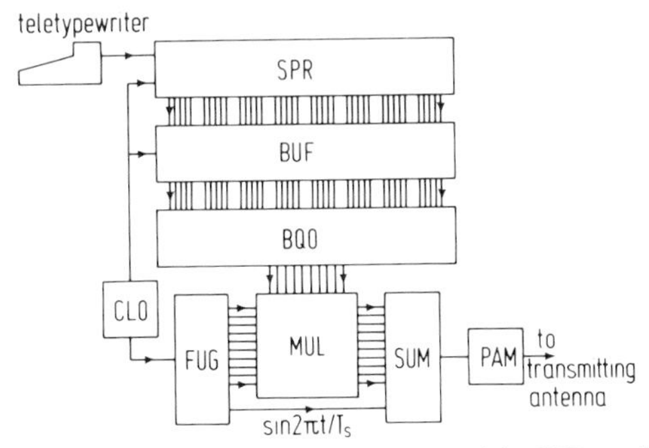

Fig. 7.5-1. Block diagram of a transmitter of a teletype link. SPR, series-parallel register; BUF, buffer; BQO, binary, quaternary, octonary converter; CLO, clock; FUG, function generator; MUL, multiplier; SUM, summer; PAM, power amplifier.

MUL with the coefficients +1 or −1 for binary transmission, with +1, +1/3, −1/3 or −1 for quaternary transmission, and with +1, +5/7, +3/7, +1/7, −1/7, −3/7, −5/7 or −1 for octonary transmission. The multiplied signals are summed in the summer SUM and fed via the power amplifier PAM to the radiating antenna. This radiating antenna is in essence a long wire with ground return, which is nevertheless short compared with the wavelengths of radiated signals.

The block diagram of a receiver is shown in Fig. 7.5-2. The input amplifier IAM with a bandwidth from about 0.1 to 100 Hz amplifies the input signal to the 1-V level. An analog-to-digital converter samples the signals at the rate of about 60 samples/s for octonary signals, 120 samples/s for quaternary signals, and 240 samples/s for binary signals. These different sampling rates may easily be implemented without any problems of aliasing if an integrate-and-dump circuit is used with the analog-to-digital converter that integrates over intervals of duration 1/60 s, 1/120 s, or 1/240 s.

The digital signals are fed to the timing circuit TIM and the resonator RES; these circuits will be discussed presently in some detail. At the moment we assume that the binary, quaternary or octonary coefficients used for multiplication in the transmitter are obtained at the times T_s, $2T_s$, $3T_s$, ... at the output terminals of the resonator RES. These coefficients are sampled and held during a time of duration T_s in the sample-and-hold circuit SAH. Note that a sample-and-hold circuit for digital signals is a storage register. The converter OQB converts the octonary or quaternary numbers into binary numbers, which represent the transmitted information in the form used by the teletypewriter. These signals are fed via a buffer BUF to a parallel-series register PSR and from there to the tele-

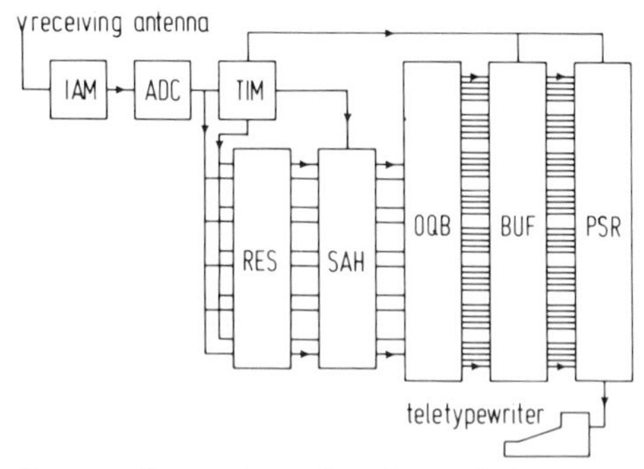

FIG. 7.5-2. Block diagram of a receiver of a teletype link. IAM, input amplifier; ADC, analog-to-digital converter; TIM, timing circuit; RES, resonators; SAH, sample-and-hold circuit; OQB, octonary, quaternary, binary converter; BUF, buffer; PSR, parallel-series register.

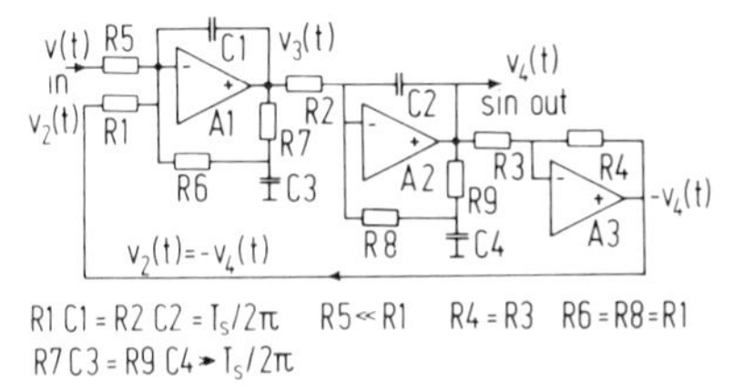

FIG. 7.5-3. Analog circuit of a filter for the synchronization signal sin $2\pi t/T_s$.

typewriter. Start-and-stop pulses required by the teletypewriter can readily be inserted in the parallel-series converter.

Except for the timing circuit TIM and the resonator RES, the receiver uses only standard circuits. Let us turn to the two nonstandard circuits. The timing circuit TIM must extract the synchronization signal sin $2\pi t/T_s$, which is a periodic sinusoidal signal with a frequency of 3.33, 1.67 or 0.833 Hz according to Fig. 7.3-7. Figure 7.5-3 shows an analog circuit suitable for the task. This circuit implements the differential equation of sinusoidal functions for a periodic force function sin $2\pi t/T_s$. The resistors R6, R7 and R8, R9 provide a dc feedback. The voltage $v_4(t)$ is the wanted synchronization signal sin $2\pi t/T_s$. The circuit resonates with sin $2\pi t/T_s$. Only the nonlinearity of the amplifiers prevents an unlimited increase of the amplitude of the output signal, but we will correct this deficiency in the digital circuit to be derived from Fig. 7.5-3.

The resonators RES in Fig. 7.5-2 are similar to the circuit of Fig. 7.5-3 but they are resonating with sinusoidal *pulses* rather than *periodic* sinusoidal functions. Figure 7.5-4 shows the modified circuit. The switches s1 and s2 have been added together with the resistors R6 and R7. The dc feedback loops of Fig. 7.5-3 are not needed now. The operation of the circuit is best demonstrated by the oscillograms of Fig. 1.2-4. A sinusoidal pulse (a) with one cycle of duration T_s is fed to the input terminal; as mentioned before, this pulse is produced by digital circuits as a step function. The output voltage $v_3(t)$ at the terminal "cos" is shown by the trace b,

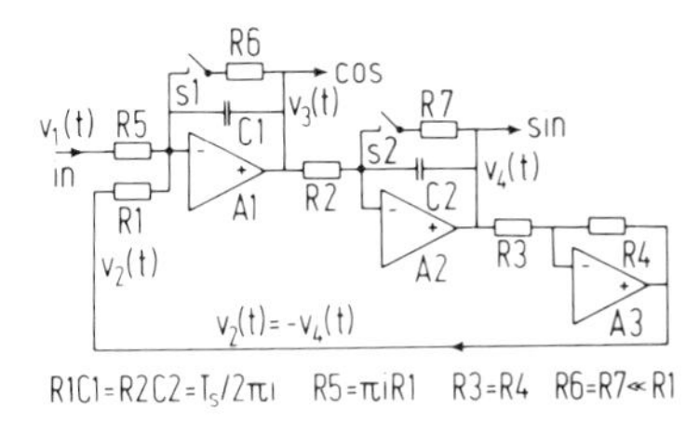

FIG. 7.5-4. Analog circuit of a resonator for sine and cosine pulses.

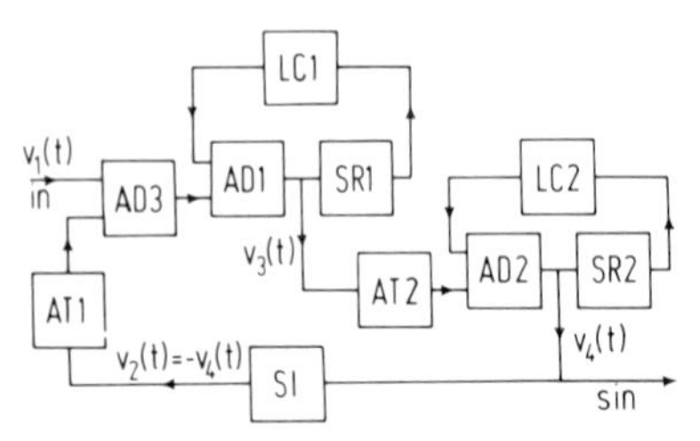

FIG. 7.5-5. Digital version of the circuit of Fig. 7.5-3. AD, adder; SR, shift register; LC, loss controller; AT, attenuator; SI, sign reverser.

while the output voltage $v_4(t)$ at the terminal "sin" is shown by the trace c. The traces begin at the time $t = 0$. At the time $t = T_s$ the trace b reaches zero, since the input pulse is a sinusoidal and not a cosinusoidal pulse; the trace c reaches a positive value representing the amplitude of the sinusoidal pulse of trace a. Immediately after the time $t = T_s$ the switches s1 and s2 in Fig. 7.5-4 are momentarily closed and the circuit is reset to receive the next pulse. This explains the jump back to zero of trace c in Fig. 1.2-4.

Trace d in Fig. 1.2-4 shows a cosinusoidal pulse as input voltage $v_1(t)$, trace e the output voltages $v_3(t)$, and trace f $v_4(t)$. Now $v_4(t)$ is zero at $t = T_s$ while $v_3(t)$ represents the amplitude of the cosinusoidal pulse.

The performance of the circuit of Fig. 7.5-4 set for pulses with $i = 128$ cycles in the interval of duration T_s is shown by the oscillogram of Fig. 1.2-5. A sinusoidal input voltage $v_1(t) = V \sin 2\pi t/T_s$ with $i = 128$ produces the output voltage $v_4(t)$ shown by trace a, while sinusoidal input voltages with $i = 129$ and 130 cycles in the interval $0 \leqq t \leqq T_s$ produce the traces b and c.

Figure 7.5-5 shows the digital implementation of the circuit of Fig. 7.5-3. The digital input signal $v_1(t)$ and the signal $v_2(t)$ arriving via the feedback loop are summed in the adder AD3. The adder AD1, shift register SR1, and the feedback loop via the loss controller LC1 implement the first integrator of Fig. 7.5-3. The output signal $v_3(t)$ has a much larger value than the input signal from adder AD3 to adder AD1. The value is determined by the loss controller LC1, which will be discussed presently. The attenuator AT2 reduces $v_3(t)$ to the level of the signal at the output terminal of the adder AD3. The adder AD2, the shift register SR2, and the feedback loop via the loss controller LC2 implement the second integrator of Fig. 7.5-3. The output signal $v_4(t)$ is fed via a sign reverser SI, which is a gate transforming 0 into 1 and 1 into 0, and the attenuator AT1 to the adder AD3. The sign reverser SI implements the amplifier A3 with the resistors R3 and R4 in Fig. 7.5-3.

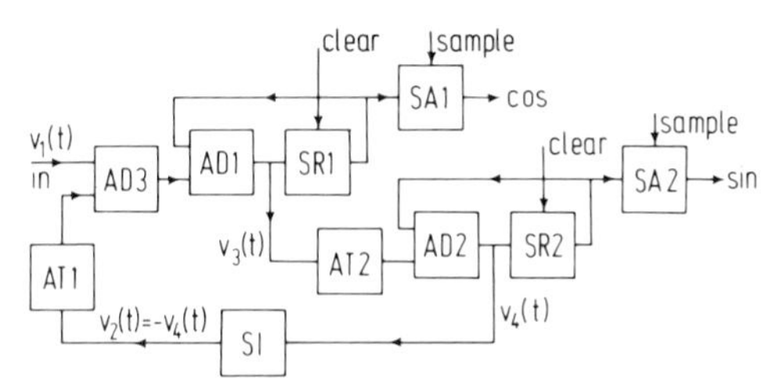

FIG. 7.5-6. Digital version of the resonator of Fig. 7.5-4. AD, adder; SR, shift register; AT, attenuator; SI, sign reverser; SA, sampling circuit.

The adders AD are standard circuits. The shift registers SR are parallel shift registers with two stages. The loss controllers LC are unusual circuits. Without them, the numbers circulating in the feedback loops of the two integrators in Fig. 7.5-5 could increase indefinitely like trace a in Fig. 1.2-5. Eventually, even the largest storage register SR would overflow. In an analog circuit as shown in Fig. 7.5-3 one prevents this unlimited increase by introducing losses that increase with the amplitude of the circulating signal. A digital equivalent of such a loss controller is a multiplier that multiplies by 1 if the numbers fed to it are small, and by $q < 1$ if the numbers are large. However, a real multiplier would be too expensive. A practical loss controller is shown in Fig. 4.4-2, and it is discussed there.

The digital attenuators AT in Fig. 7.5-5 are simpler than analog attenuators. If we want to attenuate the character at the output terminal of the adder AD1 by a factor 2^{-7}, we simply have to wire the terminals for the five highest digits to the input terminals for the five lowest digits of the adder AD2 as shown in Fig. 4.4-3.

The digital equivalent of the analog circuit of Fig. 7.5-4 is shown in Fig. 7.5-6. The loss controllers LC of Fig. 7.5-5 are not needed because the circuit is reset at multiples of T_s. The switches s1 and s2 with their resistors in Fig. 7.5-4 are replaced by a "clear" pulse fed to the shift registers SR.

7.6 Comparison of Various Systems

We have tacitly made some simplifying assumptions in the preceding sections in order to avoid obscuring the principles by the details. These simplifications will now be corrected. Furthermore, we will use data from the projects Sanguine and Seafarer for comparison. The reference frequency f' in Eqs. (7.3-2) and (7.3-3) will be chosen as

$$f' = 45 \text{ Hz} \tag{1}$$

The transmission rate currently publicized for Seafarer is 10 bits/min and the average antenna input power is 7 MW. These numbers hold for an antenna site on the Canadian Shield, such as on the upper peninsula of Michigan; antenna sites at other locations require more power. The energy E_b' per bit of information thus equals $7 \times 60/10$ or

$$E_b' = 42 \text{ MWs/bit} \tag{2}$$

We denote with η_{AF} the gain of the radiating antenna referred to the frequency f'. This gain is given by Eq. (7.3-2) for an antenna in free space or above a conducting plane in free space; the index letter F is added to characterize this antenna:

$$\eta_{AF} = 20 \log(f/f') \tag{3}$$

A positive value of η_{AF} indicates that more power is radiated for a given input power, or less input power is needed for a required radiated power.

The power radiated by an antenna in free space varies proportionate to f^2, but the power radiated by an antenna in a cavity between two conducting surfaces varies proportionate to f (Wait, 1970). The gain η_{AF} of Eq. (3) is in this case replaced by the gain η_{AC}:

$$\eta_{AC} = 10 \log(f/f') = \eta_{AF}/2 \tag{4}$$

There is some question whether Eq. (3) or Eq. (4) should be used. Some authors (Keiser, 1974; Rowe, 1974) use the free-space antenna, while others (Galejs, 1971; Ta-Shing Chu, 1974) point out that the cavity antenna between conducting surfaces should be used. The difficulty is caused by the fact that the ionosphere creates a cavity, but not one with good conducting surfaces. We will thus use Eq. (3) as the worst case and Eq. (4) as the best case, the actual antenna gain being somewhere in between.

Next comes the loss of the wave due to absorption during propagation in the Earth–ionosphere cavity as function of frequency. This is an extremely difficult topic (Wait, 1970; Galejs, 1971; Pappert and Moler, 1974). We approximate this loss in the range from about 1 to 50 Hz by $\beta(f)$,

$$\beta(f) = 0.09 \times 10^{-6}\sqrt{f} \quad [\text{dB/m}] \tag{5}$$

where the units of f are hertz. For $f = 5$ Hz one obtains $\beta(f) = 0.2$ dB/Mm (Mm = megameter = 1000 km), which is in agreement with measurements. For $f = 50$ Hz one obtains $\beta(f) = 0.64$ dB/Mm. This is lower than the observed values of 0.7 to 0.9 dB/Mm. However, using a median observed value of 0.8 dB/Mm, we obtain a difference $0.8 - 0.64 = 0.16$ dB/Mm, which yields a difference of the loss at 10 Mm of only 1.6 dB. Furthermore, this result will be on the "safe side," meaning that it will actually be better than calculated. Of more importance than the difference between 0.8 and 0.64 dB/Mm are the Schumann resonances at about 8, 14, and 20 Hz. We will have to deal with them separately (Schumann, 1952a,b, 1957).

Instead of the absolute loss defined by Eq. (5) we want the relative loss η_D (in decibels) at a distance D (in meters) referred to the reference frequency f':

$$\eta_D = [\beta(f') - \beta(f)]D = 0.09 \times 10^{-6} D(\sqrt{f'} - \sqrt{f}) \tag{6}$$

A positive value of η_D indicates that less power is lost and less radiated power is thus needed.

The loss due to geometric spreading of the power of the wave is another difficult matter. At first glance one might think that the geometry of the Earth–ionosphere cavity is the same for all sinusoidal waves with a wavelength much longer than the distance of the ionosphere from the ground. However, some of the wavelengths of interest are on the order of the circumference of the earth, which causes the previously mentioned Schumann resonances. Furthermore, the loss of a wave with a frequency of 5 Hz due to absorption is only $0.2 \times 40 = 8$ dB after having run around the Earth along a great circle. Hence, there will be interference effects due to the superposition of waves reaching the same point via different great circle routes. However, both the Schumann resonances as well as the interference effects due to multi-great-circle-route transmission—or multipath transmission—affect certain frequencies only. According to Fig. 7.3-7 our signals occupy a continuous frequency band with large relative bandwidth, e.g., from 3.33 to 23.33 Hz or from 1.67 to 11.67 Hz. As a result, the interference effects due to multipath transmission at any given receiver location will be averaged essentially in the same way as in short-wave communications using frequency diversity with equal gain summation (Harmuth, 1972, p. 308). The Schumann resonances will enhance certain frequency components of the signals. By ignoring this enhancement we will calculate a somewhat smaller signal power than the actual one and be on the safe side. Interference caused by lightning discharges will be enhanced by the Schumann resonances in the same way as a signal, and the net effect on the signal-to-noise ratio will be zero.

To reach the submarine the wave has to go through the interface between the air and the water. The electric field strength just below the waterline varies proportionate to $\sqrt{f}$, while the magnetic field strength does not vary with frequency.[1] Hence, the power transferred through the

[1] See Eqs. (34) and (16) of Rowe (1974). The attenuation due to reflection at the air–water boundary will be discussed in more detail in Section 7.8.

air–water boundary will vary like $\sqrt{f}$, and we define the following air–water boundary loss:

$$\eta_{\mathrm{I}} = 10 \log(f/f')^{1/2} \tag{7}$$

A positive value of η_{I} indicates that more power passes into the water and less power needs thus to be radiated.

After the electromagnetic wave has entered the water it is again subject to losses due to absorption and geometric spreading. The geometric spreading can be ignored since a plane wave enters the water for all practical purposes. The attenuation due to absorption is given by Eq. (7.2-2). Note that Eqs. (7.2-2) and (5) have the same form but the factor 10^{-6} makes a huge difference. We define the relative loss η_d (in decibels) in the water at a depth d (in meters) referred to the frequency f':

$$\eta_d = [\alpha(f') - \alpha(f)]d = 0.0345d(\sqrt{f'} - \sqrt{f}) \tag{8}$$

We add now the various relative gains and losses to obtain the relative gain or loss from the input power fed to the radiating antenna to the input power reaching the receiving antenna. Several cases must be distinguished.

(1) *Radiating antenna in free space.* As pointed out before, the assumption of a radiating antenna in free space is the worst case limit. We sum Eqs. (3), (6), (7), and (8). The sum is denoted η_{W},

$$\begin{aligned} \eta_{\mathrm{W}} &= \eta_{\mathrm{AF}} + \eta_D + \eta_{\mathrm{I}} + \eta_d \\ &= 25 \log(f/f') + (0.09 \times 10^{-6}D \\ &\quad + 0.0345d)(\sqrt{f'} - \sqrt{f}) \quad [\mathrm{dB}] \end{aligned} \tag{9}$$

where the units of D and d are meters, and those of f and f' are hertz.

(2) *Radiating antenna in a cavity with good conducting surfaces.* This is the best-case limit. We sum Eqs. (4), (6), (7), and (8). The sum is denoted η_{B}:

$$\begin{aligned} \eta_{\mathrm{B}} &= \eta_{\mathrm{AC}} + \eta_D + \eta_{\mathrm{I}} + \eta_d \\ &= 15 \log(f/f') + (0.09 \times 10^{-6}D \\ &\quad + 0.0345d)(\sqrt{f'} - \sqrt{f}) \quad [\mathrm{dB}] \end{aligned} \tag{10}$$

(3) In the calculations of Section 7.4 that led to the curves of Fig. 7.4-1, we assumed a radiating antenna in free space; the propagation loss in the Earth–ionosphere cavity and the loss at the air–water boundary were ignored. Hence, we have to sum Eqs. (3) and (8) for this case. The sum is denoted η_0:

$$\eta_0 = \eta_{AF} + \eta_d$$
$$= 20 \log(f/f') + 0.0345d(\sqrt{f'} - \sqrt{f}) \quad [\mathrm{dB}] \tag{11}$$

The value of η_0 is between the worst-case limit η_W of Eq. (9) and the best-case limit η_B of Eq. (10).

We now apply our results to the signal transmission according to Sanguine.[1] First, we consider a very unsophisticated transmission scheme to see the order of magnitude of the possible improvements. According to Fig. 7.4-1, curve b, the best frequency band to reach a depth of 300 m is from about 1 to about 8 Hz. A transmission rate of 10 bits/min can be achieved by sending binary pulses with a duration of 6 s. This is shown in Fig. 7.1-1, line b, for sinusoidal pulses with one cycle and a duration of 20 ms. If we increased the duration of these pulses from 20 ms to 6 s we would have pulses with most of the energy around 1/6 Hz. This is below the preferred band of 1 Hz $< f <$ 8 Hz. We may use instead pulses of 6 s duration but with 12 cycles as shown on Top of Fig. 7.6-1. The frequency energy spectrum of this pulse is shown in Fig. 7.6-1b; most of the energy is concentrated around 2 Hz. We insert $f' = 45$ Hz, $f = 2$ Hz, and $d =$ 300 m into Eqs. (9)–(11). The distance D of propagation in the earth–ionosphere cavity is chosen 10^7 m = 10,000 km, since the geometric spreading of the power yields the lowest power density at this distance,

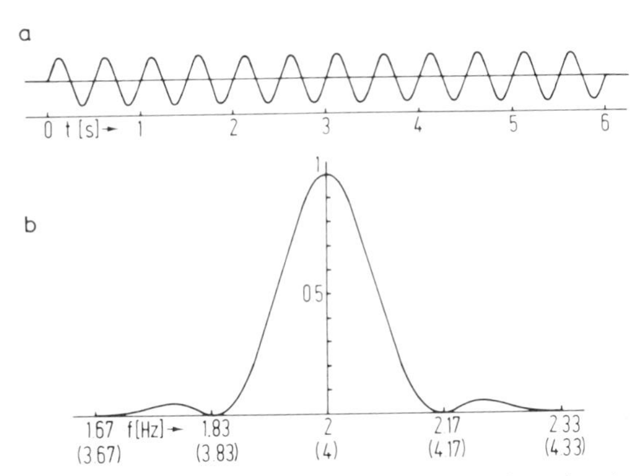

FIG. 7.6-1. Sinusoidal pulse with 12 cycles in an interval of 6-s duration (a) and its frequency power spectrum centered at 2 Hz (b). The same power spectrum but centered at 4 Hz applies to a pulse with 24 cycles in an interval of 6-s duration.

[1] See the special issue on Extremely Low-Frequency Communication of *IEEE Trans. Commun.* **22** (1974).

TABLE 7.6-1
SAVINGS OF AN UNSOPHISTICATED TRANSMISSION SYSTEM COMPARED WITH A SANGUINE SYSTEM REQUIRING $E_b' = 42$ MWs/bit AND AN AVERAGE SIGNAL POWER $P_s' = 7$ MW

	E_b'/E_b	E_b (kWs/bit)	P_s (kW)
Center frequency 2 Hz:			
$\eta_W = 25.8$ dB	375	111.8	18.6
$\eta_0 = 27.8$ dB	596	70.5	11.8
$\eta_B = 39.3$ dB	8467	5.0	0.83
Center frequency 4 Hz:			
$\eta_W = 26.7$ dB	466	90.0	15.0
$\eta_0 = 27.7$ dB	589	71.2	11.9
$\eta_B = 37.2$ dB	5248	8.0	1.3
Center frequency 6 Hz:			
$\eta_W = 26.0$ dB	401	104	17.3
$\eta_0 = 26.6$ dB	455	92.4	15.4
$\eta_B = 34.8$ dB	3009	14.0	2.3

which equals about one quarter of the earth's circumference. The results are shown in Table 7.6-1 under the heading "center frequency 2 Hz." The entry $\eta_W = 25.8$ dB means that the signal power or energy fed to the radiating antenna can be reduced by 25.8 dB to yield the same power at the receiving antenna in the worst case. This reduction of the required signal power or energy per bit is shown nonlogarithmic as the factor $E_b'/E_b = 375$ in the second column. The energy per bit, E_b, is obtained by dividing $E_b' = 42$ MWs/bit of Eq. (2) by 375. The average signal power P_s is finally found by dividing E_b by the signal duration of 6 s. Equivalent explanations apply to the rows η_0 and η_B.

There is a considerable difference between the worst-case limit η_W and the best-case limit η_B. A look at curve b in Fig. 7.4-1 shows that 2 Hz is close to the sharp drop of this curve for lower frequencies. We will expect that the worst-case limit and the best-case limit will be closer together if we use a higher center frequency. If we replace the pulse in Fig. 7.6-1a by one with 24 cycles in the time interval of 6-s duration we get the same energy spectrum as in Fig. 7.6-1b, but with the center frequency shifted to 4 Hz. The resulting values are shown in Table 7.6-1 under the heading "center frequency 4 Hz." One may see that η_W has increased by 0.9 dB from 25.8 to 26.7 dB, while η_0 has remained almost unchanged and η_B has decreased by 2.1 dB.

We may replace the pulse in Fig. 7.6-1a by one having 36 cycles in the interval of 6-s duration. The energy spectrum of Fig. 7.6.1b is then shifted to the center frequency 6 Hz. The resulting values are shown in Table 7.6-1 under the heading "center frequency 6 Hz." Here, η_W has dropped by 0.7 dB from the value for 4 Hz and η_B has dropped by 2.4 dB, which indicates that 6 Hz is definitely too high a center frequency if the ionosphere is anywhere close to being a conducting surface.

One may see from Table 7.6-1 that the power required at the input of the radiating antenna to produce a certain average power at the reception antenna can be reduced by the factor E_b/E_b' if one does not use a carrier with a frequency of 45 Hz. In other words, some two to three orders of magnitude less power are needed. Since η_0 is between the worst case η_W and the best case η_B, we will use η_0 and the approximations of Section 7.4 from here on.

To see what one can do with the saved signal power, let us consider the three teletype systems transmitting 33.3 bits/s with binary, quaternary or octonary digits according to Fig. 7.3-7 and 7.1-4. Representative energies and signal powers are given in Table 7.6-2. One must be careful to use these numbers as a guide only. For instance, the "highest frequency" in Table 7.6-2 is not used at all according to Fig. 7.3-7; the center frequency of the fastest-varying signals is 20, 10, or 5 Hz. Hence, the signal powers shown for "highest frequency" in Table 7.6-2 are an upper limit. The average signal power $\overline{P}_s$ in the last column is the average of the signal powers P_s for the highest and the lowest frequency. This is a gross approximation, but it is here more important to obtain an overview than to derive very accurate numbers.

TABLE 7.6-2

THREE TELETYPE SYSTEMS TRANSMITTING 33.3 bits/s TO A DEPTH d = 300 m COMPARED WITH A SANGUINE SYSTEM TRANSMITTING 0.167 bits/s (=10 bits/min) AND REQUIRING AN ENERGY E_b' = 42 MWs/bit OR A POWER P_s' = 7 MW

	$\eta_0 - \delta$ (dB)	E_b'/E_b	E_b (MWs/bit)	P_s (MW)	$\overline{P}_s$ (MW)
Binary signals, δ = 0 dB:					
Highest frequency f = 23.33 Hz	13.73	23.6	1.78	59.3	30.8
Lowest frequency f = 3.33 Hz	27.93	620.4	0.0677	2.26	
Quaternary signals, δ = 6.98 dB:					
Highest frequency f = 11.67 Hz	15.37	34.5	1.22	40.6	26.6
Lowest frequency f = 1.67 Hz	20.46	111.2	0.378	12.6	
Octonary signals, δ = 16.23 dB:					
Highest frequency f = 5.833 Hz	10.45	11.1	3.78	126	149
Lowest frequency f = 0.833 Hz	9.10	8.1	5.16	172	

TABLE 7.6-3

ENERGY PER BIT AND AVERAGE SIGNAL POWER REQUIRED BY TWO TELETYPE SYSTEMS TRANSMITTING 33.3 bits/s TO DEPTHS OF d = 300, 250, AND 200 m

	d = 300 m			d = 250 m			d = 200 m		
	E_b (MWs/bit)	P_s (MW)	$\overline{P}_s$ (MW)	E_b (MWs/bit)	P_s (MW)	$\overline{P}_s$ (MW)	E_b (MWs/bit)	P_s (MW)	$\overline{P}_s$ (MW)
Binary signals:									
Highest frequency f = 23.33 Hz	1.78	59.3	30.8	0.261	8.71	4.90	0.0383	1.28	0.905
Lowest frequency f = 3.33 Hz	0.0677	2.26		0.0328	1.09		0.0159	0.529	
Quaternary signals:									
Highest frequency f = 11.67 Hz	1.22	40.6	26.6	0.314	10.46	9.00	0.0809	2.69	3.60
Lowest frequency f = 1.67 Hz	0.378	12.6		0.226	7.55		0.136	4.52	

TABLE 7.6-4

ENERGY PER BIT AND AVERAGE SIGNAL POWER REQUIRED BY VARIOUS TELETYPE SYSTEMS TRANSMITTING AT DIFFERENT DATA RATES TO A DEPTH OF 300 m

	Data rate (bits/s)	Highest frequency (Hz)	E_b (MWs/bit)	P_s (MW)	Lowest frequency (Hz)	E_b (MWs/bit)	P_s (MW)	$\overline{P}_s$ (MW)
Binary signals	33.3	23.33	1.78	59.3	3.33	0.0677	2.26	30.8
	25	17.50	0.677	16.9	2.50	0.0672	1.68	9.3
	20	14.00	0.369	7.38	2.00	0.0705	1.41	4.4
	16.67	11.67	0.244	4.07	1.67	0.0757	1.26	2.7
Quaternary signals	33.3	11.67	1.22	40.6	1.67	0.378	12.59	26.6
	25	8.75	0.728	18.2	1.25	0.445	11.12	14.7
	20	7.00	0.541	10.81	1.00	0.524	10.49	10.6
	16.67	5.83	0.449	7.48	0.833	0.614	10.23	8.9

TABLE 7.6-5

ENERGY PER BIT AND AVERAGE SIGNAL POWER REQUIRED BY VARIOUS TELETYPE SYSTEMS TRANSMITTING AT DIFFERENT DATA RATES TO A DEPTH OF 400 m

	Data rate (bits/s)	Highest frequency (Hz)	E_b (MWs/bit)	P_s (MW)	Lowest frequency (Hz)	E_b (MWs/bit)	P_s (MW)	$\overline{P}_s$ (MW)
Binary signals	16.67	11.67	3.68	61.3	1.67	0.211	3.52	32.4
	11.11	7.78	1.13	12.6	1.111	0.224	2.48	7.5
	8.33	5.83	0.614	5.11	0.833	0.254	2.11	3.6
	6.67	4.67	0.426	2.84	0.667	0.292	1.95	2.4
Quaternary signals	16.67	5.83	3.06	51.0	0.833	1.268	21.14	36.1
	11.11	3.89	1.68	18.7	0.667	1.458	16.21	17.5
	8.33	2.92	1.29	10.8	0.417	2.167	18.05	14.4
	6.67	2.33	1.14	7.6	0.333	2.729	18.20	12.9

We see from Table 7.6-2 that teletype transmission to a depth of about 300 m is within our capabilities. The quaternary signals require the lowest signal power $\overline{P}_s = 26.6$ MW, but the difference to binary signals is small. According to Fig. 7.3-10 one might expect quaternary signals to do much better and octonary signals to do best, but Fig. 7.6-10 only refers to the absorption in the water, while Table 7.6-2 includes the efficiency of the radiating antenna, the attenuation in the Earth–ionosphere cavity, and the attenuation of the electric field strength at the air–sea boundary.

Let us pursue teletype transmission a little further. Table 7.6-3 shows the signal power required to reach a depth of 300, 250, and 200 m with binary and quaternary signals. Binary signals reach 250 m with an average antenna input power of $\overline{P}_s = 4.9$ MW, which is below the projected power of 7 MW for Seafarer.

Table 7.6-4 shows how the required average signal power $\overline{P}_s$ is reduced if the data rate of a teletype link is reduced. An available radiator input power of 7 MW will be sufficient to reach a depth of 300 m if the data rate is between 20 and 25 bits/s and binary signals are used. Quaternary signals require more power for all data rates except the one with 33.3 bits/s.

Table 7.6-5 shows required signal powers and various data rates to reach a depth of 400 m. A power of 7 MW will be sufficient for a data rate somewhat less than 11.11 bits/s. The quaternary signals are doing very poorly. This is due to two causes: (a) The low data rates work in favor of binary signals; (b) the operating frequency band for the quaternary signals is located at too low frequencies.

The numbers in Tables 7.6-2 to 7.6-5 applied to η_0 of Eq. (11). It may be seen from Table 7.6-1 that the best case η_B requires only one tenth the energy per bit required for case η_0. Hence, there remains a reasonable possibility that quaternary signals are better than they appear to be according to Tables 7.6-2–7.6-5. Later on we will see another reason for the inclusion of quaternary and octonary signals into this study, when it is shown that the available bandwidth for a link *from* a submarine *to* a surface station is only about 1 Hz.

7.7 Local and Distant Noise

We have seen in the preceding section that considerably less input power to the radiating antenna is needed to provide a certain field strength at the receiving antenna of the submarine if one foregoes the usual sinusoidal carrier. Before one can interpret this reduction of required power as an actual advantage, one must investigate where the dominant noise comes from.

If the dominant noise is produced in the vicinity of the reception antenna, a higher signal power at the antenna means a realizable improvement. Thermal noise produced in the water or in the reception antenna is an example of such *local noise*. Other examples are the flow noise or the propeller shaft discharges produced by a moving submarine.

A completely different situation is encountered if the dominant noise is produced in the atmosphere by electric discharges. This *distant noise* is attenuated by the seawater like a signal. The noise will benefit from the reduced attenuation at lower frequencies just like the signal, and the signal-to-noise ratio remains unchanged.

At the present the local noise is the dominant one.[1] We are thus assured that discarding the sinusoidal carrier will bring a substantial improvement. However, we have computed increases of the signal power by some 20–30 dB. Only if the local noise dominates the distant noise by that much can one actually realize the gain. This will be the case if the reception antenna is mounted directly on the submarine. A trailing antenna that is sufficiently far from the submarine to avoid all the local noise produced by the submarine will not permit realization of the gain since the noise produced by the submarine is almost eliminated in this case. Of course, a trailing antenna is undesirable since it limits the speed of the submarine to about one-quarter of its full speed, limits maneuverability, weakens the hull, etc.

To get some theoretical insight into noise in the band $3 \text{ Hz} < f < 30 \text{ Hz}$ produced by electric discharges in the atmosphere, let us start with the data published in the book by Galejs (1971). About 100 lightning flashes are received per second (p. 64); most of the energy is contained in the so-called return stroke, which is about 1 ms long (p. 41). A rectangular pulse of 1 ms duration and its frequency power spectrum are shown in Fig. 7.7-1a. The power spectrum is constant in the band $3 \text{ Hz} < f < 30 \text{ Hz}$ for all practical purposes.

Let us assume the amplitude of the pulse produced by the lightning stroke equals 100 times the largest signal amplitude. One may then clip at the value of the largest signal amplitude, and reduce the amplitude of the lightning pulse to 1% without affecting the signal. The amplitude spectrum of this clipped pulse in the band $4 \text{ Hz} < f < 22 \text{ Hz}$ is shown by the solid line of Fig. 7.7-1b.

Let the pulse of Fig. 7.7-1a propagate 322 m through seawater. Equation (7.2-2) yields an absorption loss of 40 dB at 13 Hz for this case. The

[1] Practical experience with the systems Sanguine and Seafarer has shown that the noise limiting the performance is local noise, as proved by the cooling used for the input circuits. Such cooling would, of course, have no effect on distant noise.

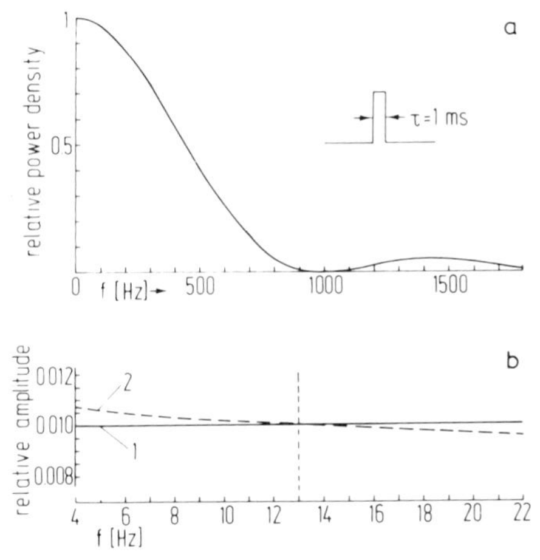

FIG. 7.7-1. (a) Rectangular pulse and its frequency power spectrum $(\sin \pi\tau f)^2/(\pi\tau f)^2$ in the band $0 \leqq f \leqq 1800$ Hz. (b) The amplitude spectrum in the band 4 Hz $\leqq f \leqq$ 22 Hz is shown for two cases: (1) The amplitude was clipped at 1% of its original value. (2) The pulse was attenuated by absorption in seawater, yielding an attenuation of 40 dB at 13 Hz, corresponding to a depth of 322 m.

amplitude diagram is represented by the dashed line in Fig. 7.7-1b. The difference between amplitude clipping and attenuation by absorption is insignificant, and this will hold even more so for depths of less than 322 m.

If 100 pulses with a duration of 1 ms are received randomly per second we have 9 ms of undisturbed signal for every 1 ms of disturbance. The power of the disturbance cannot exceed the peak signal power if clipping is used. Hence, about 90% of the received energy is due to the undisturbed signal and 10% is due to the disturbance, implying a signal-to-noise energy or power ratio of about 10 dB. Let us observe that pulses according to Fig. 7.7-1a having a duration of 1 ms can theoretically be eliminated completely from signals of duration T_s = 300 ms and more—according to Fig. 7.3-7—that consist of superpositions of sine and cosine pulses transmitted in parallel. It is not even necessary that the disturbing pulses have the form shown in Fig. 7.7-1a as long as they vary much faster than the sine and cosine pulses of the signal. The advantages of slowly varying signals transmitted in parallel for combating pulse-type disturbances and the use of amplitude clipping have been investigated in more detail in the literature (Harmuth, 1972, pp. 334–340). The conclusion here is that disturbing pulses of 1 ms duration arriving at a rate of 100 pulses/s pose theoretically no problem to the reception of signals of a duration of 300 ms or more consisting of a superposition of about 10 sine and cosine

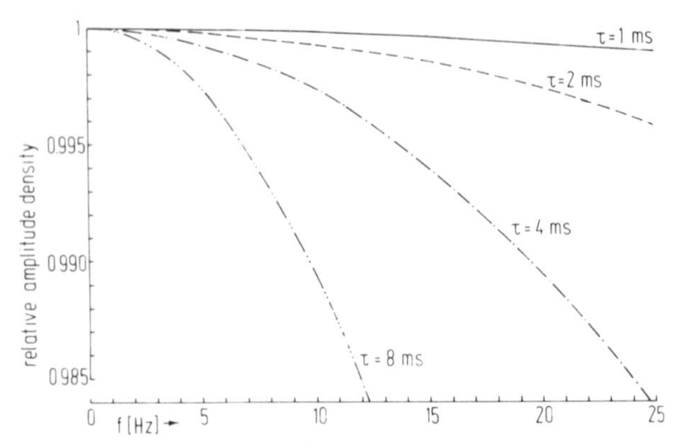

FIG. 7.7-2. Amplitude spectra $(\sin \pi\tau f)/\pi\tau f$ of rectangular pulses with duration τ for τ = 1, 2, 4, 8 ms.

pulses according to Fig. 7.3-7. The practical problems are not too great either since pulses of a duration of 1 ms and more are sufficiently slow to be handled by clipping circuits as well as by more sophisticated circuits.

The signals according to Fig. 7.3-7 occupy the frequency band below 23.33 Hz, but the individual sine and cosine pulses occupy only sections of this band. We have acted as if they all were affected equally by short disturbing pulses. This will be so only if the power or amplitude spectrum of the disturbing pulses is essentially constant in that band. The amplitude spectra of rectangular pulses according to Fig. 7.7-1a, but with various durations τ, are plotted in Fig. 7.7-2 for the band $0 \leqq f \leqq 25$ Hz. The curve for $\tau = 4$ ms drops from 1 at $f = 0$ to 0.985 at $f = 24$ Hz, which is a drop of 1.5%. Hence, pulses with a duration of less than 4 ms have a practically constant amplitude spectrum in this band. Since powerful pulses produced by lightning are almost always shorter than 4 ms we do not need to carry the investigation any further at this time.

7.8 Radiation from a Submarine

The wave impedance Z_w of seawater is given by the formula

$$Z_w = [i\omega\mu_0/(i\omega\epsilon + \sigma)]^{1/2} \doteqdot 0.0014\sqrt{f}\sqrt{i}, \quad \omega\epsilon \ll \sigma \tag{1}$$

It is general practice to use the wave impedance of empty space, $Z_0 = (\mu_0/\epsilon_0)^{1/2} \doteqdot 377\ \Omega$ for air. We will follow this practice for the moment, but we will show later on that completely different results are obtained by using a wave impedance that allows for the ionization of the air.

Let an electromagnetic wave propagate vertically to the surface of the

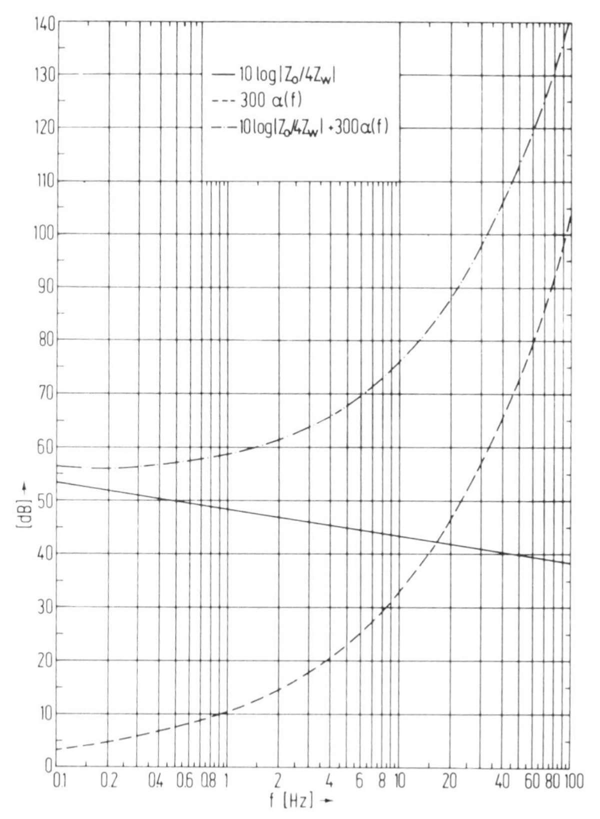

FIG. 7.8-1. Attenuation of an electromagnetic wave due to the reflection at the water–air boundary (solid line) and due to the absorption in a layer of water 300 m thick (dashed line) as function of the frequency f. The dash–dot line shows the sum of the attenuations due to reflection and absorption.

sea from the water to the air. The electric field strength E_w in the water becomes the field strength E_a in the air:

$$E_a = 2E_w Z_0/(Z_w + Z_0) \doteqdot 2E_w \tag{2}$$

The electric field strength just above the surface of the sea is twice the field strength just below the surface. The magnetic field strength H_w in the water, on the other hand, is reduced by orders of magnitude to the field strength H_a in the air:

$$H_a = 2H_w Z_w/(Z_w + Z_0) \doteqdot 7.45 \times 10^{-5}\sqrt{f}\sqrt{i} \tag{3}$$

The ratio[1] of the power of the wave in the air to that in the water follows from Eqs. (2) and (3),

$$|E_a H_a|/E_w H_w \doteqdot 4|Z_w|/Z_0 \doteqdot 1.49 \times 10^{-5}\sqrt{f} \tag{4}$$

where the units of f are hertz.

The attenuation $-10 \log(4|Z_w|/Z_0)$ is plotted as a function of f in Fig. 7.8-1. Its value is about 48 dB at 1 Hz and it drops to about 44 dB at 20 Hz. The attenuation $300\alpha(f)$ due to absorption by seawater is also plotted in Fig. 7.8-1 for a depth of 300 m according to Eq. (7.2-2). Furthermore, the sum of these two attenuations is plotted.[2]

At the typical frequencies of 40–80 Hz considered for Project Sanguine one obtains attenuations of 105–130 dB from Fig. 7.8-1; the attenuation due to geometrical spreading in the water is not yet included in these figures. The attenuation of 58 dB at 1 Hz is better but still too high, since the attenuated signal has to compete with the static discharges of lightning that have passed neither through the water nor through the water–air boundary. We have seen in the previous section that the most powerful lightning discharges can be suppressed by amplitude clipping but neither the clipping nor other processing methods at the receiver will suffice to permit communications from a submarine to a surface station unless we either radiate a very powerful electromagnetic wave from the submarine or find a way around the attenuations shown in Fig. 7.8-1.

Since the wavelength in water is substantially shorter than in the air, it has been suggested that efficient and powerful radiation is possible by using a resonant dipole as radiator. Let us investigate this suggestion. The wavelength λ in water follows from Eq. (7.2-3),

$$\lambda = v(f)/f = (4\pi/f\sigma\mu_0)^{1/2} \doteqdot 1581/\sqrt{f} \quad [\mathrm{m}] \tag{5}$$

where the units of f are hertz.

Figure 7.8-2 shows the wavelength λ and the magnitude $|Z_w|$ of the wave impedance of water as function of the frequency in the range 0.1 to 100 Hz. At 15 Hz the wavelength equals about 300 m. This is a reasonable number for a resonant dipole of length $\lambda/2$ or λ. A submarine at a

---

[1] A wave with vertical incidence propagating from the air into the water suffers the same attenuation, but the roles of the electric and magnetic field strengths are reversed. One obtains in this case $H_w = 2H_a$ and $E_w = 2E_a Z_w/Z_0$. Both of these formulas differ by a factor 2 from Rowe's formulas (34) and (16), but Rowe (1974) calculates the transition of a wave in the Earth–ionosphere duct into the water rather than a wave with vertical incidence. Hence, there is no contradiction.

[2] In reality, conditions are worse than suggested by these plots. If the incidence of the wavefront on the surface of the sea is not almost exactly vertical, the wave will be totally reflected.

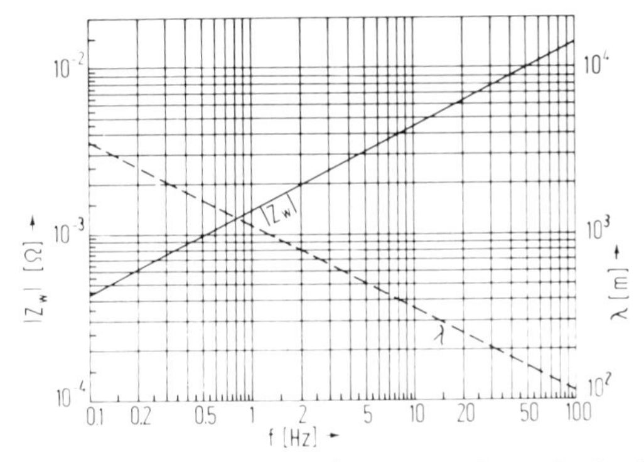

FIG. 7.8-2. Wavelength λ of electromagnetic waves and magnitude of the wave impedance Z_w in seawater as function of the frequency f.

depth of 300 m has a distance of one wavelength from the water–air boundary. The attenuation due to absorption equals 40 dB for 300 m according to Fig. 7.8-1 and 80 dB for a wave that has propagated to the surface and returned. Hence, one will be reasonably confident that this boundary has no influence on the radiation efficiency. At 15 Hz we have a reflection loss of 42 dB according to Fig. 7.8-1 or a total loss of 82 dB. This value is too high to be practically acceptable.

Let us try a lower frequency. The wavelength becomes 1000 m at about 1.3 Hz, which makes the use of a λ/2 resonant dipole difficult but still possible. The submarine is in this case only the distance λ/3 from the surface. However, the attenuation due to absorption equals 12 dB for 300 m and 24 dB for 600 m. Hence, the reaction of the surface on the efficiency of radiation should still be small.

An alternative to a towed cable used as a resonant dipole is to use the submarine itself as radiator. This is shown in Fig. 7.8-3. The power source PS drives a current through a cable to an electrode at the bow of the submarine that flows through the water to the stern, and returns to the power source via a second electrode and a cable. The outside of the submarine

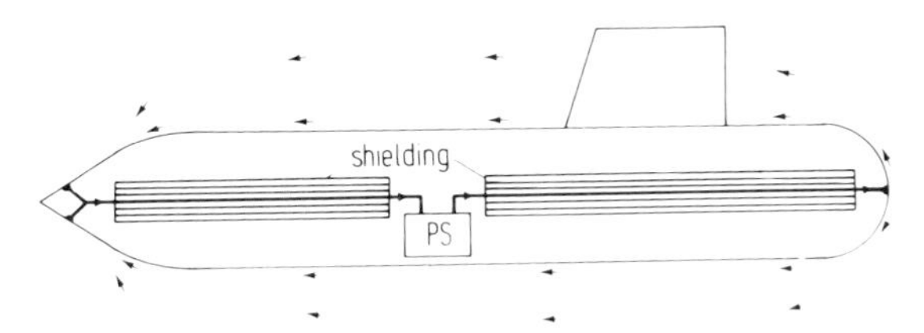

FIG. 7.8-3. Use of a submarine as an electrically short radiator. L/λ equals about 1/6 for 1 Hz and a length L = 170 m of the submarine. PS, power source.

must be insulated with paint to make the current flow through the water rather than through the hull of the submarine; theoretically, one could use the hull instead of the cable to carry the current, if the power source PS is placed close to one of the two electrodes. This arrangement is mechanically better than a towed resonant dipole but the large antenna current[1] must flow through the power source PS, whereas it would flow only in the antenna in the case of a resonnant dipole.

Let us investigate the frequency band around 1 Hz. It is below the frequencies of importance for Schumann resonances. The usual theory of communications would require about 100 cycles of a wave with a frequency of 1 Hz to transmit one digit, or a time of 100 s/digit. This is unacceptably slow. However, our method of using nonperiodic waves does away with the requirement of a narrow relative bandwidth. An absolute bandwidth of 1 Hz permits a transmission rate of up to 2 digits/s according to the Nyquist limit. We can use bands like 1 Hz $\leqq f \leqq$ 2 Hz or 0.5 Hz $\leqq f \leqq$ 1.5 Hz to transmit 2 digits/s.

Let us return to Fig. 7.8-1. The absorption loss at 1 Hz is about 10 dB. This is an acceptable value. The reflection loss of 48 dB is the problem. It appears that it is possible to eliminate the reflection loss, while no one has yet found a way to reduce the absorption loss by any other means than a reduction of the frequency. Figure 7.8-1 shows that reflection and absorption losses are equal at about 17 Hz. Thus the frequency band below 17 Hz will be of primary interest. Since the absorption losses drop with the frequency to zero, one would like to operate at the lowest possible frequency; this lowest possible frequency is determined by the desired transmission rate of information. For simplicity, we will assume that we operate in a band 1-Hz wide centered at 1 Hz.

We have so far assumed that the wave impedance of air equals that of empty space. A more realistic model of the air takes the free electrons into account that are produced by ionization due to a variety of causes. The propagation factor k and the wave impedance Z are generally defined as follows:

$$k = [-i\omega\mu(i\omega\epsilon + \sigma)]^{1/2}, \qquad Z = \omega\mu/k \tag{6}$$

We use for μ the free-space value $\mu_0 = 4\pi \times 10^{-7}$ H/m. For ϵ and σ we use values derived by Huxley (1937, 1938, 1940; Wagner, 1953). For simplicity, the magnetic field of the Earth is ignored,

[1] Preliminary calculations indicate that currents in the order of 10–100 kA will be needed, even though the required power will not be high compared with the engine power of a submarine. Currents of this order are routine in large electric power stations and in the chemical industry, but not in electrical communications.

$$\epsilon = \epsilon_0 - \frac{2Ne^2}{3m(\nu^2 + \omega^2)}\left(\frac{1}{2} + \frac{\omega^2}{\nu^2 + \omega^2}\right) \tag{7}$$

$$\sigma = \frac{2Ne^2}{3m}\frac{\nu}{\nu^2 + \omega^2}\left(1 + \frac{\omega^2}{\nu^2 + \omega^2}\right) \tag{8}$$

where $\epsilon_0 = 8.854 \times 10^{-12}$ F/m; N is the electron density $[\mathrm{m}^{-3}]$; $e = 1.60 \times 10^{-19}$ [As] is the charge of the electron; $m = 9.11 \times 10^{-31}$ [kg] is the mass of the electron; $\nu = 1/\tau$ is the collision frequency of an electron $[\mathrm{sec}^{-1}]$; τ is the average time between collisions [sec]; and $\omega = 2\pi f$.

For $\omega \gg \nu$ one obtains from Eqs. (7) and (8) the following simplified formulas that are generally used for the study of the influence of the ionosphere on electromagnetic waves:

$$\epsilon = \epsilon_0 - Ne^2/m\omega^2, \qquad \sigma = 4Ne^2\nu/3m\omega^2, \qquad \omega \gg \nu \tag{9}$$

We cannot use these approximations since we are interested in frequencies of about 1 Hz. The relation $\nu \ll \omega = 6.28$ holds for altitudes of 300 km and more, but one will suspect that a wave with a frequency of 1 Hz will not penetrate the ionosphere that far. For an altitude of 200 km one has $\tau = 7 \times 10^{-2}$ s or $\nu \doteqdot 15\ \mathrm{s}^{-1}$, and for 150 km $\tau = 3 \times 10^{-3}$ s or $\nu \doteqdot 333\ \mathrm{s}^{-1}$. For altitudes up to somewhat more than 150 km we may thus assume $\nu \gg \omega$ and obtain the following simplifications of Eqs. (7) and (8):

$$\epsilon = \epsilon_0 - Ne^2/3m\nu^2, \qquad \sigma = 2Ne^2/3m\nu, \qquad \nu \gg \omega \tag{10}$$

Note the very different form of Eq. (10) from the usual Eq. (9).

Introduction of Eq. (10) into Eq. (6) yields the following formulas for k and Z:

$$k = \{-i\omega\mu_0[i\omega(\epsilon_0 - Ne^2/3m\nu^2) + 2Ne^2/3m\nu]\}^{1/2} \tag{11}$$

$$Z = (i\omega\mu_0)^{1/2}[i\omega(\epsilon_0 - Ne^2/3m\nu^2) + 2Ne^2/3m\nu]^{-1/2} \tag{12}$$

For water we have the following values k_{w} and Z_w,

$$\begin{aligned} k_{\mathrm{w}} &= [-i\omega\mu_{\mathrm{w}}(i\omega\epsilon_{\mathrm{w}} + \sigma)]^{1/2} \doteqdot (-i\omega\mu_0/\sigma)^{1/2} \\ &\doteqdot 5.620 \times 10^{-3}\sqrt{f}\sqrt{-i} \quad [\mathrm{m}^{-1}] \end{aligned} \tag{13}$$

$$Z_{\mathrm{w}} = \omega\mu_{\mathrm{w}}/k_{\mathrm{w}} \doteqdot (\mathrm{i}\omega\mu_0/\sigma)^{1/2} = 1.405 \times 10^{-3}\sqrt{f}\sqrt{i} \quad [\mathrm{V/A}]$$

where the units of f are hertz.

Figure 7.8-4a shows Z_{w}, Z_0, and Z with ϵ inserted from Eq. (9) for $\sigma \ll \omega|\epsilon|$. In order to have an efficient transfer of electromagnetic power from water to air we must not have a phase difference between the wave impedances. Both Z_0 and Z in Fig. 7.8-4a have a phase difference of $\pi/4$

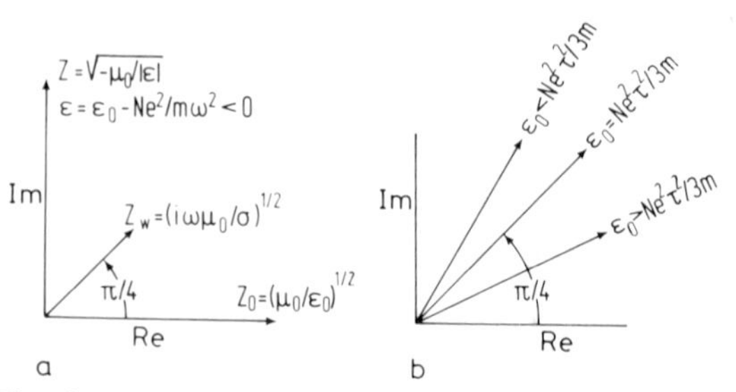

FIG. 7.8-4. (a) The phases of the wave impedance Z_0 for vacuum, Z_w for water, and Z for a wave in ionized air below the critical frequency. (b) Phases of Z in ionized air for very low frequencies $\omega \ll \nu$.

relative to Z_w. The question is whether it is possible to have a wave impedance Z in ionized air with the phase angle $\pi/4$.

Consider the wave impedance Z of Eq. (12). For $\epsilon_0 = Ne^2/3m\nu^2$ it has the phase $\pi/4$ just like Z_w for water; the magnitudes of Z and Z_w are very different. For smaller values of ϵ_0 the phase angle becomes larger than $\pi/4$, while it decreases toward zero for larger values of ϵ_0. Hence, the condition

$$Z_w/Z = n = \text{real} \tag{14}$$

is satisfied for the following value of the electron density N per cubic meter,

$$N = \frac{3m\nu^2\epsilon_0}{e^2} = \frac{3m\epsilon_0}{e^2\tau^2} = 9.45 \times 10^{-4}\tau^{-2} \quad [\text{m}^{-3}] \tag{15}$$

where the units of τ are seconds.

The conductivity σ is obtained for this case by inserting Eq. (15) into Eq. (8):

$$\sigma = 2\epsilon_0/\tau = 1.77 \times 10^{-11}\tau^{-1} \quad [\text{A/Vm}] \tag{16}$$

Let us see at which altitude Eq. (15) may be satisfied. The average time τ between collisions at various altitudes is found in books on atmospheric physics (Siedentopf, 1950). It is shown in the second column of Table 7.8-1. The third column lists the required electron density according to Eq. (15), while the observed density for the E- and F_1-layer are shown in the fourth column. One may see that somewhere in the lower reaches of the F_1-layer the observed value of N equals the required value of N. Unfortunately, the approximation of Eq. (10) loses its validity at about the

TABLE 7.8-1

ALTITUDE AT WHICH THE CONDITION $Z/Z_w = n =$ REAL IS SATISFIED. THE OBSERVED ELECTRON DENSITY N MUST EQUAL THE REQUIRED DENSITY $N = 3m\epsilon_0/e^2\tau^2$

Altitude h (km)	Average time between collisions τ (s)	Required electron density $3m\epsilon_0/e^2\tau^2$ (m^{-3})	Observed electron density N (m^{-3})	Name of ionized layer	Conductivity $\sigma = 2\epsilon_0/\tau$ (A/Vm)	Wave impedance Z(V/A)	Refraction index $n = Z_w/Z$	Brewster angle φ_B	Critical angle φ_C
0	1.2×10^{-10}	6.56×10^{16}							
10	5×10^{-10}	3.78×10^{15}							
20	2×10^{-9}	2.36×10^{14}							
40	4×10^{-8}	5.91×10^{11}							
60	4×10^{-7}	5.91×10^{9}		D					
80	1×10^{-6}	9.45×10^{8}							
100	8×10^{-5}	1.47×10^{5}	10–25	E					
120	4×10^{-4}	5910							
150	3×10^{-3}	105			5.9×10^{-9}	$36.6(if)^{1/2}$	3.84×10^{-5}	7.9″	7.9″
200	7×10^{-2}	19	20–30	F_1	2.49×10^{-8}	$17.8(if)^{1/2}$	7.89×10^{-5}	16.3″	16.3″

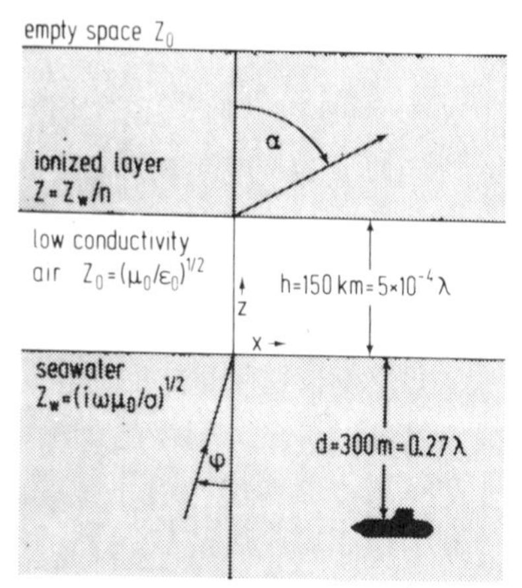

FIG. 7.8-5. Radiation from a submarine to an ionized layer with a very thin layer of low-conductivity air in between.

same altitude. The more correct formulas of Eqs. (7) and (8) permit a correct but lengthier calculation which we will forego here.

The F-layer exists during day and night, and any radio transmission depending on it should work around the clock. The E-layer depends strongly on the elevation of the sun, but Table 7.8-1 shows that its electron density is much too low to have an effect on our problem.

Table 7.8-1 lists for the altitudes of 150 and 200 km the conductivity σ according to Eq. (16), the wave impedance Z according to Eq. (12), and the refraction index n referred to seawater according to Eq. (14).

Let us now see how the F-layer can help to get an electromagnetic wave out of the water. Figure 7.8-5 shows a submarine at a depth of $d = 300$ m, which corresponds to about 0.27λ at a frequency of 1 Hz according to Fig. 7.8-2. We had observed before that the attenuation by absorption for a round-trip submarine-surface-submarine equals 20 dB at 1 Hz. Hence, the surface appears far away electrically. The F-layer is shown in Fig. 7.8-5 at a rather low altitude of 150 km but this value yields the convenient distance $5 \times 10^{-4}\lambda$ in terms of wavelength. This is such a short distance that we can ignore in first approximation the air of low conductivity below the F-layer, and this would still be true for the F_2-layer at an altitude of 300 km $= 10^{-3}\lambda$.

Let a wavefront from the submarine hit the surface of the sea with vertical incidence, which means an angle of incidence $\varphi = 0$. The electric and magnetic field strengths produced in the F-layer have the following value:

$$E_F = 2E_w(1 + Z_w/Z)^{-1} = 2E_w$$
$$H_F = 2H_w(1 + Z/Z_w)^{-1} = 2H_w Z_w/Z \qquad (17)$$

The ratio

$$E_F H_F/E_w H_w \doteqdot 4Z_w/Z \doteqdot 1.54 \times 10^{-4} \qquad (18)$$

follows for $h = 150$ km from the eighth column of Table 7.8-1. It translates into a reflection loss of 38 dB, which is not much better than the loss of 48 dB at $f = 1$ Hz in Fig. 7.8-1. However, since the ratio $Z_w/Z = n$ is now real, we have the phenomenon of the Brewster angle φ_B. A plane wave polarized in the plane of incidence will pass without reflection from the water to the F-layer if the angle of incidence φ in Fig. 7.8-5 equals the Brewster angle φ_B. The value of φ_B is given by the following relation:

$$\cos \varphi_B = 1/(1 + n^2)^{1/2} \qquad (19)$$

Note that n can be larger or smaller than 1, as long as it is real. Since n is very small according to Table 7.8-1, we approximate Eq. (19) by a series and obtain the following value for φ_B:

$$\varphi_B = n \quad [\text{rad}] \quad \text{or} \quad \varphi_B = 180n/\pi \quad [\text{deg}] \qquad (20)$$

The values 7.9'' and 16.3'' for $h = 150$ and 200 km are obtained and shown in Table 7.8-1. Hence, if the angle of incidence φ in Fig. 7.8-5 is about 10 seconds of arc, a properly polarized wave will get out of the water without attenuation by reflection. The attenuation due to absorption remains, but it amounts to only 10 dB for $f = 1$ Hz and $d = 300$ m according to Fig. 7.8-1. Furthermore, the loss due to geometric spreading remains. This is an exceedingly difficult problem to solve theoretically since the wavefronts produced by the radiating submarine are not nearly planar when they reach the surface of the sea, and the surface of the sea is not nearly planar due to the sea waves.

It may be much easier to obtain data on the geometric spreading experimentally. Any ship produces fairly strong currents due to the electrolytic action of the seawater. A ship's propeller rotating at 60 rpm should produce a periodic wave with a period of 1 s. This wave is, of course, not sinusoidal but only periodic. We have discussed selective receivers for general periodic waves in Section 4. Such receivers could be used to measure the electromagnetic field produced, and thus solve the geometric spreading problem experimentally. One could also use a receiver for sinusoidal waves with a period of 1 s but the signal-to-noise ratio would be much worse—since only one Fourier component of the signal would be used—and the radiation might be obscured by noise.

Let us return to Fig. 7.8-5. The angle α follows from Snell's law:

$$\sin \varphi / \sin \alpha = n \tag{21}$$

The largest possible value of α is $\pi/2$. From this follows the critical angle φ_C:

$$\sin \varphi_C \doteqdot \varphi_C = n \quad [\text{rad}] \tag{22}$$

A comparison with Eq. (20) shows that the Brewster angle and the critical angle have practically the same value. This means that essentially all the energy of the wave comes almost vertically out of the seawater but it flows parallel to the surface of the Earth when it reaches the F-layer.

The F-layer helps us to get the wave out of the water. What happens then? First, the wave can propagate in the F-layer. Its long wavelength makes the absorption losses quite small. According to Table 7.8-1 the conductivity σ is of the order of 10^{-8} A/Vm in the important section of the F-layer, while σ equals about 4 A/Vm in seawater. Furthermore, the wave can enter the thin layer of low conductivity air in Fig. 7.8-5 from the F-layer more readily than from the sea. The reason is that the magnitude of the wave impedance Z in Table 7.8-1 differs only by about a factor 10 from the wave impedance $Z_0 = 377\ \Omega$ of empty space or low conductivity air, while the magnitude of the impedance Z_w of seawater differs by a factor 10^5 for $f = 1$ Hz according to Eq. (13). There is little point carrying this qualitative argumentation any further. For reliable results one must solve Maxwell's equation for the four layers with impedances Z_w, Z_0, Z_w/n, and Z_0 of Fig. 7.8-5. The layers "seawater" and "empty space" are thick compared with the wavelengths in these media, while the layers "low-conductivity air" and "ionized layer" are very thin. The problem of one thin layer between two thick layers, all having wave impedances with real value, was investigated successfully and in much detail by Schaefer and Gross (1910; Schaefer, 1950); this work can be used as a starting point for the more general problem encountered here.

References

Ahmed, N., and Rao, K. (1974). "Orthogonal Transforms for Digital Signal Processing." Springer-Verlag, Berlin and New York.

Ahmed, N., Schreiber, H., and Lopresti, P. V. (1973). On notation and definition of terms related to a class of complete orthogonal functions. *IEEE Trans. Electromagn. Compat.* **15,** 75–80.

Altschuter, E. E., Falcone, V. J., and Wulfsberg, K. N. (1968). Atmospheric effects on propagation at millimeter wavelengths. *IEEE Spectrum,* July, 83–90.

Anacker, W. (1979). Computing at 4 degrees Kelvin. *IEEE Spectrum,* May, 26–37.

Annan, A. P., and Davis, J. L. (1976). Impulse radar sounding in permafrost. *Radio Sci.* **II,** No. 4, 383–394.

Barker, R. H. (1953). Group synchronizing of binary digital systems. *In* "Communication Theory" (W. Jackson, ed.). Academic Press, New York.

Barrows, M. L. (1974). Surface impedance and the efficiency of horizontal-dipole extremely low frequency (ELF) antenna arrays. *IEEE Trans. Commun.* **22,** 399–401.

Barton, D. K. (1964). "Radar Systems Analysis." Prentice Hall, Englewood Cliffs, New Jersey.

Barton, D. K., ed. (1975). "Radars—Pulse Compression," Vol. 3 (collection of 17 papers). Artech House, Dedham, Massachusetts.

Barton, D. K. (1976). Low angle tracking. *Microwave J.* **19,** No. 12, 19–24.

Beauchamp, K. G. (1975). "Walsh Functions and Their Applications." Academic Press, New York.

Becker, R., and Sauter, F. (1964). "Theorie der Elektrizität," 18th ed., Vol. 1. Teubner, Stuttgart.

Bennet, C. L., and Ross, G. F. (1978). Time-domain electromagnetics and its applications. *Proc. IEEE* **66,** 299–318.

Bertram, C. L., Campbell, K. J., and Sandler, S. S. (1972). Locating large masses of ground ice with an impulse radar system. *Proc. 8th Int. Symp. Remote Sens.,* pp. 241–260. University of Michigan, Ann Arbor.

Blake, L. V. (1970). Prediction of radar range. *In* "Radar Handbook" (M. I. Skolnik, ed.). McGraw-Hill, New York.

Bljumin, S. L., and Shtirin, A. M. (1974). On sequency ordered Walsh functions and their generalization. (In Russian.) *Probl. Peredachi Inf.* **10,** No. 3, 116–117.

Bljumin, S. L., and Shtirin, A. M. (1977). On the generation of Walsh functions and their generalization. (In Russian.) *Matematika* **2,** 6–8.

Bljumin, S. L., and Trachtman, A. M. (1977). Discrete Hilbert transformation in finite intervals (in Russian). *Radiotekh. Elektron.* **22,** No. 7, 1390–1398.

Blyth, J. D. M., ed. (1953). Ice islands in the Arctic Ocean. *Polar Rec.* **6,** No. 45, 684–687.

Booker, H. G. (1946). Slot aerials and their relation to complementary wire aerials (Babinet's principle). *J. Inst. Electr. Eng., Part 3A* **93,** No. 4, 620–626.

Born, M. (1962). "Einstein's Theory of Relativity." Dover, New York.

Born, M. (1964). "Die Relativitätstheorie Einsteins," 4th ed. Springer-Verlag, Berlin and New York.

Burrows, M. L. (1974). Surface impedance and the efficiency of horizontal-dipole extremely low frequency (ELF) antenna arrays. *IEEE. Trans. Commun.* **22,** 339–401.

Campbell, K. J., and Orange, K. J. (1974). A continuous profile of sea ice and freshwater ice thickness by impulse radar. *Polar Rec.* **17,** 31–41.

Caspers, J. W. (1970). Bistatic and multistatic radar. *In* "Radar Handbook" (M. I. Skolnik, ed.). McGraw-Hill, New York.

CCIR (1975). "XIIIth Plenary Assembly," Vol. 1: "Spectrum Utilization and Monitoring" (Study Group I). International Telecommunications Union, Geneva.

Chang, D. C., and Wait, J. R. (1974). Extremely low frequency (ELF) propagation along a horizontal wire located above or buried in the Earth. *IEEE Trans. Commun.* **22,** 421–427.

Chapman, J. C. (1976). Experimental results with a Walsh wave radiator. *IEEE Natl. Telecom. Conf. Symp. Rec.* pp. 44.2-1–44.2-3.

Chen, C. C. (1975). "Attenuation of Electromagnetic Waves by Haze, Fog, Clouds, and Rain, *Rep. R-1694-PR.* Rand Corp., Santa Monica California.

Chow, W. F. (1964). "Principles of Tunnel Diode Circuits." Wiley, New York.

Close, C. M. (1966). "The Analysis of Linear Circuits." McGraw-Hill, New York.

Cook, C. E. (1960). Pulse compression, key to more efficient radar transmission. *Proc. IRE* **48,** 310–316.

Cook, C. E., and Bernfeld, M. (1967). "Radar Signals." Academic Press, New York.

Cook, J. C. (1960). Proposed monocycle-pulse very high frequency radar for airborne ice and snow measurement. *Trans. AIEE Commun. Electron.* **79,** 588–594.

Cook J. C. (1970). An airborne, ground penetrating radar. *In* "Conference on Electromagnetic Exploration of the Moon" (W. I. Linlor, ed.). Mono Book, Baltimore, Maryland.

Cook, J. C. (1973). Radar exploration through rock in advance of mining. *Trans. Soc. Min. Eng. AIME* **254,** 140–146.

Cook, J. C. (1975). Radar transparencies of mine and tunnel rocks. *Geophysics* **40,** 865–885.

Cuccia, C. L. (1952). "Harmonics, Sidebands, and Transients in Communication Engineering." McGraw-Hill, New York.

Cutrona, L. J. (1970). Synthetic aperture radar. *In* "Radar Handbook" (M. I. Skolnik, ed.). McGraw-Hill, New York.

Cutrona, L. J., and Hall, G. O. (1962). A comparison of techniques for achieving fine azimuth resolution. *IEEE Trans. Mil. Electron.* **6,** 119–121.

Davidson, D., Macklin, D. N., and Vozoff, K. (1974). Resistivity surveying as an aid in Sanguine site selection. *IEEE Trans. Commun.* **22,** 389–393.

Davis, J. L., Scott, W. J., Morey, R. M., and Annan, A. P. (1976). Impulse radar experiments on permafrost near Tuktoyaktut, Northwest Territories. *Can. J. Earth Sci.* **13,** 1584–1590.

Davis, J. R., Baker, D. J., Shelton, J. P., and Ament, W. S. (1979). Some physical constraints on the use of "carrier-free" waveforms in radio-wave transmission systems. *Proc. IEEE* **67,** 884–890.

Deirmendjian, D. (1975). Far infrared and submillimeter wave attenuation by clouds and rain. *J. Appl. Meteorol.* **14,** 1584–1593.

Deley, G. W. (1970). Waveform design. *In* "Radar Handbook" (M. I. Skolnik, ed.). McGraw-Hill, New York.

Djadjunov, N. G., and Senin, A. I. (1977). "Orthogonal and Quasiorthogonal Signals." Svjas, Moscow (in Russian).

Duckworth, K. (1970). Electromagnetic depth sounding applied to mining problems. *Geophysics* **35,** 1086–1098.

Durant, W. (1939). "The Story of Civilization," Vol. II: "The Life of Greece." Simon & Schuster, New York.

Edelman, P. E. (1920). "Experimental Wireless Stations." N. W. Henley, New York.

Erickson, C. W. (1961). "Clutter Cancellation in Auto Correlation Functions by Binary Sequence Pairing," AD 446 146. Natl. Tech. Inf. Serv., Springfield, Virginia.

Falcone, V. J., and Abreu, L. W. (1979). Atmospheric attenuation of millimeter and submillimeter waves. *EASCON 79 Rec.* pp. 36–41.

Farnett, E. C., Howard, T. B., and Stevens, G. H. (1970). Pulse compression radar. *In* "Radar Handbook" (M. I. Skolnik, ed.). McGraw-Hill, New York.

Fessenden, C. T., and Cheng, D. H. S. (1974). Development of a trailing wire E-field submarine antenna for extremely low frequency (ELF) reception. *IEEE Trans. Commun.* **22,** 428–437.

Foner, S., and Schwartz, B. B., eds. (1974). "Superconducting Machines and Devices" (NATO Adv. Study Inst. Ser.) Plenum, New York.

Fritzsche, G. (1977). "Informationsübertragung." VEB Technik, Berlin.

Galejs, J. (1971). On the propagation of long waves to large distances. *Proc. IEEE* **59,** 1635–1636.

Golay, M. J. (1961). Complementary series. *IRE Trans. Inf. Theory* **7,** 82–87.

Graeme, J. G. (1973). "Applications of Operational Amplifiers—Third Generation Techniques." McGraw-Hill, New York.

Graeme, J. G., Tobey, G. E., and Huelsman, L. P. (1971). "Operational Amplifiers." McGraw-Hill, New York.

Haggarty, R. D., Hart, L. A., and O'Leary, G. C. (1975). A 10 000 to 1 pulse compression filter using a tapped delay line linear filter synthesis technique. *EASCON 1968 Rec.* pp. 216–221; reprinted *in* "Radars" (D. K. Barton, ed.), Vol. 3, Artech House, Dedham, Massachusetts.

Harger, R. O. (1970). "Synthetic Aperture Radar Systems: Theory and Design." Academic Press, New York.

Harmuth, H. F. (1960a). Radio communication with orthogonal time functions. *Trans. AIEE Commun. Electron.* **79,** 221–228.

Harmuth, H. F. (1960b). On the transmission of information by orthogonal time functions. *Trans. AIEE Commun. Electron.* **79,** 248–255.

Harmuth, H. F. (1969). "Transmission of Information by Orthogonal Functions." Springer-Verlag, Berlin and New York (Russian transl., Svas, Moscow, 1975).

Harmuth, H. F. (1972). "Transmission of Information by Orthogonal Functions," 2nd ed. Springer-Verlag, Berlin and New York.

Harmuth, H. F. (1977a). "Sequency Theory: Foundations and Applications." Academic Press, New York (Russian transl., Mir, Moscow, 1980).

Harmuth, H. F. (1977b). Selective reception of periodic electromagnetic waves with general time variation. *IEEE Trans. Electromagn. Compat.* **19,** 137–144.

Harmuth, H. F. (1978a). Frequency sharing and spread spectrum transmission with large relative bandwidth. *IEEE Trans. Electromagn. Compat.* **20,** 232–239.

Harmuth, H. F. (1978b). Low angle tracking by carrier-free radar. *IEEE Trans. Electromagn. Compat.* **20,** 419–425.

Harmuth, H. F. (1978c). Synthetic aperture radar based on nonsinusoidal functions. I. Moving radar and stationary arrays in one or two dimensions. *IEEE Trans. Electromagn. Compat.* **20,** 426–435.

Harmuth, H. F. (1978d). Radio signals with large relative bandwidth for over-the-horizon radar and spread spectrum communications. *IEEE Trans. Electromagn. Compact.* **20,** 501–512.

Harmuth, H. F. (1979a). "Acoustic Imaging with Electronic Circuits." Academic Press, New York.

Harmuth, H. F. (1979b). Synthetic aperture radar based on nonsinusoidal functions. II.

Pulse compression, contrast, resolution and Doppler shift. *IEEE Trans. Electromagn. Compat.* **21,** 40–49.

Harmuth, H. F. (1979c). Synthetic aperture radar based on nonsinusoidal functions. III. Beamforming by means of the Doppler effect. *IEEE Trans. Electromagn. Compat.* **21,** 122–131.

Harmuth, H. F. (1979d). Synthetic aperture radar based on nonsinusoidal functions. IV. Tracking radar and beam rider. *IEEE Trans. Electromagn. Compat.* **21,** 245–253.

Harmuth, H. F. (1980a). Synthetic aperture radar based on nonsinusoidal functions. V. Look-down radar. *IEEE Trans. Electromagn. Compat.* **22,** 45–53.

Harmuth, H. F. (1980b). Synthetic aperture radar based on nonsinusoidal functions. VI. Pulse position and pulse shape coding. *IEEE Trans. Electromagn. Compat.* **22,** 93–106.

Harmuth, H. F. (1980c). Synthetic aperture radar based on nonsinusoidal functions. VII. Thumbtack ambiguity function. *IEEE Trans. Electromagn. Compat.* **22,** 181–190.

Harmuth, H. F. (1980d). Synthetic aperture radar based on nonsinusoidal functions. VIII. Velocity and acceleration processing. *IEEE Trans. Electromagn. Compat.* **22,** 308–319.

Harmuth, H. F. (1980e). Angular resolution of sensor arrays for signals with a bandwidth larger than zero. *Proc. 1980 IEEE Int.Symp. Electromagn. Compat., Baltimore* (80CH1538-8EMC), pp. 316–320.

Harmuth, H. F. (1980f). Nonsinusoidal waves—when to use them. *Proc. 1980 IEEE Int. Symp. Electromagn. Compat., Baltimore* (80CH1538-8EMC), pp. 1–8.

Harmuth, H. F. (1981a). Synthetic aperture radar based on nonsinusoidal functions. IX. Array beam forming. *IEEE Trans. Electromagn. Compat.* **23,** 20–27.

Harmuth, H. F. (1981b). Synthetic aperture radar based on nonsinusoidal functions. X. Array gain, planar arrays, multiple signals. *IEEE Trans. Electromagn. Compat.* **23** (in press).

Harmuth, H. F. (1981c). Fundamental limits for radio signals with large bandwidth. *IEEE Trans. Electromagn. Compat.* **23,** 37–43.

Harrison, C. H. (1970). Reconstruction of subglacial relief from radio echo sounding records. *Geophysics* **35,** 1099–1115.

Hartley, R. V. L. (1928). Transmission of information. *Bell Syst. Tech. J.* **7,** 535–563.

Hertz, H. (1893). "Electric Waves." Macmillan, London (reprinted Dover, New York, 1962).

Horvat, V. (1969). Underwater radio wave transmission. *In* "Handbook of Ocean and Underwater Engineering" (J. J. Myers, ed.). McGraw-Hill, New York.

Hund, A. (1942). "Frequency Modulation." McGraw-Hill, New York.

Huxley, L. G. H. (1937). Motions of electrons in magnetic fields and alternating electric fields. *Philos. Mag.* **23,** 442–464.

Huxley, L. G. H. (1938). Propagation of electromagnetic waves in an ionized atmosphere. *Philos. Mag.* **25,** 148–159.

Huxley, L. G. H. (1940). Propagation of electromagnetic waves in an atmosphere containing free electrons. *Philos. Mag.* **29,** 313–329.

Jaroslavskij, P. L. (1979). "Introduction to Digital Image Processing." Sovietskoe Radio, Moscow (in Russian).

Jasik, H. (1961). "Antenna Engineering Handbook." McGraw-Hill, New York.

Johnson, C. K. (1963). "Analog Computer Techniques." McGraw-Hill, New York.

Jordan, E. C. (1950). "Electromagnetic Waves and Radiating Systems." Prentice-Hall, Englewood Cliffs, New Jersey.

Karpovsky, M. G. (1976). "Finite Orthogonal Series in the Design of Digital Devices." Wiley, New York.

Keiser, B. E. (1974). Early development of the project Sanguine radiating system. *IEEE Trans. Commun.* **22,** 364–371.

Kirkpatrick, G. R. (1974). "Final Engineering Report on Angular Accuracy Improvement." Reprinted *in* "Monopulse Radar" (D. K. Barton, ed.), Vol. 1. Artech House, Dedham, Massachusetts.

Klauder, J. R., Price, A. C., Darlington, S., and Albersheim, W. J. (1960). The theory and design of chirp radar. *Bell Syst. Tech. J.* **39,** 745–808.

Korn, G. A., and Korn, T. M. (1964). "Electronic Analog and Hybrid Computers." McGraw-Hill, New York.

Kotel'nikov, V. A. (1947). "The Theory of Optimum Noise Immunity" (in Russian) (English transl.: McGraw-Hill, New York, 1959).

Kovacs, A., and Abele, G. (1974). Crevasse detection using an impulse radar system. *Antarct. J. U.S.* **9,** No. 4, 177–178.

Kovacs, A., and Gow, A. J. (1975). Brine infiltration in the McMurdo Ice Shelf, McMurdo Sound, Antarctica. *J. Geophys. Res.* **80,** 1957–1961.

Kovacs, A., and Gow, A. J. (1977). Subsurface measurements of the Ross Ice Shelf, McMurdo Sound, Antarctica. *Antarct. J. U.S.* **12,** No. 4, 137–140.

Kovacs, A., and Morey, R. M. (1978). Radar anisotropy of sea ice due to preferred azimuthal orientation of the horizontal c axes of ice crystals. *J. Geophys. Res.* **83,** 6037–6046.

Kovacs, A., and Morey, R. M. (1979a). Anisotropic properties of sea ice in the 50 to 150 MHz range. *J. Geophys. Res.* **84,** 5749–5759.

Kovacs, A., and Morey, R. M. (1979b). Remote detection of massive ice in permafrost along the Alyeska pipeline and the pump station feeder pipeline. *Proc. Conf. Pipelines Adverse Environ., 1979* pp. 268–279. Am. Soc. Civ. Eng., New York.

Kovacs, A., and Morey, R. M. (1979c). Remote detection of a freshwater pool off the Sagavanirktok River delta, Alaska. *Arctic* **32,** 161–164.

Kovaly, J. J. (1972). Radar techniques for planetary mapping with orbiting vehicle. *Ann. N.Y. Acad. Sci.* **187,** 154–176.

Kovaly, J. J., ed. (1976). "Synthetic Aperture Radar." Artech House, Dedham, Maschusetts.

Labunets, W. G., and Sitnikov, O. P. (1976). Generalized fast Fourier transform on arbitrary finite Abelian groups. *In* "Harmonic Analysis of Groups in Abstract Systems Theory," pp. 24–43. Polytech. Inst. S. M. Kirova, Sverdlovsk, USSR (in Russian).

Lange, F. H. (1971). "Signale und Systeme." VEB Technik, Berlin GDR.

Livius, T. (17). "The War with Hannibal" (in Latin). Slavelabor Duplicators, Rome (English transl. Penguin Books, Hormondsworth, Middlesex, England 1965).

McClenon, D. (1976). ELF communications. *Nav. Eng. J.,* Aug., 33–40.

Maqusi, M. (1981). "Applied Walsh Analysis and its Applications." Heyden, Philadelphia, Pennsylvania.

Marconi, G. (1901). Apparatus for wireless telegraphy. U.S. Patent 676,332.

Marconi, G. (1904). Apparatus for wireless telegraphy. U.S. Patent 763,772.

Maxwell, J. C. (1981). "A Treatise on Electricity and Magnetism." Oxford Univ. Press (Clarendon), London and New York (reprinted: Dover, New York, 1954).

Moffat, D. L., and Puskar, R. J. (1976). A subsurface electromagnetic pulse radar. *Geophysics* **41,** 506–518.

Mooney, D. H., and Skillman, W. A. (1970). Pulse-Doppler radar. *In* "Radar Handbook" (M. I. Skolnik, ed.). McGraw-Hill, New York.

Moore, R. P. (1979). Environmental factors affecting the development and use of millimeter systems for naval applications. *EASCON 79 Rec.* pp. 4–11.

Morey, R. M. (1974). Continuous subsurface profiling by impulse radar. *In* "Subsurface Exploration for Underground Excavation and Heavy Construction," pp. 213–232. Am. Soc. Civ. Eng., New York.

Morey, R., and Kovacs, A. (1977). Detection of moisture in construction materials. *US Army Cold Reg. Res. Eng. Lab. Rep.* **77-25.** Hanover, New Hampshire.

Newhouse, V. L. (1975). "Applied Superconductivity." Academic Press, New York.

Nicholson, A. M., and Ross, G. F. (1975). A new radar concept for short range application. *Proc. IEEE Int. Radar Conf.*, pp. 146–151.

Pappert, R. A., and Moler, W. F. (1974). Propagation theory and calculations at lower extremely low frequencies (ELF). *IEEE Trans. Commun.* **22,** 438–451.

Porcello, L. J., Jordan, R. L., Zelenka, J. S., Adams, G. F., Phillips, R. G., Brown, W. E., Ward, S. H., and Jackson, P. L. (1974). The Apollo lunar sounder radar system. *Proc. IEEE* **62,** 769–783.

Ramp, H. O., and Wingrove, E. R. (1961). Principles of pulse compression. *IRE Trans. Mil. Electron.* **5,** 109–116.

Reutov, A. P., and Mikhaylov, B. A., eds. (1970). "Sidelooking Radar" (in Russian). Sovietskoe Radio, Moscow (English transl.: AD 787 070, Natl. Tech. Inf. Serv., Springfield, Virginia.

Rhodes, D. R. (1974). "Synthesis of Planar Antenna Sources." Oxford Univ. Press (Clarendon), London and New York.

Ridenour, L. N. (1947). "Radar Systems Engineering." McGraw-Hill, New York.

Rihaczek, A. W. (1969). "Principles of High Resolution Radar." McGraw-Hill, New York.

Rodgers, A. E., and Ingalls, R. P. (1970). Radar mapping of Venus with interferometric resolution of the range-Doppler ambiguity. *Radio Sci.* **5,** No. 2, 425–433.

Ross, G. F. (1974). BARBI, a new radar concept for precollision sensing. *Soc. Auto Eng.* [*Tech. Pap.*] **740574,** 141–152.

Rossiter, J. R., and Butt, K. A. (1979). "Remote Estimation of the Properties of Sea Ice," Beaufort Sea Field Trip Rep. March 1979. Centre for Cold Ocean Resources Engineering, Memorial University of Newfoundland, St. John's.

Rossiter, J. R., and Gustajtis, K. A. (1978) Iceberg sounding by impulse radar. *Nature* (*London*) **271** (No. 5640), 48–50.

Rossiter, J. R., and Gustajtis, K. A. (1979). Determination of iceberg underwater shape with impulse radar. *Desalination* **29,** 99–107.

Rossiter, J. R., Narod, B. B., and Clarke, G. K. C. (1979). Airborne radar sounding of arctic icebergs. *Proc. Int. Conf. Port Ocean Eng. Under Arct. Cond., 5th. Trondheim, Norway,* pp. 289–305.

Rowe, H. E. (1974). Extremely low frequency (ELF) communications to submarines. *IEEE Trans. Commun.* 371–385.

Rumsey, V. H. (1966). "Frequency Independent Antennas." Academic Press, New York.

Sakrison, D. J. (1970). "Notes on Analog Communication." Van Nostrand-Reinholds, Princeton, New Jersey.

Sano, H. and Tanada, Y. (1973). Logical Walsh functions (in Japanese). *Trans. Inst. Elec. Commun. Eng. Japan* **56-D** (9), 531–532.

Saunders, W. K. (1970). CW and FM radar. *In* "Radar Handbook" (M. I. Skolnik, ed.). McGraw-Hill, New York.

Schaefer, C. (1950). "Einführung in die theoretische Physik," 2nd ed., Vol. 3. De Gruyter, Berlin.

Schaefer, C., and Gross, G. (1910). Totalreflektion. *Ann. Phys.* (*Leipzig*) **32,** 648–672.

Schelkunoff, S. A. (1943). "Electromagnetic Waves." Van Nostrand-Reinhold, Princeton, New Jersey.

Schelkunoff, S. A. (1952). "Advanced Antenna Theory." Wiley, New York.

Schrader, W. W. (1970). MTI radar. *In* "Radar Handbook" (M. I. Skolnik, ed.). McGraw-Hill, New York.

Schreiber, H. H. (1976). Communications with Walsh functions. *1976 IEEE Electromagn. Compat. Symp. Rec.* pp. 258–263 (IEEE Cat. No. 76CH1104-9EMC).

Schumann, W. O. (1952a). Über die strahlungslosen Eigenschwingungen einer leitenden Kugel, die von einer Luftschicht und einer Ionosphärenhülle umgeben ist. *Z. Naturforsch., Teil A* **7,** 149–154.

Schumann, W. O. (1952b). Über die Dämpfung der elektromagnetischen Eigenschwingungen des Systems Erde-Luft-Ionosphäre. *Z. Naturforsch., Teil A* **7,** 250–252.

Schumann, W. O. (1957). Über elektrische Eigenschwingungen des Hohlraums Erde-Luft-Ionosphäre erregt durch Blitzentladungen. *Z. Angew. Phys.* **9,** 373–378.

Schwartz, M., Bennett, W. R., and Stein, S. (1966). "Communication Systems and Techniques." McGraw-Hill, New York.

Setzer, D. E. (1970). Computed transmission through rain at microwave and visible frequencies. *Bell Syst. Tech. J.* **49,** 1873–1892.

Siedentopf, H. (1950). "Grundriss der Astrophysik." Teubner, Stuttgart.

Sitnikov, O. P. (1976). Harmonic analysis of groups in abstract systems theory. *In* "Harmonic Analysis of Groups in Abstract Systems Theory," pp. 5–23. Polytech. Inst. S. M. Kirova, Sverdlovsk, USSR (in Russian).

Skolnik, M. I., ed. (1970). "Radar Handbook." McGraw-Hill, New York.

Smith, C. E. (1966). "Log Periodic Antenna Design Handbook." Smith Electronics, Cleveland, Ohio.

Smith, C. E., Butler, C. M., and Umashankar, K. R. (1979). Characteristics of a wire biconical antenna. *Microwave J.* **22,** No. 9, 37–40.

Sobol, I. M. (1969). "Multidimensional Approximation Formulas and Haar Functions." Nauka, Moscow (in Russian).

Solymar, L. (1972). "Superconducting Tunneling Applications." Wiley, New York.

Soroko, L. M. (1976). Dyadic derivative (in Russian). *Communication of the Joint Institute for Nuclear Research Dubna.* P11-9725, 24 pages.

Spetner, L. M. (1974). Radiation of arbitrary electromagnetic waveforms. *In* "Applications of Walsh Functions and Sequency Theory" (G. F. Sandy and H. H. Schreiber, eds.), pp. 249–274 (IEEE Cat. No. 74CH0861-5EMC).

Stratton, J. A. (1941). "Electromagnetic Theory." McGraw-Hill, New York.

Tanada, Y. and Sano, H. (1974). Walsh spectra of time-shifted waves (in Japanese). *Trans. Inst. Elec. Commun. Eng. Japan* **57-D** (8), 503–504.

Tanada, Y. and Sano, H. (1976). A hybrid Walsh waveform analyzer (in Japanese). *Trans. Inst. Elec. Commun. Eng. Japan* **57-D** (2), 101–108.

Tanada, Y. and Sano, H. (1978). Linear filtering of time signals using the Walsh transform (in Japanese). *Trans. Inst. Elec. Commun. Eng. Japan* **61-A** (6), 596–603.

Tanada, Y. and Sano, H. (1979). A Walsh waveform analyzer and its applications to filtering pulse signals (in English). *Mem. School Eng., Okayama Univ.,* Okayama-shi 700, Vol. 13, Feb., pp. 163–180.

Ta-Shing Chu (1974). The radiation of extremely low frequency (ELF) waves. *IEEE Trans. Commun.* **22,** 386–388.

Taub, H., and Schilling D. L. (1971). "Principles of Communication Systems." McGraw-Hill, New York.

Thomas, L., Jr. (1965). Scientists ride ice islands on arctic odysseys. *Natl. Geogr.* **128,** 670–691.

Trachtman, A. M., and Trachtman, V. A. (1973). The frequency of Walsh functions. *Telecommun. Radio Eng. (USSR), Part 2* **28,** No. 12, 56–58.

Trachtman, A. M., and Trachtman, V. A. (1975). "Basic Theory of Discrete Signals in Finite Intervals." Sovietskoe Radio, Moscow (in Russian).

Trachtman, V. A. (1973). Factorization matrix of Walsh functions with Paley and sequency ordering. (In Russian.) *Radiotekh. Elektron.* **18,** No. 12, 2521–2528.

U.S.A. (1977a). "Review of Band Utilization of Frequencies Between 40 and 3000 GHz," US CCIR Study Groups, Doc. USSG 1A/33. US Dept. of Commerce, Off. Telecommun., Inst. Telecommun., Boulder, Colorado.

U.S.A. (1977b). "Frequency Sharing by Means of Nonsinusoidal Radio Waves," US CCIR Study Groups, Doc. USSG 1A/49. US Dept. of Commerce. Off. Telecommun., Inst. Telecommun., Boulder, Colorado.

U.S.A. (1977c). "Need for Frequency Bands for the Radiolocation Service Between 40 and 300 GHz," US CCIR Study Groups, Doc USSG 1A/55. US Dept. of Commerce, Off. Telecommun., Inst. Telecommun., Boulder, Colorado.

Wagner, K. W. (1953). "Elektromagnetische Wellen." Birkhäuser, Basel.

Wait, J. R. (1970). "Electromagnetic Waves in Stratified Media," 2nd ed. Pergamon, Oxford.

Wallis, W. D., Street, A. P., and Wallis, J. S. (1972). "Combinatorics: Room Squares, Sum-Free Sets, Hadamard Matrices," Lect. Notes Math. No. 292. Springer-Verlag, Berlin and New York.

Walsh, J. L. (1923). A closed set of normal orthogonal functions. *Am. J. Math.* **45,** 5–24.

Whalen, R. J. (1979). Millimeter wave radar. *Microwave J.* **22,** No. 8, 16.

White, W. D. (1974). Low-angle tracking in the presence of multipath. *IEEE Trans. Aerosp. Electron. Syst.* **10,** 835–852.

White, W. D. (1976). Double null technique for low angle tracking. *Microwave J.* **19,** No. 12, 35–38.

Willwerth, F. G., and Kupiec, I. (1976). Array aperture sampling techniques for multipath compensation. *Microwave J.* **19,** No. 6, 37–39.

Wiltse, J. C., Jr. (1978). Millimeter waves—they are alive and healthy. *Microwave J.* **21,** No. 8, 16–18.

Wiltse, J. C., Jr. (1979). Millimeter wave technology and applications. *Microwave J.* **22,** No. 8, 39–42.

Woodward, P. M. (1953). "Probability and Information Theory, with Applications to Radar." McGraw-Hill, New York.

Zhevakin, S. A., and Naumov, A. P. (1967). The propagation of centimeter, millimeter, and submillimeter radio waves in the Earth's atmosphere. *Radiophys. Quantum Electron. (Engl. Transl.)* **10,** 678–694; Russian original in *Izv. vuz. Radiofiz.* **10,** 1213–1243 (1967).

References from the Chinese Literature

The following papers were published in Part II of the book "Conference Proceedings of (1) the Third Annual Meeting of the Chinese Electrical Institute, Information Theory Professional Society, (2) the Chinese Communication Theory Institute—Meeting During the Formation of the Communication Theory Professional Society, (3) the Chinese Electronic Society Conference on Walsh Functions" published by the Chinese Electrical Institute, Information Theory Professional Society, Beijing 1980; all papers are in Chinese.

Chang Tong, Applications of Walsh functions in information systems, pp. 215–217.

Chen Xiaoyan, The fast transforms using micro-computers, pp. 289–291.

Chen Xueyu, An application of the method of Walsh series in optimal control, pp. 277–280.

Cheng Mingde, Sun Jing, Pan Junzhuo, and Zhang Xuding, Computer simulation of image compression by finite Walsh transforms, pp. 232–233.

Fan Changxin, Applications of Walsh functions in the transmission of information, pp. 218–221.

Hu Dekun and Wang Chengyi, Discussion on the various ordering of finite Walsh functions and their transforms using the method of linear encoding, pp. 234–235.

Hu Zheng, About the debate on Walsh functions, pp. 222–223.

Hu Zhengming, Triadic additive group and orthogonal functions $N(i,t)$, pp. 244–246.

Liu Changguo, A new design method of generating Walsh functions, pp. 287–288.

Ma Huaxiao, Logical analysis and network design of sequency conversion for Walsh functions, pp. 283–286.

Mi Zhenju, An application of Walsh functions in multiple-access communications, pp. 292–294.

Su Weiyi, The logical derivatives of n-variable functions and the logical partial differential equations, pp. 240–243.

Tang Guoxi, Image transforms, pp. 257–260.

Teng Chengsong and Jiangsu, The calculation of OTF functions using Walsh functions, pp. 271–273.

Wang Zhaohua, The Walsh function quadrature amplitude modulation system, pp. 300–302.

Wang Zhaohua and Li Zhihua, V-transform and its application to the analysis of images, pp. 250–253.

Wang Zhaohua and Li Zhihua, The analysis of PAL signals by multidimensional Walsh functions, pp. 274–276.

Wei Hongjun, Review of electromagnetic Walsh wave communications, pp. 281–282.

Wei Hongjun, On the definition and characteristics of Walsh functions, pp. 303–304.

Wu Pinjing, An optimization design method of transformed matrix, pp. 247–249.

Xiao Guozhen, Error-correcting codes and finite orthogonal expansions, pp. 261–264.

Xiao Guozhen, On the criterion of a Boolean function which is independent of some arguments, pp. 295–297.

Xu Ningshou, Analysis of step periodic waveform via block-pulse functions, pp. 226–231.

Yang Youwei, An experimental adaptive 12 channel delta-modulation system, pp. 265–267.

Zan Yulun, Application of Walsh functions in the low frequency swept frequency signal generators, pp. 254–256.

Zhang Quishan and Liu Zhongkan, Telemetry system using Walsh functions, pp. 224–225.

Zhao Xiangwen, The correlation functions of Walsh functions, pp. 268–270.

Zheng Sumin, The sampling theorem in the sequency domain, pp. 298–299.

Zheng Weixing, The generalized logical derivative and its applications, pp. 236–239.

Index

M

N

O

P

Q

R